SINGULARITIES & DYNAMICAL SYSTEMS

NORTH-HOLLAND MATHEMATICS STUDIES 103

NORTH-HOLLAND – AMSTERDAM • NEW YORK • OXFORD

SINGULARITIES & DYNAMICAL SYSTEMS

Proceedings of the International Conference on Singularities and Dynamical Systems
Heraklion, Greece, 30 August – 6 September, 1983

edited by

Spyros N. PNEVMATIKOS
University of Dijon
France

1985

NORTH-HOLLAND – AMSTERDAM • NEW YORK • OXFORD

ISBN: 0 444 87641 3

Publishers:
ELSEVIER SCIENCE PUBLISHERS B.V.
P.O. BOX 1991
1000 BZ AMSTERDAM
THE NETHERLANDS

Sole distributors for the U.S.A. and Canada:
ELSEVIER SCIENCE PUBLISHING COMPANY, INC.
52 VANDERBILT AVENUE
NEW YORK
U.S.A.

Library of Congress Cataloging in Publication Data

International Conference on Singularities and Dynamical
Systems (1983 : Hērakleion, Crete)
Singularities & dynamical systems.

(North-Holland mathematics studies ; 103)
English and French.
1. Differentiable dynamical systems--Congresses.
2. Singularities (Mathematics) I. Pnevmatikos,
Spyros N., 1950- . II. Title. III. Title:
Singularities and dynamical systems.
QA614.8.I58 1983 515.3'5 84-18774
ISBN 0-444-87641-3 (U.S.)

PRINTED IN THE NETHERLANDS

PREFACE

This volume is an account of the lectures delivered in the International Conference on "Singularities and Dynamical Systems", organized in Heraklion from August 30th to September 6th 1983 by the University of Crete and the University of Dijon. The main purpose of the Conference was to create conditions of scientific contact between mathematicians and physicists who have singularities and dynamical systems as common interests.

The main themes dealt with at the meeting, and contained in this volume, are devoted to recent progress in the following topics:

* the global study of dynamics generated by diffeomorphisms or foliations,
* the local study of the singularities of differential equations of real and complex fields,
* the singularities of symplectic geometry, contact geometry, and Riemannian geometry,
* the singularities of functions and complex hypersurfaces,
* the bifurcations in dynamical systems and the appearance of chaos,
* the study of some specific dynamical systems, nonlinear differential equations, and solitons.

The Conference was sponsored by the Ministry of Research and Technology, and the Ministry of Civilization and Sciences of Greece. In addition, the French governement, through its Embassy in Athens, provided some support for participants. We are grateful to these organizations for their interest and assistance.

The scientific preparations for, the organization at, and the publications following this Conference are the result of a close cooperation with:

Dominique Cerveau, University of Dijon, Department of Mathematics,

Jean-François Mattei, University of Toulouse, Department of Mathematics,

Robert Moussu, University of Dijon, Department of Mathematics,

Stephanos Pnevmatikos, University of Dijon, Department of Physics,

Robert Roussarie, University of Dijon, Department of Mathematics.

I am pleased to express my gratitude to the authors for their contributions, to the referees for their helpful comments, to all persons who assisted me with the many details of running the Conference and preparing this volume, and in particular to North-Holland Publishing Company.

Spyros N. Pnevmatikos

Heraklion 1984

CONTENTS

* * *

Singularities & Dynamical Systems
S.N. Pnevmatikos (editor)

LA MARCHE AU CHAOS VUE PAR UN TOPOLOGUE

René Thom

I.H.E.S.
France

En topologie différentielle, la relation d'équivalence la plus générale est celle définie par un feuilletage (ρ) de codimension k dans une variété lisse paracompacte M^n de dimension n. En toute généralité, on devrait accepter que le feulletage puisse admettre des singularités génériques. Malheuresement, la théorie des singularités de feuilletages est chargée de beaucoup d'obscurité; les singularités peuvent dépendre de la manière (via formes différentielles ou champs de vecteurs) par laquelle on définit le feuilletage; et les singularités génériques ne sont pas connues en dehors des cas les plus simples;(basse codimension). Pour ces raisons, on supposera ici le feuilletage (ρ) partout régulier. On désignera par (M^n/ρ) l'espace quotient de M^n par la relation d'équivalence ρ ($x \underset{\rho}{\sim} y$ si x et y sont dans la même feuille).

Reprenant une idée proposée autrefois [1], on va pratiquer sur M^n la théorie de Morse. On se donnera sur M^n une fonction de Morse réelle $F: M^n \longrightarrow \mathbb{R}$ propre et positive. F atteint alors son minimum en un point minimisant m_o avec $F(m_o) = \mu_o > 0$ qu'on supposera unique. Le principe est alors, à toute valeur $c \in \mathbb{R}^+$, d'associer la variété $F \leq c$, et d'étudier comment varie l'espace quotient $(F \leq c/\rho)$, où la relation (ρ) est définie par restriction à la variété à bord $F \leq c$ dans M. On fera en plus les hypothèses de généricité naturelles que voici:

-1.En tout point critique ν de F, le cône tangent (cône quadratique) à la variété de niveau $F^{-1}(F(\nu))$ est en *position générale* par rapport à la ρ-feuille contenant ν.

-2.Pour presque toute valeur de c, la variété bord $F^{-1}(c)$ est en position générale par rapport au feuilletage. Ceci veut dire que, en général, les submersions locales $\pi_i: U_i \longrightarrow \mathbb{R}^k$ qui définissent le feuilletage sont *génériques* sur l'hyperplan $F = c$. Il y a donc un lieu critique $\Gamma(c) = F^{-1}(c)$ où ces applications locales sont critiques, lieu qui est en général de codimension (n-k) dans $F^{-1}(c)$, de corang à la source : $(n-1)-(k-1) = n-k$, ayant des singularités de codimension $2(n-k-1)$.

Ceci nous permet de définir divers types de points critiques:

1. Les points critiques ν_c de la fonction de Morse F. Les valeurs $c_i = F(\nu_c)$ seront dites *critiques ordinaires*.

2. Les valeurs σ pour lesquelles, sur $F = \sigma$, les submersions locales de ρ, $\pi_i : U_i \longrightarrow \mathbb{R}^k$, cessent d'être génériques. On supposera ces singularités de codimension un, associées à un point critique $\tilde{\mu}$ isolé, telles que le chemin défini par (c) franchissant σ se relève en un point de contact $\mu(c)$ qui défini localement une traversée transversale de la strate associée à la singularité $\pi_i | F = \sigma$. On appellera ces valeurs *σ-critiques*.

Les valeurs critiques ordinaires et les valeurs σ-critiques sont isolées. Toutefois, ces valeurs ne suffisent pas à construire une théorie de Morse. Il y a lieu d'y associer d'autres types de valeurs critiques.

- *Valeurs τ-critiques*. Pour chaque valeur (c) on a un lieu critique $\Gamma(c)$ dans $F = c$. On peut alors *saturer* ce lieu par la relation ρ dans $F \leqq c$; on obtient ainsi un ensemble $\tilde{\Gamma}(c)$, génériquement un ensemble stratifié (ou une image continue d'ensemble stratifié). Il y a lieu alors de considérer l'intersection $(F = c) \cap \tilde{\Gamma}(c)$; à l'extérieur de $\Gamma(c)$ cette intersection est en général *transversale* (en un sens généralisé, si $\Gamma(c)$ a des singularités). On appellera *τ-valeur critique* toute valeur pour laquelle cette intersection $(F = \tau) \cap \tilde{\Gamma}(\tau)$ cesse d'être transversale.

Ici encore, on peut définir la notion de défaut de transversalité de codimension un, et on supposera que les valeurs (τ) ainsi obtenues sont effectivement transversalement obtenues lors de la variation de (c) à travers la valeur (τ). Le dernier type de valeurs critiques à considérer est celui qui dans le cas classique des flots, correspond à des intersections non transversales de variétés stables et instables dans la théorie de la bifurcation. Pour cela il sera nécessaire d'admettre pour notre espace quotient $(F \leq c)/\rho$, une structure d'espace stratifié *non séparé*. Par exemple, le quotient d'un espace $\mathbb{R}^n$ par un endomorphisme linéaire hyperbolique T, $\mathbb{R}^n = \mathbb{R}^u + \mathbb{R}^s$, est un espace de quatre strates: U^{u+s} de la forme $S^{u-1} \times S^{s-1} \times \mathbb{R}^2$, domaine fondamental par T restreint au complémentaire $\mathbb{R}^n - \mathbb{R}^u \cup \mathbb{R}^s$; les variétés stables $S^{u-1} \times \mathbb{R} = \mathbb{R}^u$, $S^{s-1} \times \mathbb{R} = \mathbb{R}^s$ et l'origine 0. La seule différence avec l'axiomatique usuelle des ensembles stratifiés est que les applications d'attachement $\overset{X}{\underset{Y}{}} < Z$ ne satisfont plus à la condition de séparation lorsque les strates bord sont disjointes (non incidentes l'une sur l'autre).

Dès qu'il existe une feuille compacte φ dans notre variété $F \leq c$, on peut s'intéresser aux feuilles qui contiennent (φ) dans leur adhérence. Elles forment un ensemble $W(\rho)$ (généralisant variété stable et instable), qui dans les "bons" cas est un ensemble stratifié. On est alors amené, pour toute strate Δ du quotient $(F\leq c)/\rho$, même non séparé, à lui associer l'ensemble $W(\Delta)$, qui peut être une strate de la stratification - ou un ensemble infiniment plus pathologique.

On appellera *valeur ζ-critique* toute valeur c pour laquelle l'intersection d'une strate de la forme $W(\Delta)$ avec $F = c$ est non transversale, ou pour laquelle les ensembles $W_c(\Delta_i)$ cessent de se couper transversalement, ou encore toute valeur c à la traversée de laquelle un ensemble $W(\Delta)$ cesse d'être stratifié par suite d'une complexification de l'holonomie de la feuille (Δ). En particulier, les valeurs critiques de F restreintes aux feuilles (Δ) sont des valeurs ζ-critiques. Ces définitions étant données, on observe qu'au départ pour c voisin du minimum μ_o, $c = \mu_o + \varepsilon$, $F \leq c$ est une boule, et ρ y est quasi linéaire. On a donc un bon quotient $(F\leq c)/\rho$ qui est une $(n-k)$ boule. Toute feuille est simplement connexe. Tant qu'on n'a pas construit de feuilles non simplement connexes, la théorie de stabilité de Reeb permet d'affirmer que toute feuille des $F \leq c$ a un voisinage saturé, et par suite le quotient reste un ensemble de Hausdorff qui est stratifié selon le mode des *variétés branchues* de R. Williams [2] (compactifiées). Ceci se produira tant qu'on n'aura pas franchi une certaine valeur critique (ordinaire, ou σ, τ) pour laquelle apparaîtra une feuille compacte à holonomie non triviale.

On peut conjecturer que génériquement, on obtient une feuille compacte à holonomie hyperbolique. On notera que ceci exige des contacts d'indice 1 pour créer du π_1 (en général par des valeurs τ-critiques).

Avant cette valeur c*, pour $c < c^*$, l'application $(F\leq c) \to (F\leq c)/\rho$ est une application stratifiée sans éclatement. Elle présente donc la stabilité topologique pour tout intervalle $]c_1, c_2[< c^*$ ne contenant aucune valeur critique. En général cette valeur c* est précédée d'une suite convergente de valeurs c_i qui sont τ-critiques. Et on peut voir que le quotient $(c^*-c_i)/(c^*-c_{i+1})$ tend vers une limite (comme dans le phénomène de Feigenbaum).

Considérons le "pli" $\Gamma(c)$ sur $F = c$. Le saturé $\tilde{\Gamma}(c)$ va couper $F = c$

en un autre lieu $\psi(c)$; quand c tend vers c_i , $\psi(c_i)$ aura un contact ordinaire avec $\Gamma(c_i)$ en un point γ_i , ce qui augmente la feuille saturée $\tilde{\Gamma}(c_i)$. Alors le saturé $\tilde{\Gamma}(c_{i+1})$, pour $c = c_i+\varepsilon$, va contenir un morceau issu d'un voisinage à droite de γ_i, lequel va créer plus tard un nouveau contact (γ_{i+1}) avec $\Gamma(c_{i+1})$ pour une valeur c_{i+1}. D'où la suite des segments γ_i , γ_{i+1} , associés à des figures approximativement semblables, qui convergent vers un point $\tilde{\gamma}$ de valeur c^*.

Il est vraisemblable que la création d'une feuille à holonomie non triviale débute génériquement par une feuille à holonomie hyperbolique. Ceci est une exigence de la théorie de Morse classique, et pourrait être une formulation d'un "*closing lemma*" généralisé.

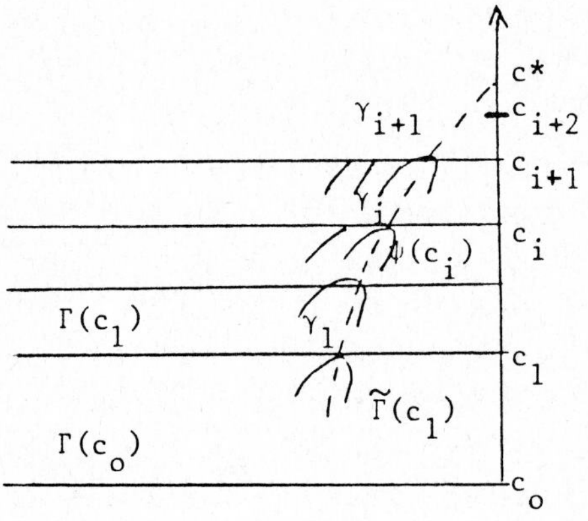

Note: La coordonnée verticale au voisinage des points γ_i signifie non la variable c, mais une coordonnée locale transverse à $\Gamma(c)$ dans $F^{-1}(c)$.

Après la valeur c^*, on a un quotient stratifié non séparé; on peut alors aller jusqu'à une valeur c^{**} , où cette propriété cesse. Il est vraisemblable que dans l'intervalle $c^* < F < c^{**}$, il n'y a pas stabilité topologique, car les applications d'attachement des stratifiés non séparés donnent naissance à des diagrammes d'applications divergentes. Si $\begin{matrix} X \\ Y \end{matrix} < Z$, on a $Z \rightrightarrows \begin{matrix} X \\ Y \end{matrix}$ diagramme divergent, topologiquement instable selon la théorie de Dufour [3]. Toutefois, dans certains cas, la stabilité topologique peut subsister. Ainsi, s'il n'existe qu'un nombre fini de strates (Δ_i) qui sont des feuilles hyperboliques compactes, si les variétés $W(\Delta_i)$ se coupent transversalement, avec une condition du type (No-Cycle) qui assure que toutes les autres feuilles sont non compactes à holonomie triviale, on peut avoir un théorème de stabilité topologique généralisant celui des systèmes de Morse-Smale. Ceci est dû au fait que dans le cas hyperbolique, les applications d'attachement qui se présentent lors des incidences multiples telles que $\genfrac{}{}{0pt}{}{X}{Y} < Z$ ont leur fibres transversales, ce qui donne la raison profonde de la stabilité topologique des points singuliers hyperboliques énoncée par le théorème de Hartmann.

Au-delà pour $F > c^{**}$, on a un quotient non séparé, non stratifié ; c'est ce qu'on appelle - par abus de langage sans doute - le "chaos". Soit le schéma général:

	μ_o	c^*	c^{**}	c
(D)	Espace quotient stratifié séparé	Espace quotient stratifié non séparé	⋮	"chaos"

Le seul fait qu'on ait le théorème de stabilité structurelle des difféomorphismes d'Anosov, montre que dans certains cas, on peut dire quelque chose de ces quotients non séparés. Très souvent, on a à considérer un couple d'endomorphismes φ_1, ψ_1 sur un espace, définis à la conjugaison près. C'est le cas par exemple des holonomies non hyperboliques, définis par un homéomorphisme $\pi_1(F) \longrightarrow GL(k)$ dont il serait bien intéressant de caractériser la structure générique pour un $\pi_1(F)$ donné. Le diagramme (D) suscite la question suivante:les bifurcations de l'endomorphisme quadratique $x \longrightarrow x^2+\lambda$ dans sa partie utile $\lambda < \frac{1}{4}$ ne présentent que la partie $c^*\, c^{**}$ du schéma (D). Cela est dû au fait que pour $\lambda = 0$ on a déjà un endomorphisme global (et non local comme dans la théorie de Morse).

Terminons par deux simples remarques. L'espace de phase de l'oscillateur linéaire, associé à l'hamiltonien $H = p^2 + q^2$, a un quotient séparé. En introduisant du frottement, ou de la dissipation, le quotient devient non séparé, car le centre devient foyer.

On observera d'ailleurs que dans la suite de Feigenbaum $x \longrightarrow x^2 + \lambda$ le quotient $Q(\lambda)$ reste isomorphe à lui-même à travers les bifurcations qui précèdent la valeur de chaos c^{**}, seule varie l'application $\rho : \mathbb{R} \longrightarrow Q(\lambda)$ où le quotient $Q(\lambda)$ est formé de trois strates: deux points O_i $(i=1,2)$ et un cercle S^1 où les strates points O_1, O_2, compactifient les bouts du revêtement universel $\mathbb{R}$ du cercle S^1.

Cette propriété d'invariance du quotient avant la valeur c^{**} est sans doute exceptionnelle. Par ailleurs il est probable que les déformations d'un endomorphisme n'engendrent pas nécessairement toutes les déformations génériques de la structure feuilletée globale que l'endomorphisme engendre par ses itérations.

REFERENCES

[1] R. Thom, Généralisation de la théorie de Morse aux variétés feuilletées. Annales de l'Institut Fourier, XIV (1964) p.173-189.

[2] R.F. Williams, Expanding attractors. Publ. IHES n°43 (1974) p. 169 - 203.

[3] J.-P. Dufour, Diagrammes d'applications différentiables. Thèse Université de Montpellier 1979.

* René THOM
Institut des Hautes Etudes Scientifiques.
35,Route de Chartres
91440 Bures-sur-Yvette
FRANCE

* * *

Singularities & Dynamical Systems
S.N. Pnevmatikos (editor)

HAMILTONIAN - LIKE PHENOMENA IN SADDLE - NODE BIFURCATIONS OF INVARIANT CURVES FOR PLANE DIFFEOMORPHISMS

Alain Chenciner

Université Paris VII
France

I present here a short report on my recent work on degenerate Hopf bifurcation of plane diffeomorphisms; for the proofs, the reader can consult [1],[2],[3]. The talk I gave in Crete stressed a particular result [2] concerning the existence of Aubry - Mather invariant sets. I shall discuss more here results connected with invariant curves, particularly a very new theorem asserting the existence of "many" "good" saddle-node bifurcations of invariant curves contained in a generic two-parameter family "unfolding" a local diffeomorphism possessing an elliptic fixed point of formal codimension two.

We consider (generic) two parameter families of smooth local diffeomorphisms $P_{\mu,a}$ of $\mathbb{R}^2,0$ and write them as perturbations of normal forms; more precisely, provided the derivative $DP_{0,0}(0)$ is conjugate to a rotation by an angle $2\pi\omega_0$ where ω_0 is not a rational number of denominator smaller than $2n+3$, one can choose $z,\bar{z}$ coordinates in $\mathbb{R}^2$ such that

$$P_{\mu,a}(z) = N_{\mu,a}(z) + O(|z|^{2n+3}),$$

$$N_{\mu,a}(z) = z(1+f(\mu,a|z|^2))e^{2\pi i g(\mu,a,|z|^2)},$$

where $f(\mu,a,X)$ and $g(\mu,a,X)$ are polynomials of degree n in the last variable with real coefficients depending smoothly on the parameters μ,a, which can be written

$$f(\mu,a,X) = \mu+aX+a_2(\mu,a)X+\ldots+a_n(\mu,a)X^n,$$

$$g(\mu,a,X) = b_0(\mu,a)+b_1(\mu,a)X+\ldots+b_n(\mu,a)X^n,$$

and satisfy the following assumptions:

$$a_2(0,0)\neq 0,\ b_1(0,0)\neq 0,\ b_1(0,0)-2a_2(0,0)\frac{\partial b_0}{\partial a}(0,0)\neq 0.$$

The bifurcation diagram of the family of normal forms and the associated dynamics are represented on fig.1 in the case where $a_2(0,0)<0$:

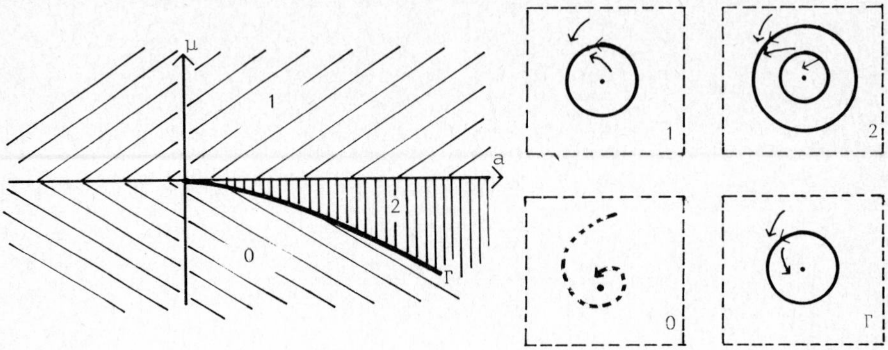

Fig.1

The following theorems are true as soon as $n \geqq 15$ (i.e. one must avoid the 33 first resonances). We say that $P_{\mu,a}$ looks like a normal form $N_{\mu',a'}$ if in a uniform (i.e. independant of μ,a) neighborhood Ω of 0, $P_{\mu,a}$ and $N_{\mu',a'}$ have the same number (0,1,or 2) of invariant curves and the same decomposition of Ω into basins of attraction or repulsion of 0 and the invariant curves.

THEOREM 1. For a generic family $P_{\mu,a}$ the set $\mathcal{N}$ of values of the parameters μ,a where $P_{\mu,a}$ looks like a normal form contains the complement of an infinite number of "bubbles" arranged in a string along Γ (fig.2). The bubbles are pinched on both sides and the closure $\tilde{\Gamma}$ of their contact points $\tilde{\gamma}_\omega$ is a Cantor set corresponding to diffeomorphisms $P_{\tilde{\gamma}_\omega}$ which look like a normal form N_{γ_ω} with $\gamma_\omega \in \Gamma$. In fact one even gets that the dynamics of $P_{\tilde{\gamma}_\omega}$ and N_{γ_ω} on their unique (saddle-node) invariant curve are conjugate.

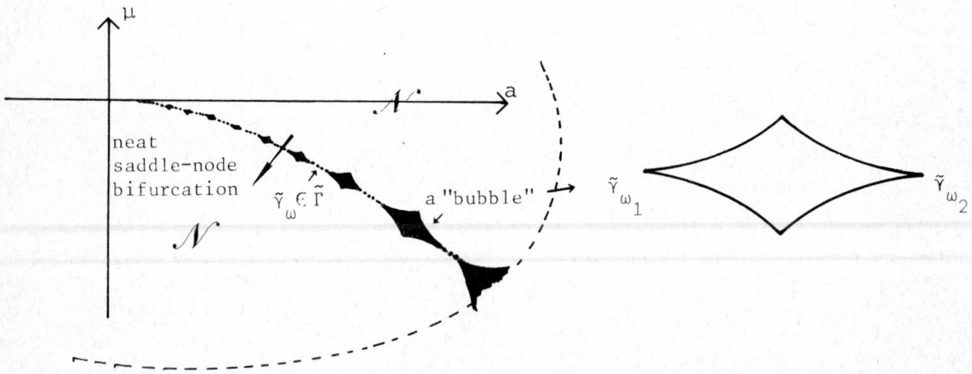

Fig.2

COROLLARY. *Almost every local one-parameter subfamily going through $P_{\tilde{\gamma}_\omega}$ with $\tilde{\gamma}_\omega \in \tilde{\Gamma}$ undergoes a neat saddle-node bifurcation of invariant curves as would be the case if one dealt with generic families of vector-fields.*

Note that $\tilde{\Gamma}$ replaces Γ in the same sense that the Cantor set of invariant curves given by K.A.M. theorem in the conservative setting replaces the one-parameter family of invariant curves of the associated normal form. Similarly, the bubbles play the role of Birkhoff's domains of instability:this last statement is made precise in the following theorem

THEOREM 2. *Generically the bubbles are open, that is the region $\mathcal{E}$ of the (μ,a) plane where $P_{\mu,a}$ does not look like a normal form contains an infinite number of disjoint open sets. In this region one can prove the existence of values of (μ,a) such that $P_{\mu,a}$ possesses ordered periodic orbits with associated homoclinic points and also of values of (μ,a) such that $P_{\mu,a}$ leaves invariant an Aubry-Mather Cantor set.*

The proof of Theorem 1 is divided into three parts whose first two are detailed in [1] and the last will appear in the published version of [1]:

1. One defines a horned neighborhood V of Γ (fig.3) such that, for any (μ,a) outside V, the normal attraction (or repulsion) of an invariant circle of $N_{\mu,a}$ is strong enough to resist the perturbation $P_{\mu,a} - N_{\mu,a} = 0(|z|^{2n+3})$. One can then prove that $P_{\mu,a}$ looks like $N_{\mu,a}$. V is defined by the apparent contour on the (μ,a) plane of the surfaces $f(\mu,a,r^2)+r^{2n} = 0$ and $f(\mu,a,r^2)-r^{2(n-3)} = 0$ and the techniques of proof are the same as in the classical Hopf bifurcation theorem for diffeomorphisms (for instance Hadamard's graph transform method). One can think of V (resp. the complement $\mathcal{H}$ of V) as the elliptic (resp. the hyperbolic) domain.

2. In V one uses techniques analogues to the ones used in K.A.M. theory (implicit function theorem in "good" Fréchet spaces of smooth functions):we first enrich the bifurcation diagram of fig.1 with the information relative to rotation numbers on invariant circles. On figure 3 we have depicted (in the case $b_1(0,0)>0$ and $\frac{\partial b_0}{\partial a}(0,0)>0$) the curves C_ω defined by $C_\omega = \{(\mu,a),\ N_{\mu,a}$ has an invariant circle on which it is a rotation by an angle $2\pi\omega$, i.e. of rotation number $\omega\}$. Note that Γ is the envelope of the C_ω's for $\omega > \omega_0$.

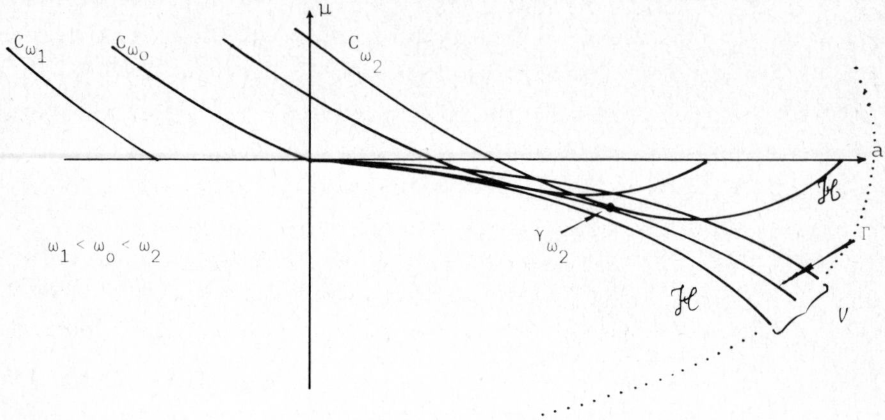

Fig. 3

The strategy of the second part (and also of the proof of theorem 2) is to understand the analogue $\tilde{C}_\omega$ of C_ω for the generic family $P_{\mu,a}$. Let $\tilde{C}_\omega = \{(\mu,a), P_{\mu,a}$ possesses an Aubry-Mather invariant set of rotation number $\omega\}$. Recall (4) that such a set is either an ordered periodic orbit, or an invariant curve, or an invariant Cantor set.

DEFINITION. Call an irrational number ω *"good" if* $\exists\, C>0, \beta>0$ *, such that*

1) For any rational number $\frac{p}{q}$ *one has* $|\omega - \frac{p}{q}| \geq \frac{C\,|\omega - \omega_0|}{|q|^{2+\beta}}$

2) $|\omega - \omega_0| \leq \varepsilon(C,\beta)$ *(*$\varepsilon(C,\beta)$ *depends on* C,β*, and the family* $P_{\mu,a}$*).*

PROPOSITION. If ω *is "good",* $\tilde{C}_\omega$ *is a smooth curve near* C_ω *coresponding to values of* (μ,a) *such that* $P_{\mu,a}$ *has an invariant curve, close to a circle, on which the induced diffeomorphism is smoothly conjugate to the rotation of rotation number* ω*. Moreover, there exists a unique point* $\tilde{\gamma}_\omega \in \tilde{C}_\omega$ *playing the same role as* γ_ω *for* C_ω *, i.e. the invariant curve of* $P_{\tilde{\gamma}_\omega}$ *is saddle-node, if* $(\mu,a) \in \tilde{C}_\omega$ *is on the left (resp. the right) of* $\tilde{\gamma}_\omega$ *, the invariant curve of* $P_{\mu,a}$ *is a normally hyperbolic attractor (resp. repellor).*

The set of all points $\tilde{\gamma}_\omega$ defines the Cantor set $\tilde{\Gamma}$. The proof uses Rüssmann's idea of "translated" curves, the preservation of area being here replaced by the existence of a translation parameter.

3) One proves easily that $P_{\tilde{\gamma}_\omega}$ looks like N_{γ_ω}. One then undertakes a new bifurcation study from the diffeomorphism $P_{\tilde{\gamma}_\omega}$:using the diophantine property of ω one can write normal forms for $P_{\mu,a}$, (μ,a) clo-

se to $\tilde{\gamma}_\omega$, in the neighborhood of the invariant curve of $P_{\tilde{\gamma}_\omega}$ which , at least for (μ,a) on $\tilde{C}_\omega$, look like normal forms around a fixed point. Outside $\tilde{C}_\omega$, the existence of a translation parameter forbids a total suppression of the angle dependance in the first terms of the normal forms, but Lipschitz estimates coming from the implicit function theorem in part 2) allow us to make this dependance very small. One is now ready to prove, using the same method as in 1), that if (μ,a) is near enough to $\tilde{\gamma}_\omega$ (more precisely in a box $\mathcal{D}_\omega$ defined by $|\mu-\mu_\omega|\leq c|\omega-\omega_o|^3$, $|a-a_\omega|\leq c|\omega-\omega_o|^2$), $P_{\mu,a}$ looks like a normal form with 0 (resp.2) invariant curves in the region $\tilde{A}_\omega$ (resp. $\tilde{B}_\omega$) depicted on figure 4.

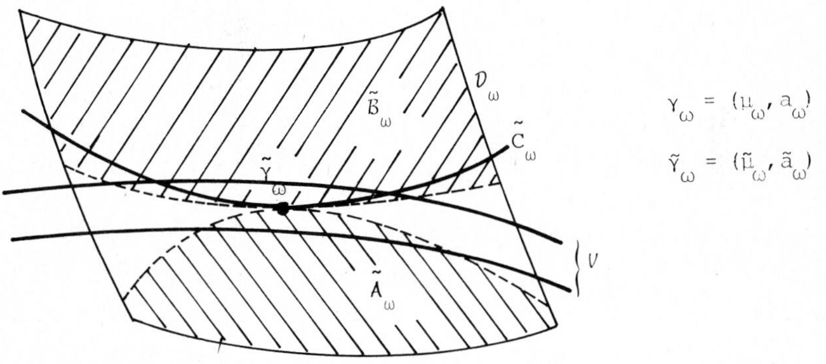

Fig.4

As we said before, the proof of theorem 2 relies on a study of the sets $\tilde{C}_\omega$ for bad ω's. Outside $\mathcal{V}$ the generic structure of $\tilde{C}_\omega$ is easy enough to understand through methods analogous to the ones used in classical Hopf bifurcation theory of diffeomorphisms: as $C_\omega\cap\mathcal{H}$, $\tilde{C}_\omega\cap\mathcal{H}$ has one component or two connected components corresponding respectively to attracting and repelling invariant curves. If ω is irrational, $\tilde{C}_\omega\cap\mathcal{H}$ is a Lipschitz curve (with lipschitz constant independant of ω); if ω is rational, $\tilde{C}_\omega\cap\mathcal{H}$ is generically a "fat" region bounded by Lipschitz curves, analogous to Arnold tongues (locking of rational frequencies as physicists say).

In order to understand the structure of $\tilde{C}_\omega\cap\mathcal{V}$ we work again in the "square" $\mathcal{D}_\omega$ whose definition makes sense for any ω. We prove that for any continuous path $c:[0,1]\longrightarrow\mathcal{D}_\omega$ going from the lower to the upper part of the boundary of $\mathcal{D}_\omega$, there exists a $t_o\in[0,1]$ such that $P_{c(t_o)}\in\tilde{C}_\omega$. As always in this kind of problem, Lipschitz estimates attached to ordered periodic orbits of monotone twist maps of an annulus allow us to reduce the irrational case to the rational one. For rational ω, the proof is a combination of a dissipative version of Birkhoff's idea for proving Poincaré's last geometric theorem and of Birkhoff-Aubry-Percival-Mather variational approach to ordered periodic orbits in the conservative case. The conclusion is that for any $\omega > \omega_o$ near enough to ω_o, the two connected components of $\tilde{C}_\omega \cap \mathcal{H}$ are connected in $\tilde{C}_\omega$.

Finally, rational ω's can also be good or bad: let ρ_ω be the radius of the unique (saddle-node) invariant circle of N_{γ_ω}; We say that $\omega = p/q$ is "good" if for (μ,a) near γ_ω, $P^q_{\mu,a}$ is still a perturbation of $N^q_{\mu,a}$ in a neighborhood of the circle $|z| = \rho_\omega$, i.e. if q is not too big with respect to $|\omega_o - \frac{p}{q}|^{-1}$.

If $\omega = \frac{p}{q}$ is "good", one can prove that $\tilde{C}_\omega$ is a connected fat region bounded by two smooth curves (fig.5). Moreover, there is an open region $\tilde{\gamma}_\omega$ contained in the interior of $\tilde{C}_\omega$ such that if $(\mu,a) \in \tilde{\gamma}_\omega$, $P_{\mu,a}$ has an ordered hyperbolic periodic orbit of rotation number ω with homoclinic intersections on both sides of stable and unstable manifolds (fig.6). Such a periodic orbit certainly does not lie on an invariant curve and the associated homoclinic tangle should be considered as the analogue of the saddle-node invariant curve of the corresponding normal form N_{γ_ω}.

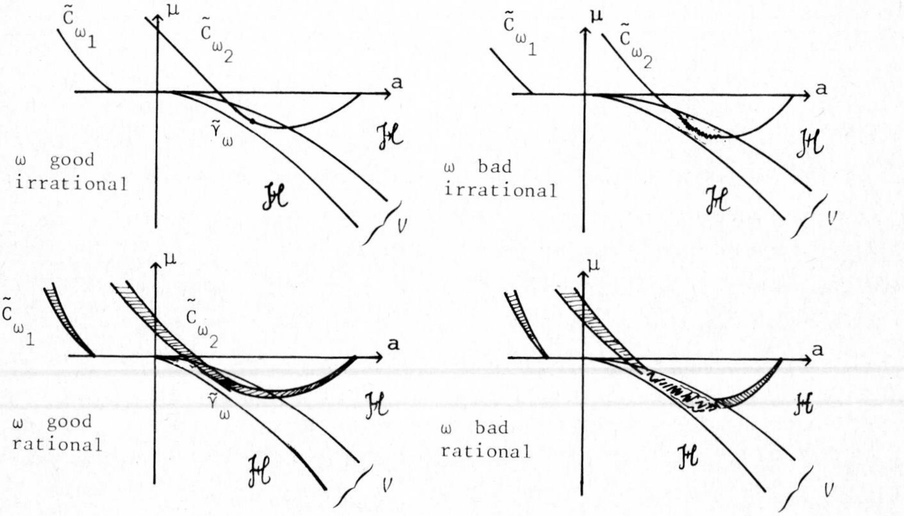

Fig. 5

The proof relies on the idea of "souvenir of nearby resonances", the periodic orbits being obtained by bifurcation from a resonant situation ($\omega = \omega_o = \frac{p}{q}$). As in Zehnder's work on the conservative case the phenomena just described are generic even for analytic families in a very fine topology.

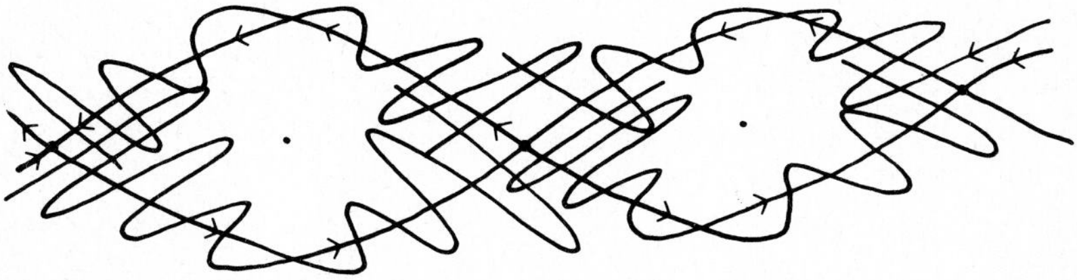

Fig. 6

Finally, a simple argument of limit shows that there exists indeed bad irrational numbers ω and values of (μ,a) on $\tilde{C}_\omega$ such that $P_{\mu,a}$ has no invariant curve cutting each radius in exactly one point and having rotation number ω. The invariant circle of a corresponding normal form $N_{\mu,a}$, $(\mu,a) \in C_\omega$, has been changed into an invariant Cantor set of the same rotation number. It is not unreasonable to think that these invariant Cantor sets lie on Birkoff-like attractors.

REFERENCES

[1] A. Chenciner, Bifurcations de points fixes elliptiques: 1-Courbes invariantes. Preprint Université Paris VII (to appear in: Publ. Math. I.H.E.S.).

[2] A. Chenciner, Bifurcations de points fixes elliptiques: 2-Orbites périodiques et ensembles de Cantor invariants.Prep. Univ.Paris7,1984. C.R.Acad.Sciences Paris 297 I (Nov. 1983), p. 465-467.

[3] A. Chenciner, Bifurcations de points fixes elliptiques: 3-Points homoclines. (en préparation),et, C.R.Acad.Sciences Paris 294 I (Fév. 1982), p. 269-272.

[4] A. Chenciner, La dynamique au voisinage d'un point fixe elliptique conservatif: de Poincaré et Birkhoff à Aubry et Mather. Séminaire Bourbaki, N° 622, Février 1984.

(March 1984)

* Alain CHENCINER
Université Paris VII
Département de Mathématiques
2 Place Jussieu
75 251 Paris , Cedex 05
FRANCE

* * *

Singularities & Dynamical Systems
S.N. Pnevmatikos (editor)
© Elsevier Science Publishers B.V. (North-Holland), 1985

THE DYNAMICS OF A DIFFEOMORPHISM AND THE RIGIDITY OF ITS CENTRALIZER

Jacob Palis

I.M.P.A.

Brasil

Our purpose here is to discuss a number of results of a work in progress with J.C. Yoccoz concerning centralizers of diffeomorphisms of a compact, connected and smooth manifold.

In general terms, the question we address is the following: given a diffeomorphism f, what can be said about its centralizer $C(f)$; i.e., the group of diffeomorphisms that commute with f ? In such a broad setting, without some restriction on the set of f's, hardly any general statement can be made about $C(f)$. For instance, depending on f, $C(f)$ is sometimes "large" and sometimes "small". The first case occurs, for example, when f embeds in a smooth flow (action of $\mathbb{R}$) as its time-one map or when f embeds in a smooth action of a even larger group such as $\mathbb{R}^k$ or $\mathbb{Z}^\ell \times \mathbb{R}^k$, ℓ and k being positive integers. On the other hand, $C(f)$ can be "small", like for instance when it is discrete: the only element that $C(f)$ contains in some neighborhood of the identity map is the identity map itself. The extreme (and important) case is when the centralizer is the smallest possible: there are diffeomorphisms that commute only with their integral powers. In such cases we say that their centralizers are trivial.

The key point of view here is that if we consider diffeomorphisms with "nice" persistent dynamical properties, we can go a long way toward the understanding of their centralizers. In fact, our results state that for many f's their centralizers are trivial. Although these results can be somewhat extended, our basic initial assumption is that the diffeomorphisms satisfy the so called Axiom A (hyperbolic nonwandering set) and the strong transversality condition, which we explain in Section 1. In this section we present the formal statements of the results. In Section 2 we mention some of the ideas involved in the proofs. Finally, in Section 3 we comment previous results and pose several open problems about centralizers in several contexts.

SECTION 1. THE STATEMENTS OF THE RESULTS.

We consider the set of all C^∞ diffeomorphisms $\mathcal{D}(M)$ of a compact C^∞ manifold M, endowed with the Whitney C^∞ topology. An element f in $\mathcal{D}(M)$ satisfies the Axiom A if

(i) its nonwandering set $\Omega(f)$ is hyperbolic

(ii) the set of periodic points $P(f)$ is dense in $\Omega(f)$.

A closed invariant subset $\Lambda \subset M$ is hyperbolic if there are a (continuous) splitting $T_\Lambda M = E^s \oplus E^u$ and a Riemannian metric for which there exists a constant $0 < \lambda < 1$ such that $\|Df_x v\| \leq \lambda \|v\|$ and $\|Df_x w\| \geq \lambda \|w\|$ for all $x \in \Lambda$, $v \in E^s_x$ and $w \in E^u_x$. If f satisfies the Axiom A, then it also satisfies the strong transversality condition if for all $x,y \in \Omega(f)$ their stable $W^s(x)$ and unstable $W^u(y)$ manifolds meet transversally.

We denote the set of f's satisfying both conditions by $\mathcal{G} = \mathcal{G}(M)$. Notice that $\mathcal{G}$ is open (and non-empty!) in $\mathcal{D}(M)$. Important particular cases are the Morse-Smale and the Anosov diffeomorphisms, denoted by MS and An. Recall that $f \in MS$ if $\Omega(f)$ consists of finitely many periodic orbits and $f \in An$ if all of M is hyperbolic in the sense above. Notice that MS is always non-empty on any ambient manifold M: the time-one map of any gradient flow on M can be approximated by an element in MS.

We now state our first results.

THEOREM 1. There is a generic (Baire second category) subset of $\mathcal{G}$ such that all of its elements have trivial centralizers.

In particular, the elements of a dense subset of $\mathcal{G}$ have trivial centralizers. It is very revelant, however, to try to obtain such a property to be also valid for <u>open</u> (not only dense) subsets of $\mathcal{G}$. For instance, using this <u>rigidity</u> (persistency of trivial centralizer), one can construct interesting structurally stable foliations with non-trivial holonomies (see [5]). In this direction we have the following results.

THEOREM 2. When $\dim M = 2$, the property of having a trivial centralizer holds for an open and dense subset of $\mathcal{G}$.

THEOREM 3. Let $\mathcal{G}^*$ be the open subset of $\mathcal{G}$ whose elements have a periodic attracting orbit. There exists an open and dense subset of $\mathcal{G}^*$ whose elements have trivial centralizers.

In particular, the same holds for an open and dense subset of MS in any compact manifold.

If we restrict our attention to roots of diffeomorphisms, we obtain as a consequence of our techniques the next corollary.

COROLLARY. For an open and dense subset of $\mathcal{G}$, its elements have no roots of any order. In particular, they do not embed in flows (continuous one-parameter group of diffeomorphisms).

Pursuing the same goal, we now restrict our attention to Anosov diffeomorphism on tori. We know that two elements in $\mathcal{G} = \mathcal{G}(T^n)$ are homotopic if and only if they induce the same isomorphism on homology. So let $f \in An$ and $\tilde{A} \in G\ell(n,\mathbb{Z})$ be the corresponding map on homology; let us denote by $A \in \mathcal{G}$ the projection (through the natural projection $\pi: \mathbb{R}^n \to T^n$) of $\tilde{A}$. By Manning [4], we know $A \in G\ell(n,\mathbb{Z})$ is hyperbolic (all eigenvalues have norm different from one) and thus its projected map A in T^n is Anosov. Moreover, there is a homeomorphism h of T^n such that $hf = Ah$.

Let $\mathcal{D}_A$ denote the connected component of $\mathcal{D}(T^n)$ containing A. Assume that the eigenvalues of $\tilde{A}$ are all simple (multiplicity one). Then, we obtain

THEOREM 4. Among the Anosov diffeomorphisms in $\mathcal{D}_A$, there is an open and dense subset whose elements have trivial centralizers.

SECTION 2. SOME IDEAS OF THE PROOFS.

A first main fact is the following

A. Rigidity Theorem. Let $f \in \mathcal{G}(M)$ and let $g_1, g_2 \in C(f)$. If $g_1 = g_2$ on an open set of M, then $g_1 = g_2$ on all of M. In particular, if $g = id$ on an open set of M then $g = id_M$.

This comes from the fact that any open set U intersects the stable manifold of some attractor Λ for f. In particular, it intersects the stable manifold of some periodic point in Λ. From that, we get $g_1 = g_2$ on all of the stable manifold of Λ. From this, it follows that $g_1 = g_2$ on the unstable manifolds of all the repellors for which the unstable manifolds intersect the stable manifold of Λ. Proceeding in this way, using the fact that M is connected, we get that $g_1 = g_2$ on a dense subset of M and thus on all of M.

The rigidity theorem lead us to consider the centralizer of f restricted to the stable manifold of an attractor. In the cases where there is no periodic attractor, we will consider both the restriction of f to the stable and the unstable manifolds of a periodic orbit in some non trivial attractor and the corresponding homoclinic structure.

B. Centralizer of a contraction. Following (A) above, we consider f restricted to the stable manifold of a periodic orbit which we may assume to be fixed and call it p. We denote the stable manifold by $W^s(p)$. We may also assume the spectrum of Df_p is different from the spectrum of Df at the other fixed points of f (open and dense property). This implies that if $g \in C(f)$ then $g(p) = p$. Again, we assume that the eigenvalues of Df_p with norm less than one $\lambda_1, \lambda_2, \ldots, \lambda_n$ are non-resonant that is $\lambda_i \neq \prod_j \lambda_j^{n_j}$, $1 \leq i,j \leq n$, $n_j \geq 0$ and $\sum_j n_j \geq 2$. We also assume $\lambda_i \neq \lambda_j$ for $i \neq j$. Again these are open and dense properties. Under these conditions, we have by [6] that $f/W^s(p)$ is linear under a C^∞ change of coordinates. In these coordinates any $g/W^s(p)$, $g \in C(f)$, is also linear [3]. From this we get, always restricting to $W^s(p)$, that $C(f)$ is isomorphic to $\mathbb{R}_*^r \times \mathbb{C}_*^s$, $r + 2s = n$. Notice that we are interested in $Z = C(f)/(f)$, where (f) is the cyclic group generated by f. If Z_1 denotes the maximal compact subgroup of Z, then $g \in Z - Z_1$ if and only if the ratios $\operatorname{Log}|\mu_i|/\operatorname{Log}|\lambda_i|$ are not all equal for $1 \leq i \leq n$, where $\mu_1, \ldots, \mu_n$ are the eigenvalues of $D_p(g/W^s(p))$. The same holds for the centralizer of $f/W^u(p)$ when p is not a sink (fixed point attractor). Moreover, since the local linearizing coordinates for $f/W^s(p)$ and $f/W^u(p)$ depend continuously on f, we can consider the above groups Z and Z_1 as being the same for all $\tilde{f}$ in a neighborhood of f in $\mathcal{G}$.

C. Global Obstructions. In view of the previous comments, we reduce the centralizer problem for $\tilde{f}$ near f in $\mathcal{G}$ to the following questions:

1. In any neighborhood U of f there is an open set V such that if $\tilde{f} \in V$ then $C(\tilde{f}) \cap Z_1 = id_M$ (Z_1 is the compact part of the centralizer group as above).

2. The same holds with respect to the full centralizer group **Z**.

From general facts about Lie groups, question 1 can be reduced to:

1'. Given $g \in Z_1$, $g \neq 1$, the set of $\tilde{f}$'s such that $g \in C(\tilde{f})$ is nowhere dense.

In all cases, starting with $f \in \mathcal{G}$, it is possible to solve question 1' by taking into account the global dynamics of f (and of $\tilde{f}$, $\tilde{f}$ near f). For instance, suppose that the fixed point p as in (B) above is attracting. Then $W^s(p)-p$ can be contained in the unstable manifold of a repellor or not. If so, there are two possibilities: the repellor is periodic or not. If it is periodic (a source), then we consider its local centralizer as before. Then, by perturbing f away from the sink and the source, we can make these "local" centralizers not to be globally compatible. In the other case, any element of the centralizer has to preserve the unstable foliation of the repellor. Again, via perturbations away from the sink and the repellor, we achieve a contradiction. Finally, we can do the same if $W^s(p)-p$ intersects the unstable manifold of a saddle-type basic set.

When p is not attracting, the solution of (1) is again possible by playing with its homoclinic structure.

In this way we can always solve question 1 in all cases, starting with f in $\mathcal{G}$. In particular, this proves the corollary of Theorem 3 in the introduction.

Notice that by solving (1) or (1'), we are already solving the centralizer problem when the ambient manifold is the circle S^1, a result previously obtained by Kopell [3].

The solution of question 2 is in general much more delicate. When the fixed point p above is attracting, we repeat the same scheme above for solving question 1. In this way, one proves Theorem 3 and its consequence for Morse-Smale diffeomorphisms.

The idea for proving Theorem 2 is the following. Since we had already treated the case where there is a periodic attracting orbit, we may suppose that the (fixed) point p for f is a saddle and $W^s(p)-p$ intersects $W^u(p)$. That is, there is a homoclinic structure for p. On the other hand, since we already took care of the compact part of the centralizer group (question 1), we may restrict our attention to elements $g \in Z-Z_1$. Thus, if μ_1 and μ_2 are the eigenvalues of Dg_p, then $\mathrm{Log}|\mu_1|/\mathrm{Log}|\lambda_1| \neq \mathrm{Log}|\mu_2| \neq \mathrm{Log}|\lambda_2|$, where λ_1 and λ_2 are the eigenvalues of Df_p (see subsection B above). We can then find integers n and m such that $\tilde{g} = f^n g^m$ has p as an attracting fixed point. Clearly $\tilde{g} \notin C(f)$ because $\tilde{g}$ can not preserve the homoclinic structure of p. And so $g \notin C(f)$, which answers question 2 since the same holds for an open set of $\tilde{f}$'s near f. With a somewhat similar argument, one can prove Theorem 1 using the homoclinic structure of f since it contains countably many points.

The "non-compact part" of Theorem 4 is more technical and we will not indicate it here.

SECTION 3. COMMENTS ON PREVIOUS RESULTS AND OPEN QUESTIONS

As mentioned before, Kopell has showed for the circle S^1 that the set of elements in MS with trivial centralizers is open and dense. Since MS is open and dense in $\mathcal{D}(S^1)$, we conclude that for an <u>open</u> and <u>dense</u> subset of $\mathcal{D}(S^1)$ its elements have <u>trivial</u> centralizers. On the other hand, it follows from remarkable results of Herman and Yoccoz that for a diffeomorphism of S^1 with an irrational rotation number which is non-Liouville the centralizer is isomorphic to the full group of rotations. Other relevant results we also obtained by them [2,7].

In higher dimensions, Anderson [1] proved that for an <u>open</u> and <u>dense</u> subset of <u>MS</u> the elements have <u>discrete</u> centralizers. This result was extended by the author of the present paper [5] to the open set $\mathcal{G}$ of diffeomorphisms satisfying the Axiom A (hyperbolic nonwandering set) and the strong transversality condition.

From the results announced here, Yoccoz and the author went

quite a long way in proving the following

CONJECTURE. For an open and dense subset of $\mathcal{G} \subset \mathcal{D}(M)$ its elements have trivial centralizers.

Still the proof of the conjecture is not yet complete. Specially, it remains open the "openess" part of the statement.

More generally, one can pose the following question which seems very difficult:

QUESTION. Is it true that for an open and dense subset of $\mathcal{D}(M)$ its elements have trivial centralizers?

Finally, we point out that most of the results above, including the ones we announce here, correspond to open questions for real analytic diffeomorphisms. The main difficulty, of course, concerns the perturbations one is allowed to perform. Similarly for volume preserving or for sympletic diffeomorphisms.

REFERENCES

[1] R.B. Anderson, Diffeomorphisms with discrete centralizer, Topology 15, (1976).

[2] M. Herman, Sur la conjugaison différentiable de difféomorphisms du cercle à des rotations, Publ. Math. IHES 49 (1979).

[3] N. Kopell, Commuting diffeomorphisms, Proc. A.M.S. Symp. 14 (1970).

[4] A. Manning, There are no new Anosov on tori, Am. J. Math. 96 (1974).

[5] J. Palis, Rigidity of the centralizers of diffeomorphisms and structural stability of suspended foliations, Proc. Conf. PUC-RJ, Lect. Notes in Math. 652 (1978), Springer-Verlag.

[6] S. Sternberg, Local contractions and a theorem of Poincaré, Am. J. Math. 79 (1957).

[7] J.C. Yoccoz, Thèse-Doctorat d'État (1984), to appear.

Jacob PALIS
Instituto de Matemática Pura e Aplicada (IMPA)
Estrada Dona Castorina 110
22460 - Jardim Botânico - Rio de Janeiro, RJ, Brasil

* * *

Singularities & Dynamical Systems
S.N. Pnevmatikos (editor)

TRANSLATION REPRESENTATIONS OF SCATTERING FOR RATIONAL MAPS*

Jean - Pierre Françoise
Université de Paris Sud
France

D. Sullivan (S) introduced a heuristic dictionary between iteration of rational maps on the Riemann sphere $\bar{\mathbb{C}}$ and Kleinian groups. In his dictionary, the dynamics of the iteration near a rationally indifferent fixed point correspond to a Kleinian group of cofinite volume. In 1972, L.D. Faddeev and B.S. Pavlov (F.P.) showed that the Lax-Phillips theory of scattering (L.P.) could be applied to the automorphic wave equation for Fuchsian groups with fundamental domain having a cusp. We define here translation representations of scattering for rational maps near a rationally indifferent fixed point. To do this, we have to use the Ecalle theory of resurgent functions and alien derivations.

I am grateful to L. Carleson and P.Jones for inviting me to participate in the workshop on iterations and for helpful discussions. I thank the Mittag-Leffler institute for providing such fine working conditions. It is a pleasure to acknowledge my indebtness to H.Epstein and D. Sullivan for introducting me to the study of rational maps. I also express my gratitude to J. Ecalle and B. Malgrange for discussions and guidance.

1. Let $f: \bar{\mathbb{C}} \to \bar{\mathbb{C}}$ be a rational map, such that $f^p \neq \mathrm{Id}$, for all $p \in \mathbb{Z}$, and let F be the Fatou-Julia set of f.

A rationaly indifferent fixed point of f is a point $t_0 \in \bar{\mathbb{C}}$ such that

$$f^p(z_0) = t_0 \ , \quad p \in \mathbb{N} \ , \quad Df^p(t_0) = e^{2\pi i k/\ell} \ , \quad k,\ell \in \mathbb{N}.$$

To simplify notations, we assume $z_0 = 0$. Hence the map $f^{p\ell}$ has for its Taylor development:

$$f^{p\ell}(z) = z + a_0 z^{n_0} + a_1 z^{n_0+1} + \dots$$

(*) *Report n°21, 1983, of Mittag-Leffler Institute.*

We get for the successive iterates

$$f^{p\ell n}(z) = z + na_0 z^{n_0} + \dots$$

from which we deduce that $\dfrac{d^{n_0} f^{p\ell n}}{dz^{n_0}}$ is not bounded; this implies that $0 \in F$.

In this paragraph, we remind the reader of some facts about the Fatou-Julia set F in a neighborhood of the point 0 .

We assume first that f itself is of the form

$$f(z) = z + z^2 + \dots$$

After an inversion $z \longrightarrow w = \frac{1}{z}$, the map f becomes

$$f: w \longrightarrow f(w) = w-1 + \psi(w) .$$

We may control ψ outside of a big disc:

$$|w| \geq R >> 0 \Longrightarrow |\psi(w)| \leq \frac{c}{|w|^2} .$$

We introduce the translation

$$f_0 : w \longrightarrow f_0(w) = w-1 .$$

The following may then be easily shown.

Lemma 1.1 There exists a unique formal series $\hat{\Phi}$ *such that:*

1.1 $$\hat{\Phi} \circ f = f_0 \circ \hat{\Phi} .$$

Fatou (F) introduced a domain which looks like this:

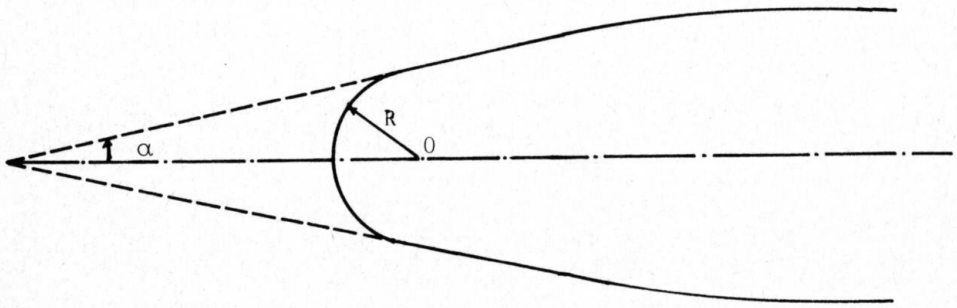

Figure 1.2

The domain's boundary is the half-circle δD_R plus two segments, which have angle α with the x - axis and two paraboloid domains whose *asymptotic direction is the* x - *axis* .

We will denote by $D_-(\alpha,R)$ the complement of this domain. It turns out that if R is large enough and α small, $D_-(\alpha,R)$ is invariant by f , and the sequence $\{f^n(w)-n\}$ converges uniformy on any compact subset of $D_-(\alpha,R)$. The limit function, $\Phi_-(w)$, is holomorphic on $D_-(\alpha,R)$ and verifies

1.3 $$\Phi_- \circ f(w) = \Phi_-(w) - 1 = f_0 \circ \Phi_-(w) .$$

Kimura (K) improved Fatou's result with

LEMMA 1.4 Φ_- *is asymptotic to* $\hat{\Phi}$ *when* $|w| \to \infty$ *in* $D_-(\alpha,R)$.

The same work may be donc Mutadis Mutandis for the branch of the inverse f^{-1} which takes 0 to 0 . Thus we find on a domain $D_+(\alpha,R)$:

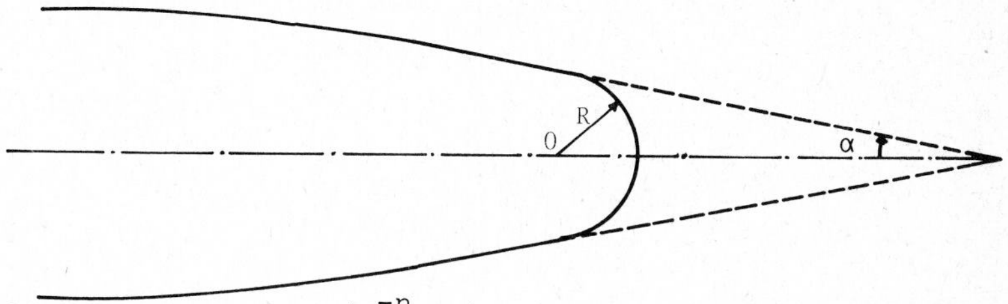

Figure 1.4

and a holomorphic map $\Phi_+ = \lim_{n\to\infty} (f^{-n} + n)$ which satisfies

1.5 $$\Phi_+ \circ f^{-1}(w) = f_0^{-1} \circ \Phi_+(w) = \Phi_+(w) + 1 .$$

By the fact that f has a simple pole at ∞, f is univalent in $D_\pm(\alpha,R)$ if R is large enough and so are the f^n and $\Phi_\pm$. We may consider the inverse Φ_-^{-1} on a $D_-(\alpha^1,R^1)$, $\alpha^1 < \alpha$, $R^1 < R$ and introduce

1.6 $$\Sigma(w) = \Phi_+ \circ \Phi_-^{-1}(w) .$$

From 1.3 and 1.5 we find out that Σ commutes with translation. A priori Σ is only defined on a domain $D_-(\alpha^1,R^1) \cap D_+(\alpha,R)$:

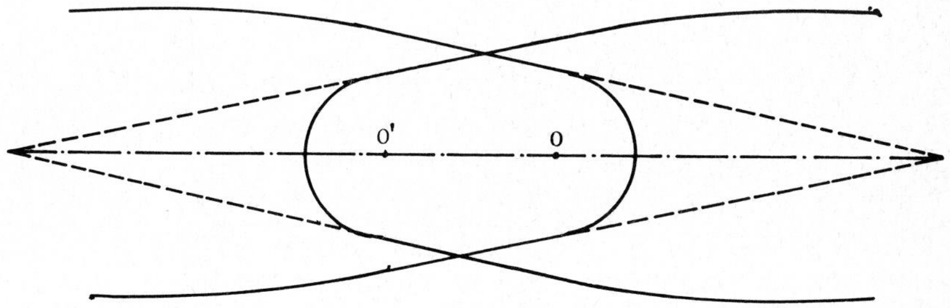

Figure 1.7

but we may extend it by using the fact that it commutes with translations , and consider it defined on the complement of a strip $|\text{Im} z| \geq A$.

Finally, we get two Fourier series

$$\Sigma_{\pm}(w) = w + \sum_{n\geq 1} c_n^{\pm} e^{\pm inw}$$

convergent for $\mathrm{Im}w > A$ and $\mathrm{Im}w < -A$ respectively such that $\Sigma_{\pm}(w)$ is flat to the identity at ∞. (by 1-4). B. Malgrange introduced this construction and proved that the data $\Sigma_{\pm}(w)$ give a *complete system of analytic invariants* for the maps we study, (M).

Let us now turn to quasi-conformal conjugacy classes with

THEOREM 1.8 Let $f: \mathbb{C},0 \to \mathbb{C},0$ *be a local germ of an analytic map of the form* $f: z \to f(z) = z + z^p + \ldots$. *Then there exists a quasi-conformal homeomorphism* $h: \mathbb{C},0 \to \mathbb{C},0$ *such that*

$$h \circ f(x) = h(z) + h(z)^p$$

This theorem was announced by Voronin (V) in a publication of Sherbakov to appear. It is actually one the key tool of Voronin construction of analytic invariants. We give the proof here to make our paper self-contained.

Proof. Assume first $p = 2$. We have

$$f(w) = w-1 + \psi(w) , \quad |\psi(w)| < \frac{c}{|w|} \quad \text{if} \quad |w| > R$$

and we choose $f_0(w) = w-1$ as a local model.
We take two circles of radius R_1, R_2 $(R_1 < R_2)$ large enough and centered at 0. The maps f_0 and f are homeomorphisms near the boundaries of the annulus $D_{R_2} - D_{R_2}$, hence we may glue them together in a quasi-conformal way into an $\tilde{f}$.

We look for an h such that $h \circ f = h-1$. Let us define h to be the identity on the imaginary axis. Now observe that outside of D_{R_2} the image by f of the imaginary axis Δ, does not intersect Δ_1 because f is nearly a translation and it has no cusps because f is univalent there. Define

$$h(z) = h(\tilde{f}^{-1}(z))-1 , \; z \in \Delta_1 .$$

We now glue together $h|_{\Delta_1}$ and $h|_{f(\Delta_1)}$ in a quasi - conformal way into h. Then we take the successive images of Δ_1 by f and f^{-1} and we get strips limited by $\Delta_n = f^n(\Delta_1)$ and $\Delta_{n+1} = f^{n-1}(\Delta_1)$ which play the role of a fundamental domain for the map f. We use the map itself to propagate the homeomorphism h to the whole sphere $\overline{\mathbb{C}}$. To be sure we get quasi-conformality at ∞, we have to check control of the derivatives of f^n. This is done easily using the majorations from Fatou (F). More precisely, after an analytic conjugacy, we may assume that

$$f(w) = w-1+\psi(w) \quad , \quad |\psi(w)| < \frac{c}{|w|^2} \quad \text{if} \quad |w| \geq R \ .$$

Hence, we get for all $n \geq 0$

$$f^n(w) = w-n+c_n(w)n^{-1}$$

where $c_n(w)$ is uniformly bounded.

We get from this that the coefficient of quasi-conformality of the homeomorphism h restricted to the strip (Δ_n, Δ_{n-1}) tends to 1.

We deal now with the general case:

$$f(z) = z + z^p + \dots \quad , \quad p > 2 \ .$$

We use the blowing-up $\pi: z \to \xi = z^{p-1}$ and consider $\pi \circ f \circ \pi^{-1} : \xi \to \xi + (p-1)\xi^2 + \dots$.

After the inversion it appears, far away from 0, to be tangent to translation by $-(p-1)$. We may thus apply the same construction as we did before. We have a fundamental domain, we glue together homeomorphisms on the boundaries of this fundamantal domain in a quasi-conformal way and we propagate it.

Finally, we find for f a quasi-conformal conjugacy to any local model, outside of 0, and we may extend it through 0.

We see that the analogy of D. Sullivan (S) is particularly striking in the case where we have a fundamental domain outside of the Fatou-Julia set.

Let us complete in some points the description of F near 0. First, for $f = z + z^2 + \dots$, from what we saw, we have an f-invariant domain $D_-(\alpha,R)$.

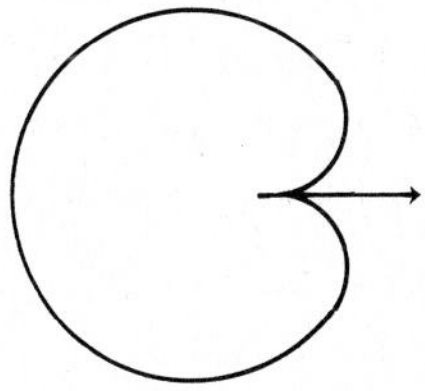

Figure 1.9

We have an f^{-1}-invariant domain $D_+(\alpha,R)$ which looks symmetric. The intersection S of the two domains is *parabolic*

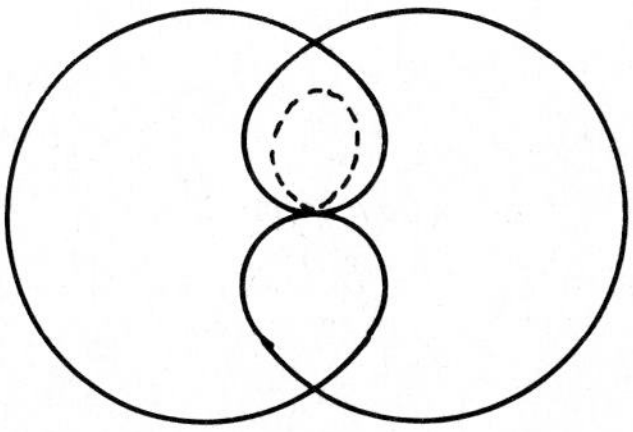

Figure 1.10

We define the *pétal* of f at 0 as the maximal parabolic domain which contains S . Its boundary will be a part of the Fatou - Julia set. We see that F cannot go inside of $D_{-}(\alpha,R)$. Hence it has a cusp at 0. Fatou proved that for $0 \leq \varepsilon \leq \frac{1}{4}$ the set F for $z^2 + \varepsilon$ is a Jordan curve. Pommerenke has pointed out that for the map $z \to z^2 + \frac{1}{4}$ one obtains an *invariant Jordan curve* which is not *quasi-circle.*

The same kind of analysis may be done using the blowing-up trick for the general case, and we find out that the Fatou-Julia set looks like a flower with p-1 pétals which are parabolic domains at each point of the periodic orbit.

To be more complete, let us note that the boundary of the pétal need not be a *Jordan curve*. It may, e.g., carry a Liouville indifferent fixed point. There always exists a *critical point* inside the flower. See the proof of Fatou (F) or Julia (J) or the book of W. Thurston about iteration of rational maps to appear (T).

When R is large enough, we get small invariant parabolic domains on which we will study the iteration of the map from the viewpoint of scattering theory. The idea that a conjugacy problem might be presented as a scattering theory was already in Nelson's book (N) and is related to S. Sternberg linearization theorem for a C^∞ vector field. It was resuggested as a formal analogy by B. Malgrange (M).

2. Let $\mathbb{D}$ be the disc and $H_1(\mathbb{D})$ be the set of holomorphic functions $\bar{\varphi}$ on $\mathbb{D}$ with growth condition $|\bar{\varphi}(z)| \leq \frac{c}{1-|z|}$ near the boundary.

Let R be a Riemann surface isomorphic to the plane $\mathbb{C} - \Omega$, where $\Omega = 2\pi i\mathbb{Z}$. Then the universal cover of R is the disc $\mathbb{D}$. Let $\pi : \mathbb{D} \to R$ be the projection. For a function $\varphi : R \to \mathbb{C}$ we set $\bar{\varphi} = \varphi \circ \pi : \mathbb{D} \to \mathbb{C}$.

Proposition 2.1 Single - valued functions $\varphi : R \to \mathbb{C}$ *such that* $\bar{\varphi} \in H_1(\mathbb{D})$ *are exactly the meromorphic functions on* $\mathbb{C}$ *having poles only on* Ω *and with each pole having order less than or equal to one and which grow at most exponentially at infinity.*

Proof.

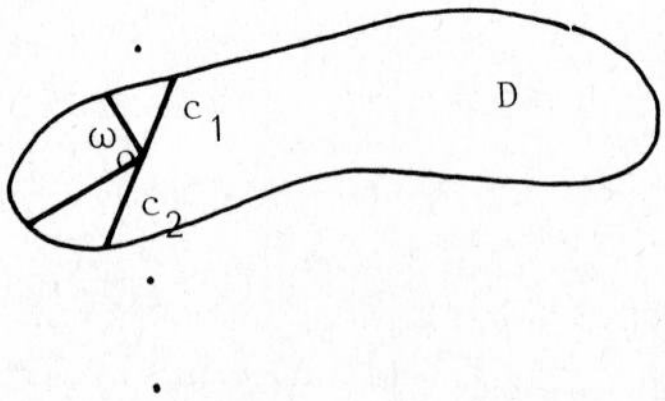

Figure 2.2

Choose a domain U with $\omega_0 \in \Omega$ inside and a cut C_1.
The map π^{-1} is univalent on $U \setminus C_1$ and hence by the Koobe one quarter theorem, the hyperbolic distance from z to ω_0 is controlled by the hyperbolic distance on the disc:

$$\delta(z,\omega_0) \leq c\ \delta(\pi^{-1}(z)\ ,\ \delta\, \mathbb{D})\ .$$

Hence we have the estimate

$$|\varphi(z)| \leq \frac{c^1}{|z-\omega_0|}\ .$$

If we draw another cut C_2, we get the same kind of estimate for any point in $U \setminus \omega_0$. This, together with the fact that f is single-valued, implies there exists at most a pole of order one at ω_0.

We introduce now $H_2(\mathbb{D})$, the set of homomorphic functions $\bar{\varphi}$ on the disc such that $|\bar{\varphi}'(z)| \leq \dfrac{c}{(1-|z|)^2}$ for some constant c.

By the same argument used in 2.1 we get:

Proposition 2.3 The holomorphic functions $\varphi : R \to \mathbb{C}$, *whose derivative* φ' *is single-valued and such that* $\bar{\varphi} \in H_2(\mathbb{D})$ *and exactly the holomorphic functions on* $\mathbb{C} - \Omega$ *with possibly a pole of order one plus a logarithmic singularity at points of* Ω *which grow at most exponentially.*

We see that the set of such functions is closed under uniform convergence. This is essential in the proof of

THEOREM 2.4 (J. Ecalle) *The Borel transform of* $\bar{\Phi}$ (up to δ,δ') *defines an analytic germ at* $0 \in \mathbb{C}$ *which extends to an holomorphic map on* $\mathbb{C} - 2\pi i\mathbb{Z}$ *with possibly a pole of order one plus a logarithmic singularity at the points* $\omega \in 2\pi i\mathbb{Z}$.

Let us recall that the Borel transform of a formal series is the $\mathbb{C}$-linear operator defined by

$$\begin{aligned} B(w) &= \delta \\ B(1) &= \delta' \qquad \text{(Dirac distributions)} \\ B\left(\frac{1}{w^n}\right) &= \frac{\xi^{n-1}}{(n-1)!}\ ,\ n \in \mathbb{N}\ ,\ n > 1\ . \end{aligned}$$

We may associate to f its vectorfield $\hat{u}$, which is the formal series such that f is the time 1 flow of $\hat{u}$, and its iterates $\hat{f}^t$, where $t \in \mathbb{R}$, which are formal series in w with polynomial coefficients in t.
The same proof as in 2.4 gives that the Borel transforms of $\hat{u}$ and of the $\hat{f}$ are holomorphic maps on $\mathbb{C} - 2\pi i\mathbb{Z}$ with possibly poles of order

one plus logarithmic parts at points $\omega \in 2\pi i\mathbb{Z}$.

3. The exponential map takes $\mathbb{C} - 2\pi i\mathbb{Z}$ onto $\bar{\mathbb{C}} - \{0,1,\infty\}$ and is a covering map. We prefer to use the spehere minus three points in the following because it has finite topological type. Its universal cover is the disc $\mathbb{D}$ and more precisely:

$$\bar{\mathbb{C}} - 0,1 = \mathbb{D}/\Gamma$$

where Γ is the Fuchsian group $\{ \frac{az+b}{cz+d} \in PSL_2(\mathbb{Z})$, where a,d are odd, b and c are even $\}$. The Fuchsian group Γ is generated by the three parabolic elements which are the direct rotations around $\{1,\infty,0\}$; noted $R,T,T,$ with the relation $RST = 1$.

Ecalle define a convolution product $*$ which lifts on $\mathbb{D}$ the usual convolution of germs of holomorphic maps at the point 1 :

$$(\bar{\varphi}_1, \bar{\varphi}_2) \to \bar{\varphi}_1 * \bar{\varphi}_2(z) = \int_1^z \bar{\varphi}_1(z/\xi)\bar{\varphi}_2(\xi)\,d\xi$$

This involve contractible symetric paths (symmetric relatively to its middle and contractible inside the class of symmetric paths) .

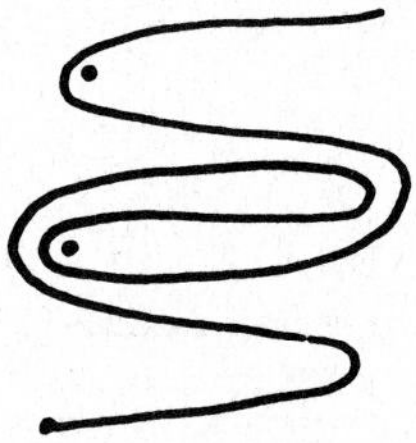

3.1 Example of a contractible symmetric path.

Now the Fuchsian group Γ operates on the algebra of holomorphic functions on the disc by composition

$$\gamma_1\bar{\varphi}_1 \to \bar{\varphi}_1 \circ \gamma^{-1} .$$

By the fact that elements in $SL_2(\mathbb{R})$ preserve hyperbolic distance we deduce that Γ operates on $H_1(\mathbb{D})$. We may extend this action into an action of the free elgebra $\mathbb{C}(\Gamma)$ of the group Γ. We consider the subalgebra $D(\Gamma) = \mathbb{C} \oplus (1-R)\mathbb{C}(\Gamma)$ generated by

$$R, R^{-1}, S_n = (1-R)S^n, \; T_n = (1-R)T^n .$$

One checks that

$$\text{i)} \quad S_n(\varphi_1 * \varphi_2) = \sum_{p=0}^{n} S_{n-p}(\bar{\varphi}_1) * S_p(\bar{\varphi}_2)$$

$$\text{ii)} \quad T_n(\bar{\varphi}_1 * \bar{\varphi}_2) = \sum_{p=0}^{n} T_{n-p}(\bar{\varphi}_1) * T_p(\bar{\varphi}_2)$$

iii) $R^{\pm}(\bar\varphi_1 * \bar\varphi_2) = R^{\pm}(\bar\varphi_1) * R^{\pm}(\bar\varphi_2)$.

Hence we build the formal operator

$$S(t) = 1 + \Sigma\, t^n S_n$$

and take its formal logarithm

3.2 $$\log S(t) = \sum_p \frac{(1-S(t))^p}{p} = \sum_n t^n S_n^* \ .$$

From i) $S(t)$ preserves $*$-product, therefore $\log S(t)$ is a derivation for the $*$-product and all the coefficients S^* are derivations. The same can be dane with $T(t) = 1 + \sum_n t^n T_n$. Hence we get a collection of derivations

$$\{R^m S_n R^{-m}\} \qquad \text{and} \qquad \{R^m T_n R^{-m}\}$$

which Ecalle calls *alien derivations*.

This relate analytic data, the convolution product, to a purely geometric one, i.e. the action of the Fuchsian group on the disc.

Note that these derivations really depend on the topology of $\mathbb{C}-2\pi i\mathbb{Z}$ (or $\bar{\mathbb{C}} - \{0,1,\infty\}$) via the fundamental group and do not depend of the uniformization.

Given a $\omega \in 2\pi i\mathbb{Z}$, there exists a distinguished element in the family $\{R^m S_n^* R^{-m}\}_{m,n} \cup \{R^m T_n^* R^{-m}\}_{m,n}$, denoted by Δ_ω , such that $\Delta_\omega \bar\varphi = 0$ for all $\bar\varphi \in H_1(\mathbb{D})$ holomorphic at ω .

Similarly, for the family $\{R^m S_n R^{-m}\}_{m,n} \cup \{R^m T_n R^{-m}\}_{m,n}$ there exists a distinguished element Δ_ω^+ (resp Δ_ω^-) such that $n > 0$ (resp $n < 0$) and such that $\Delta_\omega^{\pm} \bar\varphi = 0$ for all $\bar\varphi \in H_1(\mathbb{D})$ holomorphic at ω .

These are not derivations but have property

3.3 $$\Delta_\omega^{\pm}(\bar\varphi_1 * \bar\varphi_2) = \sum_{\omega_1+\omega_2=\omega} \Delta_{\omega_1}^{\pm}(\bar\varphi_1) * \Delta_{\omega_2}^{\pm}(\bar\varphi_2)$$

We define now a *composition product* by the "Taylor formula" (E):

3.4 $$\bar\varphi_1 \circ \bar\varphi_2(\xi) = \sum_{n=0}^{\infty} \frac{1}{n!} (\bar\varphi_1-\partial)^{*n}(\xi) * (-\xi)^{*n} * \bar\varphi_2(\xi) \ .$$

We also introduce the convolutive exponential (E) , denoted by Exp by setting

3.6 $$\operatorname{Exp} \bar\varphi = \sum_{n=0}^{\infty} \frac{1}{n!} \bar\varphi^{*n}$$

We thus deduce the formula (E)

3.6 $\Delta_\omega(\bar\varphi_1 \circ \bar\varphi_2)(\xi) = (\Delta_\omega \bar\varphi_1) \circ \bar\varphi_2 * \mathrm{Exp}-(\bar\varphi_2-\partial)(\xi) +$

$$\left((\xi * \bar\varphi_1) \circ \bar\varphi_2\right) * \Delta_\omega \bar\varphi_2(\xi)$$

4. Let us consider the set of all formal series of the form

$$\varphi(w) = w+a_0 + \sum_{i=1}^{\infty} a_i w^{-i}$$

whose Borel transform satisfies $\overline{B(\varphi)} \in H_2(\mathbb{D})$ and whose logarithm u also satisfies $\bar u \in H_2(\mathbb{D})$.
We denote by $H_3(\mathbb{D}) \subset H_2(\mathbb{D})$ the set of such $\overline{B(\varphi)}$.
Given a convergent germ $f : \mathbb{C},0 \to \mathbb{C},0$ of the form $f : z \to z+z^2+\ldots$ its iterates f^t for $t \in \mathbb{R}$ define elements in $H_3(\mathbb{D})$. Note furthermore that $H_3(\mathbb{D})$ is stable under the composition product. Hence we may introduce a one-parameter group $U(t)$ on the space $H_3(\mathbb{D})$ by setting

4.1 $U(t)\bar\varphi = \bar\varphi \circ f^t$

Now the logarithm gives obviously a translation representation of this one-parameter group on $H_2(\mathbb{D})$. We need to adapt it slightly to get.

THEOREM 4.2. There exists a couple of translation representations $R_{1,2} : H_2(\mathbb{D}) \longrightarrow \mathrm{Aff}(\mathbb{R}, H_2(\mathbb{D}))$ *of the one - parameter group 4.1 such that the associated Scattering Matrix is* $\Sigma_\pm(w)-w$.

See the looks of Lax-Phillips (L-P) to compare with their situation. Lax-Phillips introduced an abstract scattering theory and gave several examples which fit this abstract presentation , among them scattering for the acoustic equation and for dissipative hyperbolic systems. Later Faddeev and Pavlov showed that this abstract scattering theory could be applied to the automorphic wave equation for Fuchsian groups with one cusp. Then L-P extended this treatment to any Fuchsian group of cofinite volume.

We define

4.3 $R_1(\bar\varphi)(s) = \mathrm{Exp}\left[2\pi i(\bar\Phi-\partial) * \bar u^{-1} * (\log \bar\varphi + s\bar u)\right]$

$$R_2(\bar\varphi)(s) = \sum_\omega R_1(\bar\varphi)(s) * \Delta_\omega^\pm \bar\Phi$$

i) It may be easily verifief that $R_{1,2}$ are translation representations. For instance

$$R_1(\bar\varphi \circ f^t)(s) = \mathrm{Exp}\left[2\pi i(\bar\Phi-\partial) * \bar u^{-1} * (\log(\bar\varphi \circ f^t)+s\bar u)\right]$$

$$R_1(\bar{\varphi} \circ f^t)(s) = \mathrm{Exp}\left[2\pi i(\Phi-\partial) * \bar{u}^{-1} * (\log\bar{\varphi} + t\bar{u} + s\bar{u})\right]$$

4.4 $$R_2(\bar{\varphi} \circ f^t)(s) = R_1(\bar{\varphi})(s+t)$$

Now from 1.1 and 3.6 we get that there exist constants c_ω such that

4.5 $$\Delta_\omega \cdot \bar{\Phi} = c_\omega \, \mathrm{Exp}{-\omega}(\bar{\Phi}-\partial)$$

Our choices for $R_{1,2}$ rely deeply on this equation which is called by Ecalle a *Resurgence Equation* (analytic bootstrap).

ii) We deduce from 3.2 the existence of constants $c_\omega^{\pm}$ such that

4.6 $$\Delta_\omega^{\pm} \cdot \bar{\Phi} = c_\omega^{\pm} \, \mathrm{Exp}{-\omega}(\bar{\Phi}-\partial)$$

This relation would have been difficult to chech directly because of the intricacy of 3.3.

Now, as usual in Borel theory, we get a "realization" of the Borel transform by taking the Laplace transform of $\bar{\Phi}$.

It turns out that it exists and that we have:

4.7
$$\bar{\Phi}_-(w) = \int_0^{\infty} \bar{\Phi}(\xi)\, e^{+w\xi} d\xi$$
$$\bar{\Phi}_+(w) = -\int_0^{-\infty} \bar{\Phi}(\xi)\, e^{w\xi} d\xi$$

When we use properties of the Laplace transform we find that the coefficients $c^{\pm}$ are exactly those of the series $\Sigma \pm (w) - w$. (E)

Now we have by 4.5

$$R_2(\bar{\varphi})(s) = \Sigma c_\omega^{\pm} \,\mathrm{Exp}\, 2\pi i \left[(\bar{\varphi}-\partial) * \bar{u}^{-1} * (\log\bar{\varphi} + s\bar{u})\right] * \mathrm{Exp}{-\omega}(\bar{\varphi}-\partial)$$

and if we write $\omega = 2\pi i n$, we get

$$R_2(\bar{\varphi})(s) = \sum_n c_n^{\pm} R_1(\bar{\varphi})(s-n)$$

Therefore

4.8 $$R_2(\bar{\varphi})(s) = \int \Sigma\, c_n^{\pm}\, \delta_n(t)\, R_1(\bar{\varphi})(s-t)\,dt$$

Hence the scattering distribution associated to the couple $R_{1,2}$ we introduced is

$$\sum_n c_n^{\pm} \exp 2\pi i n w = \Sigma_{\pm}(w) - w .$$

* * *

* *I am glad to thank University of Crete, particulary Spyros Pnevmatikos, and the University of Dijon, for inviting me to participate in the meeting "Singularities and Dynamical Systems".*

R E F E R E N C E S

[E] *J. Ecalle*,"Les fonctions résurgentes" (en trois parties), Publ. Math. d'Orsay, 81-06.

[F.P.] *L.D. Faddeev/B.S. Pavlov*, "Scattering theory and automorphic functions", Seminar of Steklov Mathematical Institute of Leningrad, vol. 27 (1972), 161-193.

[F] *P. Fatou*, "Sur les équations fontionelles", Bull.Soc. Math. France, 47(1919), 161-271, 48(1920) 33-94, 208-316.

[J] *G. Julia*, "Mémoire sur l'itération des fonctions rationelles", I.de Math. pures et Appl. 8.1 (1918) 67-245.

[K] *T. Kimura*, "On the iteration of analytic functions", Funk. Eqvavoj 14-3(1971) 197-238.

[L.-P.] *P.D. Lax/R.S. Phillips*, "Scattering theory", Academic Press, New York 1967
"Scattering theory for automorphic functions" , Annals of Math.Studies, Princeton Univ.Press 1976.

[M] *B. Margrange*, "Travaux d'Ecalle et de Martinet-Ramis sur les Systèmes Dynamiques", Sém. Bourbaki 34, 1981-1982 no. 582, Astérisque 92-93.

[N] *E. Nelson*, "Topics in Dynamics, I:Flows" Math.Notes, Princeton Univ.Press 1970.

[S] *D. Sullivan*, Seminar on Conformal and Hyperbolic Geometry", IHES/M/82/12
"Quasi conformal homeomorphisms and Dynamics I, II, III" , IHES/1983.

[T] *W. Thurston*, "On the combinatoric of iterated rational maps" Preprint Princeton University, 1983.

[V] *S.M. Voronin*, "Classification des germes d'applications conformes $\mathbb{C},0 \to \mathbb{C},0$ targents à l'identité" Funct. Analis. et appl. 15-1 (1981) 1, 17.

* Jean-Pierre FRANÇOISE
Institut des Hautes Etudes Scientifiques
35 Route de Chartres
91440 Bures-sur-Yvette
FRANCE

Singularities & Dynamical Systems
S.N. Pnevmatikos (editor)

ON THE $T^k \to T^{k+1}$ BIFURCATION PROBLEM

Dietrich Flockerzi
Universität Würzburg
West Germany

I. INTRODUCTION

In 1979 G.R.Sell[13] and A.Chenciner and G.Iooss[1] published results describing a Hopf-type bifurcation of a k-dimensional into a (k+1)-dimensional invariant torus within a one-parameter family of differential equations or maps,respectively.The present contribution gives an extension of these results since it will not be assumed that

i) the normal spectrum of the k-dimensional torus crosses from $\mathbb{R}^-$ to $\mathbb{R}^+$ in a transversal fashion (or that the corresponding transversality condition in Theorem III.2.11 of [1] holds),

ii) the quadratic and cubic terms are the only significant nonlinearities.

The proof of our main result in Section IV is rather involved (cf. [5]). So we confine ourselves to an outline of the main ideas and put more emphasis on the derivation of the set up we use for this kind of bifurcation problem.

Just to give the flavor of the result we are aiming at we recall the essential features of the corresponding bifurcation problem in $\mathbb{R}^2$. To this end we consider an analytic one-parameter family of ordinary differential equations

$$(1.1)_\alpha \qquad \dot{x} = f_\alpha(x), \ x \in \mathbb{R}^2, \ \alpha \in I=(-\alpha_o,\alpha_o),$$

with $f_o(o)=o$ and we assume that the variational equation $\dot{x}=f_o'(o)x$ along the trivial solution x=o of the unperturbed system $(1.1)_o$ admits a family F of closed orbits C_ρ^*,i.e. we assume

A1) $f_o'(o)$ has eigenvalues $\pm i\omega_o$ with $\omega_o>o$.

Then,by the implicit function theorem,the system $(1.1)_\alpha$ possesses a stationary solution $x(\alpha)$ which we may take to be the trivial solution x=o. Here -as always- we suppose that $\alpha_o > o$ is sufficiently small. We further assume that the closed orbits C_ρ^* of F are of multiplicity m . In terms of the spectrum of $f_\alpha'(o)$ this can be formulated as an m-th order contact condition:

A2) $f'_\alpha(o)$ has eigenvalues $\alpha^m\lambda_m(\alpha)\pm i\omega(\alpha)$ for $\alpha\in I$ with $m\in\mathbb{N}$, $\lambda_m(o)\neq o$ and $\omega(o)=\omega_o$.

The assumptions A1 and A2 imply that x=o is a focus for $\dot{x}=f'_\alpha(o)x$ for $\alpha\neq o$ and a center for $\alpha=o$. Now, the effect of taking the nonlinearities of $(1.1)_\alpha$ into account should be that some of the m copies of each invariant curve C^*_ρ of F are bent to the right (supercritical bifurcation for $\alpha>o$), some to the left (subcritical bifurcation for $\alpha<o$) and some are left in place (vertical bifurcation for $\alpha=o$). A vague attractor or repellor condition like

A3) x=o is a focus of finite multiplicity for $(1.1)_o$

excludes the vertical bifurcation and implies the existence of at least one transcritical branch of bifurcating periodic solutions provided m is odd. This can be seen by purely topological arguments (Poincaré-Bendixson theorem). The number of bifurcating branches, their direction of bifurcation and the stability properties of the bifurcating periodic solutions can be determined by analytical means for any m (cf. [3]).

In the present paper we want to investigate what the corresponding hypotheses for the above indicated $T^k\to T^{k+1}$ bifurcation are that allow to answer questions involving the number of branches of (k+1)-dimensional tori bifurcating from a given k-dimensional torus, their direction of bifurcation and their stability properties. Thus we will consider a C^∞-smooth one-parameter family of differential equations

$$(1.2)_\alpha \qquad \dot{\xi} = F_\alpha(\xi),\ \xi\in\mathbb{R}^n,\ \alpha\in I=(-\alpha_o,\alpha_o),$$

possessing a smooth family of k-dimensional invariant tori M_α with $k+2\leq n$, e.g. one may think of M_α as being generated by k weakly coupled van der Pol oscillators. This family M_α takes over the role of the family $x(\alpha)$ of stationary solutions of $(1.1)_\alpha$. We will establish not-too-restrictive criteria which allow the existence of (k+1)-dimensional invariant tori $\hat{M}_\alpha$ tending as a whole to M_o for $\alpha\to o$. In Section II we give an example where just part of a 2-torus undergoes a Hopf-type bifurcation to a pinched torus. By a second example we illustrate the possibility that one part of a 2-torus can undergo such a bifurcation to a pinched torus before the rest of it does so that in the end a proper 3-torus has bifurcated.

The case of bifurcation from a periodic solution (k=1) is much easier to deal with than the case $k\geq 2$. There are essentially two ways of proving the $T^1 \to T^2$ bifurcation: first, by considering the first return map on a transversal hypersurface one is led to the Hopf bifurcation problem for maps (cf.[11]), secondly, by using Floquet theory one can reduce this bifurcation problem to the Hopf bifurcation problem from a stationary solution. For $k \geq 2$ there is no Floquet theory available, and it is known that the reduction of a quasi-periodic linear system to a linear system with constant coefficients is not always possible. Thus there is a problem in determining precisely a critical value of the parameter at which the stability of the invariant torus M_α changes in a way which allows the bifurcation of a (k+1)-dimensional torus. In the sequel we mainly have this case $k \geq 2$ in mind. In the next section we describe how the theory of the spectrum of an invariant manifold (cf.[12]) can be used to define such a critical value.

II. THE BIFURCATION SET UP

We consider system $(1.2)_\alpha$ and assume that the flow on M_α is generated by

$$(2.1)_\alpha \qquad \dot{y} = \bar{\omega} + g_\alpha(y) = \bar{\omega} + O(|\alpha|), \; y \in T^k, \; \alpha \in I,$$

where T^k denotes the standard torus of dimension k. The components of y are thus defined modulo 2π. Here it is generally not sufficient to require the existence of the k-torus just for $\alpha=0$. If one has the above bifurcation in mind M_0 cannot be normally hyperbolic and thus the continuation of M_0 to M_α for $\alpha \in I$ is not always possible (compare the remark following H1ii below). Adapting the hypotheses A1 and A2 to the present situation amounts -in somewhat loose terms- to the following assumption:

The "variational equation" of the unperturbed system $(1.2)_0$ along the torus M_0 possesses a family F of invariant (k+1)-dimensional tori M^*_ρ of "multiplicity m".

In what follows we will make precise what is meant by "variational equation" and by "multiplicity m".

The first objective is to introduce a local coordinate system near M_α via a decomposition of the tangent space of $\mathbb{R}^n$ at y into the sum of the k-dimensional tangent space $Y_\alpha(y)$ to M_α at y, the "non-critical" part $Z_\alpha(y)$ of the normal space to M_α at y and the "critical" part $X_\alpha(y)$ of the normal space which reflects the fact of M_α not being normally hyperbolic for all $\alpha \in I$, i.e. via a decomposition

$$(2.2)\qquad \mathbb{R}^n = X_\alpha(y) + Y_\alpha(y) + Z_\alpha(y) \quad \text{for all } (y,\alpha)\in T^k\times I.$$

For the Hopf-type bifurcation we are aiming at $X_\alpha(y)$ should be a 2-dimensional, smoothly varying subspace of $\mathbb{R}^n$ for all $(y,\alpha)\in T^k\times I$. In order to derive such a decomposition (2.2) we consider the linear skew-product flow

$$\Pi(\xi_o,y_o,\alpha,t) = (\Xi(y_o,\alpha,t)\xi_o, Y(y_o,\alpha,t),\alpha)$$

on $\mathbb{R}^n\times T^k\times I$ induced by $(1.2)_\alpha$. Here, Y denotes the solution of $(2.1)_\alpha$ with initial value y_o at $t=o$ and Ξ denotes the fundamental matrix solution of $\dot\xi=F'_\alpha(Y(y_o,\alpha,t))\xi$ with $\Xi(y_o,\alpha,o)=I$. As in [13,V] we denote the induced tangential and normal linear skew-product flows by Π^T and Π^N, respectively. For the notion of the spectra of such flows we refer to [12] or [13]. Then we have a decomposition as in (2.2) if the following spectral conditions hold:

S1) Π^T and Π^N have disjoint spectra for $\alpha\neq o$. (Note that o belongs to both spectra for $\alpha=o$ if $k\geq 1$.)

S2) Π^N has a "critical" spectral set $C_\alpha\subset(-\delta,\delta)$ for $\alpha\in I$ with a two-dimensional associated spectral subspace $X_\alpha(y)$ such that

$$C_o=\{o\},\qquad C_\alpha\cap\mathbb{R}^+ \neq \phi \Rightarrow C_\alpha\cap\mathbb{R}^- = \phi \quad \text{for } \alpha\neq o.$$

The remaining spectral intervals are bounded away from o by δ.

S3) $X_\alpha(y)$ varies smoothly in α and $X_o(y)$ is disjoint from $Y_o(y)$ for all $y\in T^k$.

Under these circumstances $X_o(y)$ and $Z_o(y)$ can be parametrized in a smooth way by $x\in\mathbb{R}^2$ and $z\in\mathbb{R}^{n-k-2}$, respectively. The corresponding differential equation will be 2π-periodic in each component of y and will have the following form:

$$(2.3)_\alpha\qquad \begin{aligned} \dot x &= \Omega(y)x+\alpha\Omega_1(y,\alpha)x+\alpha\Omega_2(y,\alpha)z + X(x,y,z,\alpha), & x\in\mathbb{R}^2,\\ \dot y &= \bar\omega + Y(x,y,z,\alpha), & y\in T^k,\\ \dot z &= B(y)z+\alpha B_1(y,\alpha)x+\alpha B_2(y,\alpha)z + Z(x,y,z,\alpha), & z\in\mathbb{R}^{n-k-2}, \end{aligned}$$

where X and Z are of order $O(|x|^2+|z|^2)$ and where $Y(o,y,o,\alpha)$ equals $g_\alpha(y)$. Moreover the matrix $B(y)$ in $(2.3)_\alpha$ satiesfies the following:

$$(2.4)\qquad \dot z = B(\bar\omega t+y_o)z \text{ admits an exponential dichotomy for all } y_o\in T^k$$

(cf. [5]). The considerations to follow always refer to system $(2.3)_\alpha$ with the property (2.4).

The following examples reveal the important role the nature of the rotation vector $\bar\omega$ plays and motivate our hypothesis H1 below.

Example 1 (cf. [6])

We consider the following 4-dimensional system of the form $(2.3)_\alpha$ with a rational rotation vector $\bar{\omega}=(1/2,1)^T$ and $\omega_0 \notin \mathbb{Q}$:

$$(2.5)_\alpha \quad \begin{aligned} \dot{x} &= \begin{pmatrix} o & -\omega_o \\ \omega_o & o \end{pmatrix} x + \alpha \begin{pmatrix} 1+2\cos(2y_1-y_2) & o \\ o & 1+2\cos(2y_1-y_2) \end{pmatrix} x - r^2 x \\ \dot{y}_1 &= \frac{1}{2} + \frac{\alpha}{2}\sin(2y_1-y_2), \qquad \dot{y}_2 = 1 , \end{aligned}$$

with $(x,y_1,y_2)\in\mathbb{R}^2\times T^1\times T^1$. With polar coordinates (r,θ) for x and the change of coordinates

$$(2.6) \qquad \psi = 2y_1-y_2 , \quad \phi = -y_1+y_2$$

we write $(2.5)_\alpha$ in the form

$$(2.7)_\alpha \quad \begin{aligned} \dot{r} &= \alpha(1+2\cos\psi)r - r^3, & \dot{\psi} &= \alpha\sin\psi , \\ \dot{\theta} &= \omega_o , & \dot{\phi} &= (1-\alpha\sin\psi)/2. \end{aligned}$$

The (ψ,ϕ)-system describes a flow on a 2-torus having two periodic orbits for $|\alpha|\neq o$. The set $\{r=o\}$ corresponds to the invariant 2-torus of $(2.7)_\alpha$ which is attractive for $\alpha=o$. The linearized equation of $(2.7)_o$ admits a family F of invariant 3-tori. The (r,θ)-system is independent of (ψ,ϕ) and can be analyzed by phase plane techniques. For the choice of coordinates as in Figure 1

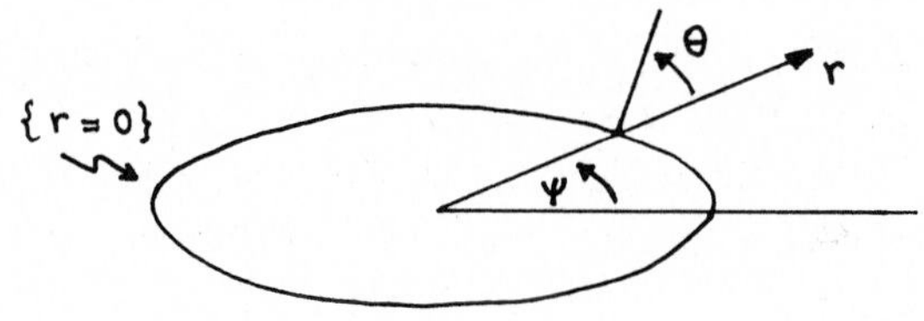

Fig.1

the two sketches in Figure 2 show the flow of the (r,θ)-system for $\alpha<o$ and $\alpha>o$:

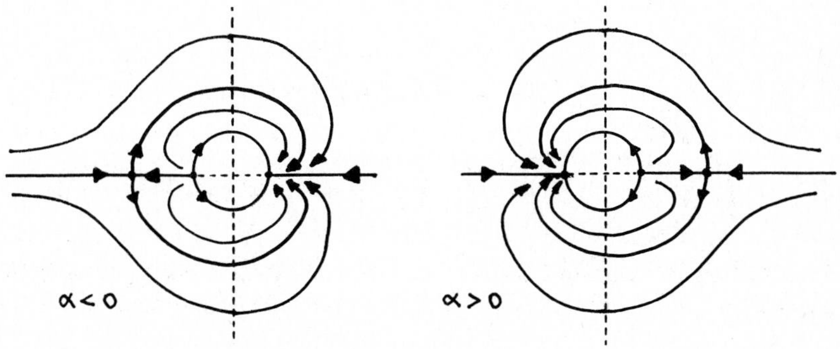

Fig.2

The corresponding invariant sets of the period map on the surface $\{y_2 = 0 \bmod 2\pi\}$ -one may think of y_2 playing the role of the time- are thus pinched tori as shown in Figure 3.

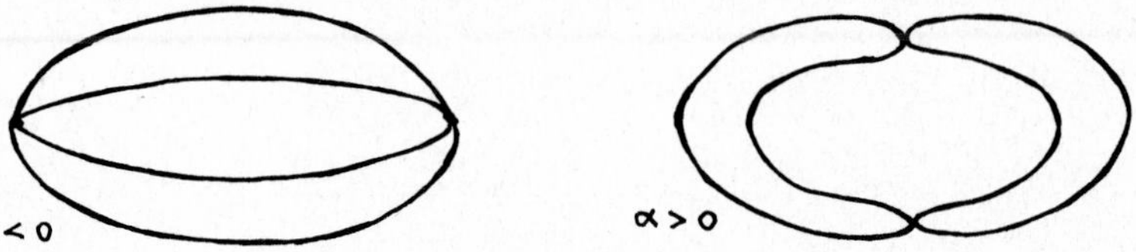

Fig.3

Thus, instead of a one-sided bifurcation to a 3-torus one has a two-sided bifurcation to invariant sets that can be visualized as three-dimensional tori which are pinched along a circle. For a more extensive study of this resonance phenomenon we refer to [6].

Example 2 (cf.[1o])
The essential features of the above example can be used to show that the $T^2 \to T^3$ bifurcation is very sensitive to perturbations. Consider

$$(2.8)_\varepsilon \qquad \begin{aligned} \dot r &= \alpha r - r^3 + \varepsilon^2 r\cos(2y_1 - y_2), & \dot\theta &= \omega_o + O(\varepsilon^2) \\ \dot y_1 &= \tfrac{1}{2} + \tfrac{1}{2}\varepsilon^2 \sin(2y_1 - y_2), & \dot y_2 &= 1, \end{aligned}$$

where ε is just an auxiliary parameter measuring the size of the perturbation. The unperturbed system $(2.8)_o$ undergoes a supercritical Hopf bifurcation from the invariant 2-torus $\{r=0\}$ to the invariant 3-torus $\{r=\sqrt{\alpha}\}$. By means of the perturbed system $(2.8)_\varepsilon$ we illustrate that part of an invariant 2-torus can undergo a Hopf bifurcation before a proper 3-torus appears. By introducing the new bifurcation parameter β via $\alpha=\varepsilon^2\beta$, the scaling $r \to \varepsilon r$ and the transformation (2.6) system $(2.8)_\varepsilon$ can be written as

$$\begin{aligned} \dot r &= \varepsilon^2 r(\beta + \cos\psi - r^2), & \dot\psi &= \varepsilon^2 \sin\psi, \\ \dot\theta &= \omega_o + O(\varepsilon^2), & \dot\phi &= \tfrac{1}{2} + O(\varepsilon^2). \end{aligned}$$

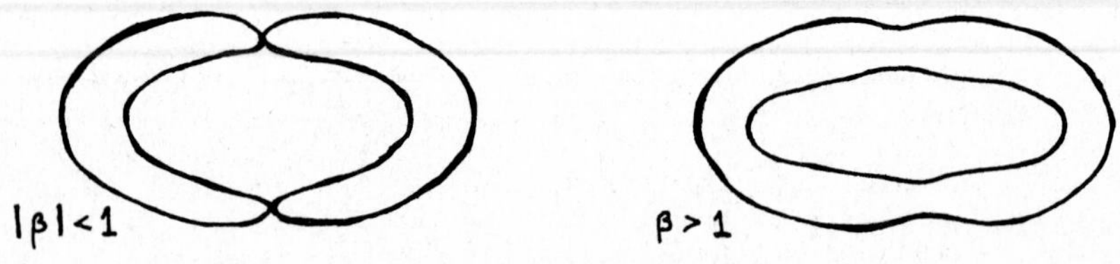

Fig.4

Thus, for increasing β, the invariant 2-torus undergoes a partial Hopf bifurcation to a pinched torus at $\beta=-1$ and a second such bifurcation -now to full 3-torus- at $\beta=1$. The invariant sets of $(2.8)_\varepsilon$ for the period map on $\{y_2=o \bmod 2\pi\}$ are shown in Figure 4.

In view of these examples we ask for a nonresonant rotation vector $\bar{\omega}$ whose components are independent over $\mathbb{Q}$. Our geometric idea of the family F requires all solutions of

$$(2.9) \qquad \dot{x} = \Omega(\bar{\omega}t+y)x \ , \ y \in T^k ,$$

to be bounded. (This is not implied by our spectral assumptions, see [9].) We therefore can take $\Omega(y)$ to be of the form $\begin{pmatrix} o & -c(y) \\ c(y) & o \end{pmatrix}$ (cf. [8]). There is no loss of generality in taking c(y) equal to its mean value provided (2.9) admits a nontrivial almost periodic solution for some $y \in T^k$, since then any solution of (2.9) will be almost periodic (for any $y \in T^k$, cf. [8]). We restrict our attention to this case and assume

$$\text{H1i)} \quad \Omega(y) = \begin{pmatrix} o & -\omega_o \\ \omega_o & o \end{pmatrix} \quad \text{for all } y \in T^k \ , \ \omega_o \neq o \ .$$

Now, the existence of the family F of (k+1)-dimensional invariant tori M^*_ρ for the linearized system of $(2.3)_o$ is obvious. We would like to add a remark: R.A.Johnson has shown in [7,8] that already for dimy=2 there need not exist a nontrivial almost periodic solution of (2.9) even if all solutions of (2.9) are bounded. The crucial point in proving the nonexistence of such a solution is the construction of an analytic function c and a square-integrable, but not continuous antiderivative γ on T^2 with the following properties:

i) c has mean value o and $\int_o^t c(\bar{\omega}s+y)ds$ is unbounded for all $y \in T^2$,

ii) $\gamma(\bar{\omega}t+y) - \gamma(y) = \int_o^t c(\bar{\omega}s+y)ds$ for all $y \in T^2$ and all $t \in \mathbb{R}$,

where $\bar{\omega}_1$ in $\bar{\omega}=(\bar{\omega}_1,1)^T$ is a suitable irrational number.

As it turns out the above assumption of nonresonance on $\bar{\omega}$ alone will not suffice for our purposes. In [1] A.Chenciner and G.Iooss have considered some cases of strong resonance where ω_o depends rationally on $\bar{\omega}$. Here we will ask for the strong version of nonresonance in form of a diophantine condition for $\omega=(\omega_o,\bar{\omega})^T$. We assume

H1ii) There exist positive constants κ_o and κ such that

ω satisfies $|\nu\cdot\omega| \geq \kappa_o|\nu|^{-\kappa}$ for all $o \neq \nu \in Z^{1+k}$

A careful analysis of our argumentation below shows that H1ii is only needed for a finite number of ν_o's (if ν denotes $(\nu_o,\bar{\nu})\in Z\times Z^k$). As A. Chenciner and G. Iooss have noted hypothesis H1ii (with $\nu_o=o, \nu_o=1$) allows the continuation of the torus M_o to the tori M_α for $\alpha\in I$ ([2]).

We now proceed to make the statement about the "multiplicity of M^*_ρ" precise. In cylindrical coordinates

$$(x,y,z) = (r,\theta,y,z) = (r,\phi,z) \in \mathbb{R}^+ \times T^{1+k} \times \mathbb{R}^{n-k-2}$$

system $(2.3)_\alpha$ is now of the form

$$\begin{aligned}
&\dot{r} = \alpha A_1(\phi,\alpha)r + \alpha A_2(\phi,\alpha)z + R(r,\phi,z,\alpha),\\
(2.1o)\quad &\dot{\phi} = \omega + \Phi(r,\phi,z,\alpha),\\
&\dot{z} = B(\phi)z + \alpha B_1(\phi,\alpha)r + \alpha B_2(\phi,\alpha)z + Q(r,\phi,z,\alpha),
\end{aligned}$$

where R and Q are of second order in (r,z). By a smooth change of variables

$$(2.11)\quad r \to r + \alpha U(\phi,\alpha)r + \alpha V(\phi,\alpha)z$$

one is led to solve partial differential equations of the form

$$(2.12)\quad u_\phi(\phi)\cdot\omega + \chi(\phi) = \bar{\chi} = \text{mean value of } \chi(\phi)$$

where χ is a known 2π-periodic function on T^{k+1}. Unless all mean values $\bar{\chi}_i$ vanish when performing (2.11) -the $\bar{\chi}_i$ can be computed recursively because of H1ii- the r-equation takes the form

$$\dot{r} = \alpha^m K_m r + \alpha^{m+1}[O(r)+O(|z|)] + O(r^2+|z|^2),\ K_m \neq o$$

(cf. [5]). We assume that the following situation prevails:

H2) There exists an $m\in\mathbb{N}$ such that $K_m \equiv \bar{\chi}_m \neq o$ whereas the mean values $\bar{\chi}_1,\ldots,\bar{\chi}_{m-1}$ are all equal to o.

In case of the Hopf bifurcation considered in Section I H2 reduces to the condition A2. The existence of such an integer m can be taken for the definition of the multiplicity of M^*_ρ. H2 is the exact formulation of the previous geometric idea and states in analytical terms that the normal spectrum of M_α, i.e. the spectral subset C_α, has an m-th order contact with the tangential spectrum at $\alpha=o$. Finally we state the assumption corresponding to A3 :

H3) M_o is vaguely hyperbolic for $(2.3)_o$ (cf. (3.3) below).

Having established the set up for our bifurcation problem we formulate our *goal* in precise terms:

Show the existence of invariant (k+1)-tori $\hat{M}_\alpha$ of the form

$$(2.13)\quad \hat{M}_\alpha = \{z=\hat{z}(r,\phi,\alpha),\ r=\hat{r}(\phi,\alpha),\ \phi\in T^{k+1}\} \quad \text{for } \pm\alpha > o$$

with appropriate functions $\hat{z}$ and $\hat{r}$ such that $\hat{M}_\alpha \to M_o$ as $\alpha \to o$ (cf.Section IV.B).

III. NORMAL FORMS FOR (2.3)

We confine ourselves to outline the derivation of the normal forms. For the details we refer to [5]. After the transformation (2.11) we make the substitutions

$$(3.1) \qquad r \to \varepsilon^m r \ , \quad z \to \varepsilon^m z \ , \quad \alpha = \varepsilon\beta$$

introducing $\varepsilon > o$ as the new bifurcation parameter. As we go along we will be led to various choices of β (in terms of ε) and thus to the exact scalings with $\varepsilon = \varepsilon(\alpha)$. In a first step we compute an approximation to the center manifold of $(2.3)_\alpha$ in the new variables (3.1) via a transformation

$$(3.2) \qquad z \to z + \varepsilon\mathcal{W}(r,\phi,\varepsilon,\beta).$$

Because of the exponential dichotomy in (2.4) the arising partial differential equations which are of the form

$$-B(\phi)w(\phi) + w_\phi(\phi)\cdot\omega + Q(\phi) = o$$

have unique smooth 2π-periodic solutions. By means of (3.2) we arrive at a system of the form

$$\begin{aligned} \dot{r} &= \varepsilon R^*(r,\phi,\varepsilon,\beta) + \varepsilon O(|z|), \\ \dot{\phi} &= \omega + O(\varepsilon), \\ \dot{z} &= B(\phi)z + \varepsilon O(|z|) + O(\varepsilon^N), \end{aligned}$$

where N is an arbirarily large finite positive integer. Invariant manifold theory tells us that the z-term in the radial equation can be considered as being a higher order term (see Section V). In the second step we replace the ϕ-dependent coefficients of the leading powers of r in R^* by their constant mean values. This can be achieved by an averaging transformation $r \to r + \varepsilon\mathcal{U}(r,\phi,\varepsilon,\beta)$ where the corresponding partial differential equations are of the form (2.12). Because of the diophantine condition H1ii they have unique smooth 2π-periodic solutions with mean value o. The resulting radial equation will then be of the form

$$\dot{r} = P(r,\varepsilon,\beta) + P_R(r,\phi,\varepsilon,\beta) + \varepsilon O(|z|) = O(\varepsilon)$$

where -because of H3- there exists an $\ell_o \in \mathbb{N}$ such that one has

$$(3.3) \qquad P(r,\varepsilon,o) = \varepsilon^{2\ell_o m} K_o r^{2\ell_o m+1}, \ K_o \neq o.$$

P_R can thus be assumed to be of higher order with respect to ε. By employing the Newton diagram for $P(r,\varepsilon,\beta)=o$ (where one marks the powers of ε on the abscissa and those of β on the ordinate) one notices that this algebraic bifurcation equation can be written as

$$(3.4) \qquad P(r,\varepsilon,\beta) = \sum_{j\in J} \varepsilon^{2\ell_j m+j}\beta^j K_j r^{2\ell_j+1} + \text{h.o.t}$$

with nonzero constants K_j and nonnegative integers ℓ_j. The set J corresponds to the pairs of ε- and β-powers on the convex polygonal line in the Newton diagram. The various slopes $-1/\Delta(h) \in \mathbb{Q}^-, h=1,\dots,H$, of Newton's polygonal line then yield the correct relation between α and ε by

$$(3.5) \qquad \alpha = \varepsilon\beta, \quad \beta = \pm\varepsilon^{\Delta(h)}, \quad h=1,\dots,H,$$

and lead to H normal representations of $(2.3)_\alpha$ in the form

$$(3.6)_\varepsilon \qquad \begin{aligned} \dot r &= \varepsilon^{\sigma(h)}[P_h^\pm(r)+o(1)] + \varepsilon O(|z|), \\ \dot\phi &= \omega + O(\varepsilon), \\ \dot z &= B(\phi)z + \varepsilon O(|z|) + O(\varepsilon^N) \end{aligned}$$

with appropriate positive rationals $\sigma(h)$. The "Newton polynomial" $P_h^\pm(r)$ is the coefficient of the lowest ε-power in $P(r,\varepsilon,\pm\varepsilon^{\Delta(h)})$ and thus no monomial.

IV. MAIN RESULT

A) Under the hypotheses H1,H2 and H3 the above H normal forms $(3.6)_\varepsilon$ can be derived from $(2.3)_\alpha$. Any bifurcating invariant (k+1)-torus of the form (2.13) will be one for one of these representations $(3.6)_\varepsilon$.

B) If P_h^+ or P_h^- has a simple positive zero ρ then there exists an invariant (k+1)-torus $\hat M_\alpha$ for $(2.3)_\alpha$ of the form

$$\{(r,\phi,z,\alpha):\ r=\hat r(\phi,\alpha),\ z=\hat z(\hat r(\phi,\alpha),\phi,\alpha)\ ,\ \phi\in T^{k+1}\}$$

for $\alpha>o$ or $\alpha<o$, respectively. The functions $\hat r$ and $\hat z$ are given by

$$\hat r(\phi,\alpha) = |\alpha|^{m(h)}(\rho+o(1)), \quad \hat z(\hat r(\phi,\alpha),\phi,\alpha) = o(|\alpha|^{m(h)})$$

with $m(h)=m/(1+\Delta(h))$. The stability property of $\hat M_\alpha$ within the center-stable manifold follows from the sign of $\frac{\partial}{\partial r}P_h^\pm(\rho)$.

C) If m in H2 is odd then there exists at least one $h\in\{1,\dots,H\}$ so that the "Newton slope" $-1/\Delta(h)$ in (3.5) corresponds to an odd ordinate segment in the Newton diagram of (3.4). If $-1/\Delta(h)$ is such a slope then at least one of the polynomials P_h^+ or P_h^- possesses an odd order positive zero. Such zeros can "in general" be reduced to simple zeros (cf.[5,Section V]: The only exception can be caused by the need of infinitely many reduction steps).

D) In the particular case m=1 there exists just the Newton slope $-1/\Delta(1)=-1/(2\ell_o-1)$. In this case either P_1^+ or P_1^- has the simple positive zero $\rho=|K_1/K_o|^{1/2\ell_o}$, and one has a unique bifurcation of an invariant (k+1)-torus either for $\alpha>o$ or for $\alpha<o$.

E) The number of positive zeros of all the Newton polynomials together does not exceed $\min(m,\ell_o)$. Thus,"in general" there are at most $\min(m,\ell_o)$ branches of bifurcating (k+1)-tori.

V. INVARIANT MANIFOLDS AND $T^k \to T^{k+1}$ BIFURCATION

We have already sketched the proof of part A in Section III.Parts C,D and E follow from a thorough discussion of the Newton diagram for (3.4). Here,we will just give an outline how the theory of invariant manifolds can be used to prove part B. We suppose that P_h^+ possesses a simple positive zero ρ,denote P_h^+ simply by P and drop the dependence on h in the sequel. If Δ denotes γ/δ in lowset terms ($\gamma>o,\delta>o$) we introduce the new parameter μ by $\varepsilon=\mu^\delta$ to avoid fractional powers of the parameter. With the translation $r\to r+\rho$ system $(3.6)_\varepsilon$ is of the form

$$(5.1)_\mu \quad \begin{aligned} \dot r &= \mu^{\delta\sigma}(P'(\rho)r+o(1)) + \mu^\delta O(|z|), \\ \dot\phi &= \omega + O(\mu^\delta), \\ \dot z &= B(\phi)z + \mu^\delta O(|z|) + O(\mu^{N\delta}). \end{aligned}$$

In order to reduce this n-dimensional system $(5.1)_\mu$ to a (k+2)-dimensional system we modify the right-hand side so it has the necessary global properties. Because of (2.4) the system

$$\dot\phi = \omega + O(\mu^\delta), \qquad \dot z = B(\phi)z$$

also admits an exponential dichotomy (cf.[12] or [4]).Thus we can define the Green's function for the modified system which enables us to reduce the n-dimensional system to a (k+2)-dimensional one on a center manifold

$$(5.2) \quad \{z=\tilde z(r,\phi,\mu)=O(\mu^{N\delta}): \phi\in T^{k+1}\}.$$

A further averaging transformation $\phi \to \phi+\mu^\delta V(r,\phi,\mu)$ -where for the recursive computation of V the diophantine condition H1ii is used once more- generates a weakly coupled system

$$(5.3) \quad \begin{aligned} \dot r &= \mu^{\delta\sigma}(P'(\rho)r+o(1)), \\ \dot\phi &= \omega + \Phi_o(r,\mu) + O(\mu^{\delta\sigma+1}) \end{aligned}$$

on the center manifold (5.2). Since Φ_o does not depend on ϕ the normal flow dominates the tangential flow for (5.3). Thus there exists an invariant manifold

(5.4) $\{r=\tilde{r}(\phi,\mu)=O(\mu) \ : \ \phi\in T^{k+1}\}$

on the manifold (5.2). All in all,we thus have arrived at the desired (k+1)-torus $\hat{M}_\alpha$ with the properties stated in Section IV.B. We note that the various scalings (3.1) and (3.5) amount to

$$\alpha=\varepsilon^{1+\Delta(h)}, \quad \varepsilon^m=\alpha^{m(h)}, \quad \mu=\alpha^{1/(\gamma+\delta)}$$

so that the functions $\hat{r}$ and $\hat{z}$ of (2.13) and Section IV.B are given by

$$\hat{r}(\phi,\alpha) = \alpha^{m(h)}(\rho+\tilde{r}(\phi,\mu)), \quad \hat{z}(r,\phi,\alpha) = \alpha^{m(h)}\tilde{z}(r,\phi,\mu).$$

REFERENCES

1. A.Chenciner and G.Iooss:Bifurcations de tores invariants,Archive Rat.Mech.Anal.69(1979)1o9-198.
2. A.Chenciner and G.Iooss:Persistance et bifurcations de tores invariants,Archive Rat.Mech,Anal.71(1979)3o1-3o7.
3. D.Flockerzi:Existence of Small Periodic Solutions of ODE's in $\mathbb{R}^2$, Archiv der Mathematik 33(1979)263-278.
4. D.Flockerzi:Weakly Nonlinear Systems and Bifurcation of Higher Dimensional Tori,in:H.W.Knobloch and K.Schmitt,Equadiff 82,Springer Lecture Note in Math.1o17(1983)185-193.
5. D.Flockerzi:Generalized Bifurcation of Higher Dimensional Tori, University of Würzburg preprint 97(1983),(to appear in the JDE).
6. D.Flockerzi:Resonance and Bifurcation of Higher Dimensional Tori, University of Würzburg preprint 1o2(1983).
7. R.A.Johnson:Measurable Subbundles in Linear Skew-Product Flows, Illinois J.Math.23(1979)183-198.
8. R.A.Johnson:Analyticity of Spectral Subbundles,JDE 35(198o) 366-387.
9. R.A.Johnson:On a Floquet Theory for Almost-Periodic Two-Dimensional Linear Systems,JDE 37(198o)184-2o5.
1o.K.R.Meyer:Tori in Resonance,Univ.of Minnesota preprint 13(1983).
11.D.Ruelle and F.Takens:On the Nature of Turbulence,Comm.Math.Phys. 2o(1971)167-192.
12.R.J.Sacker and G.R.Sell:A Spectral Theory for Linear Differential Systems,JDE 27(1978)32o-358.
13.G.R.Sell:Bifurcation of Higher Dimensional Tori,Archive Rat.Mech. Anal.69(1979)199-23o.

Dietrich Flockerzi
Math.Institut der Universität Würzburg
Am Hubland,D-8700 Würzburg,West Germany.

Singularities & Dynamical Systems
S.N. Pnevmatikos (editor)

HÖLDER CONTINUOUS PATHS AND HYPERBOLIC AUTOMORPHISMS OF T^3

M. C. Irwin
Liverpool University
England

A hyperbolic toral automorphism $f : T^n \to T^n$ is a diffeomorphism of the torus that lifts to a hyperbolic automorphism of $\mathbb{R}^n$. So we have the diagram

$$\begin{array}{ccc} \mathbb{R}^n & \xrightarrow{A} & \mathbb{R}^n \\ \pi \downarrow & & \downarrow \pi \\ T^n & \xrightarrow{f} & T^n = \mathbb{R}^n/\mathbb{Z}^n \end{array}$$

where A is hyperbolic (i.e. has no (complex) eigenvalues of modulus 1) and π is the quotient map. Thus A induces a direct sum decomposition $\mathbb{R}^n = E^s \oplus E^u$, where the *stable summand* E^s is the generalised eigenspace corresponding to eigenvalues of modulus < 1 and the *unstable component* E^u corresponds to eigenvalues of modulus > 1. They project under π to W^s and W^u, the *stable* and *unstable manifolds* of $\pi(0)$. Each of these is dense in T^n.

A subset K of T^n is *invariant* if it is compact and $f(K) = K$. On iteration f mixes up the torus very thoroughly, and this puts heavy restrictions on invariant sets K. We are here concerned with two questions: what dimensions K can have and how smooth it can be. By dimension, we mean, in the first instance, topological (covering, inductive) dimension, which is an integer. Here are some examples of invariant sets K.

EXAMPLE 1. $K = O(p)$, the orbit of p where $p \in \mathrm{Per}\, f$. There are lots of periodic orbits, since $\mathrm{Per}\, f$ is $\pi(\mathbb{Q}^n)$, see [1].

EXAMPLE 2. It is possible to obtain other 0-dimensional sets by taking closures of orbits of single points. This needs care, as there are many points whose orbit closure is the whole of T^n (see, for example, Theorem 7.14 of [6]). However if p and q are periodic points, and if r is on the stable manifold of p and the unstable manifold of q, then the orbit closure of r is the union of the orbits of p, q and r, which is certainly 0-dimensional.

EXAMPLE 3. If $\mathbb{R}^n$ has a proper A-invariant subspace with basis in $\mathbb{Z}^n$, this will project onto an f-invariant toral subgroup of T^n (necessarily of dimension > 1). Thus, for example, if f is a given hyperbolic toral automorphism of T^n, $T^n \times \{0\}$ is a toral subgroup of $T^{2n} = T^n \times T^n$ invariant under the hyperbolic automorphism $f \times f$, and $T^n \times 0(p)$ is invariant under $f \times f$ for any $p \in \text{Per } f$. A slightly more complicated example is the automorphism g of $T^n \times T^n$ given by

$$g(x, y) = (y, f(x)).$$

Here, the invariant set $(T^n \times \{0\}) \cup (\{0\} \times T^n)$ contains the toral subgroup $T^n \times \{0\}$, which is invariant under g^2 but not under g. The question of whether any power of a given hyperbolic toral automorphism has proper invariant toral subgroups has been considered by Hancock [3, 4]. Generically it does not.

There are various theorems about invariant sets K of a hyperbolic toral automorphism $f : T^n \to T^n$, mostly to the effect that there are not very many. Here are three results that are particularly relevant to us.

THEOREM (*Smale, Hirsch and Williams*) *Dim* $K \neq n - 1$.

Smale proved this for $n = 2$, and Hirsch and Williams [5] made the generalisation. Smale raised the question of whether there are 1-dimensional K for $n > 2$. The obvious way to construct such a set is to take the orbit-closure of some suitable path in T^n, but the warning in Example 2 above applies again. This is borne out by:

THEOREM (*Franks, Mañé*) *Any non-constant rectifiable path in* T^n *has orbit closure containing some coset of a toral subgroup invariant under some power of* f.

Thus, usually the orbit closure is the whole of T^n. Franks [2] proved the theorem first for C^2 curves and Mañé [9] weakened the condition to rectifiability. On the other hand:

THEOREM (*Hancock, Przytycki*) *For all* r *with* $0 < r < n - 1$, *there exists a non-constant* C^0*-map* $\delta : I^r \to T^n$ (*where* $I = [0, 1]$) *such that* $\delta(I^r)$ *has* r*-dimensional orbit-closure.*

Hancock [3] proved the case $r = 1$, $n = 3$ and Przytycki [10] made the generalisation.

These last two theorems suggest that, paradoxically, to keep the topological dimension of the orbit-closure of a curve small (namely, 1), one has to make its Hausdorff dimension large. How large is an interesting question. I have approached the question from the related viewpoint of Hölder continuity, in answer to a question raised by Hancock [4].

A map $h : X \to Y$ of metric spaces is *Hölder continuous* with exponent $(0 < \alpha \le 1)$ if for some $A > 0$ and all $x, x' \in X$

$$d(h(x), h(x')) \le Ad(x, x')^{\alpha}.$$

So Hölder continuity comes between uniform continuity and rectifiability (which essentially corresponds to $\alpha = 1$). Hancock asked what happens to the orbit-closure of Hölder continuous paths. The answer depends on α. This shows up very well in T^3. There are two cases, depending on whether or not the eigenvalues are all real. If they are all real, they are necessarily distinct, and by taking some power of f if necessary we may assume that they are all positive, with two of them greater than 1. Then, for $f : T^3 \to T^3$, we have the result:

THEOREM *If f has eigenvalues $\lambda > \mu > 1 > \nu$ then*

(i) there exists a Hölder continuous path with exponent $\log \mu / \log \lambda$ whose orbit-closure is 1-dimensional.

Moreover,

(ii) any non-constant path with larger exponent has orbit-closure T^3.

If f has complex conjugate eigenvalues then

(iii) for any $\alpha > 1$, there exists a path with exponent α whose orbit-closure is 1-dimensional.

Thus (i) improves Hancock's results, (ii) improves Mañé's result, and (iii) is also best possible, since we cannot have $\alpha = 1$, by Mañé's theorem. The proof of the theorem, together with generalisations to T^n, appears elsewhere [7, 8]. We here attempt to give some idea of the proofs of (i) and (ii).

Sketch proof of (i) Notice that there are no 3-dimensional invariant sets other than T^3, since f is topologically transitive. Also by Hirsch and Williams' theorem there are no 2-dimensional invariant sets. So a non-constant curve δ whose image stays out of some given open set U necessarily has 1-dimensional orbit closure. Following Przytycki, we construct δ as a map into the unstable manifold W^u, and the open set that we miss is a thin neighbourhood of $\pi(0_z)$ (where 0_z is the z-axis in $\mathbb{R}^3$) minus the component of W^u

in it containing $\pi(0)$.

It is more convenient to work in $\mathbb{R}^3$, so δ will be $\pi\gamma$ for some curve γ in E^u. We have to make the orbit of γ avoid $V = \pi^{-1}(U)$, which is a neighbourhood of a lattice of points in E^u, the intersection of vertical lines through the integer lattice (except (0, 0)) in the (x, y)-plane. For convenience, we map E^u linearly to $\mathbb{R}^2$ with the λ and μ eigenspaces as x_1- and x_2-axes. The idea is to start with a curve $\gamma_0(t) = (t, 0)$ and to make successive modifications to get $\gamma_r(t) = (t, g_r(t))$, $(r > 0)$, such that $\gamma_r, A(\gamma_r), \ldots, A^r(\gamma_r)$ miss V, or, equivalently γ_r misses V, $A(V), \ldots, A^{-r}(V)$. (We are not worried about negative iterates of γ_r, as they will be near 0 in E^u, and we have excluded this region from V). We aim to make γ_r Lipschitz with constant λ^r/μ^r, but also we want $\gamma = \lim_{r=\infty} \gamma_r$, $\gamma(t) = (t, g(t))$, to satisfy

$$\frac{1}{\lambda^r} < |t - t'| < \frac{1}{\lambda^{r-1}} \quad \text{implies} \quad |g(t) - g(t')| \le A(\lambda/\mu)^{r-1}|t - t'|.$$

In this case, for $|t - t'| \in \left[\frac{1}{\lambda^r}, \frac{1}{\lambda^{r-1}}\right]$, $|g(t) - g(t')|$ is bounded by $A|t - t'|^{\log\mu/\log\lambda}$, which gives the Hölder exponent for g and hence for γ.

To control the Lipschitz constants while rerouting the path γ_{r-1} to avoid the forbidden $A^{-r}(V)$, one needs to decide on a standard rerouting technique, so that changes do not occur too violently or too often. I found it easiest to consider thin diamond shaped neighbourhoods of the lattice. Then, providing γ_{r-1} is flat enough, it does not hit the diamonds too often, and does so in a standard way (passing across the diamond from left to right). It is then possible to reroute, by a piecewise linear displacement, always upwards, that only increases the slope by a factor λ/μ, and that only locally. The curve γ obtained is without self-intersections, and being a graph of a function, is still comparatively uncomplicated, despite its infinite length.

Sketch proof of (ii) Let $\gamma : I \to \mathbb{R}^3$ be non-constant and α-Hölder, for $\alpha > \frac{\log \mu}{\log \lambda}$. Choose β with

$$\alpha > \beta = \frac{\log \mu}{\log \lambda} + \varepsilon > \frac{\log \mu}{\log \lambda}.$$

We want to prove that $\pi\gamma(I)$ has orbit-closure T^3. Let U be open in T^3, and let $V = \pi^{-1}(U)$. We must show that $\gamma(I)$ intersects

$A^r(V)$ for some r. Take coordinates $x = (x_1, x_2, x_3)$ with respect to a basis of eigenvectors corresponding to λ, μ, ν respectively. For $p \in \mathbb{R}^3$, $1 \le i \le 3$, $a > 0$ and $\ell > 0$, let $C_i(p, \ell, a)$ be the closed solid circular cylinder with centre p, radius a and axis length 2ℓ in the x_i-direction. It follows from the denseness of $\pi(0x_i)$ in T^3 and the compactness of T^3 that there exist $a > 0$ and $\ell > 0$ such that, for any cylinder $C_i(p, \ell, a)$ in $\mathbb{R}^3$, any path running the length of the cylinder intersects V. We deduce that, for this ℓ and a and any $r > 0$, any path running the length of any cylinder $C_1(p, \ell\lambda^{-r}, a\mu^{-r})$ intersects $A^{-r}(V)$.

Let $\gamma = (\gamma_1, \gamma_2, \gamma_3) = (\gamma_1, \gamma_{23})$ and consider the case γ_1 not constant. By an argument in real variable analysis, there is some subinterval $J = [t_0, t_0 + c]$ of I on which

$$|\gamma_1(t) - \gamma_1(t_0)| \ge |t - t_0|^{\alpha/\beta}.$$

We may suppose that, for some sufficiently large r, c is the first positive number for which $\gamma_1(t_0 + c) = \gamma_1(t_0) + 2\ell\lambda^{-r}$. We deduce that, on J,

$$\begin{aligned} |\gamma_{23}(t) - \gamma_{23}(t_0)| &\le |\gamma(t) - \gamma(t_0)| \\ &\le A|t - t_0|^{\alpha} \\ &\le A|\gamma_1(t) - \gamma_1(t_0)|^{\beta} \\ &\le A(2\ell\lambda^{-r})^{\beta} \\ &= (2A\ell)^{\beta}\lambda^{-r\varepsilon}\mu^{-r} \\ &\le a\mu^{-r} \end{aligned}$$

for sufficiently large r. Thus $\gamma|J$ runs the length of $C\ (p, \ell\lambda^{-r}, a\mu^{-r})$ and hence intersects $A^{-r}(V)$.

If γ_1 is constant, consider the smallest rectangle with edges in the x_2- and x_3-directions containing $\gamma(I)$. If γ_2 is non-constant this rectangle becomes long (in the x_2-direction) and thin (in the x_3-direction) under A^r, $r > 0$, and so $A^r(\gamma)$ runs the length of some $C_2(p, \ell, a)$, and hence intersects V. If γ_3 is non-constant, we use a similar argument with A^r, $r < 0$ and $C_3(p, \ell, a)$. In either case, $0(\gamma)$ intersects V.

REFERENCES

[1] T. F. Banchoff & M. R. Rosen: Periodic points of Anosov diffeomorphisms, Proc. Symp. in Pure Math. 14 (1970) 17-21.

[2] J. Franks: Invariant sets of hyperbolic toral automorphisms. American J. Math. 99 (1977) 1089-1095.

[3] S. G. Hancock: Construction of invariant sets for Anosov diffeomorphisms. J. London Math. Soc. (2) 18 (1978) 339-348.

[4] S. G. Hancock: Invariant sets of Anosov diffeomorphisms. Doctoral thesis, Warwick University, 1979.

[5] M. W. Hirsch: On invariant subsets of hyperbolic sets. Essays on Topology and Related Topics (Springer-Verlag, 1970).

[6] M. C. Irwin: Smooth Dynamical Systems (Academic Press, 1980).

[7] M. C. Irwin: The orbit of a Hölder continuous path under a hyperbolic toral automorphism. Ergod. Th. and Dynam. Sys. 3 (1983).

[8] M. C. Irwin: Hölder continuous paths and hyperbolic automorphisms of T^n. (To appear).

[9] R. Mañé: Orbits of paths under hyperbolic toral automorphisms. Proc. Amer. Math. Soc. 73 (1979) 121-125.

[10] F. Przytycki: Construction of invariant sets for Anosov diffeomorphisms and hyperbolic attractors. Studia Math. 68 (1980) 199-213.

M. C. Irwin
Department of Pure Mathematics
University of Liverpool
P.O. Box 147
Liverpool L69 3BX
England.

Singularities & Dynamical Systems
S.N. Pnevmatikos (editor)
© Elsevier Science Publishers B.V. (North-Holland), 1985

A LABYRINTH AND OTHER WAYS TO LOSE ONE'S WAY

Ana Cascon* & Rémi Langevin
Université de Dijon
France

The theorem of Poincaré and Bendixon says that an orbit of a vector field on S^2 accumulates only on singularities, closed orbits or leaf cycles.

This is no more true of non orientables foliations or laminations.

We will first show, following Rosenberg, that some foliations of S^2 have leaves which can be dense or can accumulate on a transversly Cantor set.

Then we will show that the geometry of the unstable manifolds of a Morse Smale diffeomorphism of S^2 can be also much more complicate that the geometry of a vector field orbit, and provides curvature accumulation (cf infra) not unlike the unorientable foliations.

I - THE THEOREM OF POINCARE AND BENDIXON.

Let $\mathfrak{X}$ be a vector field on a manifold M. For each point m of M, let $\mathcal{O}(m)$ be the orbite of $\mathfrak{X}$ going through m. The ω-limit set of m, $\omega(m)$ is where the orbit through m accumulates when time goes to $+\infty$.

$$\omega(m) = \{x \in M \ / \ \exists t_n \to +\infty ; \ \varphi_{t_n}(m) \to x\}$$

where φ_t is the flow of the vector field $\mathfrak{X}$.

Replacing $t_n \to +\infty$ by $t_n \to -\infty$ proxides the definition of the α-limit set.

A cycle is a set $p_1 \ldots p_n = p_1$ of singularities and a family γ_i, $i \in I$, of orbits such that $\omega(\gamma_i) = p_j$ and $\alpha(\gamma_i) = p_k$.

THEOREM (Poincaré-Bendixon)

Let $\mathfrak{X}$ be a vector field of S^2 with a finite number of singularities. Then the ω-limit set of any point $m \in S^2$ can be only :

(*) Supported by CNPq, Brazil.

1) a singularity

2) a closed orbit

3) a cycle.

(see for example for detail and examples).

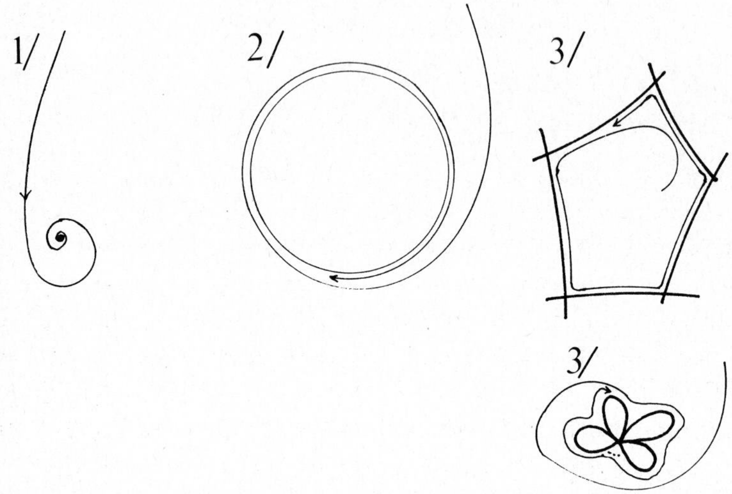

Fig. 1

II - ARATIONAL FOLIATIONS.

A singular foliation of dimension p of a manifold M is a foliation of the complement of a "small" "bad" set Σ, which is called singular set.

We will consider only the case where M is a surface, Σ a finite number of points.

1) The foliation $\mathcal{F}$ will be called arational if the neighbourhood of the singular points admits a model of one of the two following types :

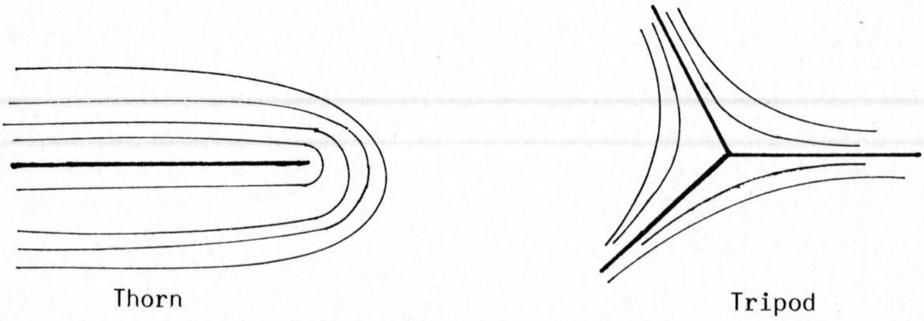

Fig. 2

that is, there is an homeomorphism between a neighbourhood of a singular point and a neighbourhood of $0 \in \mathbb{C}$ foliated by, in the case of the thorn the image of the horizontal lines by the map $z \mapsto z^2$, and in the case of the tripod the images of the levels of $(\mathrm{Re}\, z^3 = \lambda)$ by the same map $z \mapsto z^2$.

2) F has no compact leaf in the interior of M.

3) no leaf join two singular points.

4) if the boundary ∂M of M is non empty then $\mathcal{F}$ is transverse to ∂M.

Arational foliations on compact orientable surfaces have been studied by H. Rosenberg [Ro]. We will just take the opportunity of this conference in Crete to present a significant particular case of labyrinth.

A description, following Rosenberg's ideas of arational foliations of compact non orientable surfaces, has been done by the first autor [Ca].

First notice that any leaf of an arational foliation of S^2 is "infinite" as it can contain in its closure at most one singular point. If a leaf L accumulates on a regular point of a leaf ℓ it accumulates on all the points of ℓ.

Let now give an example of arational foliation of S^2 with dense leaves.

Let $\tilde{\mathcal{F}}$ be the Anosov foliation of T^2 (gived by the eigen vectors corresponding to the positive eigen-value of the linear map of matrix $\begin{pmatrix} 2 & 1 \\ 1 & 1 \end{pmatrix}$). All the leaves of $\tilde{\mathcal{F}}$ are dense.

Represent T^2 as a torus of revolution in $\mathbb{R}^3$ and take a line Δ as on fig 3 (T^2 is symetric with respect to Δ, Δ is orthogonal to the axis of revolution).

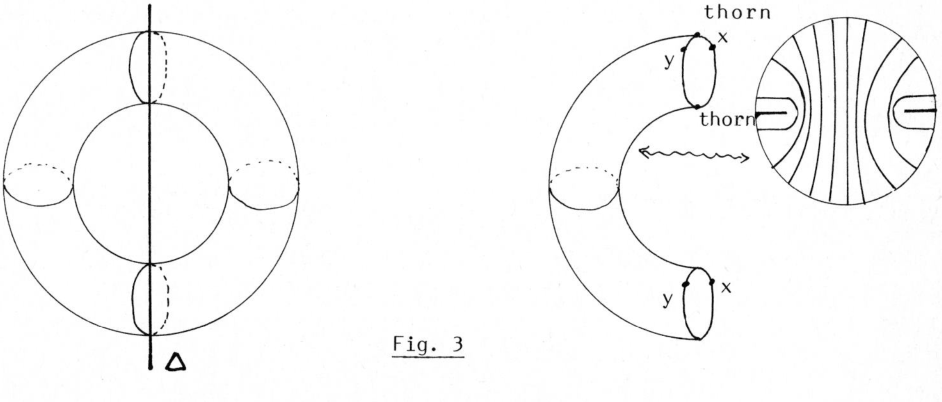

Fig. 3

The foliation $\tilde{\mathcal{F}}$ is globally invariant by the symetry with respect to Δ. The quotient is the sphere S^2 foliated with four thorns. The foliation can be obtained by gluing two discs endowed with the standard foliation of figure 4

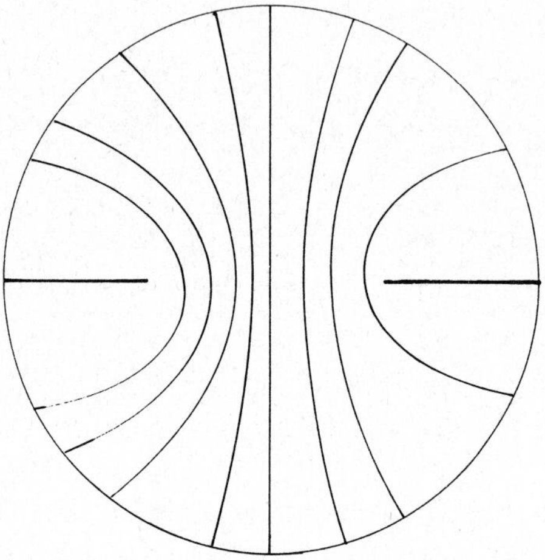

Fig. 4

by an irrational rotation. This shows that all the leaves are dense. Let now give an example of a "Denjoy-type" phenomenon.

Let D_1, D_2, D_3 three half discs foliated by concentric (half) circles. Covering with those three discs an horizontal segment, one obtains, forgetting the point A, a foliation with three thorns (the centers of the discs).

Suppose the bigger disc is of diameter one and the two smaller of diameter α and $(1-\alpha)$.

If α is irrational all the leaves interior to $D = D_1 \cup D_2 \cup D_3$ are dense in D.

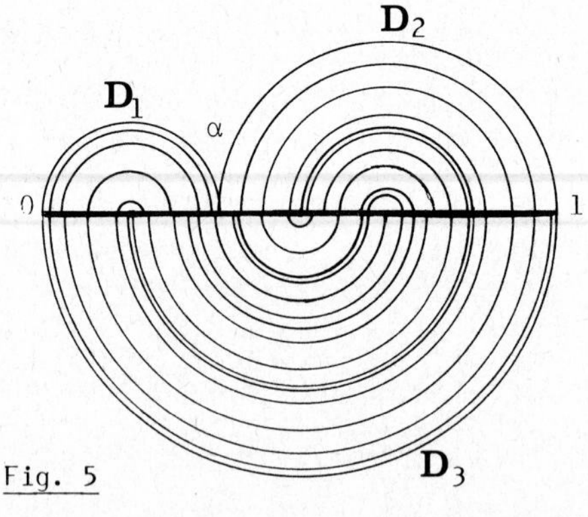

Fig. 5

We will now modify the picture of figure 5 to get an arational foliation of the disc.

Let cut open the leaf entering D at the point A and replace it by a "river" R, that is a band of width converging very rapidly (exponentially for example) to zero. The river being foliated as on fig. 6, on obtains a piece $\bar{D}$ of arational foliation.

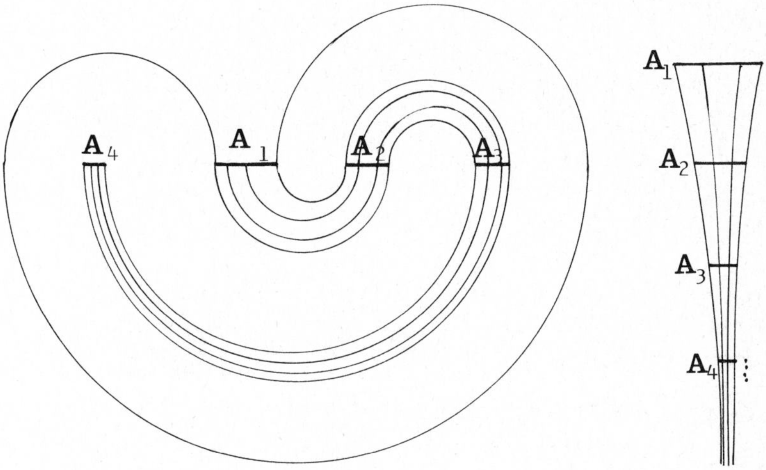

Fig. 6

The river R is dense in $\bar{D}$, and all the leaves inside R are proper, the complement $\mathcal{L}$ of the interior of R is cut by the segment I in a Cantor set. Any leaf do accumulate on $\mathcal{L}$.

This foliation can be extended to an arational foliation of a disc B by adding one tripod. (fig. 7).

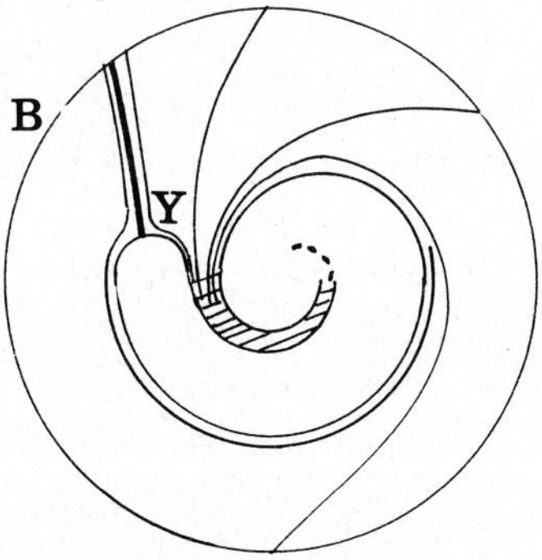

Fig. 7

Notice that one can still recognise three half discs $\tilde{D}_1, \tilde{D}_2, \tilde{D}_3$ covering almost completely the segment I ; each half disc being endowed with a foliation conjugate to the foliation by concentric (half) circles.

To obtain an arational foliation of S^2 all leaves of which accumulates on a labyrinth $\mathcal{L}$ it is enough to glue B to another disc endowed with the standard foliation of fig. 4, avoiding only to connect by a leaf a thorn and the tripod . H. Rosenberg [Ro] proves these two example are typical.

Suppose now the surface M is smooth (C^∞). When the leaves of $\mathcal{F}$ are smooth enough, say of class C^2 one can in every regular point x compute the absolute value of the geodesic curvature of the leaf of $\mathcal{F}$ going through x. Let denote by $|k|$ this function.

The singularities of the foliation are points where curvature accumulates.

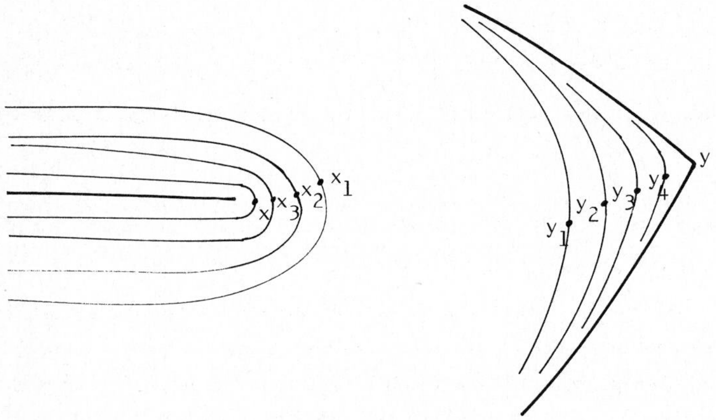

Fig. 8

On the models one can check that, restricted to rays going to the singular point, the curvature $|k|$ grows geometrically.

Points of accumulation of curvature, even if the leaves of the foliation $\mathcal{F}$ are smooth, are not automatically singular points (build a conterexample using $\frac{1}{n}$ sin nx like leaves).

It is true of a foliation of class C^2, that is such that all the foliation charts are of class C^2, as then, if a sequence x_n converges to a regular point x, the curvature $|k|(x_n)$ converge to the curvature $|k|x$.

We will see in next paragraph that the curvature function defined by instable manifolds of a Morse-Smale diffeomorphism of a sphere can also accumulate only on fixed point of the diffeomorphism.

III - MORSE-SMALE DIFFEOMORPHISMS OF A SPHERE S^2.

Let f be a diffeomorphism of a manifold M, and P an hyperbolic point of f.

The stable and unstable manifolds associated to P are defined by :

$$W^s(p) = \{x \in M \mid f^n(x) \to P,\ n \to +\infty\}$$

$$W^u(p) = \{x \in M \mid f^{-n}(x) \to P,\ n \to +\infty\}$$

Those manifolds are globally invariant by f, immersed, and as smooth as f. Our goal is to understand when those manifolds define a lamination of the surface. A first step in studying this question is to detect the points of accumulation of curvature.

We will prove that the unstable manifolds of the saddles of a Morse-Smale diffeomorphism of S^2 can provide curvature accumulation only at sinks. They form then a lamination of S^2 - {sinks of f} .

Let give first some definition (see [Pa-Me] for more details).

. The non-wandering set $\Omega(f)$ of a diffeomorphism f is defines by:

$$\Omega(f) = \{x \mid \forall v(x)\ ,\ \forall N\ \exists n > N,\ f^n(v(x)) \cap v(x) \neq \emptyset\ \}.$$

where $v(x)$ is any neighbourhood of x.

. The diffeomorphism f is Morse-Smale if it has only a finite number of periodic points, all hyperbolic, and all stable and unstable manifolds are transverse.

An immediate consequence is that, then, the manifold is the union of the unstable manifolds of the periodic points of f.

. A cycle is a sequence $P_o, P_1, \ldots, P_n = P_o$ of periodic points such that $W^u(P_i) \cap W^s(P_{i+1}) \neq \emptyset$; $i = 0,\ldots,n-1$.

When f is Morse-Smale, there are no cycles because otherwise $\Omega(f)$ would not be finite.

One has then a partial order given by the relation $P_i \leqslant P_j$ if $W^u(P_j) \cap W^s(P_i) \neq \emptyset$.

The behaviour beh (P,Q) of the pair (P,Q) of periodic points is the maximal length of a sequence $P = P_1 \geqslant P_2 \geqslant \ldots \geqslant P_k = Q$ of distinct periodic points ordered by the above relation.

Basic references on dynamical systems are [S] and [Pa - Me] .The methods we use in the chapiter are from [Pa].

Let Q be a periodic saddle of a Morse-Smale diffeomorphism f of a compact surface.

We will say that the curvature of $W^u(Q)$ accumulates at the point x if there exists a sequence (x_n) of points of $W^u(Q)$ converging to x and such that the curvature $|k|(x_n)$ (of $W^u(Q)$ at the point x_n) goes to infinity. In brief a point of accumulation of curvature is a point where the curvature of some $W^u(Q)$, Q saddle, accumulates.

THEOREM. Let f *be a Morse-Smale diffeomorphism of a closed surface. If* x *is a point of accumulation of curvature then* x *is a sink.*

The main step of our proof is to understand the geometry of the unstable manifolds in the neighbourhood of a saddle. So let recall some lemmas due to Palis and Smale concerning Morse-Smale diffeomorphisms of surfaces.

Strong λ lemma : Let P be a periodic saddle point of f. Let N be a transversal to $W^s(P)$ (that is a small transverse arc cutting $W^s(P)$ in one point). Let v(P) be a small open neighbourhood of P and $V^u(P)$ the intersection of $W^u(P)$ with v(f).Then $f^n(N)$ approches C^2 the unstable manifold of P in v(P), that is :

$$\{\forall \varepsilon > 0, \exists n_o \in \mathbb{N} \mid \forall n > n_o , f^n(N) \cap v(p) \text{ is } \varepsilon\text{-}C^2 \text{ close from } V^u(P)\}$$

where by close we mean close as submanifolds of v(P).

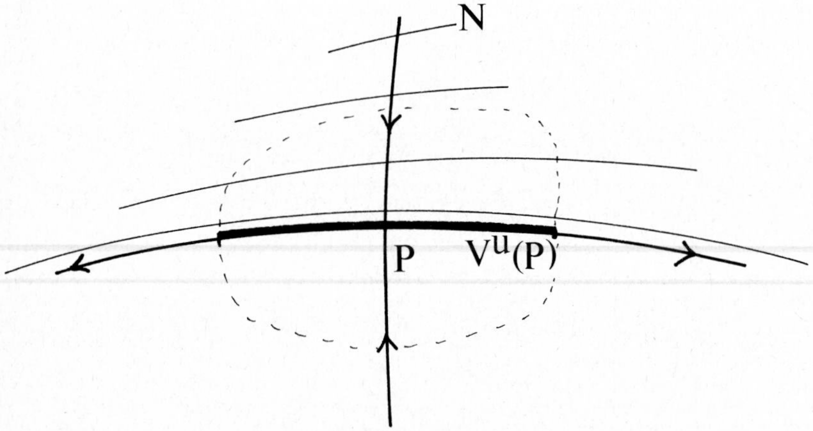

Fig. 9

Observation. . The λ-lemma as set by Palis says that $f^n(N)$ is C^1-close to $V^u(P)$. The strong version seems to be "folklore" and is known to Palis, Takens, Von Strien etc...

. A corollary of the strong λ-lemma is that no accumulation of curvature, which would have been of "weak type", can occur in the neighbourhood of a saddle.

COROLLARY. *If two singularities* S,Q *of a Morse-Smale diffeomorphism satisfy* beh $(S,Q) \neq 0$, *then* $W^u(S) \cap W^s(Q) \neq \emptyset$.

Proof.

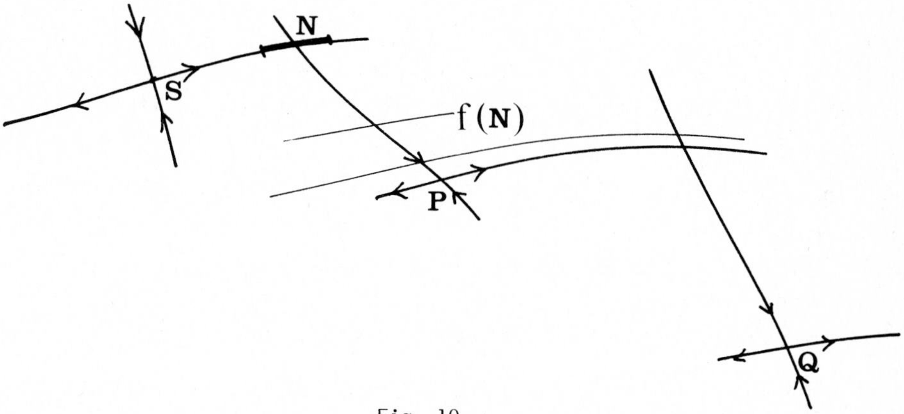

Fig. 10

LEMMA (Smale). *Suppose that* $\overline{W^u}(Q) \cap W^u(P) \neq Q$. *Then there exists a sequence* $P = P_1, P_2, \ldots, P_n = Q$ *of periodic points of* f *such that*

$$W^s(P_i) \cap W^u(P_{i+1}) \neq \emptyset \quad ; \quad 1 \leqslant i \leqslant n-1$$

The previous corollary implies then

$$W^u(Q) \cap W^s(P) \neq \emptyset .$$

LEMMA Palis [Pa] (*local lamination structure*)

Let P *be a saddle and* Q *be any periodic point. There exists a neighbourhood* V *of* P *such that if* $W^u(Q) \cap V \neq \emptyset$, *then*

$$W^u(Q) \cap V \simeq B \times A \quad (\simeq \textit{ is an homeomorphism})$$

where $B = V \cap W^u(P)$ *and* $A = V \cap W^s(P) \cap W^u(Q)$.

. notice that the strong λ-lemma does not imply that this chart is of class C^k.

COROLLARY. *For any neighbourhood $V^u(P)$ of P in $W^u(P)$ there exists a neighbourhood v of $V^u(P)$ such that the components of $W^u(Q) \cap v$ are C^k-close to $V^u(P)$.*

The corollary is obtained by iterating by f the neighbourhood v(P) introduced the strong λ-lemma.

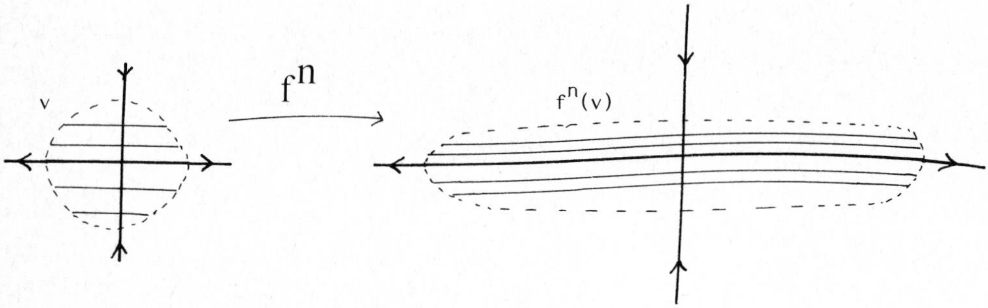

Fig. 11

Let now prove the theorem.

Let x be a point of accumulation of curvature of the unstable manifold of the saddle point Q.
As any point of the surface it should belong to the unstable manifold of some point P, and as f is Morse-Smale P is different from Q.

1) If P is a source, then $W^u(P)$, the "unstable basin of P", is an open manifold of dimension 2. Then the fact the point x belongs to $\overline{W^uQ}$ implies by Smale's lemma that $W^u(Q) \cap W^s(P) \neq \emptyset$, which is impossible because the stable manifold of P is the point P alone.

2) P cannot be a saddle because of the local lamination structure.

Then P is a sink ; $W^u(P) = P$, and so $x = P$ is a sink, which proves our theorem.

Lets end with some examples of Morse-Smale diffeomorphisms which are not embeddable of in a flow.

Example 1. Let X be a vector field on S^2 with six singularities as on fig.12 one source F, three sinks P_1, P_2, P_3 and two saddles S_1 and S_2. The two saddles are linked by the curve V, that is $W^u(S_1) - S_1$ and $W^s(S_2) - S_2$ have a common component V.

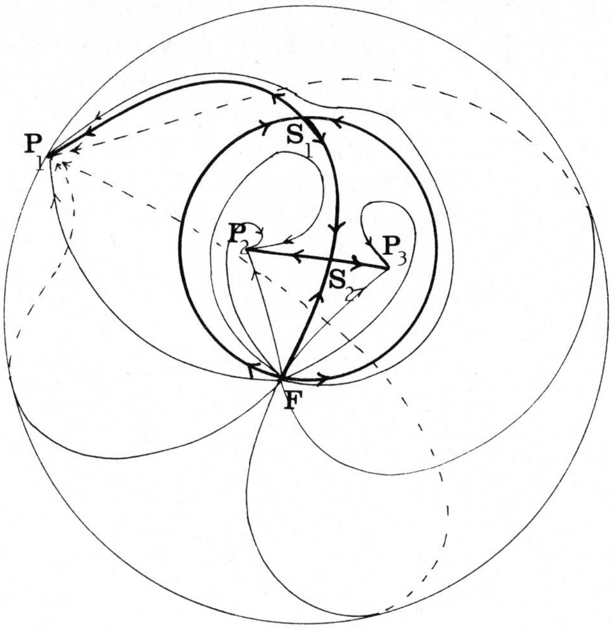

Fig. 12

Let $f(x) = \varphi_1(x)$ the diffeomorphism "time one" of the flow φ_t of the above vector field.

Let h be a diffeomorphism of S^2 of support in a neighbourdhood N of a period of f on the curve V.

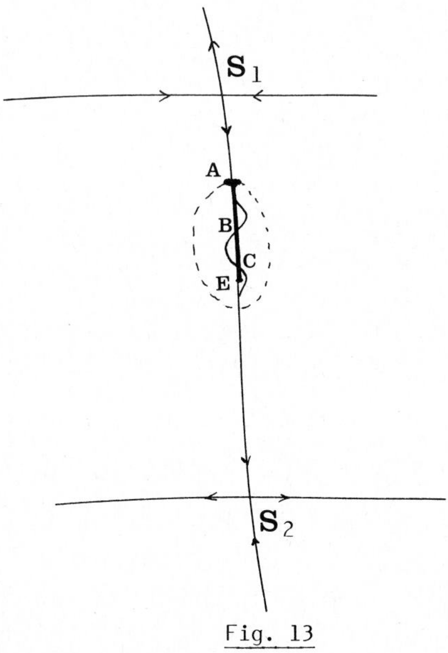

f(A) = E

Fig. 13

Denote by AE an arc a little longer than the period AC of V contained in N and choose a point B between A and E. The diffeomorphism shifts to the right between A and B, leaving A and B fixed, and to the left between the fixed points B and E.

One can check that $g = h \circ f$ is a Morse-Smale vector field of S^2, which coincide with f outside of N.

The arc S_1A still belongs to the unstable manifold of S_1 (for g) $W^u_g(S_1)$, but $W^u_g(S_1)$ leaves $W^u_f(S_1)$ in A as $W^u_g(S_1)$ is obtained by iterating S_1A by g. The unstable manifold $W^u_g(S_1)$ approches accordion-like the segment $W^u_g(S_2) \cup P_1 \cup P_2$.

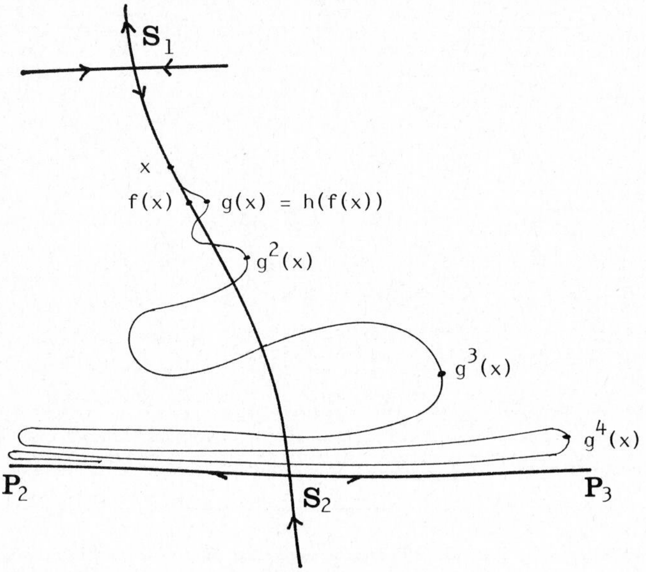

Fig. 14

One can check that the only intersection points of $W^u_g(S_1)$ and $W^s_g(S_2)$ are the points of the orbits of B and C.

The positive iterates of any other point of $W^u_g(S_1)$ still tends toward P_2 or P_3.

Symetrically $W^s_g(S_2)$ accumulates also accordion - like on $W^s_g(S_1) \cup F$, as on fig.15.

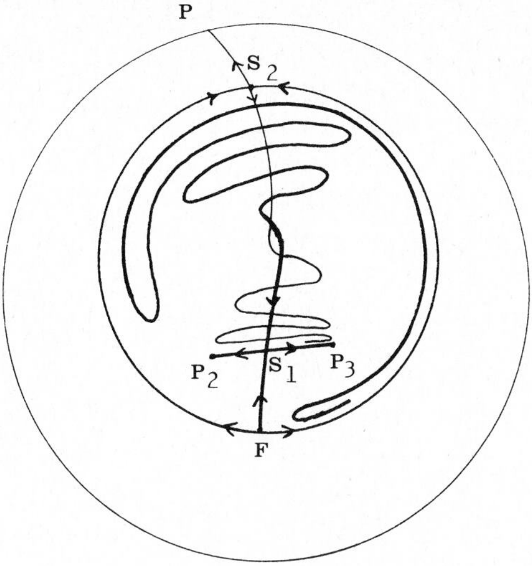

Fig. 15

Example 2. Let start from the vector field $\mathcal{X}$ on S^2 represented on fig. 13 . It has one source F, three sinks P_1, P_2, P_3 , and two saddles S_1, S_2. We of course suppose all the zeros are hyperbolic

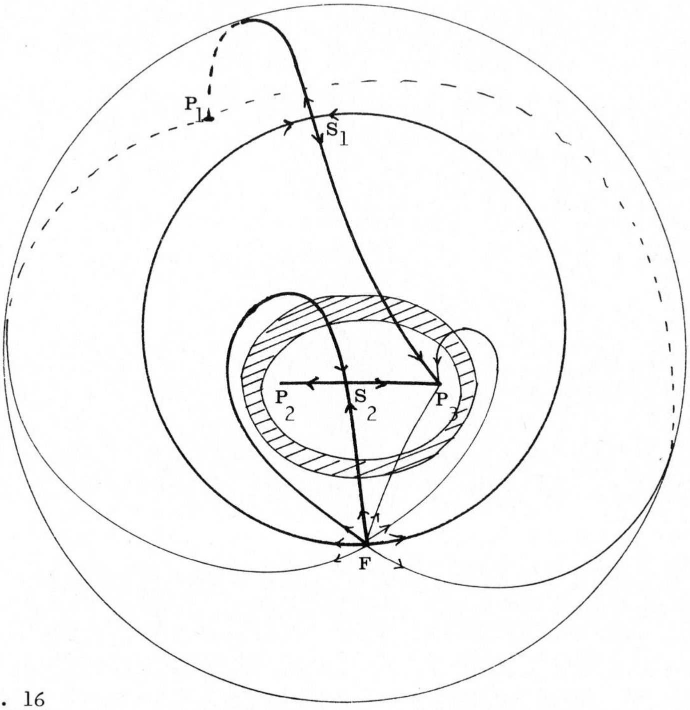

Fig. 16

Once again f is the time one flow of the vector field $\mathfrak{X}$.

The set of points attracted by the segment $I = \{W^u(S_2) \cup P_1 \cup P_2\}$, that is $\{x \mid \lim_{n \to \infty} d(f^n(x), I) = 0\}$ is a disc D so one can choose in this disc a fundamental domain, that is an anulus A such that $\partial A_1 = A_1 \cup A_2$; $A_2 = f(A_1)$, and of course $D - I = \bigcup_{n \in \mathbb{Z}} f^n(A)$.

The orbit of $\mathfrak{X}$ cut exactly once A_2 and A_1.

Let compose f with a Dehn twist in h, that is a map A of the annulus into it self such that the two boundary components are fixed by h, such that on a model annulus bounded by two concentric circles, it preserves globally the concentric circles and twists them gradually in order that the image of a radial segment "makes one turn", see fig. 17.

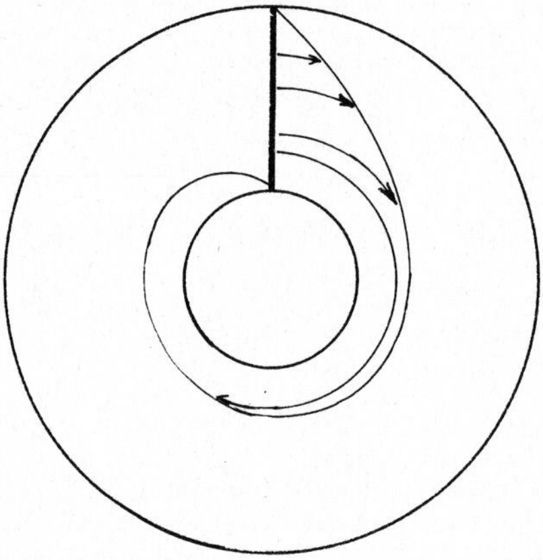

Fig. 17

One can check that $g \circ f$ is Morse-Smale.

The Dehn twist will make $W_g^u(S_1)$ turn in the annulus A and then $W_g^u(S_1)$ will have to spiral towards $W_g^u(S_2) \cup P_2 \cup P_3$. In a similar way the two components of $W_g^s(S_2)$ will spiral towards $W_g^s(S_1) \cup F$. see fig. 18 .

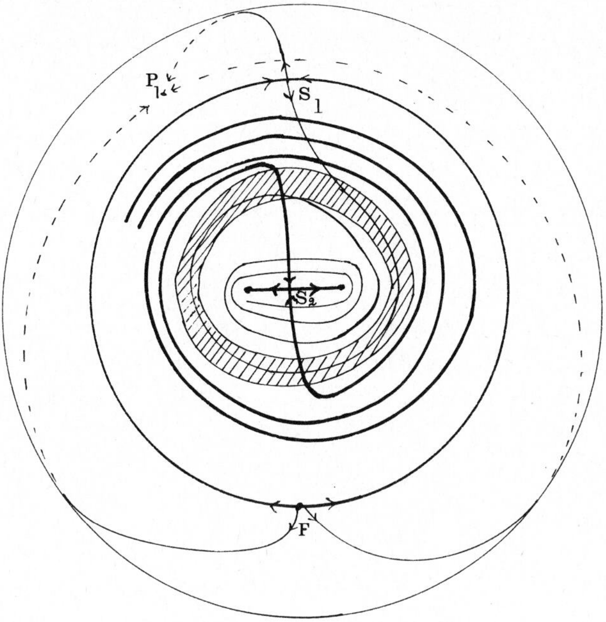

Fig 18.

Conjecture. Any Morse-Smale diffeomorphism on compact surface can be obtained from a vector feild by composing with Dehn twists and isotipies The second author thinks he can built a classification of Morse-Smale diffeomorphisms on compact surfaces which is a first step in showing this conjecture.

BIBLIOGRAPHIE

[Ca] A. CASCON : Folheações Arracionais em Superficies Compactas, Tese de Mestrado - PUC - Rio de Janeiro - Brésil.

[Pa] J. PALIS : On Morse Smale Dynamical Systems , Topology,(1969) pp 385-405

[PaMe] J. PALIS & W. de MELO : Geometric theory of Dynamical systems Springer-Verlag (1982)

[Ro] H. ROSENBERG : Labyrinths in the disc and surfaces , Annals of Maths, (1983) vol. 117 , pp. 1-33.

[S] S. SMALE : Differentiable dynamical systems , The Mathematics of the time , Springer , on BAMS 73 (1967) pp. 749-817

Ana CASCON & Rémi LANGEVIN
Laboratoire de Topologie
ERA 07 945
Département de Mathématiques
Université de Dijon B.P.138
21 004 DIJON Cedex
FRANCE

Singularities & Dynamical Systems
S.N. Pnevmatikos (editor)

MODULI OF STABILITY FOR GRADIENTS

Floris Takens
Rijksuniversiteit Groningen
The Netherlands

Generic k - parameter families of gradient vector fields are not always structurally stable.

1. INTRODUCTION

Investigations of structural stability, especially in a context of genericity, have played in important role in the study of dynamical systems in the last decades; for the definitions of genericity and structural stability see e.g. Smale [12] and Palis [7]. It was found that on (orientable) surfaces generic vector fields are structurally stable, see Peixoto [10]. In higher dimensions the situation is different: it follows from examples of Abraham and Smale [1] and Newhouse [4] that, for vector fields in dimensions greater than two, the structurally stable ones are no longer dense. On the other hand, Palis [5] and Palis-Smale [8] proved that for so called Morse - Smale vector fields the two-dimensional result could be generalized: in all dimensions, generic Morse-Smale vector fields are structurally stable.

Later, in analogy with the corresponding problem for functions, and in part inspired by the fact that generic k-parameter families of smooth functions are stable, see Looijenga [3], the problem of structural stability was taken up for generic vector fields, depending on parameters. It was not enough to restrict to Morse-Smale vector fields, since already in one-parameter families of such systems, so called moduli were found by Palis [6], Takens [15] and Van Strien [14]. These moduli are related with the existence of closed orbits and with non-real eingenvalues (of the linear part of the vector field) in singularities. Hence it was natural to restrict further investigation of structural stability of parameterized vector fields to gradients.

It turned out that generic one-parameter families of gradient vector fields are structurally stable, Palis - Takens [9]; there are also strong indications that the same holds for two-parameter families. This led us in [9] to pose the problem whether generic k-parameter families of gradients are structurally stable, or at most would only have a "finite number of moduli of stability".

In this paper I provide examples which disprove this conjecture. In fact we show that:

- it is not a generic property of k-parameter families of gradient vector fields on an n-manifold to be structurally stable, at least not if $k \geq 8$ and $n \geq 3$;
- it is not a generic property of k-parameter families of gradient vector fields on an n-manifold to have only a finite number of moduli of stability, at least not if $k \geq 16$ and $n \geq 4$.

It may be that the bounds on the dimension, and especially on the number of parameters can be sharpened, but that would probably need more complicated examples.

The author acknowledges hospitality at the Institut des Hautes Etudes Scientifiques during the preparation of this paper.

2. A THREE DIMENSIONAL EXAMPLE

We consider a C^{∞}, or highly differentiable, gradient vector field X on a 3-manifold M which satisfies the following specifications:

- X has a hyperbolic singularity in p with a 2-dimensional stable manifold such that the two contracting eigenvalues are equal (and real because X is a gradient);
- X has a hyperbolic singularity in q with a 2-dimensional unstable manifold such that the two expanding eigenvalues are equal (and real);
- each branch of $W^u(p)$ coïncides with a branch of $W^s(q)$;
- M is orientable on a neighbourhood of the loop formed by $W^u(p) = W^s(q)$.

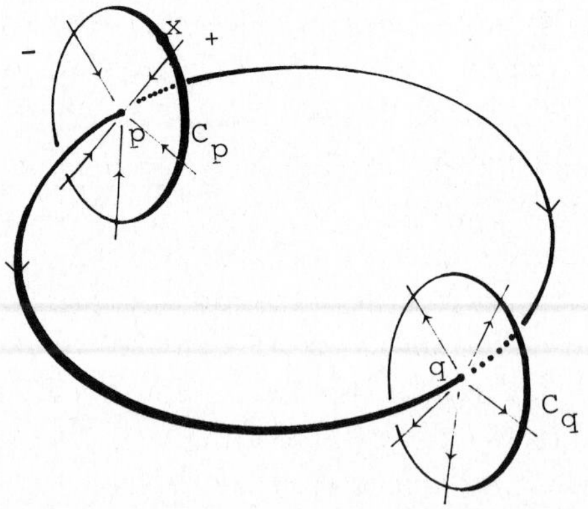

In this situation we say that the loop formed by $W^u(p)$, or $W^s(q)$, is a *false cycle* of X.

Let C_p be a fundamental domain in $W^s(p)$ and C_q be a fundamental domain in $W^u(q)$. We define two mappings P^+, and P^-, from C_p to C_q.

For $x \in C_p$, we consider a sequence of points $x_i \to x$ such that $x_i \notin W^s(p)$ and such that all x_i are on the same side of $W^s(p)$, (in the above figure, these sides are denoted by + and -). Let γ_i be positive X-integral curves starting in x_i. These integral curves pass, at least for big i, near p, then follow a branch of $W^u(p)$, (since all x_i are on the same side of $W^s(p)$, all γ_i follow the same branch) pass near q and then pass near a point of C_q. We claim that, for given $x \in C_p$ and for a given "side" of $W^s(p)$, (in wich to choose x_i), there is a unique point $y \in C_q$ for which there is a sequence $y_i \in \gamma_i$ such that $y_i \to y$. According to the side on which we took x_i, we have $y = P^+(x)$ or $y = P^-(x)$. In the next section we generalise this definition. The fact that P^+ and P^- are well defined and invertible follows from an analysis of the trajectories near a saddle point of the type X has in p or in q; this analysis is carried out in appendix 1. From this analysis it also follows that P^+ and P^- are *projective* in the following sense. Both $X|W^s(p)$ and $X|W^u(q)$ can be linearized by smooth coordinates according to Sternberg [13]. The linearizing coordinates are unique up to a linear coordinate transformations. Such linear coordinates induce an identification between points of C_p and "*directions* in $T_p(W^s(p))$" with directions we means: non-zero vectors up to multiplication with a *positive* scalar. Note that, in this way, C_p , and of course also C_q , are identified with a double covering of the real projective line. Also, we can now interprete the maps P^+ and P^- as maps from the directions in $T_p(W^s(p))$ to the directions in $T_q(W^u(p))$. P^+ and P^- are projective in the sense that there are linear maps from $T_p(W^s(p))$ to $T_q(W^u(q))$ which induce the same transformations on directions. Hence there is a linear transformation in $T_p(W^s(p))$ corresponding to $(P^-)^{-1} \circ P^+$.

From this it follows that $(P^-)^{-1} \circ P^+$ has

either four fixed points (two eigenvectors)

or two fixed points (one eigenvector)

or has no fixed points but is conjugated to a rotation (non real eigenvalues).

In the last case we speak of a *false cycle with rotation*.

Note that, if one has a false cycle, we can obtain one with rotation by modifying the flow along one of the branches of $W^u(p)$; this can even be done within the class of gradient vectorfields, e.g. see [16] . If one has a false cycle with rotation then, if one slightly perturbes the vectorfields such that the false cycle remains, then it remains as a false cycle with rotation. Furthermore, if one has a false cycle with rotation, the *rotation number* of $(P^-)^{-1} \circ P^+$ is clearly an invariant of topological equivalence.

Next we observe that in generic (k+8)-parameter families of gradient vector fields on M, the set of parameter values for which the vector field has a false cycle, is an immersed k-dimensional manifold. To see this, we have to "count the number of independent conditions" in the definition of a false cycle:

1. The condition on the eigenvalues of X at p counts for two co-dimensions: since X is a gradient, the linear part of $X|W^s(p)$ has the form

$$\begin{pmatrix} \alpha & \beta \\ \beta & \gamma \end{pmatrix}$$

the condition are $\beta=0$ and $\alpha=\gamma$.

2. For each branch of $W^u(p)$, the condition to coïncide with a branch of $W^s(q)$ also counts for two co-dimensions.

Since both of the above conditions appear twice, the false cycle occures with co-dimension 8.

THEOREM. *For k-parameter families of gradient vector fields on a 3-manifold, $k \geq 8$, it is not a generic property to be structurally stable.*

Proof. Let X_μ be a generic k-parameter family of gradients, $k \geq 8$. We may assume that there is some co-dimension 8 immersed submanifold in the μ-space where X_μ has a false cycle with rotation. The rotation number is a smooth function of μ on this manifold; we assume that for certain values $\mu_1, \mu_2, \ldots$ this rotation number has a local maximum, let $\{\sigma_1, \sigma_2, \ldots\}$ denote the rotation numbers at these maxima. Now one can make an arbitrarily small perturbation X'_μ of X_μ such that the corresponding set of rotation numbers $\{\sigma'_1, \sigma'_2, \ldots\}$ is different. Hence X_μ and X'_μ are not topologically equivalent. X is not structurally stable if it contains a false cycle._

3. RELATIONS DETERMINED BY SADDLE CONNECTIONS

We consider one aspect of the three-dimensional example in some more detail. Let X be again a vector field on a manifold M (of arbitrary dimension) which has hyperbolic singularities p and q. Let r be an orbit in the intersection of $W^u(p)$ and $W^s(q)$; such orbits are called heteroclinic orbits. Then r defines a relation, denoted R_r, between

$W^s(p)$ and $W^u(q)$:

if $x \in W^s(p)$ and $y \in W^u(q)$, then $x R_r y$ if and only if there are segments of X-integral curve $\gamma_i : [s_i, v_i] \to M$ such that $\lim \gamma_i(s_i) = x$, $\lim \gamma_i(v_i) = y$ and $\lim \operatorname{Im} \gamma_i = V^+(x) \cup r \cup V^-(y)$; this last limit is meant in the Hausdorff sense, V^+, respectively V^-, denotes the positive, respectively negative, X-orbit.

Observe that if $x, x' \in W^s(p)$ and $y, y' \in W^u(q)$, such that x, x' are on the same X-orbit and also y, y' are on the same X-orbit, then $x R_r y$ if and only if $x' R_r y'$. So R_r can be considered as a relation between orbits in $W^s(p)$ and $W^u(q)$. If one considers a fundamental domain in $W^s(p)$ as a parametrization of the orbits in $W^s(p)$ (with the exception of $\{p\}$), then one can also see R_r, as a relation between fundamental domains. In the previous section we considered a case where this relation was a map, even a homeomorphism. That was due to the conditions on the eigenvalues occuring in a false cycle.

Let now X be a vectorfield on M which has an invariant manifold $N \subset M$ which is ℓ-normally hyperbolic. This means that for each point $x \in N$, there is a splitting

$$T_x(M) = T_x(N) \oplus E^s(x) \oplus E^u(x)$$

and there are constants $\lambda_s < 0$ and $\lambda_u > 0$ such that for some Riemannian metric $\| \ \|$, we have

- for all t, $d(X_t)$ maps $E^u(x)$ to $E^u(X_t(x))$ and $E^s(x)$ to $E^s(X_t(x))$, X_t is the time t map of X;
- for $Y \in T_x(N)$, and $t \in 0$, $e^{\lambda_s . t} \| Y \| < \| dX_t(Y) \| < e^{\lambda_u . t} \| Y \|$;
- for $Y \in E^s(x)$, and $t > 0$, $\| dX_t(Y) \| < e^{\ell . \lambda_s . t} \| Y \|$;
- for $Y \in E^u(x)$, and $t > 0$, $e^{\ell . \lambda_u . t} \| Y \| < \| dX_t(Y) \|$.

We recall [2] that if N is ℓ-normally hyperbolic, N is C^ℓ and also, if X' is C^ℓ near X, X' has also an ℓ-normally hyperbolic invariant manifold.

We assume that X has two saddle points p, q in N, and that r is a heteroclinic orbit in N, connecting p and q. We want to compare the relations R_r for $X|N$ and for X (on all of M). For this we have to recall a few facts about normally hyperbolic invariant manifolds [11]. X induces a flow in the normal bundle of N, or, equivalently, in the bundle $E^u \oplus E^s$ over N. There is a continuous conjugation between this flow on the normal bundle (the normally linearized flow), and the original flow on M, at least if we restrict to a neighbourhood of the zero section respectively to a neighbourhood of N.

From this result it is clear that we have the following.

PROPOSITION. For $X, N \subset M$, p, q, r as above let $\pi_{\ell in}$ denote the projection of the normal bundle of N, and let π denote the projection from a neighbourhood of N to N, obtained by the above mentioned conjugacy, i.e., π maps X-integral curves to $(X|N)$-integral curves. For $x \in W^s(p)$ and $y \in W^u(q)$, $x R_r y$ holds if and only if, for $X|N$, $\pi(x) R_r \pi(y)$ holds.

We see that the induced projections

$$\pi|W_X^s(p) : W_X^s(p) \to W_{X|N}^s(p) = W_X^s(p) \cap N$$

and

$$\pi|W_X^u(q) : W_X^u(q) \to W_{X|N}^u(q) = W_X^u(q) \cap N$$

are important in the description of the relation R_r. Although π is in general not C^1, the induced projections, or rather the foliations whose leaves are inverse images of points, are differentiable: if X is C^k these foliations are C^{k-1}; this last fact is more or less well known, but for the sake of completeness I give a proof in appendix 2.

4. EXAMPLES IN DIMENSIONS GREATER THAN FOUR

In order to extend the theorem in section 2 to manifolds of higher dimension we carry out the following construction. Let M be a manifold of arbitrary dimension (but of dimension greater than three). Take a 3-dimensinal submanifold $N \subset M$. Choose a vector field X on N which has a false cycle with rotation. Extend X to M in such a way that N is an ℓ-normally hyperbolic invariant manifold where ℓ is choosen so that everything in section 2 is valid for C^ℓ-gradients. If M has a Riemannian metric, such a vector field can be made a gradient. Using the proposition in section 3, one can extend the arguments in section 2 to the above extended vector field on M. Thus we have

THEOREM. For k-parameter families of gradient vector fields on an n-manifold, $k \geq 8$ and $n \geq 3$, it is not a generic property to be structurally stable.

In view of that we are going to do in the next section, we describe the "relations" in the sense of the preceding section for these higher dimensional false cycles. As before, let $p, q \in N$ be the saddle points of X as above. If N has, in the normal direction, an n_s-dimensional stable bundle E^s and an n_u-dimensional unstable bundle E^u then,

$$\dim(W^s(p)) = n_s + 2$$

$$\dim(W^u(q)) = n_u + 2 .$$

From section 2 we have n_s, respectively n_u, dimensional foliations in $W^s(p)$, respectively $W^u(q)$. If we omit the exceptional leaves containing p, respectively q, and if we restrict to a fundamental domain (with dimension n_s+1, respectively n_u+1), we have a foliation of $S^1 \times D^{n_s}$, respectively $S^1 \times D^{n_u}$, whose space of leaves in both cases is S^1. The relation R_r, where r is one of the hetroclinic orbits between p and q, can now be interpreted as a diffeomorphism between the spaces of leaves in the fundamental domains of $W^s(p)$ and $W^u(q)$. If r and r' are the two hetroclinic orbits between p and q, $R_{r'}^{-1} o R_r$ defines a diffeomorphism of the space of leaves in the fundamental domain of $W^s(p)$. This diffeomorphism plays here the same role as $(p^-)^{-1} o P^+$ in section 2.

5. EXAMPLES WITH INFINITELY MANY MODULI

Let M be a manifold of dimension n>3. Let X be a gradient vector field on M which has two false cycles as constructed in the previous section; such vector fields occure in generic k-parameter families of gradient vector fields if $k \geq 16$. We denote the saddle points of the first false cycle by p and q and the saddle points of the second false cycle by $\tilde{p}$ and $\tilde{q}$. Using the notation of section 4, we have that

$$\dim (W^s(p)) = n_s+2$$

$$\dim (W^u(\tilde{q})) = \tilde{n}_u+2$$

where n_s, $\tilde{n}_u \leq n-3$. We choose the dimensions so that $n_s+2+\tilde{n}_u+2 = n+2$ or $n_s+\tilde{n}_u+2 = n$. Then we arrange X so that $W^s(p)$ and $W^u(\tilde{q})$ have a transversal intersection, necessaraly of dimension 2, which satisfies the following conditions:

- in a fundamental domain of $W^s(p)$, the (one-dimensional) trace of this intersection is a circle which intersects all the leaves of the foliation, defined by the relations R_r and $R_{r'}$, transversally and projects diffeomorphically onto the space of leaves of this foliation;
- in a fundamental domain of $W^u(\tilde{q})$, the (one-dimensional) trace of this intersection is a circle which intersects all the leaves of the foliation, defined by the relations $R_{\tilde{r}}$ and $R_{r'}$, transversally and projects diffeomorphically onto the space of leaves of this foliation.

As a consequence, the intersection of $W^s(p)$ and $W^u(\tilde{q})$ determines a diffeomorphism between the spaces (circles) of leaves in the fundamental domains of $W^s(p)$ and $W^u(\tilde{q})$.

Next we assume that both false cycles have rotation, and that their rotation numbers are irrational. In this case we have an identification of the circle of leaves in the fundamental domain of $W^s(p)$ with the standard circle $\mathbb{R}/1$ which conjugates the map defined by $(R_{r'})^{-1} \circ (R_r)$ with a "pure" rotation $\varphi \to \varphi + \sigma \bmod(1)$, σ being the rotation number of the first false cycle. This identification is unique up to a pure rotation. The same holds in a fundamental domain of $W^u(\tilde{q})$. Hence $W^s(p) \cap W^u(\tilde{q})$ determines a map from the standard circle to itself up to left and right composition with a pure rotation. This map, up to compositions with pure rotations, is an invariant for topological equivalences. The space of these maps, modulo composition with pure rotations is infinite dimensional so we find:

THEOREM. *For k-parameter families of gradient vector fields on an n-manifold, k≥16 and n≥4, it is not a generic property to have only finitely many moduli.*

* * *

* APPENDIX 1. *Orbits passing near a hyperbolic singularity.*

Let X be a smooth vector field on $\mathbb{R}^3$ with a hyperbolic singularity in the origin, and let the linear part of X in the origin be

$$\begin{pmatrix} -\lambda_1 & 0 & 0 \\ 0 & -\lambda_1 & 0 \\ 0 & 0 & \lambda_2 \end{pmatrix}$$

$\lambda_1, \lambda_2 > 0$.

Without loss of generality we may assume that (locally) the stable manifold coïncides with the x_1, x_2 plane and that the unstable manifold coïncides with the x_3-axis. As we may assume that X, restricted to the stable manifold, is linear, see Sternberg [13]. This means the X has the following form:

$$X = -\Lambda_1(x)\left(x_1\frac{\partial}{\partial x_1} + x_2\frac{\partial}{\partial x_2}\right) + \Lambda_2(x).x_3\frac{\partial}{\partial x_3} + \Lambda_3(x)\left(x_1\frac{\partial}{\partial x_2} - x_2\frac{\partial}{\partial x_1}\right)$$

with $\Lambda_1|\{x_3=0\} \equiv \lambda_1$, $\Lambda_2(0) = \lambda_2$ and $\Lambda_3|\{x_3=0\} \equiv 0$.

We want to modify our coordinates so that we also get Λ_3 zero along the x_3 axis. For this we consider the manifold V consisting of pairs (r,L), with $r \in W^u(0)$ and L a linear transformation from $T_r(\mathbb{R}^3)/\langle X(r)\rangle$ to $T_0(W^s(0))$. On V there is an induced vector field X_V whose time t map sends (r,L) to $(\varphi_t(r), d\varphi_t(0) \circ L \circ (d\varphi_t(r))^{-1})$, where φ_t stands for the time t map of X. Since $d\varphi_t(0)$ commutes with all linear maps,

X_V is zero in all points of the form $(0,L)$. So in these points, there is a 4-dimensional center manifold and a 1-dimensional unstable manifold (with eigenvalue λ_2). We consider the unstable manifold of $(0,id)$; this must be of the form $\{(r,L(r)) \mid r \in W^u(0)\}$ with $L(0) = id$.

Now we choose new x_1, x_2 coordinates, coinciding on $\{x_3=0\}$ with the old coordinates, such that $L(r)$, expressed in these new coordinates, is the identity for all $r \in W^u(0)$. Then, along the x_3 axis, Λ_3 is zero and Λ_1 is λ_1.

Next we take points $s \in W^s(0)$ and $r \in W^u(0)$ and consider integral curves which pass very close to s and r; we are interested in how they approach r. For this we introduce polar coordinates ρ, φ: $x_1 = \rho \cos\varphi$ and $x_2 = \rho \sin\varphi$. Then it follows that if $\gamma:[a,b] \to R^3$ is a segment of an X-integral curve, then

$$\frac{|\varphi(a) - \varphi(b)|}{\rho(a) - \rho(b)} = \mathcal{O}(d(\mathrm{Im}(\gamma), W^s(0) \cup W^u(0))) ,$$

where d denotes here the maximal distance from a point in $\mathrm{Im}(\gamma)$ to the set $(W^s(0) \cup W^u(0))$; this estimate holds at least for those γ which do not leave some fixed neighbourhood of 0. For segments of integral curves whose end points approach s and r, this distance $d(\mathrm{Im}(\gamma), W^s(0) \cup W^u(0))$ goes to zero, while $\rho(a) - \rho(b)$ remains bounded; hence $\varphi(a) - \varphi(b)$ goes to zero.

In other words, if $\{\gamma_i(a_i,b_i)\}$ is a sequence of X-integral curves so that $\lim \gamma_i(a_i) = s$ and $\lim \gamma_i(b_i) = r$, then

$$\lim \frac{x_1(\gamma_i(b_i))}{x_2(\gamma_i(b_i))} = \frac{x_1(s)}{x_2(s)} .$$

In this way, the integral curves near $W^s(0) \cup W^u(0)$ define a map from the set of direction in $T_o(W^s(0))$ to the directions in $T_r(\mathbb{R}^3)/\langle X(r)\rangle$ wich is independent of the coordinates and projective in the sens of section 2.

* APPENDIX 2. *Canonical projections.*

We consider a vector field X with a hyperbolic singularity. Since we want to consider only constructions inside the stable manifold, and since we know that the stable manifold is as differentiable as X, we may assume that the hyperbolic singularity is a sink, i.e. all eigenvalues of the linear part have negative real part. Since we are only interested in local constructions, we may assume X to be defined on $\mathbb{R}^n$ and the hyperbolic sink to be the origin.

Let $T_o(\mathbb{R}^n) = E_1(0) \oplus E_2(0)$ be a splitting, invariant under dX_t, such that for some number $\lambda<0$,

- if $Y \in E_1(0)$, $t>0$, $\| dX_t(Y) \| < e^{t\lambda} \| Y \|$;
- if $Y \in E_2(0)$, $t>0$, $\| dX_t(Y) \| > e^{t\lambda} \| Y \|$.

We want to construct a smooth invariant foliation such that $E_1(0)$ is a tangent plane of the leaf through the origin.

Consider the manifold consisting of pairs (x,V), with x in $\mathbb{R}^n$ and $V \subset T_x(\mathbb{R}^n)$ a linear subspace with the same dimension as $E_1(0)$. (X_t, dX_t) induces a flow in this manifold. $(0,E_1(0))$ is a hyperbolic singularity of this flow; in fact, the stable manifold of this singularity has dimension n and projects diffeomorphically to $\mathbb{R}^n$. This stable manifold defines in each point $x \in \mathbb{R}^n$ (near the origin) a linear subspace $\mathcal{D}(x) \subset T_x(\mathbb{R}^n)$. The distrubition $x \longrightarrow \mathcal{D}(x)$ is C^{k-1} if X is C^k. Finally we prove that $\mathcal{D}$ is integrable.

For this, let $\mathcal{D}_o$ be the "constant distribution", i.e., $\mathcal{D}_o(x)$ is obtained by translating $E_1(0)$ to x. We define $\mathcal{D}_t$ as $(d(X_{-t}))\mathcal{D}_o$, i.e., $\mathcal{D}_t(x) = (dX_{-t})(\mathcal{D}_o(X_t(x)))$. From the proof of the stable manifold theorem it follows that $\mathcal{D}_t$ converges in the C^1 topologie to $\mathcal{D}$. Since $\mathcal{D}_t$ is integrable for all t so is $\mathcal{D}$. So we have:

PROPOSITION. Let X be a C^k vector field with a hyperbolic singularity in p. If $E \subset T_p(W^s(p))$ is a linear subspace, invariant under dX_t, such that all contractions of dX_t in E are stronger than the induced contractions in $T_p(W^s(p))/E$, then there is a unique invariant C^{k-1}-foliation in $W^s(p)$ whose leave in p has E as tangent space.

The foliation in $W^s(p)$, defined by restriction of the projection π in section 3 is equal to the above foliation; for more informations see [2].

* * *

REFERENCES

[1] R.Abraham - S.Smale, Non genericity of ω-stability. Proc. A.M.S. Symp. in Pure Math. 14 (1970), 5-9.

[2] M.W.Hirsch - C.C.Pugh - M.Shub, Invariant manifolds. Lecture Notes in Mathematics 583 (1977), Springer-Verlag.

[3] E.Looijenga, Structural stability of smooth families of C^∞ functions. Thesis, Amsterdam, 1974.

[4] S.Newhouse, Non-density of action A(a) on S^2. Proc. A.M.S. Symp.

in Pure Math. 14(1970) p. 191-203.

[5] J.Palis, On Morse-Smale dynamical systems. Topology 8(1969) p. 385-403.

[6] J.Palis, A differentiable invariant of topological conjugacies and moduli of stability. Asterisque 51(1978) p. 335-346.

[7] J.Palis, Moduli of stability and bifurcation theory. Proc. Int. Congres Math., Helsinki 1978.

[8] J.Palis & S.Smale, Structural stability theorems. Proc. A.M.S. Symp. in Pure Math. 14(1970) p. 223-232.

[9] J.Palis & F.Takens, Stability of parametrized families of gradient vector fields. Ann. of math.,118,(1983),383-421.

[10] M.M.Peixoto, Structural stability on two dimensional manifolds. Topology 1(1962) p. 101-120.

[11] C.C.Pugh & M.Shub, Linearization of normally hyperbolic diffeomorphisms and flows. Inv. Math. 10(1970) p. 187-198.

[12] S.Smale, Differentiable dynamical systems. B.A.M.S. 73(1967) p. 747-817.

[13] S.Sternberg, Local contractions and a theorem of Poincaré. Amer. J. Math. 79(1957) p. 809-827.

[14] S.J. van Strien, One parameter families of vector fields, bifurcations near saddle-connections. Thesis, Utrecht 1982.

[15] F.Takens, Global phenomena in bifurcations of dynamical systems with simple recurrence. Jber.d.Dt.Math.-Verein. 81(1979)p.87-96.

[16] F.Takens, Mechanical systems and gradient systems; local perturbations and generic properties. Preprint R.U.G. Groningen,ZW-8206 To appear in Boletim Soc. Bras. Mat.

Floris TAKENS
University of Groningen
Department of Mathematics
P.O.B. 800
9700 AV Groningen
THE NETHERLANDS

* * *

Singularities & Dynamical Systems
S.N. Pnevmatikos (editor)

SINGULARITIES OF GRADIENT VECTOR FIELDS AND MODULI

Floris Takens
Rijksuniversiteit Groningen
The Netherlands

1. INTRODUCTION

We consider gradient vector fields, i.e., vector fields X for which there exist a Riemannian metric g and a function V such that $g(X,-)=dV$. Our considerations are mainly local, so we assume all these objects to be defined on $\mathbb{R}^n$. We assume that X has a singularity in the origin, so $dV(0) = 0$.

The analysis of these singularities was motivated by the following considerations. It is known that generic gradient vector fields on compact manifolds are structurally stable [2,4]. This is also true if the vector field depends on one parameter [5], while in low dimensions the result even remains valid with more parameters [8]. On the other hand, there is a recent example [6] showing that generic k-parameter families of gradients need not be structurally stable if $k \geq 8$. The example is based on a configuration of two saddles with two orbits of non-transverse intersection of stable and unstable manifolds and on the equality of certain eigenvalues at the saddles. This configuration leads to a so called modulus of stability. For the purpose of this paper we can define this as follows.

Let $X^g(M)$ be the space of gradient vector fields on a compact manifold M. Let $W \subset X^g(M)$ be a smooth submanifold of finite codimension and let $\mu: W \to \mathbb{R}$ be a smooth function with non-zero derivative. Then we say that (W,μ) is a *modulus (of stability)* if $\mu(X) \neq \mu(X')$ for $X, X' \in W$ implies that X and X' are not topologically equivalent. The codimension of W is also called the *codimension of the modulus*.

The example in [6] left a number of problems, like what is the lowest codimension of a modulus of stability of gradient vector fields? or, can a modulus be due to only one (isolated) singularity, or to only one orbit of non-transverse intersection of a stable and an unstable manifold? Here we deal with possible moduli of isolated singularities of gradient vector fields. To state the result we use the space J_k^n of k-jets of singularities of gradient vector fields on $\mathbb{R}^n$.

THEOREM. *For $n \geq 5$ and k sufficiently big there are a smooth submanifold $W \subset J_k^n$ and a smooth function $\mu: W \to \mathbb{R}$ with non-zero derivative such that, whenever X and X' are gradient vector fields on $\mathbb{R}^n$ with singularity in the origin and such that their k-jets $j_k(X)$ and $j_k(X')$ are in W, then, if X and X' are topologically equivalent (near the origin),*

$$\mu(j_k(X)) = \mu(j_k(X')) \quad .$$

One could call such $\mu: W \to \mathbb{R}$ with $W \subset J_k^n$ a modulus of stability for singularities of vector fields. It is not difficult to see that the conclusion of the theorem leads to the existence of a modulus of stability for gradient vector fields in the sens of the above definition.

The construction of (W,μ) is based on the following ideas. Let X be a gradient vector field on $\mathbb{R}^n$ with an isolated singularity in the origin. The stable, unstable set is denoted by W^s, W^u respectively. Both these sets consist of integral curves of X; we denote by F^s, F^u the space of X-integral curves in $W^s - 0$, $W^u - 0$. So one may consider W^s, W^u as a cone on F^s, F^u. Next we define a relation R between F^s and F^u: $x R y$ if and only if there are X-integral curves passing arbitrarily near both x and y. Then the idea is to make this relation the union of two diffeomorphisms from F^s to F^u which we denote by Φ_+ and Φ_-. This leads to a diffeomorphism $d_u = \Phi_+ . \Phi_-^{-1} : F^u \to F^u$ which is now a topological invariant in the following sense. Let X' be another gradient vector field with singularity which is topologically equivalent with the singularity of X; denote the topological equivalence by h. F'^u and d'_u are defined as above using X' instead of X. Then h induces a homeomorphism from F^u to F'^u which conjugates d_u with d'_u or with $(d'_u)^{-1}$. This follows from the "topological characterization" of d_u : if $y, y' \in F^u$ and if there is some $x \in F^s$ such that xRy and xRy', then $y = y'$ or $y = d_u(y')$ or $y' = d_u(y)$. Finally we make an example where d_u has a modulus of stability of the type considered in [1,3] .

This example suggests that we may expect all complications, known to exist for diffeomorphisms, to show up when studying isolated singularities of gradient vector fields. Though I think this expectation is correct, I was not able to prove it: in the present examples d_u is quite special, in fact, d_u is very close to the time one map of a gradient vector field.

Finally we should mention that in many cases, like in [6], moduli of stability were used to show that generic k-parameter families need not be structurally stable if k is greater than or equal the codimension of the modulus. Such an argument however always makes use of some extra structure. For example in [6] this conclusion is based on the fact that if we denote the modulus constructed there by (W',μ') and if $X'\notin W'$ but X' is near W', then X' is not topologically equivalent with any $X\in W'$. For our present modulus of stability it is not clear whether there is such extra structure; hence we cannot conclude to new types of instability of generic parametrised families of gradient vector fields.

2. CONSTRUCTION OF THE MODULUS.

A. *Blowing up.*

First we recall the blowing up construction. Let X be a vector field on $\mathbb{R}^n$ with $X(0)=0$. Then there is an induced vector field $\tilde{X}$ on $S^{n-1}\times\mathbb{R}$ such that $\Phi_*(\tilde{X})=X$, where $\Phi(w,r)=r.w$ (we identify S^{n-1} with the unit sphere in $\mathbb{R}^n$), e.g. see [7]. $\tilde{X}$ is said to be obtained by blowing up X. If the ℓ-jet of X is zero in the origin, $\ell\geq 1$, then $\tilde{X}$ is zero in the points of $S^{n-1}\times\{0\}$ and we can in fact devide by r^ℓ; the resulting vector field we denote by $\bar{X}=r^{-\ell}.\tilde{X}$.

We shall apply this method to singularities of gradient vector fields. So let V_0 be a homogeneous polynomial of degree k. We denote grad V_0 by X_0 and the corresponding vector fields on $S^{n-1}\times\mathbb{R}$ by $\tilde{X}_0$ and $\bar{X}_0=r^{-k+2}.\tilde{X}_0$. Then $\bar{X}_0|S^{n-1}\times\{0\}$ is the gradient of $V_0|S^{n-1}$ (again we identify S^{n-1} with the unit sphere in $\mathbb{R}^n$). We shall take V_0 so that the critical points of $V_0|S^{n-1}$ are either non-degenerate or are part of critical manifolds where the second derivative of $V_0|S^{n-1}$, normal to these critical manifolds is non-degenerate. Also we shall take V_0 so that in the critical points of $V_0|S^{n-1}$, V_0 is nowhere zero.

In this situation the stable set W^s is a cone on the union of the stable manifolds in $S^{n-1}\times\{0\}$ of those singularities of $\bar{X}_0|S^{n-1}\times\{0\}$ where $V_0|S^{n-1}$ is negative (where we now identify both S^{n-1} and $S^{n-1}\times\{0\}$ with the unit sphere in $\mathbb{R}^n$). The unstable set W^u is the union of the unstable manifolds in $S^{n-1}\times\{0\}$ of those singularities of $\bar{X}_0|S^{n-1}\times\{0\}$ where $V_0|S^{n-1}$ is positive. If we add to V_0 a function V_1 with k-jet zero to obtain $V=V_0+V_1$, then the stable and unstable sets of grad V_0 and grad V are homeomorphic. Also $\bar{X}_0|S^{n-1}\times\{0\}$ is equal to $\bar{X}|S^{n-1}\times\{0\}$; $\bar{X}$ is the vector field obtained from grad V by blowing up and deviding by r^{k-2}.

B. *Construction of V_o.*

As announced, V_o will be a homogeneous polynomial. In the construction we first make a polynomial $\tilde{V}$ such that $\tilde{V}|S^{n-1}$ has the required properties and then choose V_o so that $V_o|S^{n-1} = \tilde{V}|S^{n-1}$. For this last construction we need that $\tilde{V}(x) = \tilde{V}(-x)$ for all x in S^{n-1} or that $\tilde{V}(x)=-\tilde{V}(-x)$ for all x in S^{n-1}. For this reason we shall make $\tilde{V}$ so that $\tilde{V}(-x)=\tilde{V}(x)$.

A first step in the construction of V is the construction of a polynomial $v_1:\mathbb{R}^2\to\mathbb{R}$ such that

- $v_1(x_1,x_2) = v_1(\pm x_1,\pm x_2)$ for all $(x_1,x_2)\in\mathbb{R}^2$;

- we associate to v_1 the phase portrait of the following vector field: extend v_1 tot $\mathbb{R}^n$ by $v_1(x_1,x_2,\dots,x_n) = v_1(x_1,x_2)$, then take $\mathrm{grad}(v_1|S^{n-1})$ and project the integral curves on $\mathbb{R}^2$ (note that the result is independent of n); we require that this phase portrait has the following form:

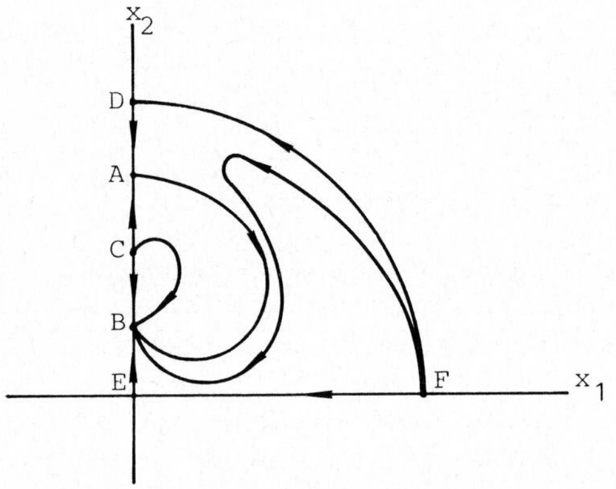

in the other quadrants, the phase portrait is determined by symmetry;

- from the phase portrait it follows that

$v_1(F) < v_1(E) < v_1(B)$,

$v_1(F) < v_1(D) < v_1(A) < v_1(B)$,

$v_1(C) < v_1(A)$;

we alsoassumme that $v_1(B)$ is positive and $v_1(A)$, $v_1(C)$, $v_1(D)$, $v_1(E)$ and $v_1(F)$ are negative;

- $v_1|S^{n-1}$ has critical points corresponding to D and F and has critical submanifolds corresponding to A, B, C and E, these points and submanifolds in S^{n-1} will also be denoted by A,...,F; we assume that $v_1|S^{n-1}$ has nondegenerate 2nd derivative normal to A,B,C and E.

We observe that for each $\lambda_3,\dots,\lambda_n$, not all zero, the 3-sphere $\{(x_1,\dots,x_n)\in S^{n-1}\,|\,x_3:\dots:x_n=\lambda_3:\dots:\lambda_n\}$ is an invariant submanifold for grad $(v_1|S^{n-1})$. A consequence of this is that for each point $a\in A$ (as subset of S^{n-1}) the two branches of the one-dimensional unstable manifold of a approach B in the same point. We shall distroy this last property by adding to v_1 a perturbation of the following form

$$\tilde{V}(x_1,\dots,x_n)=v_1(x_1,x_2)+\varepsilon.x_1^3.v_{2,\eta}(x_3,\dots,x_n),$$

where $v_{2,\eta}(x_3,\dots,x_n)$ is a homogeneous polynomial, depending on $\eta\in\mathbb{R}^2$ and satisfying $v_{2,\eta}(-x)=-v_{2,\eta}(x)$; we shall come back to the definition of $v_{2,\eta}$. $\tilde{V}|S^{n-1}$ has the same critical points and critical submanifolds as $v_1|S^{n-1}$. For ε shall we still have that for each point $a\in A$ the two branches of $W^u(a)$ approach B, but now in different points. This defines, for ε small, two diffeomorphisms $\Phi_{+,\varepsilon,\eta}$, $\Phi_{-,\varepsilon,\eta}: A\longrightarrow B$: $\Phi_{+,\varepsilon,\eta}(a)$, $\Phi_{-,\varepsilon,\eta}(a)$ is the limit point of the branch of $W^u(a)$ in $x_1\geqq 0$, $x_1\leqq 0$. If we identify A and B with S^{n-3} using

$$(x_3,\dots,x_n)\longrightarrow\frac{1}{\sum_{i\geq 3}x_i^2}(x_3,\dots,x_n),$$

then we see from the formula for $\tilde{V}$ that

$$\frac{\partial}{\partial\varepsilon}\Phi_{\pm,\varepsilon,\eta}(a)\Big|_{\varepsilon=0}=\pm c.\mathrm{grad}(v_{2,\eta}|S^{n-3})(a)$$

for some constant c. From this it also follows that

$$\frac{\partial}{\partial\varepsilon}(\Phi_{+,\varepsilon,\eta}\cdot\Phi^{-1}_{-,\varepsilon,\eta}(a)\Big|_{\varepsilon=0}=2c.\mathrm{grad}(v_{2,\eta}|S^{n-3})(a).$$

Before we analyse the situation further we want to show that $\Phi_{+,\varepsilon,\eta}\cdot\Phi^{-1}_{-,\varepsilon,\eta}$ can be interpreted as the map $d_u: F^u\longrightarrow F^u$ mentioned in the introduction.

C. *Analysis of grad* V_0.

From the preceding constructions and remarks it follows that the unstable set of grad V_o is the cone on B or, more precisely, on two copies of the (n-3)-sphere, one in $\{x_2>0\}$ and one in $\{x_2<0\}$. The stable set is more complicated: we have to take the union of the stable manifolds of the singularities where $V_o|S^{n-1}$ is negative. This means, in the (x_1,x_2)-plane description: the curve C-A-D-F-E, including the endpoints. In S^{n-1} this corresponds with the two closed (n-2)-discs for the line C-A-D (one in $\{x_2>0\}$ and one in $\{x_2<0\}$), four curves for D-F (one in each quadrant), and an (n-2)-sphere for E-F. The stable set is the cone on all this. Because of the symmetry we may restrict our attention to $\{x_1 \geq 0\}$.

Now we consider the relation R between F^s and F^u defined in the introduction. We shall make use of both vector fields $X_0 = \text{grad}\, V_0$ and $\bar{X}_0$, obtained from X_0 by blowing up and deviding by the appropriate power of r. We shall often identify X integral curves with the corresponding $\bar{X}$ integral curves in $S^n \times \{r<0\}$. The sets A and B in S^{n-1} will be identified with the corresponding subsets in $S^{n-1}\times\{0\}$.

Restricting to $\{x_2 \geqslant 0\}$, F^u is an (n-3)-sphere which we can identify with B: each $\bar{X}_0$ integral curve in F^u approaches, for $t\to-\infty$, a unique point in $B\subset S^{n-1}\times\{0\}$. F^s, also restricted to $\{x_2 \geqslant 0\}$ consists of an (n-2)-disc, two curves and an (n-2)-sphere. We restrict our attention to the part of F^s which consists of the interior of the (n-2)-disc, minus the point where the two curves are attached; this part of F^s we denote by $\tilde{F}^s$. There is a canonical projection $\pi\colon \tilde{F}^s \to A \subset S^{n-1}\times\{0\}$: each $\bar{X}_0$ integral curve in $\tilde{F}^s$ has a unique limit point in A for $t \to +\infty$.

Using the fact that A and B, as submanifolds of $S^{n-1}\times\{0\}$, are normally hyperbolic invariant manifolds of $\bar{X}_0$ (this time not restricted to $S^{n-1}\times\{0\}$) we see that an orbit x in $\tilde{F}^s$ is related with an orbit y in F^u if and only if $\Phi_{+,\varepsilon,\eta}(\pi(x))=y$ or $\Phi_{-,\varepsilon,\eta}(\pi(x))=y$. This means that $\Phi_{+,\varepsilon,\eta}\cdot\Phi^{-1}_{-,\varepsilon,\eta}$ is indeed $d_u F^u \to F^u$ as defined in the introduction (except that we had here the further complications that $\tilde{F}^s \subsetneqq F^s$ and that $\dim \tilde{F}^s = \dim F^u + 1$) in the sense that for each y, y' in F^u such that for some $x \in \tilde{F}^s$ we have x Ry and x Ry′, $y=y'$, or $y=d_u(y')$, or $y'=d_u(y)$.

D. The choice of $v_{2,\eta}$ and construction of the modulus.

We choose $v_{2,\eta}$ so that grad $(v_{2,\eta}|S^{n-3})$ is a two-parameter family of gradient vector fields on S^{n-3} wich satisfies the following specifications:

- for some line ℓ in the η-plane, $\text{grad}(v_{2,\eta}|S^{n-3})$ has an orbit of tangency of a stable and an unstable manifold of hyperbolic singularities; transverse to this line the tangency unfolds generically in the sense of [5];

- along ℓ the modulus (of topological conjugacy) defined in [2] is not constant.

For example in the case n=5 (so $S^{n-3}=S^2$) one might take $v_{2,\eta}$ such that $\text{grad}(v_{2,\eta}|S^2)$ has the following local phase portraits:

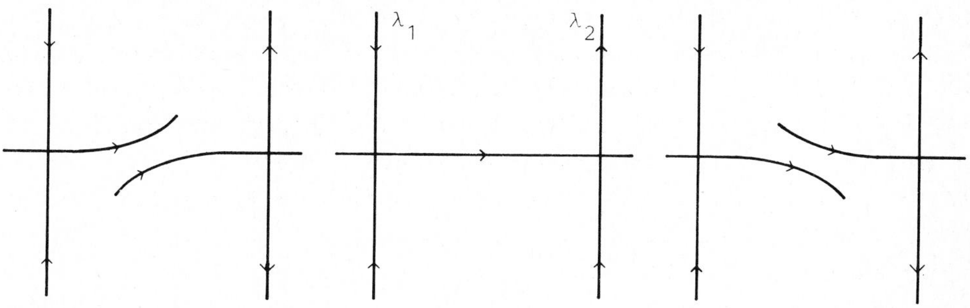

If λ_1 and λ_2 are the contracting and expanding eigenvalues as indicated, then the above mentioned modulus is λ_1/λ_2.

If we now take ε sufficiently small, then also $d_{u,\varepsilon,\eta} = \Phi_{+,\varepsilon,\eta} \cdot \Phi^{-1}_{-,\varepsilon,\eta}$ has two hyperbolic fixed points which have a tangency for η on some line L near ℓ. Along this line L the modulus, associated with this tangency, is again not constant. We fix ε and only consider $\eta \in L$. This gives a one-parameter family of homogeneous polynomials $V_{0,s}$ ($s \in \mathbb{R}$ is the parameter) say of degree k. Now we take $W \subset J^n_k$ as the set of k-jets of $\{V_{0,s}\}_{s\in\mathbb{R}}$, so $\dim W = 1$, and as map $\mu : W \to \mathbb{R}$ we take the modulus associated with the tangency.

This is indeed a modulus as announced because if V is any smooth function with the same k-jet as $V_{0,s}$ then, as we mentioned in the section on "blowing up", the unstable sets of grad V and grad $V_{0,s}$ are homeomorphic and, by construction the maps d_u for grad V and grad $V_{0,s}$ are conjugate. This means that the modulus of the tangency is also the same in both cases.

REFERENCES

[1] S. Newhouse & J. Palis & F. Takens, Stable families of diffeomorphisms. Publ. Math. I.H.E.S. 58 (1983) p. 5-71.

[2] J. Palis, On Morse - Smale dynamical systems. Topology 8 (1969) p. 385 - 403.

[3] J. Palis, A differentiable invariant of topological conjugacies. Asterisque 51 (1978) p. 335 - 346.

[4] J. Palis & S. Smale, Structural stability theorems. Proc. A.M.S.

Symp. in Pure Math. 14 (1970) p. 223-232.

[5] J. Palis & F. Takens, Stability of parametrized families of gradient vector fields. Ann. of Math. 118 (1983), 383-421.

[6] F. Takens, Moduli of stability for gradients: generic k-parameter families of gradient vector fields are not always structurally stable. In this volume.

[7] F. Takens, Singularities of vector fields. Publ. Math. I.H.E.S. 43 (1974) p. 47-100.

Floris TAKENS
University of Groningen
Department of Mathematics
P.O.B. 800
9700 AV Groningen
THE NETHERLANDS

* * *

Singularities & Dynamical Systems
S.N. Pnevmatikos (editor)

GEOMETRY OF TRIPLES OF VECTOR FIELDS IN $\mathbb{R}^4$

Piotr Mormul & Robert Roussarie
Université de Dijon
France

1. INTRODUCTION

We indicate by $X = \{X_1,\ldots,X_k\}$ a k-uple of C^∞ vector fields in $\mathbb{R}^n$. Two k-uples $X = \{X_1,\ldots,X_k\}$ and $Y = \{Y_1,\ldots,Y_k\}$ are equivalent if and only if there exist a C^∞ diffeomorphism g of $\mathbb{R}^n$ and field of C^∞ $k \times k$-matrices : (f_{ij}) such that :

$$Y_i = g_*\left(\sum_{j=1}^{k} f_{ij}X_j\right) \quad \text{for } i = 1,\ldots,k.$$

Interest for k-uples of vector fields comes from control theory. Here, we only want to speak about the singularities of these objects. A general theory for singularities of k-uples was initiated in the article of B. Jakubczyk and F. Przytycki [2] and next continued by the first author [4] , [5] , [6] . Our aim is just to show how some geometrical considerations may help the study. More precisely we establish for the triples of vector fields in $\mathbb{R}^4$ the following result :

THEOREM. *A generic triple of vector fields in $\mathbb{R}^4$ is locally equivalent outside a union of dimension 1 manifolds, to the triple $\{\frac{\partial}{\partial x},\frac{\partial}{\partial y},\frac{\partial}{\partial z}+x\frac{\partial}{\partial w}\}$ or the triple $\{\frac{\partial}{\partial x},\frac{\partial}{\partial y},x\frac{\partial}{\partial z}-y\frac{\partial}{\partial w}\}$, in local coordinates (x,y,z,w).*

The existence of local explicit models implies that the germs of generic triple of vector fields are stable, in an evident sense, for points of codimension ≤ 2. We show below that the first model in the theorem corresponds to the regular case (codimension 0 - stratum of singularity) and the second one to a singularity stratum of codimension 2.

2. THE SINGULARITIES OF k-UPLES OF VECTOR FIELDS.

We begin with some definitions coming from the general theory of Jakubczyk and Przytycki. Let L_0X the modulus of vector fields generated by a given k-uple of vector fields in $\mathbb{R}^N$,X. More generally, for $1 \leq i$, the modulus $L_i X$ is defined as the modulus generated by $L_{i-1}X$ and its Lie brackets. These module are used by Jakubczyk and Przytycki to define the singular subsets of X in the following manner:

$$S_{(i,j)}X = \{ m \in \mathbb{R}^n | \dim L_i X(m) = j\}$$

for each couple of integers (i,j), $i \geq -1$, $j \geq 0$. ($L_{-1}X(m)$ means $T_m\mathbb{R}^n$).

If $S_{(i,j)}$ is a submanifold, you can go on and define, for a second couple (i',j') :

$$S_{(i,j)(i',j')}X=\{ m \in S_{(i,j)}X | \dim \left[L_i X(m) \cap T_m S_{(i,j)}\right] +j' = \dim L_{i'}X(m)\}$$

and so on, you may define possibly singular sets : S_I for $I = (i_1,j_1)\dots(i_\ell,j_\ell)$, and more generally subjets as $(S_I \cap S_j)_K$ for systems of couples I,J,K. (If U is a modulus of vector fields, and $m \in \mathbb{R}^n$, we note $U(m) = \{ v(m) \in T_m\mathbb{R}^n | v \in U\}$.).

Using similar methods to those used in the Boardman theory [1] , B. Jakubczyk, F. Przytycki and next, P. Mormul have shown that these singular subsets generically exist and are submanifolds.

In the following, we limit ourselve to triple of vector fields in $\mathbb{R}^4$. For this case, a complete description of generic singularities was given by P. Mormul [4], [5] .

It is summed up in the following figures 1,2 taken in [5] . (For two singular sets A,B, $A \to B$ means : $A \subset \bar{B}$). For simplicity we named the differents singular sets and their closure : $M_0, M_1, M_2, M_3, M_{10}, M_{20}, M_{30}, M_{21}, M_{31}, M_2 \cap M_3$; the correspondance with the complete name is given in figure 1.

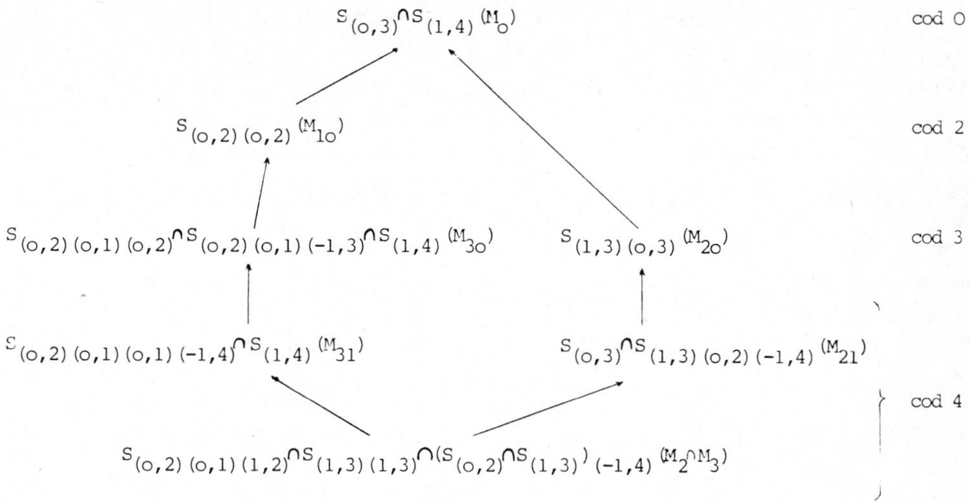

Figure 1

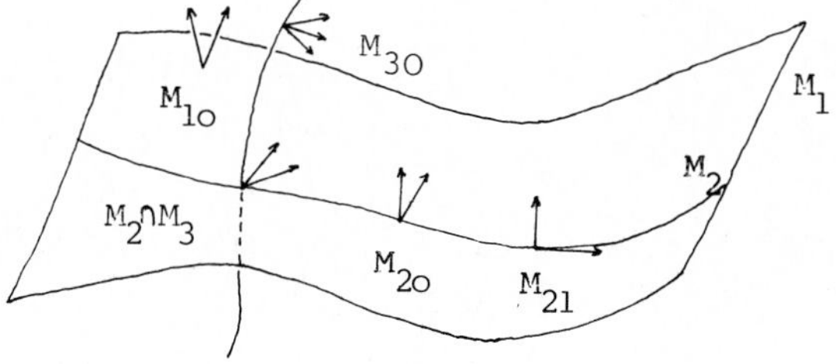

Figure 2

For a generic triple $X = \{X_1, X_2, X_3\}$ there exist submanifolds M_1 (dimension 2), M_3 (dimension 1), M_2 (dimension 1) $\subset M_1$ such that the different singular sets are (see figure 2) :

1. $M_o = \mathbb{R}^4 \setminus M_1 \cup M_3$: at $m \in M_o$, $\dim X(m) = 3$ and $\dim L_1 X(m) = 4$.
2. $M_{1o} = M_1 \setminus M_2$: $\dim X(m) = 2$, $X(m)$ is transversal to $T_m M_1$ and $\dim L_1\ X(m) = 4$.
3. M_{2o} (open intervals on M_2) : $\dim X(m) = 2$, $\dim (X(m) \cap T_m M_1) = 1$, $X(m) \cap T_m M_1$ transversal to M_3, and some other generic independant conditions (see figure 1).
4. M_{3o} (open intervals on M_3) : $\dim X(m) = 3$, $\dim L_1\ X(m) = 3$, $X(m)$ transversal to M_3, and some other generic conditions.
5. M_{21} (isolated points on M_2) : $\dim (X(m) \cap T_m M_1) = 1$, $X(m) \cap T_m M_1$ is tangent to M_3 and some other generic conditions.
6. M_{31} (isolated points on M_3) : $\dim X(m) = 3$, $\dim L_1 X(m) = 3$, $X(m)$ tangent to M_2 and some other generic conditions.
7. $M_2 \cap M_3 = M_1 \cap M_3$ (isolated points) : for the definition, see figure 1.

   It is easy to find algebraic examples of each singularities [5]. Below, we show that the germs of generic triples at points of M_o and M_{1o}, admit the unique models given in the theorem of introduction.

3. THE MODEL FOR REGULAR POINTS (M_o).

For the germ at $m_o \in M_o$, the given model is an easy consequence of the Darboux theorem for contact structure. Because we need the geometry of the regular stratum M_o to study the other strata M_{1o}, we indicate how to deduce the demonstration from the Darboux theorem.

Outside the manifold M_1 , $\dim X = 3$ and X defines a Pfaffian structure (a field of 3-planes). Around each $m_o \notin M_1$, the equivalence of triple is the same as equivalence of 3-plane fields (via diffeomorphisms of $\mathbb{R}^4$). So there is no risk of confusion to note also by X, the field of planes $m \to X(m)$. This field may also be given as the field of Kernels of a 1-form Ω , non-zero around m_o, and defined up to a multiplicative, non-zero function.

Now $\dim L_1 X(m) = 4 \Longleftrightarrow \Omega \wedge d\Omega\ (m) \neq 0$. So $\mathrm{Ker}\,\Omega \wedge d\Omega$ is a field of lines

(depending only on X). Choose a non-zero vector field U, parametrizing locally this field. (For example, if dV is a volume-form, let U defined by $U \lrcorner dV = \Omega \wedge d\Omega$).
Now it is easy to see that : $\Omega(U) = 0$ or equivalently : $U(m) \in X(m)$ for each m near m_o ; $\Omega \wedge L_U \Omega \equiv 0$ or equivalently : $[U,X] \subset X$. This last property means that X is invariant by the U-flow. So, if W is a flow-box of U around x_o, diffeomorphic to $\mathbb{R}^4$, there exists, on the quotient $\tilde{W}$ of W by the trajectories of U, (diffeomorphic to $\mathbb{R}^3$), a 2-plane fields $\tilde{X}$ such that X is the counter-image of $\tilde{X}$ (This means that if $\pi : W \to \tilde{W}$ is the natural projection and $m \in W$, then $X(m) = d\pi^{-1}(m)(\tilde{X}(\pi(m)))$. Now the condition $\Omega \wedge d\Omega \neq 0$ implies that X is defined by a form $\tilde{\Omega} = xdz - dw$ for coordinates (x,z,w) on $\tilde{W}$ (theorem of Darboux for $\tilde{X}$) ; we can also take $\Omega = xdz - dw$ for coordinates (x,y,z,w) on W from where it results easily that X is equivalent to the tripe $\{\frac{\partial}{\partial x}, \frac{\partial}{\partial y}, \frac{\partial}{\partial z} + x\frac{\partial}{\partial w}\}$.

Remarque 1: You can visualize $\tilde{W}$ by taking a transversal section to U ; $\tilde{X}$ is then given by the intersection of $\tilde{W}$ with X.

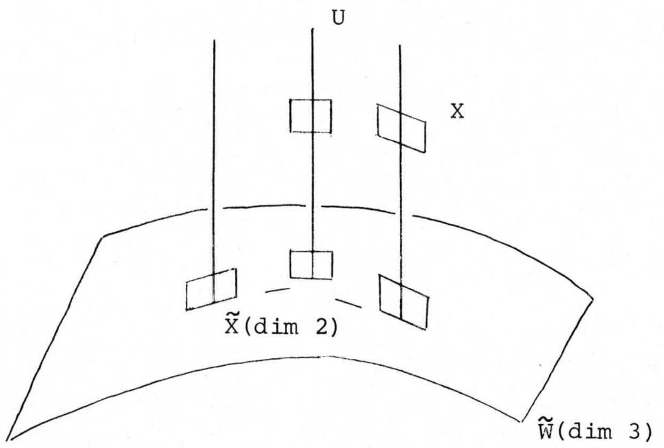

Figure 3

Remark 2 : You can notice that the 2-plane field $\Delta = \{\frac{\partial}{\partial x}, \frac{\partial}{\partial y}\}$ is an integrable field tangent to X ; $\Delta = \pi^{-1}\tilde{\Delta}$ with $\tilde{\Delta} = \{\frac{\partial}{\partial x}\}$ a field of lines, tangent to $\tilde{X}$ (A "Legendre line field"). Inversely, every integrable 2-plane field Δ in W, tangent to X is equal to the counter-image by π of Legendre field of lines of $\tilde{X}$.

To see this result, suppose that V_1 and V_2 are 2 vector fields generating Δ . We say that U is tangent to Δ .

Otherwise, at a point m, U(m), $V_1(m)$, $V_2(m)$ are independant and the relations $[U,V_1](m) \in X(m)$, $[U,V_2](m) \in X(m)$, $[V_1,V_2](m) \in \Delta(m)$ imply that dim $L_1X(m) = 3$, contrairely to the hypothesis. Now, because Δ is integrable, we have $[U,\Delta] \subset \Delta$ which implies the existence of a Legendre line field $\tilde{\Delta}$, quotient of Δ .

4. THE MODEL FOR CODIMENSION 2-POINTS (M_{1o}).

Around a point $m_o \in M_1$ the triple X is no longer equivalent to a Pfaffian structure. But, locally one can associate a 1-form Ω to X as in the following :

PROPOSITION 2. Let $m_o \in M_1$; there exists a 1-form Ω defined in a neighborhood of m_o such that :

a) $\Omega \equiv 0$ on M_1, $\Omega \neq 0$ and $X = \operatorname{Ker}\Omega$ outside M_1.

b) If $m_o \in M_{1o}$, one can find local coordinates (x,y,z,w) such that $M_1 = \{x=y=0\}$, $X(m) = \{\frac{\partial}{\partial x}, \frac{\partial}{\partial y}\}$ for $m \in M_1$ and $\Omega = xdz - ydw \bmod \mathfrak{J}(x,y)^2$ (where $\mathfrak{J}(x,y)$ is the ideal generated by the functions x, y).

Proof. a) Let $m_o \in M_1$. We can choose X_1, X_2, X_3 such that X_1 and X_2 are independant around m_o. Next suppose that coordinates (x,y,z,w) are chosen around m_o such that the projections of X_1, X_2 on the (x,y)-plane are independant. Changing X_1, X_2 by combinations of these 2 vector fields, you can suppose that the projections of X_1 and X_2 on the (x,y)-plane are respectively $\frac{\partial}{\partial x}$ and $\frac{\partial}{\partial y}$. Then, adding a combination of X_1 and X_2 to X_3, you have $X_3 = f\frac{\partial}{\partial z} + g\frac{\partial}{\partial w}$ for some C^∞ functions f,g. The local equation of M_1 is given by f=g=0 and so, f and g are independant functions around m_o.

At this point, we have obtain that the 3 vector-fields X_1, X_2, X_3 can be written :

$$(1)\quad \begin{cases} X_1 = \frac{\partial}{\partial x} + \frac{\partial}{\partial z} + * \frac{\partial}{\partial w} \\ X_2 = \frac{\partial}{\partial y} + * \frac{\partial}{\partial z} + * \frac{\partial}{\partial w} \\ X_3 = f \frac{\partial}{\partial z} + g \frac{\partial}{\partial w} . \end{cases}$$

Where the symbols $*$ repesent some C^∞ functions.
Let now $\Omega = \alpha dx + \beta dy + \gamma dz + \delta dw$ a 1-form such that $\Omega(X_1) = \Omega(X_2) = \Omega(X_3) \equiv 0$ around m_o. These equations are equivalent to the system in $\alpha, \beta, \gamma, \delta$:

$$\begin{cases} \alpha + * \gamma + * \delta = 0 \\ \beta + * \gamma + * \delta = 0 \\ \gamma f + \delta g = 0 \end{cases}$$

The third equation must be solved by $\gamma = -g$, $\delta = f$ and the two first equations give α and β. We have that $\alpha, \beta \in \mathfrak{J}(f,g)$, the ideal generated by f,g. Clearly, the form Ω verifies the point a in the proposition.

b) Suppose now that $m_o \in M_{1o}$. We can choose the local coordinates (x,y,z,w) such that $M_1 = \{x = y = 0\}$ around m_o and that $X_1 = \frac{\partial}{\partial x}, X_2 = \frac{\partial}{\partial y}$ on M_1.

It follows from this, that you can write X_1, X_2, X_3 by the formula (1) with the functions $*$ in the ideal $\mathfrak{J}(x,y) = \mathfrak{J}(f,g)$. The form Ω is equal to :

$$\Omega = gdz - fdw \mod \mathfrak{J}(x,y)^2$$

(mod $\mathfrak{J}(x.y)^2$ means that the function coefficients of the rest are in the ideal $\mathfrak{J}(x,y)^2$).
If you choose (f,g,z,w) as new coordinates, we obtain :

$$\Omega = xdz - ydw \mod \mathfrak{J}(x,y)^2$$

For the form Ω defined in proposition 2, we define the vector field U by $U \lrcorner dV = \Omega \wedge d\Omega$, where dV is a volume form. Around a point $m_o \in M_{1o}$, in the system of coodinates (x,y,z,w) given at the point b, and write $dV = dx \wedge dy \wedge dz \wedge dw$ we have :

$$U = x\frac{\partial}{\partial x} + y\frac{\partial}{\partial y} \mod \mathfrak{J}(x,y)^2.$$

So the vector field U admis locally M_1 as a normally hyperbolic set of zeros.

At a point $m \in M_{1o}$, near m_o, the unstable space $E^u(m)$ is just the 2-plane $X(m)$ (equal to the (x,y)-plane of coordinates). From the theory of stable/unstable manifolds it follows that U, is tangent in a neighborhood of m_o, to the unstable C^∞ foliation $\mathfrak{F}$. This foliation is the unique one tangent to the unstable space $X(m)$ at each point

$m \in M_{1o}$ and tangent everywhere to the field of lines generated by U. This last affirmation shows that $\mathcal{F}$ depends only on the triple X, and not on the choice of Ω. A more precise link between X and $\mathcal{F}$ is given in the following theorem :

THEOREM 3. Let $m_o \in M_{1o}$. The foliation $\mathcal{F}$ defined above in a neighborhood of m_o is tangent everywhere to X. Locally around m_o, it is the unique 2-foliation with this property.

Proof. The unicity of the foliation follows from the unicity of the unstable foliation for a normally hyperbolic set. So, we have only to prove that $\mathcal{F}$ is tangent everywhere to X. To show this point, we introduce normal polar coordinates (ρ, θ) : $x = \rho \cos\theta$, $y = \rho \sin\theta$.
Let φ the application $(\rho,\theta,z,w) \to (\rho\cos\theta, \rho\sin\theta, z, w)$. From $\Omega = xdz - ydw \mod J(x,y)^2$ we have :

$$\bar{\Omega} = \varphi^* \Omega = \rho(\cos\theta dz - \sin\theta dw) + O(\rho^2)$$

and so : $$\frac{1}{\rho}\bar{\Omega} = \cos\theta dz - \sin\theta dw + O(\rho).$$

This form $\frac{1}{\rho}\bar{\Omega}$ is regular for $\rho = 0$.
The plane field $\bar{X} = \varphi^{-1}X$, defined for $\rho \neq 0$, has a C^∞ continuation for $\rho = 0$. Let also $\bar{U} = \varphi^{-1}U$, $\bar{\mathcal{F}} = \varphi^{-1}\mathcal{F}$. The vector field $\frac{1}{\rho}\bar{U}$ has a C^∞ continuation by $\frac{1}{\rho}\bar{U} = \frac{\partial}{\partial\rho}$ for $\rho = 0$ and the foliation $\bar{\mathcal{F}}$ also. Now we notice that the vector field $\frac{1}{\rho}\bar{U}$ cut the manifolds $V_a = \{\rho = a\}$, transversally. The form $\frac{1}{\rho}\bar{\Omega}$ induce contact structures on these manifolds. In particular, for $\rho = 0$, we have the contact structure :
$\cos\theta dz - \sin\theta dw$ on $V_o = S^1 \times \mathbb{R}^2$ with coordinates (θ, z, w).

Now $\bar{\mathcal{F}}$ cut transversally V_o along the circles $\{z = C^t, w = C^t\}$, which are Legendre curves for the induced contact structure on V_o. (Because $\mathcal{F}$ is tangent to the (x,y)-plane along M_1); and next, $\bar{\mathcal{F}}$ is tangent everywhere to $\bar{U}$. (Because $\mathcal{F}$ is every where tangent to U). It follows from the remark 2 above that $\bar{\mathcal{F}}$ is tangent everywhere to $\bar{X}$ and then, that $\mathcal{F}$ is tangent everywhere to X.

Corollary 4. Around each point $m_o \in M_{1o}$, the triple is equivalent to the triple : $\{\frac{\partial}{\partial x}, \frac{\partial}{\partial y}, x\frac{\partial}{\partial z} - y\frac{\partial}{\partial w}\}$.

Proof. Choose coordinates (x,y,z,w) such that the foliation $\mathcal{F}$ is given by the planes $z = C^t$, $w = C^t$, and the vector-fields X_1, X_2, X_3 are equal to :

$$X_1 = \frac{\partial}{\partial x}, \quad X_2 = \frac{\partial}{\partial y}, \quad X_3 = f\frac{\partial}{\partial z} + g\frac{\partial}{\partial w}$$ with $J(f,g) = J(x,y)$. Let now the local diffeomorphism $G(x,y,z,w) = (f,-g,z,w)$.

We have :

$$G_*(X_1) = \frac{\partial f}{\partial x}\frac{\partial}{\partial x} - \frac{\partial g}{\partial x}\cdot\frac{\partial}{\partial y}$$

$$G_*(X_2) = + \frac{\partial f}{\partial y}\frac{\partial}{\partial x} - \frac{\partial g}{\partial y}\frac{\partial}{\partial y}$$

and $G_*(X_3) = x\frac{\partial}{\partial z} - y\frac{\partial}{\partial w} \bmod \{\frac{\partial}{\partial x}, \frac{\partial}{\partial y}\}$

Now, taking combinations of $G_*(X_1)$, $G_*(X_2)$ and adding to $G_*(X_3)$ a combination of these 2 vector fields, we obtain the desired triple.

Remark 3. In the classical coordinates (q_1,q_2,p_1,p_2) of the cotangent bundle $T^*\mathbb{R}^2$, the above model of triple may be writen :

$$\{\frac{\partial}{\partial p_1}, \frac{\partial}{\partial p_2}, p_2\frac{\partial}{\partial q_1} - p_1\frac{\partial}{\partial q_2}\}.$$

The singular set M_{1o} is then the zero-section of $T^*\mathbb{R}^2$. The foliation $\mathcal{F}$, is the foliation by the fibers. We may take Ω to be $p_1dq_1 + p_2dq_2$, the canonical 1-form, and $U = p_1\frac{\partial}{\partial p_1} + p_2\frac{\partial}{\partial p_2}$, the radial vector field.

We don't know if the theorem 2 and the above remark may be generalized to every dimension and if it is possible to interpret the canonical 1-form 1-form $\Sigma p_i dq_i$ of $T^*\mathbb{R}^n$, for $n \geq 3$ in the frame of singularities of k-uples of vector fields ((n+1)-uples ?).

5. FINAL REMARKS.

1) Outside M_1, the triple is equivalent to a Pfaffian structure and the singularities M_{3o}, M_{31} have been already described by J. Martinet [3] . In particular, it follows from is work, that the germs of triple at points of M_{3o}, depend at least on a functional modulus (functions of M_{3o} in $\mathbb{R}$, depending of the 2-jet of the triple along M_{3o}).

2) Take now a point $m_o \in M_{2o}$. As we have seen above we can choose a defining form Ω as in proposition 2. The associated vector field U is locally normally expanding on one side A of M_2 in M_1 and locally contracting on the other side B. Next, it is easy to see that there exist an open set W, with $m_o \in \overline{W}$, such that trajectory of U throw each point of W, goes from A to B (Precisely, we mean that the α-limit set of the trajectory is a point of A and the ω-limit of the trajectory is a point of B). Now, let $\mathcal{F}(A)$ and $\mathcal{F}(B)$ the foliations defined in the theorem 3 around the points of A and B and extended

along the trajectories of U. These two foliations are defined on W and it is likely that in general they don't coïncide,(but we have not demonstrate this point until now), and that the germs of triples at points of M_{2o} depend on fonctional modulus.

BIBLIOGRAPHY

[1] J.M. BOARDMAN : Singularities of differentiable maps, Publ. Math. IHES 33 (1967) 21-57.

[2] B. JAKUBCZYK, P. PRZYTYCKI : Singularities of k-uples of vector-fields, Preprint 158 (1978). Inst. of Math, Polish Academy of Science ; to appear in : Dissertationes Mathematicae.

[3] J. MARTINET : Sur les singularités des formes différentielles, Ann. Inst. Fourier 20-1 (1970), 95-178.

[4] P. MORMUL : Singularities and invariants sets of germs of triples of vector fields on $\mathbb{R}^4$, Master Thesis (1979), in Polish.

[5] P. MORMUL : Singularities of triples of vector fields on $\mathbb{R}^4$, Preprint (1982). Institut of Math., Polish Academy of Science.

[6] P. MORMUL : Classification of singularities of k-uples of vector-fields ; local models and stratifications in space of germs of k-uples of fields, Doctorat Thesis, to appear.

* Robert ROUSSARIE
Laboratoire de Topologie
ERA CNRS 945
Universite de Dijon,
Bâtiment Mirande, BP 138
21004 - DIJON - FRANCE

* Piotr MORMUL
Institute of Mathematics,
Polish Academy of Science
Sniadeckich 8,
00-950 - WARSAW - POLAND.

Singularities & Dynamical Systems
S.N. Pnevmatikos (editor)

CYCLES LIMITES, ETUDE LOCALE

Jean - Pierre Françoise

I.H.E.S.

France

INTRODUCTION

Cette étude se compose d'une suite de variations sur des sujets classiques de dynamique qualitative du plan, qui ont rapport aux cycles limites. On y envisage particulièrement un théorème de N. Bautin [2] sur l'existence d'une borne du nombre de *cycles limites* qui peuvent naître localement par perturbation d'un centre pour les champs polynomiaux de degré deux.

Le contrôle local du nombre des composantes connexes des fibres d'un morphisme *sous-analytique* qui ressort des travaux récents de R. M. Hard [7] et B. Teissier [11] permet d'en étendre l'énoncé au degré quelconque et pour un "jet initial" de degré quelconque. Il n'est pas toutefois possible d'expliciter la valeur de la borne qu'on obtient comme fonction du degré du champ. Le théorème général de géométrie analytique qu'on utilise se démontre par récurrence et on n'en dispose pas d'une preuve constructive.

La notion d'ensemble sous-analytique nous paraît bien adaptée aux problèmes que l'on rencontre dans la dynamique qualitative des champs analytiques du plan [5].

Pour la singularité de Poincaré, on construit dans les articles classiques une fonction de Lyapounov qui permet de décider le problème du centre-foyer. Dans la pratique, cette fonction de Lyapounov est d'importance, ainsi les exemples à quatre cycles de Shi-Song-Ling et Wang-Ming-Shu [10],[12],[13] commencent par son calcul.

On indique comment cette construction peut se faire dans le cas d'une singularité plus dégénérée à l'aide des *séries caractéristiques* que l'on trouve à partir du théorème de M. Sebastiani [9] et que l'on a déjà utilisées pour des problèmes de systèmes dynamiques analytiques [4].

Le problème de la finitude du nombre des cycles limites passe par la considération de la transformation de retour des graphiques les plus généraux. Nous signalons enfin une construction de principe des dérivées successives de cette transformation par un *algorithme du type de Godbillon-Vey* qui est l'extension à tous les ordres de la *formule de Poincaré* sur la dérivée première.

Dans toute la suite on considère indifféremment que le feulletage du plan est défini soit par un champ soit par une 1-forme. Le terme de *singularité* est utilisé pour désigner les zéros communs des composantes.

Je remercie l'IHES de m'avoir accordé d'excellentes conditions de travail durant mon séjour. J'exprime toute ma reconnaissance à B. Teissier pour les passionnants éclaircissements qu'il m'a apportés sur les outils de géométrie analytique que j'ai utilisés.

C'est avec grand plaisir que je remercie l'Université de Crète, en particulier Sp. Pnevmatikos, et l'Université de Dijon, pour l'invitation à participer au congrès " Singularités et Systèmes Dynamiques".

1. GENERALISATION D'UN THEOREME DE N. BAUTIN

Soit

$$X = f(x,y)\frac{\partial}{\partial x} + g(x,y)\frac{\partial}{\partial y}$$

un champ de vecteurs analytique sur un ouvert U de $\mathbb{R}^n$ et qui admet $0 \in U$ comme seule singularité dans U.

Ecrivons l'équation des trajectoires de X au voisinage de 0 sous la forme

$$\frac{dx}{dy} = \frac{f_k(x,y) + \dots + f_j(x,y) + \dots}{g_m(x,y) + \dots + g_\ell(x,y) + \dots}$$

où l'on fait apparaître f_j et g_ℓ les composantes homogènes de f et de g. Nous allons noter R^k (resp. R^k_n) l'ensemble des champs de vecteurs (resp. des n-jets de champs) du type ci-dessus et pour lesquels

$$k = m \quad \text{et} \quad yf_k - xg_m \neq 0 \text{ si } (x,y) \neq (0,0).$$

Autrement dit, nous considérons les champs de vecteurs pour lesquels, après un éclatement de l'origine, l'équation des trajectoires devient

$$\frac{dr}{d\varphi} = \frac{r\,[\,N_k(\varphi) + r\,N_{k+1}(\varphi) + \dots\,]}{D_k(\varphi) + r\,D_{k+1}(\varphi) + \dots}$$

avec $D_k(\varphi) \neq 0$ pour $\varphi \in [0,2\pi]$. Donc pour ceux-là l'équation se désingularise au bout d'un éclatement et la transformation de retour du cycle $r = 0$ est analytique. Il ne peut y avoir, dans ce cas, d'accumulation d'une suite discrète de trajectoires périodiques sur le cercle $r=0$ que si on a un centre.

Nous allons considérer un $X_o \in \mathbb{R}^k_n$ et une perturbation $X_{\underline{\lambda}} \in \mathbb{R}^k_n$ de X_o et examiner le nombre de cycles limites qui naissent avec une telle perturbation dans un voisinage de l'origine.

Comme nous l'avons vu, X_o possède une transformation de retour analytique en r : $r \longrightarrow L(r)$, par rapport à $\varphi = 0$, sur un voisinage de $r = 0$. Nous allons supposer que le paramétrage $\underline{\lambda}$ de la perturbation varie sur un voisinage V de $\underline{\lambda} = 0$ et nous choisissons un voisinage U de $0 \in \mathbb{R}^2$ tel que, pour tout $\underline{\lambda} \in V$, $X_{\underline{\lambda}}$ possède une transformation de retour $L(r,\underline{\lambda})$ sur U. Par transversalité L dépend analytiquement de $\underline{\lambda}$. Notons $\varepsilon = \max\{r/(x,y) \in \{\varphi = 0\} \cap U\}$.

Les trajectoires périodiques de $X_{\underline{\lambda}}$ contenues dans le voisinage U de 0 prescrit sont données par les solutions de $L(r,\underline{\lambda}) - r = 0$. Nous avons ainsi à considérer l'ensemble analytique

$$\Sigma = \{(r,\underline{\lambda}) \in [0,\varepsilon] \times \bar{V} : L(r,\underline{\lambda}) - r = 0\},$$

sa projection propre $p : \Sigma \longrightarrow p(\Sigma) \subset \bar{V}$ sur l'espace des paramètres et les fibres $p^{-1}(\underline{\lambda})$ sur un voisinage de $\underline{\lambda} = 0$ dans l'ensemble sous-analytique $p(\Sigma)$.

Si $L(r,0) - r \not\equiv 0$, donc si X_o présente un foyer au voisinage de $0 \in \mathbb{R}^2$ on peut appliquer le théorème de préparation de Weierstrass et on trouve tout de suite que la cardinalité de la fibre $p^{-1}(\underline{\lambda})$ est bornée au voisinage de $\underline{\lambda} = 0$.

Il n'y a donc pas de difficulté à perturber un foyer. Maintenant, supposons que $\underline{\lambda} = 0$ a un centre en 0. Dans ce cas on ne peut plus appliquer le théorème de Weierstrass. Mais en général on peut invoquer le fait suivant [7],[11].

Si Σ est un ensemble sous-analytique et $p : \Sigma \longrightarrow \mathbb{R}^j$ un morphisme sous-analytique *propre*. Pour tout $x_o \in p(\Sigma)$, il existe un voisinage $U(x_o)$ de x_o et un entier N tels que toute fibre $p^{-1}(x)$ $x \in U(x_o)$, a moins de N composantes connexes.

On en déduit donc la

PROPOSITION 1. *Soit $X_o \in \mathbb{R}^k_n$, il existe une borne $B(n)$ du nombre de cycles limites qui peuvent naître avec une petite perturbation $X_{\underline{\lambda}} \in \mathbb{R}^k_n$ dans un voisinage de 0.*

N. Bautin a donné une borne explicite $B(2) = 3$ pour le cas de degré deux et $k = 1$. L'argument général que nous avons employé ne peut nous donner une bonne explicite.

On aurait pu tout aussi bien reprendre une partie de l'argumentation de N. Bautin qui s'appuie sur le caractère Nœthérien des anneaux locaux de fonctions analytiques, mais nous aurions dû faire des majorations.

Du reste, ce serait une facon de démontrer le résultat général de Hard - Teissier dans ce cas particulier et nous préférons utiliser un argument plus conceptuel.

Des énoncés généraux de géométrie analytique [7] on peut ainsi déduire la

PROPOSITION 2. Soit $C_s \subset p(\Sigma)$ *l'ensemble des points* $\underline{\lambda}$ *tels que* $\# p^{-1}(\lambda) = s$; *soit* $C_0 \subset p(\Sigma)$ *l'ensemble des points* $\underline{\lambda}$ *tels que* $\dim p^{-1}(\underline{\lambda}) = 1$. *Ces ensembles* C_s *et* C_0 *sont sous-analytiques.*

Et par exemple le

COROLLAIRE. Si une suite convergente de points de C_0 *a pour limite* $\underline{\lambda}_1 \in C_0$, *il existe une courbe continue tracée dans* C_0 *qui est adhérente à* $\underline{\lambda}_1$.

Autrement dit une "déformation discrète" de centres implique l'existence d'une "déformation continue".

2. UNE EXTENSION DES SERIES DE LYAPUNOV

Soit X un germe de champ de vecteur de $(\mathbb{R}^2, 0)$ tel que

$$X(0) = 0 \quad , \quad j_1(X) = x \frac{\partial}{\partial y} - y \frac{\partial}{\partial x} \quad ,$$

nous dirons que X a une singularité de Poincaré en 0.

Pour de tels champs, H.Poincaré a démontré l'existence d'une série formelle f et d'une série $\psi(f_o)$ en puissance de $f_o = x^2 + y^2$ telles que

$$X.f = \psi(f_o) = \sum_{i=1}^{\infty} \psi_i f_o^i \quad .$$

Ces séries ne sont pas uniquement définies mais le coefficient ψ_i de ψ est défini modulo l'idéal engendré par les coefficients précédents $(\psi_1, \ldots, \psi_{i-1})$. Nous allons donner une preuve de ce résultat qui peut s'étendre à n'importe quelle f_o quasi-homogène. Nous envisagerons en particulier le cas du foyer dégénéré de Lyapunov.

Soit $\mathcal{O}$ l'anneau local des germes de fonctions holomorphes en $0 \in \mathbb{C}^2$.

PROPOSITION 3. Soit f_0 *un germe quasi-homogène à singularité isolée en* $0 \in \mathbb{C}^2$ *de nombre de Milnor* μ *et* ω *un germe de 2-forme. Soit* $\{x^\alpha y^\beta\}$ μ *monômes qui engendrent une base de* $\mathcal{O}/J(f_0)$ *où* $J(f_0)$ *est l'idéal jacobien de* f_0, *alors il existe une unique décomposition*

$$\omega = \sum_{\alpha,\beta} x^\alpha y^\beta \, \psi_{\alpha\beta}(f_0)\, dx \wedge dy + df_0 \wedge dg$$

Cette proposition est un cas particulier d'un théorème général qui s'énonce pour toutes les singularités isolées en n'importe quelle dimension et qui est dû à H.Brieskorn et M. Sebastiani [3],[9].

Dans le cas d'un germe qusi-homogène on dispose d'un algorithme de construction explicite des séries $\psi_{\alpha\beta}$ et de la fonction g. Si f et f_o sont des complexifications de germes analytiques réels comme dans le cas qui nous occupe, alors les séries caractéristiques $\psi_{\alpha\beta}$ sont à coefficients réels.

PROPOSITION 4 (H. Poincaré). *Notons* $f_o = x^2 + y^2$ *et* ω *une 1-forme tangente à* df_o, *il existe* $f = f_o + \ldots$ *et* $\psi(f_o)$ *telles que*

$$\omega \wedge df = \psi(f_o)\, dx \wedge dy \ .$$

Preuve. Nous allons construire par récurrence les composantes homogènes de f notées f_k au moyen de la proposition 3. Ainsi, il existe $\psi_1(f_o)$ et g_1 telles que

$$\omega \wedge df_o = \psi_1(f_o)dx \wedge dy + df_o \wedge dg_1$$

et nous posons $f_1 = -g_1$. Supposons connues $f_1,\ldots,f_k$ alors on construit $\psi_{k+1}(f_o)$ et g_{k+1} par

$$\omega_{k+1} \wedge df_o + \sum_{i+j=k+1} \omega_i \wedge df_j = \psi_{k+1}(f_o)dx \wedge dy + df_o \wedge dg_{k+1} \ ,$$

et on pose

$$f_{k+1} = -g_{k+1} \ .$$

A chaque étape de la récurrence, il y a un arbitraire dans le choix du g_k défini à un h_k près tel que $df_o \wedge dh_k = 0$ et donc $h_k = h_k(f_o)$. Cet arbitraire se répercute sur ψ_k par une combinaison linéaire des $(\psi_1, \ldots,\psi_{k+1})$.

Si donc on considère une famille de champs qui dépend analytiquement de paramètres $\underline{\lambda}$ et a comme jet d'ordre un $x\,\partial/\partial y - y\,\partial/\partial x$. Par exemple, on peut s'intéresser aux champs polynomiaux et prendre pour $\underline{\lambda}$ les coefficients de leurs composantes. Par transversalité, les fonctions $\psi_k(f_o, \underline{\lambda})$ vont dépendre analytiquement de leurs arguments et à chaque étape de la récurrence l'idéal I_k engendré par $(\psi_1,\ldots,\psi_k)$ sera bien défini.

Dans le cas des champs polynomiaux de degré inférieur à n, la longueur de la chaîne des idéaux $I_1 \supset I_2 \supset \ \ldots \supset I_k \supset \ldots$ est d'intérêt pour les exemples de construction de cycles limites [10],[12],[13].

La méthode de démonstration que nous avons proposée s'étend de suite en la

PROPOSITION 5. Soit f_0 *un germe quasi-homogène à singularité isolée en* $0 \in \mathbb{C}^2$ *et de nombre de Milnor* μ *et* ω *1-forme tangente à* df_0, *il existe* μ *séries* $\psi_{\alpha\beta}(f_0)$ *et une série formelle* f *telles que*

$$\omega \wedge df = \sum_{\alpha,\beta} \psi_{\alpha\beta}(f_0)\, x^\alpha y^\beta\, dx \wedge dy \quad .$$

De plus pour chaque (α,β) les séries $\psi^k_{\alpha\beta}$ sont bien définies à une combinaison linéaire près des $\psi^1_{\alpha\beta},\ldots,\psi^{k-1}_{\alpha\beta}$.

Nous avons cette fois-ci pour une famille analytique de champs, μ chaînes d'idéaux $(\psi^1_{\alpha\beta}(f_0,\underline{\lambda}),\ldots,\psi^k_{\alpha\beta}(f_0,\underline{\lambda}))$ et μ longueurs de chaînes qui pourraient se relier à la borne du paragraphe 1 dans des cas particuliers et en préciser son calcul.

Dans le cas de la singularité de Lyapunov

$$\omega = ydy + \ldots\ldots$$

on peut démontrer [8] qu'il existe un système de coordonnées analytiques (X,Y) tel qu'à une unité près $\omega = d(Y^2 + X^m) + \ldots$ pour un certain entier m.

Notre argument général s'applique donc à ce cas et on peut associer à ω , $\mu = m-1$ séries et une série f telles que

$$\omega \wedge df = \sum_{j=0}^{m-2} X^j\, \psi_j(f_0)\, dX \wedge dY$$

où

$$f_0 = Y^2 + X^m.$$

3. LES DERIVEES SUCCESSIVES DE LA TRANSFORMATION DE POINCARE

Soit

$$\omega = f(x,y)dx + g(x,y)dy$$

une 1-forme analytique sur un ouvert U de $\mathbb{R}^2$ et dont les points singuliers sont $P_1,\ldots,P_s$. Il existe une forme $\bar{\omega}_1$ dont les coefficients sont des fractions rationnelles à pôles en $P_1,\ldots,P_s$ telle que $d\omega = \omega \wedge \bar{\omega}_1$. Une telle forme est unique modulo une forme multiple de ω. Par exemple, nous pouvons prendre

$$\bar{\omega}_1 = \frac{fdy - gdy}{f^2 + g^2}\left(\frac{\partial g}{\partial x} - \frac{\partial f}{\partial y}\right).$$

Soit $\gamma_0 \subset \mathbb{R}^2$ un ensemble invariant pour $\omega = 0$ qui possède une transformation de retour L relative à une section transverse (σ).

Par exemple, γ_0 peut être un point singulier, un cycle ou un graphique, c'est-à-dire la réunion de trajectoires reliant des points singuliers.

Soit P un point régulier au voisinage de γ_o qui n'appartient pas à γ_o. La fonction

$$P \longrightarrow \psi(P) = \exp \int_{P_o}^{P} \bar{\omega}_1$$

où l'intégration est effectuée le long de la trajectoire de $\omega = 0$ entre P et P_o le premier point d'intersection avec (σ), ne dépend que de ω et de la section (σ). Elle est analytique hors de (σ) et discontinue le long de cette section. De la définition de ψ, il résulte que hors de (σ) :

$$d(\psi\omega) = 0 .$$

Autrement dit ψ est un facteur intégrant de ω. Nous allons voir la relation entre ce facteur ψ et la transformation L.

Soit t un paramétrage analytique de la section (σ) et γ_t la portion de l'orbite comprise entre deux points d'intersection consécutifs avec (σ). En un point $t = t_o$ qui correspond à une trajectoire périodique : $L(t_o) = t_o$, la formule de la divergence de Poincaré s'écrit

$$L'(t)\Big|_{t=t_o} = \psi^{-1}(t_o) = \exp - \int_{\gamma_{t_o}} \bar{\omega}_1 .$$

Nous nous occupons ici d'écrire des expressions pour les dérivées successives. Pour cela nous allons préciser le choix de la section (σ). Nous avons déjà introduit la 1-forme $\bar{\omega}_1 = \iota_Y d\omega = \theta_Y \omega$ où Y est le champ

$$Y = \frac{f}{f^2+g^2}\frac{\partial}{\partial x} + \frac{g}{f^2+g^2}\frac{\partial}{\partial y}$$

qui vérifie $\omega(Y) = 1$. Nous allons choisir comme section transverse (σ) une courbe intégrale de Y. Avec ce choix on trouve qu'en n'importe quel point $t \in (\sigma)$

$$L'(t) = \exp - \int_{\gamma_t} \bar{\omega}_1 = \exp \int_{\gamma_t} \theta_Y \omega$$

où t désigne le temps compté le long de la trajectoire de Y qu'est (σ) à partir de $(\sigma) \cap \gamma_o$.

On trouve les dérivées successives de L à partir de la

PROPOSITION 6. *Soit Ω une forme proportionnelle à $\iota_Y dx \wedge dy$, la dérivée*

$$\frac{d^n}{dt^n} \int_{\gamma_t} \Omega \quad \textit{est égale à} \quad \int_{\gamma_t} (\iota_{\bar{Y}} d)^n \Omega = \int_{\gamma_t} \theta^n_{\bar{Y}} \Omega$$

où $\bar{Y} = \psi^{-1} Y$.

La preuve géométrique de cette proposition est facile. En effet le champ $\bar{Y}$ a un flot qui envoie les feuilles de $\omega = 0$ restreintes au complémentaire de (σ) sur des feuilles de $\omega = 0$.

Les formes $\omega_n = \theta_Y^n \omega$ sont les formes successives de l'algorithme de Godbillon-Vey [6]. Ce ne sont donc pas simplement ces formes qui apparaissent dans les dérivées successives mais les formes, $\bar{\omega}_n = \theta_{\bar{Y}}^{n-1} \bar{\omega}_1$, dérivées de Lie de ω par le champ Y divisé par le facteur intégrant.

Lorsque $t \neq 0$, il est parfaitement clair que les intégrales $\int_{\gamma_t} \omega_n$ existent pour tout n.

Dans le cas où γ_0 est un point singulier ou un graphique, nous pouvons encore utiliser la même notation $\int_{\gamma_0} \bar{\omega}_n$ mais à condition de lire cette fois-ci l'intégrale au sens d'Allendoerfer-Eells [1]. C'est-à-dire que si D_{t_0} (pour un quelconque $t_0 > 0$ fixé) désigne le domaine bordé par γ_t, γ_o et (σ)

$$\int_{\gamma_0} \bar{\omega}_n \overset{def}{=} \iint_{D_{t_0}} d\bar{\omega}_n + \int_{\gamma_{t_0}} \bar{\omega}_n .$$

Avec cette convention d'écriture, les intégrales $\int_{\gamma_0} \bar{\omega}_k$, $k=1,\ldots,n$ existent si et seulement si la transformation de rétour L de γ_0 est de classe C^n et on a encore

$$\frac{d^{k-1}}{dt^{k-1}} \int_{\gamma_0} \omega_1 = \int_{\gamma_0} \bar{\omega}_k , \quad k = 1,\ldots,n .$$

Pour un graphique ou un point singulier il se peut, bien sûr, que la transformation existe et ne soit que de classe C^1 en $t = 0$.

REFERENCES

[1] C.B. Allendoerfer & J. Eells, On the cohomology of smooth manifolds. Com. Math. Helvetici 32 (1958) p. 165 - 179.

[2] N. Bautin, On the Number of Limit Cycles... . Math 56,30,72(1952) p. 181-196.

[3] H. Brieskorn, Die Monodromie der isolierten Singularitäten von Hyperflächen. Manuscripta Math. 2(1970) p. 103-161.

[4] J.-P. Francoise, Modèle local simultané d'une fonction et d'une forme de volume. Astérisque 59-60(1978) p. 119.130.

[5] J.-P. Francoise & C.C. Pugh, Déformations de cycles limites. Prépublication de l' IHES, M/82/62.

[6] C. Godbillon & J. Vey, Un invariant des feuilletages de codimension un. C.R.Acad Sciences Paris, 12 Juillet 1971, p. 92-93.

[7] R.M. Hardt, Stratification of real analytic mappings and images. Inventiones Math. 28 (1975) p. 193-208.

[8] R. Moussu, Symétrie et forme normale des centres et foyers dégénérés. Prépublication de l'Université de Dijon.

[9] M. Sebastiani, Preuve d'une conjecture de Brieskorn. Manuscripts Math. 2 (1970) p. 301-308.

[10] Shi Songling, A method of constructing cycles without contact around a week focus. Journal of Diff. Equations 41,3 (1981).

[11] B. Teissier, Sur trois questions de finitude en géométrie analytique réelle. Prépublication de l'Ecole Polytechnique de Paris, A paraître dans Acta Matematica.

[12] Wang-Ming Shu, A quadratic differential system having four limit cycles. Acta Matematica Sinica 22,6 (1979).

[13] G. Wanner, On Shi's counter example for the 16th Hilbert problem. Prépublication de l'Université de Genève.

Jean-Pierre FRANÇOISE
Institut des Hautes Etudes
Scientifiques.
35 Route de Chartres
91 440 Bures-sur-Yvette
FRANCE

Singularities & Dynamical Systems
S.N. Pnevmatikos (editor)

ANALYTIC CLASSIFICATION OF RESONANT SADDLES AND FOCI

Jean Martinet & Jean-Pierre Ramis

Université de Strasbourg

France

In this work, we apply our results of [17] to the study of resonant analytic saddles or foci in the real plane: we classify exhaustively, up to local analytic transformations of $\mathbb{R}^2$, the phase portraits near zero of the differential systems:

(1) Saddles $\begin{cases} \dot{x} = qx + \dots \\ \dot{y} = -py + \dots \end{cases}$ (2) Foci $\begin{cases} \dot{x} = y + \dots \\ \dot{y} = -x + \dots \end{cases}$

(p,q are relatively prime positive integers and, in both cases, the dots denote convergent series of order at least two)

The *resonance* means that, at first order, the phase portraits are defined by polynomials ($x^p y^q$ for saddles, x^2+y^2 for foci). But, in general, as is well known, (1) or (2) are not *linearizable*; in case (2) for instance, linearizability means that one has a center instead of a focus; in this case, a finite number of tests on the Taylor series provides an *order* k at which the system "essentially" departs from the linear one. We deal here only with such systems, which are often called *"weak" saddles or foci of order k*; the adjective "weak" makes obvious sense for foci: the integral curves spiral to the singular point very slowly, in contrast with "strong" foci (non resonant ones), for which they approach it exponentially fast; one can give a similar interpretation for weak saddles, by considering the complexification of the system.

To get a better understanding of the analytic structure of such singular points may be an important step in some problems about limit cycles: weak foci and singular cycles (made up of separatrices joining weak saddles) seem to play a prominent role in the global theory of algebraic differential systems in the plane, as "organizing centers" of major bifurcations.

Anyway, resonant singular points have been studied for a long time starting with Poincaré, Liapounov, Dulac in particular. Recently, Roussarie [21] and Takens [22] have solved the classification problem in the differentiable case, for saddles and foci respectively: resonant systems which are formally equivalent are differentiably equivalent. The analytic situation is quite different, as was pointed out by Bryuno [5] .

We have solved in [17] the classification problem for resonant systems in the *complex plane* $\mathbb{C}^2$. We describe the (germ of) "leaf space" Σ of a resonant system: if $k \in \mathbb{N}$ is the order, Σ is a non-Hausdorff complex analytic manifold of dimension 1; it is a "necklace" formed by 2k "beads" (each one a Riemann sphere), each bead sticking to the next one by means of a *germ* of analytic diffeomorphism; thus Σ may be characterized by 2k germs of diffeomorphisms of $(\mathbb{C},0)$, the attaching maps ϕ_i. Resonant systems having the same formal invariants are classified analytically by their leaf spaces.

Now, if we consider the complexification of a *real* resonant system, the antiholomorphic involution σ ($\sigma(x,y) = (\bar{x},\bar{y})$) endows the leaf space Σ with an antiholomorphic involution that we denote also by σ.

For a *real saddle*, the involution σ exchanges two halves of the necklace Σ, leaving fixed two of the "attaching points"; if one has $\sigma(i)=j$ for two attaching points i and j, then $\sigma\phi_i\sigma = \phi_j^{-1}$ where ϕ_i and ϕ_j are the attaching diffeomorphisms (σ reverses orientation along the necklace): Σ is then characterized by k+1 attaching maps. The "space of real leaves" corresponds to two germs of real curves, each one at a fixed attaching point.

For a *weak focus*, the involution σ still exchanges two halves of Σ; no attaching point is fixed, but two of the beads are globally invariant. We still have $\sigma\phi_i\sigma = \phi_j^{-1}$ if $\sigma(i)=j$: Σ is now characterized by k independant attaching maps. The space of real leaves corresponds to the "equatorial circle" of one of the fixed beads. In this case, an interesting consequence is the following fact: *any real analytic germ* $P(r)=-r+ar^{2k+1}+\dots$ $(a\neq 0)$ is the "Poincaré map" of a weak focus of order k (i.e the map defined on a ray through O by a half-turn along the orbits of the system); its conjugacy class characterizes the weak focus, up to analytic equivalence.

The paper is organized as follows.

In the first part, we recall in some details the *formal* classification of resonant *differential forms*: its features are quite different from the analogous theory for vector fields, and interesting by themselves; in particular, the normal forms appear very naturally as simple *meromorphic closed forms* (see 1.3).

The second part is a résumé of the main results of [17] ,explaining the nature of the analytic invariants of resonant complex systems.

The last two parts are devoted to the study of saddles and foci.

We cannot end this presentation without mentioning the papers of Markhashov [10] [11] [12] [13] [14] [15] on resonant analytic saddles or foci in the real domain. The results of Markhashov differ completely from ours: he proved in [10] [12] [13] [14] [15] that an analytic focus is always trivial (i.e analytically reducible to its normal form); this, of course, is false: beside the present paper, for a critic of Markhashov's proof, see Bryuno [4] p.727-728, [3] p.848, [5] appendix (it seems that Basov [2] was the first to prove the divergence of the normalization for certain foci). For saddles, Markhashov claims an analytic classification in [10] [11]. It seems that he has only obtained a *formal* one (Bryuno [3] p.848). In [11] he asserts that any real *analytic* saddle is analytically equivalent to an *algebraic* one; in fact the problem remains open (and we think that this assertion is probably false).

1. CANONICAL NORMALIZATION OF RESONANT DIFFERENTIAL EQUATIONS

1.1. TRANSVERSE FIBRATIONS

We start with a germ of equation at O in $\mathbb{C}^2$:

$$(1.1) \qquad \omega = py(1+\dots)dx + qx(1+\dots)dy = 0$$

(p and q are relatively prime positive integers); the separatrices of this equation are the coordinate axis.

In 1923, Dulac [8] shows that, given arbitrary large positive integers r,s, there exists analytic coordinates in which (1.1) becomes (up to a unity):

$$(1.2) \qquad \omega = py(1+P(u)+x^r y^s a(x,y))dx + qxdy = 0$$

where $u = x^p y^q$, P is a polynomial of degree $\leqslant n$, $P(0)=0$, $r > pn, s > qn$, and a is an analytic function. This result can be proved easily, using an identification process.

If, for any r,s, one gets P=0, equation (1.1) is *formally linearizable*. In this case, it is well known that (1.1) is *analytically linearizable* (Bryuno [6] ,Mattei-Moussu [18]). Thus, we are going to consider the situation in which $P \neq 0$ for some r,s. In this case, we start, using only the first non zero term in P, with an equation:

$$(1.3) \qquad \omega = (pydx + qxdy) + u^k(\alpha + xya)ydx = 0$$
$$\alpha \neq 0, \quad u = x^p y^q$$

The integer $k \geqslant 1$ will be called the *order* of ω.

Let us give a geometrical meaning to (1.3). Notice first that:

(1.4) $\omega\wedge(pydx+qxdy) = qxyu^k(\alpha + xya)dx\wedge dy.$

This means that the local foliation (by complex curves) $\mathcal{F}_\omega$ of $\mathbb{C}^2-\{0\}$ defined by (1.3) is, outside the separatrices $\{x=0\}$ and $\{y=0\}$, *transverse* to the "linear foliation" $pydx+qxdy = 0$ ($u=$ cst); the integer k represents the "order of contact" of these foliations along their common separatrices.

We are going to study the pairs (ω,π), where ω is of the form (1.3) and $\pi : \mathbb{C}^2 \to \mathbb{C}$ is the *singular fibration* defined by:

$$\pi(x,y) = u = x^p y^q$$

Remark 1.1.1.

The fibration π, which is transverse to $\mathcal{F}_\omega$ in the sense of (1.4), is not canonical. Let us look under which condition a function

$$v = x^p y^q \exp h(x,y)$$

will define a fibration having a similar property. We have, in view of (1.3):

$$\omega\wedge dv/v = \omega\wedge(pdx/x + qdy/y + dh)$$
$$= (pydx + qxdy)\wedge dh + u^k(\alpha + xya)(q + y\partial h/\partial y)dx\wedge dy$$

This equality will be of the form (1.4) if and only if:

$$(pydx + qxdy)\wedge dh = xyu^k b(x,y)dx\wedge dy$$

This means that $h(x,y) = Q(u) \mod xyu^k$, where Q is a polynomial in u of degree k. Thus, *the fibration* π *is determined modulo* xyu^k, *and may be changed arbitrarily at higher order.*

1.2. INTEGRATING FACTORS

From this point, we shall need to use *formal (divergent) series in two variables*. They will be of a particular type that we describe first.

We define $\hat{B}\in\mathbb{C}[[x,y]]$ as the *ring of formal series*

$$\hat{f} = \sum_{r,s>0} a_{r,s}x^r y^s$$

such that all the series in one variable $\Sigma_r\, a_{r,s}x^r$ and $\Sigma_s\, a_{r,s}y^s$ *converge on* $|x|<\rho, |y|<\rho$ *for some* $\rho>0$ *(*ρ *may depend on* $\hat{f}$*).*

Now, if we fix two relatively prime, positive integers p,q, it is easy to check that any $\hat{f}\in\hat{B}$ can be written as:

$$\hat{f} = \sum_{n>0} f_n(x,y)u^n \qquad (u = x^p y^q)$$

where the functions f_n are analytic on a common polydisk in $\mathbb{C}^2$. (The coefficients f_n are not unique, but there is a canonical choice, if one factors out the powers of $u = x^p y^q$ in $\hat{f}$.)

PROPOSITION 1.2.1. The analytic form

$$(1.3)\qquad \omega = (pydx+qxdy) + u^k(\alpha + xya)ydx \qquad (\alpha \neq 0)$$

has a formal integrating factor $\hat{F} = xyu^k\exp(-\hat{f})$, $\hat{f}\in\hat{B}$. *It is unique provided that* $\hat{f}(0)=0$, *and one has:*

$$\hat{f} = P(u) \bmod xyu^k$$

where P is a polynomial of degree k in u.

Proof. Recall that $\hat{F}$ is an integrating factor of a differential form ω if $\omega/\hat{F}$ is a closed form (Cerveau-Mattei [7]). From (1.3) we get:

$$\omega/xyu^k = du/u^{k+1} + \alpha dx/x + \omega'$$

where ω' has analytic coefficients vanishing at O. Hence $\hat{F}=xyu^k e^{-\hat{f}}$ is an integrating factor if:

$$d[e^{\hat{f}}du/u^{k+1} + e^{\hat{f}}\alpha dx/x + e^{\hat{f}}\omega'] = 0$$

which is equivalent to:

$$(1.5)\quad d\hat{f}\wedge(pydx+qxdy) + \alpha yu^k d\hat{f}\wedge dx + xyu^k(d\omega' + d\hat{f}\wedge\omega') = 0$$

Setting $\hat{f} = \sum_{r,s\geq 0} b_{r,s}x^r y^s$, (1.5) gives a system of linear equations in the coefficients $b_{r,s}$. Notice that:

- If (r,s) is not a multiple of (p,q), (1.5) determines $b_{r,s}$ from $b_{i,j}$ with $i\leq r-pk$ and $j\leq s-qk$, and from the coefficients of ω', the degrees of which satisfy the same inequalities.
- If $(r,s) = (dp,dq)$ $(d\geq 1)$, (1.5) determines $b_{dp,dq}$ from the $b_{i,j}$ and the coefficients of ω', such that $(i,j)<(dp,dq)$ (i.e $i\leq dp$, $j\leq dq$ and at least one of these inequalities is strict).

These remarks suffice to prove that $\hat{f}$ is unique and that:

$$\hat{f} = P(u) \bmod xyu^k$$

To show that $\hat{f}\in\hat{B}$, it is enough to notice that the series $\sum_r b_{r,s}x^r$ and $\sum_s b_{r,s}y^s$ are solutions of *regular differential equations of order one*; an easy induction proves that these series have a common disk of convergence ∎

Now, consider again a differential form (1.3), and write its integrating factor $\hat{F}$ as:

$$\hat{F} = xyu^k/(1+Q(u))(1+xyu^k\hat{f})$$

where Q is a polynomial of degree $\leq k$, $Q(0)=0$, and $\hat{f}\in\hat{B}$.

An easy computation shows that:

$$\omega/\hat{F} - (1+Q(u))du/u^{k+1} - \alpha dx/x = \hat{\omega}'$$

is a *closed form, with coefficients in* $\hat{B}$ ($\hat{\omega}'$ has no polar part); thus

$$\hat{\omega}' = d\hat{H} \qquad \hat{H}\in\hat{B},\ \hat{H}(0)=0$$

We reformulate this, with obvious changes of notations, in the following:

COROLLARY 1.2.2. Let ω be an analytic form (1.3), with integrating factor $\hat{F}$. Then:

(1.6) $\quad \omega/\hat{F} = (1+Q(u))du/u^{k+1} + \alpha dx/x + \beta dy/y + d\hat{H}$

with $Q(u) = \alpha_1 u + \ldots + \alpha_{k-1}u^{k-1}$; $\alpha, \beta \in \mathbb{C}$; $p\beta - q\alpha \neq 0$, $\hat{H} \in \hat{B}$, $\hat{H}(0) = 0$.

1.3. NORMALIZATION

We are going to put (1.6) into a normal form, keeping the fibration π unchanged. From now on , we shall denote by $\hat{D}_\pi$ the *group of formal transformations*:

$$\hat{\Phi} \left| \begin{array}{l} x' = x \exp q\hat{\phi}(x,y) \\ y' = y \exp -p\hat{\phi}(x,y) \end{array} \right. \qquad \hat{\phi} \in \hat{B}, \quad \hat{\phi}(0) = 0$$

These transformations are characterized by the properties $\pi \circ \hat{\Phi} = \pi$, and $D\hat{\Phi}(0) = Id$. We shall denote by $D_\pi \subset \hat{D}_\pi$ the subgroup consisting of *analytic diffeomorphisms.*

Now, consider the form (1.6) and the formal transformation:

$$\hat{\Phi} : (x,y) \rightarrow (x \exp \hat{\phi}_1(x,y), y \exp \hat{\phi}_2(x,y))$$

where $\hat{\phi}_1$ and $\hat{\phi}_2$ are the elements of $\hat{B}$ defined by:

$$p\hat{\phi}_1 + q\hat{\phi}_2 = 0 \quad \text{and} \quad \alpha\hat{\phi}_1 + \beta\hat{\phi}_2 = \hat{H}$$

It is obvious that:

$$\pi \circ \hat{\Phi} = \pi \quad , \quad \omega/\hat{F} = \hat{\Phi}\, \omega_0 \quad \text{with} \quad \omega_0 = (1+Q(u))du/u^{k+1} + \alpha dx/x + \beta dy/y.$$

Summarizing the results we have obtained up to now, we have the following:

THEOREM 1.3.1. Let ω be a resonant analytic form of order k, together with a transverse fibration π.

(i) ω has a unique integrating factor $\hat{F} \in \hat{B}$.

(ii) The pair $(\omega/\hat{F}, \pi)$ may be normalized into:

$$\omega_0 = (1+Q(u))du/u^{k+1} + \alpha dx/x + \beta dy/y \;, \quad \pi(x,y) = u = x^p y^q$$
$$(p\beta - q\alpha \neq 0)$$

by means of a unique normalizing transformation $\hat{\Phi} \in \hat{D}_\pi$.

The unicity in (ii) is obvious: the identity mapping is the unique element of $\hat{D}_\pi$ which preserves the above normal form ∎

The meromorphic normal form ω_0 may be further simplified, in the following way:

1) The differential equation:

$$(1+Q(u))du/u^{k+1} = dv/v^{k+1}$$

defines a unique holomorphic function $u \rightarrow v(u)$, $v(0)=0$, $v'(0)=1$ (i.e $v = u \exp \psi(u)$, $\psi(0)=0$). Then, the analytic change of coordinates in $\mathbb{C}^2$:

$$\Psi : (x,y) \rightarrow (x \exp \psi_1(u), y \exp \psi_2(u))$$

where $\alpha\psi_1 + \beta\psi_2 = 0$ and $p\psi_1 + q\psi_2 = \psi$
transforms obviously ω_o into:

$$dv/v^{k+1} + \alpha dx/x + \beta dy/y$$

and preserves the fibration π (but moves the fibers: $\pi \circ \Psi = v \circ \pi$).

2) The complex numbers α and β are the *residues* of the closed form ω_o on the separatrices $\{x=0\}$ and $\{y=0\}$. Actually, only the ratio β/α is geometrically interesting: through a linear change of coordinates $(x,y) \mapsto (ax,by)$, one may always reduce (in the complex domain) to the case where $p\beta - q\alpha = 1$.

In the sequel, we shall be interested in the pairs $(\mathcal{F}_\omega, \pi)$, where $\mathcal{F}_\omega$ is the local foliation of $(\mathbb{C}^2,0)$ defined by a resonant equation $\omega=0$. In view of the above results, *we shall consider only the analytic pairs* $(\mathcal{F}_\omega,\pi)$ *which are formally reducible, through a unique element of* $\hat{\mathcal{D}}_\pi$, *to one of the normal forms:*

$$(1.7)\quad \left| \begin{array}{l} \mathcal{F}_o : \omega_o = py\,dx + qx\,dy + u^k(\alpha y\,dx + \beta x\,dy) \quad p\beta - q\alpha \neq 0 \\ \pi : \pi(x,y) = u = x^p y^q . \end{array} \right.$$

Remarks.

1) If the integrating factor $\hat{F}$ of a resonant form is analytic, then the normalizing transformation $\hat{\Phi}$ is itself analytic: this is obvious from our computations. This fact was first noticed by Cerveau-Mattei ([7],p.129-134), with a different approach.

2) It is interesting to compare the formal normalization of resonant forms ω and resonant vector fiels X. In the latter case, the key is the (formal) Jordan decomposition:

$$X = \hat{S} + \hat{N} \qquad [\hat{S},\hat{N}] = 0$$

where $\hat{S}$ is a semi-simple (diagonalizable) vector field, and $\hat{N}$ a nilpotent one.

In the former case, as we have seen, the key is the (formal) decomposition:

$$\omega = \hat{F}.\hat{\omega}_1$$

where $\hat{F}$ is the integrating factor, and $\hat{\omega}_1$ a meromorphic closed form.

1.4. APPLICATION TO REAL RESONANT DIFFERENTIAL EQUATIONS

We consider a real analytic differential equation, defined in a neighborhood of 0 in $\mathbb{R}^2(x,y)$, as a *complex equation which is invariant under the (antiholomorphic) involution* $\sigma:(x,y) \to (\bar{x},\bar{y})$.

If $\omega=0$ is a real resonant equation, the involution σ must transform each separatrix into a separatrix. Therefore, two cases are to be considered.

Saddles: each separatrix is σ*-invariant (i.e is a real curve).*

The linear part of ω may be written, in real coordinates, as

$$pydx + qxdy$$

(there is no restriction on the values of the integers p,q)

We consider only saddles having a finite order k (i.e which are not linearizable). The integrating factor $\hat{F}$, being unique, is *real*. Moreover, we may choose a *real transverse fibration* π:

Dulac's theorem quoted in 1.1 shows the existence of a complex fibration; it is real modulo $xy(x^p y^q)^k$ (being unique modulo this ideal, following Remark 1.1.1); as it may be modified arbitrarily inside this ideal, it can be choosen real.

We then proceed as in 1.3, and obtain:

Each analytic pair $(\mathcal{F}_\omega, \pi)$ (a saddle of order k, with a transverse fibration π) transforms, through a unique real $\hat{\Phi} \in \hat{\mathcal{D}}_\pi$, into a real normal form:

$$(1.8)\quad \left| \begin{array}{l} \mathcal{F}_0: \quad \omega_0 = (1+Q(u))(pydx+qxdy)+u^k(\alpha ydx+\beta xdy) = 0 \\ \pi(x,y) = u = x^p y^q \end{array} \right.$$

where Q is a polynomial of degree <k, Q(0)=0; $\alpha, \beta \in R$, $p\beta - q\alpha \neq 0$.

Using, as in 1.3, a (real) change of coordinate in the basis of π, and (real) linear transformations $(x,y) \to (ax,by)$, we may assume that, in the normal form (1.8):

$$Q = 0 \ , \ p\beta - q\alpha = 1 \ (k \text{ odd}), \ p\beta - q\alpha = \pm 1 \ (k \text{ even})$$

Foci: the separatrices are exchanged by σ.

The symetry σ implies that p=q= 1; the linear part of ω is:

$$xdx + ydy$$

in a convenient real coordinate system: the singularity is either a *center or a focus*; infinite order corresponds to a center (the result of Bryuno quoted in 1.1 is a generalization of the famous theorem of Poincaré-Liapounov); we consider only equations of finite order k, that is *weak foci of order k*.

It is of course convenient to use complex coordinates (ξ, η) such that the (complex) separatrices are $\xi=0$, $\eta=0$, the involution being defined by $\sigma(\xi,\eta) = (\bar{\eta}, \bar{\xi})$; the real plane is then $\eta = \bar{\xi}$.

Using the same argument as in the previous case, one proves that a weak focus admits a real transverse fibration π(it is a Liapounov function). The integrating factor being real too, we obtain as in 1.3

Each analytic pair $(\mathcal{F}_\omega, \pi)$ (a weak focus of order k, with a real transverse fibration) transforms, through a unique $\hat{\Phi} \in \hat{\mathcal{D}}_\pi$, into a real normal form:

$$(1.9)\quad \left|\begin{array}{l}\mathcal{F}_0:\ \omega_0 = (1+Q(u))du + u^k(\alpha\eta d\xi + \bar{\alpha}\xi d\eta) = 0\\ \pi(\xi,\eta) = u = \xi\eta\end{array}\right.$$

where Q is a polynomial of degree $<k$, $Q(0)=0;\alpha\in\mathbb{C}$, $\bar{\alpha}-\alpha \neq 0$.

As in the previous case, we loose no generality in assuming that $Q=0$. Moreover, by using linear transformations $(\xi,\eta) \to (a\xi,\bar{a}\eta)$, and possibly $(\xi,\eta) \to (\eta,\xi)$ (these are real transformations), we may assume that:

$$\bar{\alpha}-\alpha = i$$

With these simplifications, the normal form (1.9) reads, in real coordinates (x,y) $(\xi = x+iy)$:

$$\mathcal{F}_0:\ \omega_0 = (1+\lambda u^k)(xdx+ydy) + \frac{u^k}{2}(xdy-ydx) = 0$$
$$\pi(x,y) = u = \xi\bar{\xi} = x^2+y^2 \qquad \lambda = \mathrm{Re}\,\alpha\ .$$

2. ANALYTIC INVARIANTS (COMPLEX DOMAIN)

2.1. *INTRODUCTION*

In part one, we have described the *formal invariants* (p,q,k,α,β) of a pair $(\mathcal{F}_\omega,\pi)$, where ω is a resonant form of finite order k, and π is a transverse fibration. Here, we consider the set of analytic pairs $(\mathcal{F}_\omega,\pi)$ with given formal invariants, and we intend to classify these pairs up to analytic diffeomorphisms of $(\mathbb{C}^2,0)$; that is, we look for their *analytic invariants.*

Using the results of section one, we may formulate this problem as follows. Fix a normal form:

$$(1.7)\quad \left|\begin{array}{l}\mathcal{F}_0:\ \omega_0 = pydx+qxdy + u^k(\alpha ydx+\beta xdy) = 0 \qquad (p\beta-q\alpha = 1)\\ \pi(x,y) = u = x^p y^q\end{array}\right.$$

Consider the analytic pairs $(\mathcal{F}_\omega,\pi)$ (π being the same as in (1.7)) which are $\hat{D}_\pi$-equivalent to (1.7); each of these corresponds to a unique normalizing transformation $\hat{\Phi}_\omega \in \hat{D}_\pi$:

$$\omega \wedge \hat{\Phi}_\omega^{*}\,\omega_0 = 0$$

(caution: $\hat{\Phi}_\omega^{*}\,\omega_0$ is not, in general, an analytic form, but only the product of an analytic form by a formal series $\hat{h}$, with $\hat{h}(0)\neq 0$; see [17].)

Obviously, the pairs $(\mathcal{F}_\omega,\pi)$ and $(\mathcal{F}_{\omega'},\pi)$ are analytically equivalent if and only if:

$$\hat{\Phi}_\omega^{-1}\circ\hat{\Phi}_{\omega'} \in D_\pi \quad \text{(the group of } \textit{analytic diffeomorphisms}\text{)}$$

Thus, our problem will be solved if we are able to *characterize the classes $\hat{\Phi}.D_\pi$ (i.e elements of the quotient space $\hat{D}_\pi/D_\pi$) which have the property: $\mathcal{F} = \hat{\Phi}^{-1}(\mathcal{F}_0)$ is an analytic foliation.*

The solution we give to this problem rests upon a very natural "geometrical" interpretation of the elements of $\hat{D}_\pi$, which we are going to sketch now; the following ideas generalize a viewpoint which is classical in the analytic theory of differential equations; for more details, see [16] and [17].

2.2. FORMAL DIFFEOMORPHISMS AND COCYCLES

We recall that $\hat{\Phi}\varepsilon\hat{D}_\pi$ is defined by $(x,y) \to (x \exp q\hat{\phi}, y \exp -p\hat{\phi}), \hat{\phi}\varepsilon\hat{B}$, i.e $\hat{\phi} = \Sigma\, \phi_n(x,y)(x^p y^q)^n = \Sigma\, \phi_n u^n$, where the ϕ_n are analytic on a common polydisk $\Delta \subset \mathbb{C}^2$ (Δ will be fixed, for simplicity, throughout this paragraph).

We shall consider systematically the elements of $\hat{B}$ as the *restrictions to the surface* $\Sigma \subset \mathbb{C}(u)\times\mathbb{C}^2(x,y)$, *defined by* $u=x^p y^q$, *of formal series in u with coefficients analytic on* $\Delta\subset\mathbb{C}^2(x,y)$; in this context, π will denote the projection $(u,x,y) \to u$ from $\mathbb{C}^3$ to $\mathbb{C}^2$. In particular every $\hat{\Phi}\varepsilon\, \hat{D}_\pi$ will be thought of as the restriction to Σ of the formal diffeomorphism $(u,x,y) \to (u,\phi(u,x,y))$ of $\mathbb{C}^3$ along $\{0\}\times\Delta$.

Now, we sketch the main ideas.

1) Sectorial diffeomorphisms.

Let $U = \{u\varepsilon\mathbb{C} \mid |u|<r, \theta< \operatorname{Arg} u<\theta'\}$ be a *sector* in $\mathbb{C}(u)$, with vertex 0, and set $\tilde{U} = U\times\Delta\subset\mathbb{C}^3$. Any $\hat{\phi}\varepsilon\hat{B}$ is the *Taylor series, along* $\{0\}\times\Delta$, *of a function* ϕ *which is analytic in the interior of* $\tilde{U}$: this is a consequence of Borel-Ritt's Theorem; thus, we have a surjective homomorphism:

$$B(U) \to \hat{B} \qquad\qquad \phi \to \hat{\phi}$$

where $B(U)$ is the ring of functions which are analytic inside $\tilde{U}$, and are C^∞ along $\{0\}\times\Delta$ (a part of the boundary of $\tilde{U}$).

Example 1. The interested reader should do the following fundamental exercise; show that the integral:

$$\phi(u) = \int_L \frac{\exp -t/u}{1-t}\,\frac{dt}{u}$$

($L \neq \mathbb{R}^+$ is a half-line going from 0 to ∞)

defines an element of $B(U)$ (U = half-plane with bissector L), and its Taylor series at 0 is $\hat{\phi} = \Sigma\, n!u^n$ (this integral is the Borel's sum of the series $\hat{\phi}$) ∎

We have $\hat{\phi}= \hat{\phi}'$ if $\phi-\phi'$ is *infinitely flat along* $\{0\}\times\Delta$: it vanishes, with all its partial derivatives with respect to u, for u=0. We denote by $B^\infty(U)\subset B(U)$ the ring of *flat functions*.

Now, we define $D_\pi(U)$ as the *group* (under composition of mappings) of transformations:

$$\Phi : (u,x,y)\,\varepsilon U \to (u,x\, e^{q\phi}, y\, e^{-p\phi}) \qquad \phi\, \varepsilon B(U)$$

(actually, to consider $D_\pi(U)$ as a transformation group, one must allow the radius and aperture of U to vary slightly, but this is an easily overcomed technical difficulty)

Similarly, $D^\infty_\pi(U) \subset D_\pi(U)$ will be the subgroup defined by the condition $\phi \in B^\infty(U)$, i.e Φ is *infinitely flat with respect to the identity mapping along* $\{0\}\times\Delta$.

Obviously, we have:

$$\hat{D}_\pi = D_\pi(U)/D^\infty_\pi(U)$$

Thus, we may consider a formal $\hat{\Phi}$ as the Taylor series along $\{0\}\times\Delta$ of a "true transformation" of a substantial domain: we have given some flesh to a ghost!

2) The characteristic cocycle of $\hat{\Phi} \in \hat{D}_\pi$.

Cover a neighborhood of $0 \in \mathbb{C}(u)$ with a finite number of sectors U_0, $U_1, \ldots$ ($U_0 \cap U_1 \neq 0, U_1 \cap U_2 \neq 0, \ldots$, but the intersection of any three sec tors is empty, putting aside 0). We fix such a covering $\mathcal{U}$.

Given $\hat{\Phi} \in \hat{D}_\pi$, write $\hat{\Phi} = \hat{\Phi}_0 = \hat{\Phi}_1 = \ldots$, where $\Phi_i \in D_\pi(U_i)$ is defined up to composition by an element of $D^\infty_\pi(U_i)$. The transformations $\Phi_0 \circ \Phi_1^{-1}$, $\Phi_1 \circ \Phi_2^{-1}$, ... belong obviously to $D^\infty_\pi(U_0 \cap U_1)$, $D^\infty_\pi(U_1 \cap U_2), \ldots$: they define *a Čech-cocycle:*

$$c(\Phi) \in C^1(\mathcal{U}; D^\infty_\pi)$$

If one chooses other representatives Φ'_i of $\hat{\Phi}$, the cocycle $c(\Phi')$ will be *cohomologous* to $c(\Phi)$, because they are related by an element of $C^0(\mathcal{U}; D^\infty_\pi)$. This means that $\hat{\Phi}$ *determines a unique element of the cohomology space:*

$$[\hat{\Phi}] \in H^1(\mathcal{U}; D^\infty_\pi)$$

Notice that this is *non-abelian* cohomology: this construction is similar to those of bundle theory.

It is easy to see that $[\hat{\Phi}] = [\hat{\Phi}']$ if and only if $\hat{\Phi}' = \hat{\Phi} \circ \Psi$, $\Psi \in D_\pi$; this means that each class $\hat{\Phi}.D_\pi \subset \hat{D}_\pi$ is *characterized* by an element of $H^1(\mathcal{U}; D^\infty_\pi)$, which we shall call the *characteristic class* (or cocycle, with a slight abuse) of $\hat{\Phi}.D_\pi$.

We have proved in [17] the fundamental result:

Each element of $H^1(\mathcal{U}; D^\infty_\pi)$ *is the characteristic class of an element of* $\hat{D}_\pi / D_\pi$.

Example 2. Consider the covering consisting of two sectors $U_0 = \mathbb{C} - \mathbb{R}^+.i$ $U_1 = \mathbb{C} - \mathbb{R}^-.i$, and $F \in D^\infty_\pi(U_0 \cap U_1)$ ($\pi(x,y) = xy$) defined by:

$$F: (u,x,y) \to (u, x \exp f(u), y \exp -f(u))$$

$$f(u) = \exp -1/u \text{ if } \operatorname{Re} u > 0,\ f(u) = 0 \text{ if } \operatorname{Re} u < 0.$$

We still denote by F the corresponding cohomology class. Prove that the Cauchy-Heine integral:

$$\phi_j(u) = \frac{1}{2i\pi} \int_{L_j} \frac{e^{-1/t}}{t-u}\, dt \qquad j=0,1;\ u \notin L_j$$

($L_j = OA_j$ is a segment, Re $A_j > 0$ is very small, Im $A_0 > 0$ and Im $A_1 < 0$) defines $\Phi_j \varepsilon D_\pi(U_j): (u,x,y) \to (u, x\exp\phi_j(u), y\exp-\phi_j(u))$ such that:

$$\Phi_0 \circ \Phi_1^{-1} = F$$

Show that $\hat{\phi}_j = \Sigma\, n!u^n$ ■

Finally, we have, for each covering $\mathcal{U}$ of $(\mathbb{C},0)$ by a finite number of sectors, a *natural identification of* $\hat{D}_\pi/D_\pi$ *with* $H^1(\mathcal{U};D_\pi^\infty)$. This is the main tool for solving our problem about foliations.

2.3. ANALYTIC INVARIANTS

We go back to the data $(\mathcal{F}_0,\pi)$ considered in 2.1. Our main result can be stated (loosely) as follows:

Given $\hat{\Phi}\varepsilon\hat{D}_\pi$, $\hat{\Phi}^{-1}(\mathcal{F}_0)$ is an analytic foliation if and only if, for a convenient covering $\mathcal{U}$ (which we are going to describe), $\hat{\Phi}$ is characterized by a cocycle in $C^1(\mathcal{U};D_\pi^\infty)$ which preserves the foliation $\mathcal{F}_0$ (this means that the cocycle preserves the restriction of $\mathcal{F}_0$ to each domain $U_0 \cap U_1$, $U_1 \cap U_2, \ldots$).

1) *The foliation* $\mathcal{F}_0$.

We describe it on a domain $\tilde{U} = \pi^{-1}(U) \subset \mathbb{C}^2(x,y)$, where $U \subset \mathbb{C}(u)$ is any sector with vertex 0: each leaf of $\mathcal{F}_0$ in $\tilde{U}$ (we put aside the separatrices x=0, y=0, which lie in $\pi^{-1}(0)$) is a one-to-one covering of U through π, as $\mathcal{F}_0$ and π are transverse. More precisely, write:

$$\omega_0/xyu^k = du/u^{k+1} + \alpha dx/x + \beta dy/y = (1+\lambda u^k)du/u^{k+1} + m dx/x + n dy/y$$

(where m,n are *integers*, $pn-qm=1$, $\lambda = n\alpha - m\beta$; these choices are consistent because we assume (see 1.3) $p\beta - q\alpha = 1$; λ mod $\mathbb{Z}$ does not depend on the choice of m,n)

It follows that $\mathcal{F}_0|\tilde{U}$ is defined by the *first integral*:

$$H(u,x,y) = x^m y^n K(u) \quad \text{with} \quad K(u) = u^\lambda \exp -1/ku^k$$

These definitions depend on U in general, because u^λ requires the choice of a determination of Log u on U (we often use the principal determination).

The leaf $H = c\varepsilon\mathbb{C}$ has then, for each $u\varepsilon U$, a unique point $(x,y)\varepsilon\tilde{U}$ such that $\pi(x,y)=u$: it is determined by $x^p y^q = u$, $x^m y^n = c/K(u)$ (recall that $pn-qm=1$).

A transformation $F: (x,y) \to (x\exp q\phi, y\exp -p\phi)$ preserves $\mathcal{F}_0|\tilde{U}$ if and only if $H \circ F$ is a function of H (this means that ϕ is a function of H).

Now, the point is that $H \varepsilon B^{\infty}(U)$ (i.e is flat) if $\mathrm{Re}\, u^k > 0$ on U.

In this case, we have a non-trivial subgroup:

$$D^{\infty}_{\pi}(U;\mathcal{F}_o) \subset D^{\infty}_{\pi}(U)$$

consisting of transformations which preserve $\mathcal{F}_o$; each one is uniquely defined by a relation:

$$H \circ F = f \circ H$$

where $f(z) = z + a_2 z^2 + \ldots$ is any local diffeomorphism of $\mathbb{C}$ at 0, tangent to the identity.

Similarly, if $\mathrm{Re}\, u^k < 0$ on U, we have an analogous result, using $1/H = H^{-1}$ instead of H.

If U contains a line on which $\mathrm{Re}\, u^k = 0$, it is easily seen that *no element of $D_{\pi}(U)$ (flat or not) preserves $\mathcal{F}_o$, but the identity mapping.*

The lines $\mathrm{Re}\, u^k = 0$ will be called the *Stokes lines of $\mathcal{F}_o$*.

2) *Conclusions*

Define $\mathcal{U}(\mathcal{F}_o)$ as the covering of $\mathbb{C}(u)$ such that the sectors $U_o \cap U_1$, $U_1 \cap U_2, \ldots$ are limited by two consecutive Stokes lines of $\mathcal{F}_o$. For instance, if $k=1$, it consists of the two sectors $U_o = \mathbb{C} \setminus \mathbb{R}^+ i$ and $U_1 = \mathbb{C} \setminus \mathbb{R}^- i$; in general, it is made up of $2k$ sectors of aperture $2\pi/k$.

It is now possible to state precisely our result about the analytic classification of the pairs $(\mathcal{F}_{\omega}, \pi)$; see [17].

THEOREM 2.4. *Let $(\mathcal{F}_o, \pi)$ be a normal form (1.7), $\mathcal{U} = \mathcal{U}(\mathcal{F}_o)$ the corresponding covering of $\mathbb{C}(u)$.*

(i) *One has:*

$$H^1(\mathcal{U}; D^{\infty}_{\pi}(\mathcal{F}_o)) = C^1(\mathcal{U}; D^{\infty}_{\pi}(\mathcal{F}_o))$$

Each cocycle is defined by $2k$ local analytic diffeomorphisms of $\mathbb{C}$ at 0, tangent to the identity.

(ii) *These cocycles classify the pairs $(\mathcal{F}_{\omega}, \pi)$ up to analytic equivalence: given $\mathcal{F}_{\omega}$, the characteristic class of its normalizing transformation $\hat{\Phi} \varepsilon \hat{D}_{\pi}$ is represented by a unique cocycle in $C^1(\mathcal{U}; D^{\infty}_{\pi}(\mathcal{F}_o))$; every cocycle defines an analytic foliation $\mathcal{F}_{\omega}$, up to analytic equivalence.*

This foliation is obtained, geometrically, by gluing $\mathcal{F}_o|\tilde{U}_o$ to $\mathcal{F}_o|\tilde{U}_1$, $\mathcal{F}_o|\tilde{U}_1$ to $\mathcal{F}_o|\tilde{U}_2, \ldots$; the transition mappings are the terms of the cocycle. This viewpoint enables us to describe the "leaf space" as a "necklace" of $2k$ Riemann spheres (see [17] and the introduction)

2.5. SUMMABILITY

We still fix a normal form $(\mathcal{F}_o, \pi)$; let $F = (F_o, F_1, \ldots) \varepsilon C^1(\mathcal{U}; D^{\infty}_{\pi}(\mathcal{F}_o))$ be one of the corresponding cocycles; let $\hat{\Phi} \varepsilon \hat{D}_{\pi}$ be such that $[\hat{\Phi}] = F$, and set $\mathcal{F} = \hat{\Phi}^{-1}(\mathcal{F}_0)$. Then, there is a *unique* family $\Phi=(\Phi_0,\Phi_1,\ldots)$, $\Phi_i \varepsilon D_\pi(U_i)$, such that:

$$\hat{\Phi}_i = \hat{\Phi} \quad \text{and} \quad \mathcal{F}|\tilde{U}_i = \Phi_i^{-1}(\mathcal{F}_0|\tilde{U}_i)$$

Moreover, $[\hat{\Phi}] = F$ means that:

$$\Phi_0 \circ \Phi_1^{-1} = F_0,\ \Phi_1 \circ \Phi_2^{-1} = F_1, \ldots, \text{ on } \tilde{U}_0 \cap \tilde{U}_1, \tilde{U}_1 \cap \tilde{U}_2, \ldots$$

The main features of F are:

(i) The F_i's are defined over sectors of aperture π/k.

(ii) Writing $F_i:(u,x,y) \to (u,x \exp q f_i \circ H, y \exp -p f_i \circ H)$ (see 2.3), $f_i \circ H$ is *exponentially flat of order* k: $||f_i \circ H|| = o(\exp -1/|u|^k)$

These properties ensure that $\hat{\Phi} : (u,x,y) \to (u,x \exp q\phi, y \exp -p\phi)$ *is* *k-summable*, as a consequence of Ramis' theory (see [20]). Let us explain the meaning of this fundamental property, for k=1 (Borel-summability).

Essentially, knowing the formal transformation $\hat{\Phi}$, one can *compute* Φ_0,Φ_1 (i.e the normalizing transformations on $\tilde{U}_0,\tilde{U}_1$) as follows.

1) Write, as in 2.2:

$$\hat{\phi} = \Sigma\ \phi_n(x,y)u^n \qquad (x,y)\ \varepsilon\ \Delta$$

Define $\quad \bar{\phi}(t,x,y) = \Sigma\ \phi_n(x,y)t^n/n!$ (inverse Borel-transform of ϕ)

Then, this series has a *non-zero radius of convergence* (in t) and defines, by analytic continuation, a function $\bar{\phi}$ which is holomorphic on $\Omega\times\Delta$, where:

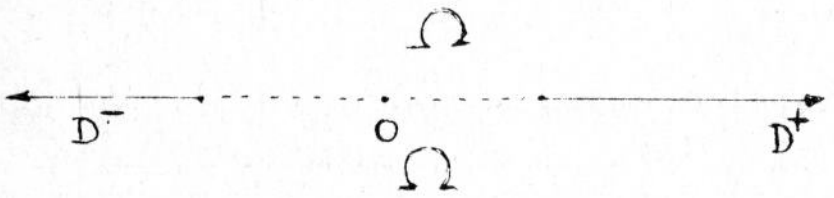

$\Omega = \mathbb{C} \setminus (D^+ \cup D^-)$

(the half lines D^+ and D^- lie on the bissectors $\mathbb{R}^+$ and $\mathbb{R}^-$ of $U_0 \cap U_1$)

Moreover, $\bar{\phi}$ has *exponential growth* (at most) when $t \to \infty$.

Remark. Ecalle's theory of resurgent functions gives much more information on $\bar{\phi}$: the cocycle F defines precisely the location and the nature of the singularities of $\bar{\phi}$ with respect to t; in general, $\bar{\phi}$ is analytic on the universal covering of $\mathbb{C}\setminus\mathbb{Z}$ (see [9]).

2) For each half line L starting from O in $\mathbb{C}(t)$, $L \neq \mathbb{R}^+$ or $\mathbb{R}^-$, set:

$$\phi_L(u,x,y) = \int_L \bar{\phi}(t,x,y) \exp -t/u . dt/u \quad \text{(Borel transform of } \bar{\phi})$$

and $\quad \Phi_L: (u,x,y) \to (u,x \exp q\phi, y \exp -p\phi)$

It is not hard to check that (see example 2, in 2.2):

- ϕ_L is analytic on $U_L \times \Delta$ (U_L = open half plane with bissector L), and $\hat{\phi}_L = \hat{\phi}$; thus $\Phi_L \varepsilon D_\pi(U)$, and $\hat{\Phi}_L = \hat{\Phi}$.
- When L moves in the lower (resp. upper) half plane, the Φ_L glue together and define $\Phi_0 \varepsilon D_\pi(U_0)$ (resp. $\Phi_1 \varepsilon D_\pi(U_1)$), with $\hat{\Phi}_0 = \hat{\Phi}_1 = \hat{\Phi}$.

The transformations Φ_0, Φ_1 *are precisely the (unique) normalizing transformations, on* $\tilde{U}_0$ *and* $\tilde{U}_1$, *of* $\mathcal{F} = \hat{\Phi}^{-1}(\mathcal{F}_0)$.

They are called the (Borel) sums of $\hat{\Phi}$; the lines $\mathbb{R}^+$ and $\mathbb{R}^-$ are the *singular lines of* $\mathcal{F}_0$: they are the bissectors of two consecutive Stokes lines.

In the general case, when k>1, the procedure which leads to $(\Phi_0, \Phi_1, \ldots)$ knowing $\hat{\Phi}$ is quite similar: one replaces in 1) the factorial by $\Gamma(1+n/k)$ (inverse Leroy-transform); in 2), exp-t/u.dt/u by $kt^{k-1}\exp(-t^k/u^k).dt/u^k$. (For more details, see [16].)

3. REAL DOMAIN: RESONANT (OR WEAK) SADDLES

In this chapter, we use the previous results to obtain the classification of *real analytic saddles* which are equivalent, through a *real* $\hat{\Phi}\varepsilon\hat{D}_\pi$, to the *real normal form*:

$$(1.8)\quad \left|\begin{array}{l} \mathcal{F}_0:\ \omega_0 = pydx+qxdy+u^k(\alpha ydx+\beta xdy) = 0 \qquad \alpha,\beta\varepsilon R \\ \pi(x,y)=u=x^py^q \end{array}\right.$$

$$p\beta-q\alpha=1 \ (k \text{ odd}) \qquad p\beta-q\alpha=\pm 1 \ (k \text{ even})$$

As in 2.3, the *complex foliation* $\mathcal{F}_0$ is defined by the first integral:

$$(3.1)\qquad H = x^m y^n K(u) \qquad \text{where} \quad K(u) = u^\lambda \exp -1/ku^k$$

$$\lambda = \pm(n\alpha - m\beta)$$

The Stokes lines of $\mathcal{F}_0$, in $\mathbb{C}(u)$, are defined by Re $u^k=0$, and the singular lines by Im $u^k=0$. Notice that the real plane $\mathbb{R}^2 \subset \mathbb{C}^2$ lies in $\pi^{-1}(\mathbb{R})$, and $\mathbb{R}$ consists of two singular lines.

The analytic weak saddles $(\mathcal{F},\pi)$ of order k and invariants α,β are in one-to-one correspondance with their normalizing transformations, from the results of 1.4. Thus, they are classified, up to real analytic equivalence, by the cocycles:

$$F \ \varepsilon\ C^1(\mathcal{U};D_\pi^\infty(\mathcal{F}_0))$$

such that $F = [\hat{\Phi}]$ for a *real* $\hat{\Phi}\varepsilon\ \hat{D}_\pi$.

We are going to study these cocycles in details, assuming k=1 for simplicity.

In this case, the covering $\mathcal{U}$ of $\mathbb{C}(u)$ consists of two sectors U_0, U_1 with bissectors $\mathbb{R}^-.i$, $\mathbb{R}^+.i$, and aperture 2π. We set:

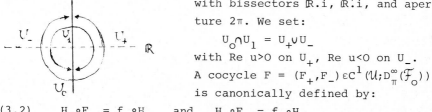

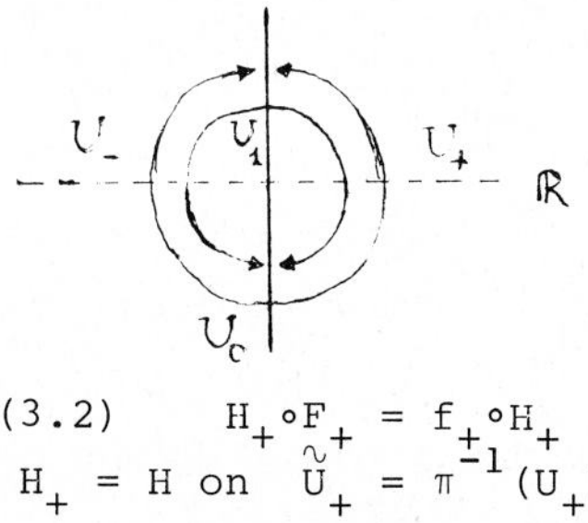

$$U_0 \cap U_1 = U_+ \cup U_-$$

with Re u>0 on U_+, Re u<0 on U_-.

A cocycle $F = (F_+,F_-)\,\varepsilon C^1(\mathcal{U};D_\pi^\infty(\mathcal{F}_0))$ is canonically defined by:

$$(3.2)\qquad H_+\circ F_+ = f_+\circ H_+ \quad \text{and} \quad H_-\circ F_- = f_-\circ H_-$$

where $H_+ = H$ on $\tilde{U}_+ = \pi^{-1}(U_+)$, $H_- = 1/H$ on $\tilde{U}_- = \pi^{-1}(U_-)$, and f_+, f_- are local diffeomorphisms of $(\mathbb{C},0)$, tangent to the identity:

(3.3) $f_+(z)=z+\sum_{n\geq 2} a_n z^n \qquad f_-(z)=z+\sum_{n\geq 2} b_n z^n$

In what follows, we shall always denote by σ the antiholomorphic involution defined by conjugation ($\sigma(x,y)=(\bar{x},\bar{y})$ in $\mathbb{C}^2$, $\sigma(u)=\bar{u}$ in $\mathbb{C}(u)$, etc...). Notice that σ exchanges the sectors U_0 and U_1 in $\mathbb{C}(u)$, $\tilde{U}_0$ and $\tilde{U}_1$ in $\mathbb{C}^2$, but it leaves invariant each sector U_+, U_-.

PROPOSITION 3.1. A cocycle $F=(F_+,F_-)\in C^1(\mathcal{U};\mathcal{D}_\pi^\infty(\mathcal{F}_0))$ *is characteristic of a real transformation* $\hat{\Phi}\in\hat{\mathcal{D}}_\pi$ *if and only if:*

(3.4) $\sigma\circ F\circ\sigma = F^{-1}$ (*i.e* $\sigma\circ F_+\circ\sigma = F_+^{-1}$, $\sigma\circ F_-\circ\sigma = F_-^{-1}$)

Proof. a) Necessity.

Let $\hat{\Phi}\in\hat{D}_\pi$ be a real transformation, with characteristic class $F=(F_+,F_-)\in C^1(\mathcal{U};D_\pi^\infty(\mathcal{F}_0))$; let Φ_0, Φ_1 be the "sums" of $\hat{\Phi}$ on $\tilde{U}_0$, $\tilde{U}_1$. We have obviously:

$$\sigma\circ\Phi_0\circ\sigma = \Phi_1 \quad \text{i.e} \quad \overline{\Phi_0(\bar{x},\bar{y})} = \Phi_1(x,y)$$

since σ transforms $\tilde{U}_0$ into $\tilde{U}_1$, and $\hat{\Phi}$ is real (see the summation formulas in 2.5). From this follows immediately:

$$\sigma\circ F_+\circ\sigma = \sigma\circ\Phi_0\circ\Phi_1^{-1}\circ\sigma = \Phi_1\circ\Phi_0^{-1} = F_+^{-1} \quad \text{on } \tilde{U}_+$$

and, similarly: $\sigma\circ F_-\circ\sigma = F_-^{-1}$ on $\tilde{U}_-$.

b) Sufficiency.

Consider an $F\in C^1(\mathcal{U};D_\pi^\infty(\mathcal{F}_0))$ satisfying condition (3.4); take any $\hat{\Phi}\in\hat{D}_\pi$ with characteristic class F; then $\sigma\hat{\Phi}\sigma\in\hat{D}_\pi$ has obviously the same characteristic class, in view of (3.4); therefore:

$$\sigma\circ\hat{\Phi}\circ\sigma = \hat{\Phi}\circ G$$

where $G\in D_\pi$ is *analytic*; this equality gives $G\circ\sigma = \Phi^{-1}\circ\sigma\circ\Phi$, which shows that $G\circ\sigma$ is an *antiholomorphic involution of* $\mathbb{C}^2$ at O. Assume that we have shown:

(3.5) $G\circ\sigma = G'\circ\sigma\circ G'^{-1}$ with $G'\in D_\pi$

It follows that $\Phi' = \Phi\circ G'\in\hat{D}_\pi$ is real; indeed:

$$\sigma\circ\hat{\Phi}'\circ\sigma = \sigma\circ\hat{\Phi}\circ\sigma\circ\sigma\circ G'\circ\sigma = \hat{\Phi}\circ G\circ\sigma\circ G'\circ\sigma = \hat{\Phi}\circ G' = \hat{\Phi}'$$

Thus, we have found a real diffeomorphism with characteristic class F.

Proof of (3.5). It is well known that any antiholomorphic involution of $\mathbb{C}^n$ at O is conjugate to the canonical involution $\sigma(x)=\bar{x}$ through a holomorphic diffeomorphism ([1],p.16; see too [19],p.263-264). From this we may write here, in $\mathbb{C}^2$:

$$G\circ\sigma = G_1\circ\sigma\circ G_1^{-1}$$

where G_1 is holomorphic, with $DG_1(O)=I$ (because $DG(O)=I$), and defined up to composition on the right by a real analytic diffeomorphism. Having chosen such a G_1, notice that the function $\pi' = \pi\circ G_1$ is *real*;

indeed $\pi \circ G \circ \sigma \circ G_1 = \pi \circ G_1 \circ \sigma$; but $\pi \circ G \circ \sigma = \sigma \circ \pi$ since $G \in D_\pi$; hence $\sigma \circ \pi \circ G_1 = \pi \circ G_1 \circ \sigma$.

Set $G_1(x,y) = (x+\xi_1(x,y), y+\eta_1(x,y))$; then we have:

$$\pi'(x,y) = (x+\xi_1)^p (y+\eta_1)^q$$

As π' is real, the curves $x+\xi_1 = 0$ and $y+\eta_1 = 0$ are real and analytic; thus we have:

$$x+\xi_1 = a(x,y)(x+\xi) \qquad y+\eta_1 = b(x,y)(y+\eta)$$

where ξ and η are real analytic, of order >2 at O, and a,b are analytic, with $a(0)=b(0)=1$, $a^p b^q$ *real*. It is then easy to define a real analytic G_2 such that:

$$\pi' = \pi \circ G_2$$

We have then $\pi \circ G_1 = \pi \circ G_2$; the diffeomorphism $G' = G_1 \circ G_2^{-1}$ belongs to D_π and gives the expected relation (3.5) ∎

This proposition, added to the results of sections 1 and 2, proves the following:

THEOREM 3.2. *The real analytic weak saddles of order one, with formal invariants $(p,q,k=1,\alpha,\beta)$, are characterized, up to real analytic equivalence, by the elements $F \in C^1(\mathcal{U}; D_\pi^\infty(\mathcal{F}_0))$ such that $\sigma \circ F \circ \sigma = F^{-1}$, where:*

$$\mathcal{F}_0 : \quad \omega_0 = py\,dx + qx\,dy + u\,(\alpha y\,dx + \beta x\,dy) = 0$$
$$\pi(x,y) = u = x^p y^q \;;\quad \sigma(x,y) = (\bar{x}, \bar{y})$$

The "reality condition" (3.4) may be interpreted easily in terms of the local diffeomorphisms introduced in (3.2) and (3.3). Assuming that H_+ and H_- are defined by the principal determination of Log u on U_+ and U_-, one has:

$$\sigma H_+ \sigma = H_+ \qquad \text{and} \qquad \sigma H_- \sigma = e^{-2i\pi\lambda} H_-$$

It follows that $F=(F_+,F_-)$ will satisfy condition (3.4) if :

$$\sigma f_+ \sigma = f_+^{-1} \qquad \text{and} \qquad e^{2i\pi\lambda} \sigma f_- \sigma\, e^{-2i\pi\lambda} = f_-^{-1}$$

The first of these relations means that $\sigma \circ f_+$ is an antiholomorphic involution of $\mathbb{C}$ at O; thus, from the result cited in the proof of Prop.3.1, we have:

$$\sigma \circ f_+ = g_+ \circ \sigma \circ g_+^{-1}$$

where g_+ is a local diffeomorphism of $(\mathbb{C},0)$, tangent to the identity; finally we obtain:

$$f_+ = \sigma g_+ \sigma g_+^{-1} = [\sigma, g_+] \quad \text{(commutator of } \sigma \text{ and } g_+)$$

Similarly, one proves that:

$$f_- = e^{2i\pi\lambda} [\sigma, g_-]$$

where g_- is any local diffeomorphism of $(\mathbb{C},0)$, with linear part $e^{i\pi\lambda}$.

CONCLUSION. As in the complex case, we obtain a "huge" moduli space for real analytic resonant saddles; the "trivial" invariants p,q,k (k=1 here),α,β, being fixed, we have one equivalence class of analytic equations for each choice of two convergent series:

$$g_+(z) = z + \sum_{n\geq 2} a_n z^n \quad \text{and} \quad g_-(z) = e^{i\pi\lambda} z + \sum_{n\geq 2} b_n z^n.$$

3.3. EXAMPLES

The "explicit" computation of the analytic invariants of a weak saddle is a difficult problem (Ecalle's theory of resurgent functions gives, in principle, a way of computing them; see [9]). We shall present here a two-parameters family for which we can compute the invariants, using one of our previous results ([16],p.154-158). The idea is to construct non trivial resonant saddles by blowing up Riccati equations with an irregular singularity.

We consider the family:

$$\omega_{a,b} = (1+ax+\lambda u-buy)du-xudy = 0 \qquad (u=\pi(x,y)= xy)$$

where a,b,λ are real parameters.

The equation $\omega_{a,b}=0$ is formally equivalent to the normal form:

$$\omega_0 = (1+\lambda u)du-xudy = 0 \quad (p=q=1;\ k=1)$$

through a unique formal transformation $\hat{\Phi} \varepsilon \hat{D}_\pi$:

$$\hat{\Phi} \quad \begin{vmatrix} x' = x \exp \hat{\phi} \\ y' = y \exp -\hat{\phi} \end{vmatrix}$$

These differential equations can be obtained from the Riccati equations:

$$\xi^2 d\eta/d\xi = (1+\lambda\xi)\eta + a\xi - b\xi\eta^2$$

by means of the Hopf blowing up: $\xi = xy, \eta = y$.

The computations we have made in the above reference show that the invariants of the saddle $\omega_{a,b}=0$ are, with the notations of 3.2:

$$f_+(z) = z/(1+\mu z) \qquad \text{and} \qquad f_-(z) = z/(1+\nu z)$$

$$\mu = -2i\pi a/\Gamma(1+\alpha)\Gamma(1+\beta) \qquad \nu = -2i\pi b e^{i\pi\lambda}/\Gamma(1-\alpha)\Gamma(1-\beta)$$

α,β being the roots of $\alpha^2+\lambda\alpha-ab = 0$.

3.4. REMARKS ON THE GENERAL CASE (k ≥ 2); THE LEAF SPACE.

The covering $\mathcal{U}$ of $\mathbb{C}(u)$ relative to the normal forms (1.8) of order k consist of 2k sectors with aperture $2\pi/k$, and bissectors the *Stokes lines* $\mathrm{Re}\, u^k = 0$.

The reality condition on the characteristic classes $F \varepsilon C^1(\mathcal{U}; D^\infty_\pi(\mathcal{F}_0))$ is still:

$$(3.4) \qquad \sigma F \sigma = F^{-1}$$

In the plane $\mathbb{C}(u)$, σ is the reflexion with axis $\mathbb{R}$, a *singular line* of $\mathcal{F}_0$ which bissects two of the intersections $U_i \cap U_{i+1}$: these sectors are invariant through σ; the other 2k-2 sectors $U_i \cap U_{i+1}$ are associated in pairs by σ. Extending the procedure used for k=1, we may conclude:

The cocycles $F \in C^1(\mathcal{U}; \mathcal{D}^\infty_\pi(\mathcal{F}_0))$ verifying (3.4) (and thus the equivalence classes of weak saddles of order k, with fixed formal invariants) are defined by the choice of k+1 analytic diffeomorphisms of $(\mathbb{C},0)$; one of them has linear part $z \to e^{i\pi\lambda}z$, the others are tangent to the identity, and they are otherwise arbitrary.

In our paper [17], we have shown that the resonant complex foliations are characterized by their "leaf spaces", which we describe as "necklaces of Riemann spheres" (see the Introduction). In particular, the complex foliation defined by a saddle of order k gives rise to a necklace Σ of 2k spheres. The reality condition means that Σ is equipped with an antiholomorphic involution σ: this involution exchanges two halves of the necklace, and fixes two of the 2k "attaching points". The submanifold of real leaves consists of the fixed points of σ (two "small arcs").

3.5. C^∞-NORMALIZATION

Let $\omega = 0$ be a resonant analytic saddle, formally equivalent to a normal form $\omega_0 = 0$. Roussarie has shown [21], in the more general case where ω has C^∞ coefficients, that this equation is equivalent to $\omega_0 = 0$ through a C^∞ diffeomorphism at 0 in $\mathbb{R}^2$. Actually, he proves still more: he classifies, up to conjugacy, germs of C^∞ vector fields (i.e, he takes into account the parametrization of the integral curves by the time). This is a difficult result; one has mainly to show that any *infinitely flat perturbation of* ω_0 is equivalent to ω_0, by means of a C^∞ diffeomorphism which is infinitely flat with respect to the identity map of $\mathbb{R}^2$.

We shall show how, in the case of *analytic* equations, our theory allows to define a local C^∞ diffeomorphism Ψ of $\mathbb{R}^2$ at 0, which transforms the foliation $\mathcal{F}_\omega$ into $\mathcal{F}_0$. We shall do this, for simplicity, when the order k is one.

Let $\hat{\Phi} \in \hat{D}_\pi$ be the *real formal* normalizing transformation, which takes $\mathcal{F}_\omega$ to $\mathcal{F}_0$, and let $F \in C^1(\mathcal{U}; D^\infty_\pi(\mathcal{F}_0))$ be its characteristic class, which satisfies (3.4): $\sigma F \sigma = F^{-1}$.

The sums Φ_0 and Φ_1 of $\hat{\Phi}$ on the domains $\tilde{U}_0, \tilde{U}_1$ transform $\mathcal{F}_\omega$ into $\mathcal{F}_0$; the real plane $\mathbb{R}^2$ is contained in each of these domains; but it is *not invariant* through any of these transformations (otherwise, we would have $\sigma\Phi_o\sigma = \Phi_o = \Phi_1$, i.e $\hat{\Phi}$ would be analytic, and there is nothing to prove). There are two ways of overcoming this difficulty.

1) Set

$$\Phi_i \left| \begin{array}{l} x' = x \exp q\phi_i \\ y' = y \exp{-p\phi_i} \end{array} \right. \qquad (i = 0,1)$$

The "reality" of $\hat{\Phi}$ means that $\phi_1 = \bar{\phi}_o$.

The foliation $\mathcal{F}_\omega$ is then defined, in $\tilde{U}_o \cap \tilde{U}_1$, by any one of the two closed meromorphic forms (because $\omega \wedge \Phi_i^* \omega_o = 0$, i=0,1;see 1.2.2):

$$du/u^2 + \alpha dx/x + \beta dy/y + d\phi_i = 0 \quad (i=0,1)$$

Therefore, it is defined as well, in $\mathbb{R}^2$, by the real equation:

$$du/u^2 + \alpha dx/x + \beta dy/y + d\psi = 0$$

where $\psi = (\phi_o + \phi_1)/2 = \mathrm{Re}\ \phi_o = \mathrm{Re}\ \phi_1$.

The fuction ψ is C^∞ in a neighborhood of 0 in $\mathbb{R}^2$: it is analytic (as ϕ_o and ϕ_1) outside the coordinate axis, and it admits $\hat{\phi}$ as its Taylor series at each point of these axis. Then the real C^∞ diffeomorphism:

$$\Psi \left| \begin{array}{l} x' = x \exp q\psi \\ y' = y \exp{-p\psi} \end{array} \right.$$

answers obviously the question.

2) Using 3.1 and 3.2, one may write the characteristic cocycle of $\hat{\Phi}$ as:

$$F = \sigma G \sigma G^{-1} = [\sigma, G]$$

where $G = (G_+, G_-)$, and G_+, G_- are elements of $D_\pi^\infty(U_\pm)$ which *preserve* the foliation $\mathcal{F}_o$. A straightforward computation shows then that $G_+^{-1} \circ \Phi_1$ preserves the two quadrants of $\mathbb{R}^2$ defined by $x^p y^q = u > 0$, and $G_-^{-1} \circ \Phi_o$ preserves the two quadrants where $x^p y^q = u < 0$. The pair $(G_+^{-1} \circ \Phi_1, G_-^{-1} \circ \Phi_o)$ thus defines a C^∞ local diffeomorphism of $\mathbb{R}^2$ at 0: it is still analytic outside the coordinate axis, and admits $\hat{\Phi}$ as its Taylor series at each point of these axis. This diffeomorphism answers the question.

Notice, moreover, that the C^∞ normalizing transformation obtained here belongs obviously to the Gevrey class of order 2.

4. REAL DOMAIN: WEAK FOCI

We consider a normal form:

$$(1.9) \left| \begin{array}{l} \mathcal{F}_o: \ \omega_o = (1 + \lambda u^k)(xdx + ydy) + \frac{u^k}{2}(xdy - ydx) = 0 \quad (\lambda \in \mathbb{R}) \\ \pi(x,y) = u = x^2 + y^2 \end{array} \right.$$

which represents a weak focus of order k in $\mathbb{R}^2(x,y)$, with a transverse fibration π, made up of concentric circles (each leaf of $\mathcal{F}_o$ is a spiral which crosses each circle). In $\mathbb{C}^2$, with complex coordinates:

$$\xi = x+iy \qquad \eta = x-iy$$

this normal form reads:

$$\omega_o = du + u^k(\alpha\eta d\xi + \bar{\alpha}\xi d\eta) = 0 \qquad \alpha = \lambda - i/2$$
$$\pi(\xi,\eta) = u = \xi\eta$$

The foliation $\mathcal{F}_o$ is defined by the first integral:

(4.1) $\quad H = \eta u^{-(i\lambda+1/2)} \exp -1/iku^k$

(write $\quad \omega_o/i\xi\eta u^k = du/u^{k+1} - (i\lambda+1/2)du/u + d\eta/\eta$)

Therefore, the Stokes lines, in $\mathbb{C}(u)$, of the complex foliation $\mathcal{F}_o$ are defined by Re $iu^k = 0$, and the singular lines by Im $iu^k = 0$. Notice that $\mathbb{R}^2 \subset \mathbb{C}^2$ $(\eta=\bar{\xi})$ lies in $\pi^{-1}(\mathbb{R}^+)$, that is in the inverse image of a Stokes line; notice to that H has modulus one in $\mathbb{R}^2$.

The analytic weak foci $(\mathcal{F},\pi)$ of order k and invariant λ are in one-to-one correspondance with their normalizing transformations, from the result of 1.4. Thus, they are classified, up to analytic equivalence, by the cocycles $F \in C^1(\mathcal{U};D^\infty_\pi(\mathcal{F}_o))$ such that $F = [\hat{\Phi}]$ for a *real* $\hat{\Phi} \in \hat{D}_\pi$.

We are going to describe these cocycles in some details for k=1. Of course, the "reality condition" will be stated, as in the saddle case, by means of the involution of $\mathbb{C}^2$ which corresponds to our real plane, that is $\sigma(\xi,\eta) = (\bar{\eta},\bar{\xi})$, $\sigma(u) = \bar{u}$.

For k=1, the covering $\mathcal{U} = \mathcal{U}(\mathcal{F}_o)$ of $\mathbb{C}(u)$ consists of the two sectors U_0, U_1 with aperture 2π and bissectors $\mathbb{R}^-, \mathbb{R}^+$. We denote by U_+, U_- the components of $U_0 \cap U_1$ on which Im u > 0 and Im u < 0.

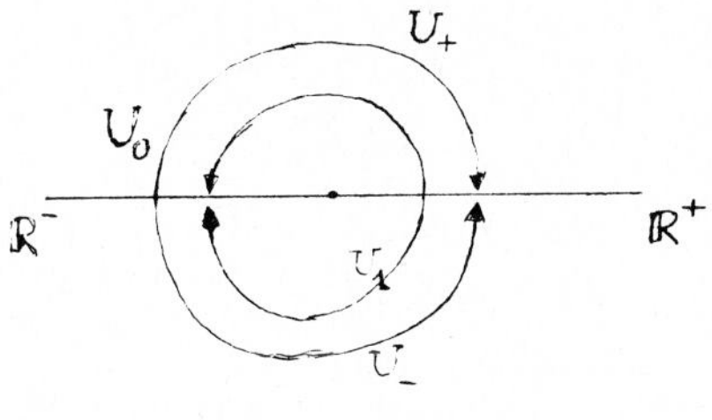

Notice that σ leaves invariant each of the U_i's, but exchanges U_+ and U_-. The elements of $C^1(\mathcal{U};D^\infty_\pi(\mathcal{F}_o))$ are the pairs $F = (F_+,F_-)$ defined on U_+,U_- by:

(4.2) $\quad H_+ \circ F_+ = f_+ \circ H_+ \qquad H_- \circ F_- = f_- \circ H_-$

where $H_+ = 1/H$ on U_+, $H_- = H$ on U_- (we use flat integrals), and f_+ f_- are local analytic diffeomorphisms of $(\mathbb{C},0)$, tangent to the identity.

PROPOSITION 4.1. *The cocycle* $F \in C^1(\mathcal{U};D^\infty_\pi(\mathcal{F}_o))$ *is characteristic of a real transformation* $\hat{\Phi} \in \hat{D}_\pi$ *if and only if:*

(4.3) $\quad \sigma \circ F_+ \circ \sigma = F_-^{-1} \qquad$ (*i.e* $\sigma \circ F \circ \sigma = F^{-1}$)

The proof is analogous to the one of Proposition 3.1 and we shall omit it. Notice only that if $\hat{\Phi} \in \hat{D}_\pi$ is real (that is $\sigma\hat{\Phi}\sigma = \hat{\Phi}$) and summable with sums Φ_0, Φ_1 on $\tilde{U}_0, \tilde{U}_1$, then we have $\sigma\Phi_i\sigma = \Phi_i$ (i=0,1); indeed σ preserves U_i, and the definition of the sum (see 2.5) gives immediately these equalities. Condition (4.3) follows obviously ■

Let us now describe more precisely the cocycles having property (4.3), in terms of the one dimensional diffeomorphisms f_+, f_-. To do this, we assume that H_+ and H_- are defined by the principal determination of Log u ($0 < \text{Arg } u < \pi$ on U_+, $-\pi < \text{Arg } u < 0$ on U_-). Then, we have:

$$\sigma H_+ \sigma = H_-$$

and it follows that F will satisfy (4.3) if and only if:

$$\sigma f_+ \sigma = f_-^{-1}$$

This means that, having chosen $f_+(z) = z + \sum_{n>2} a_n z^n$ (any convergent series), f_- will be defined by its inverse:

$$f_-^{-1}(z) = z + \sum_{n>2} \bar{a}_n z^n = \bar{f}_+(z).$$

Using these natural conventions, we have proved:

THEOREM 4.2. The analytic classes of weak foci with formal invariants $k=1, \lambda \in \mathbb{R}$, are in one-to-one correspondance with the local analytic diffeomorphisms of $(\mathbb{C},0)$:

$$f(z) = z + \sum_{n\geqslant 2} a_n z^n$$

As in the saddle case, our computations of [16] allow us to give explicitely the "characteristic diffeomorphism" of a family of weak foci.

PROPOSITION 4.3. The weak focus of order one:

$$\omega_{a,b} = (1+a(x^2-y^2)+2bxy)(xdx+ydy)+\frac{u}{2}(xdy-ydx) = 0 \quad (u=x^2+y^2)$$

has formal invariants $k=1$, $\lambda=0$; its analytic invariant is the local diffeomorphism of $(\mathbb{C},0)$ defined by:

$$f_+(z) = z/\sqrt{1+\alpha z^2} \quad , \alpha = -2e^{i\theta}\sin\pi\sqrt{a^2+b^2}, \quad \theta = \text{Arg}(a+bi)$$

Proof. The "blowing down" ρ (a two sheeted covering) defined by:

$$v = u/2i \qquad t = 2i(x-iy)/(x+iy) = 2i\eta/\xi$$

transforms the pair $(\mathcal{F}_{a,b}, \pi)$ into the Riccati equation:

$$\omega' = v^2dt-(t-4\beta v+\bar{\beta}vt^2)dv = 0 \qquad (\beta = (a-bi)/2)$$

and the projection $(v,t) \to v$.

The foliation associated to this Riccati equation is formally equivalent to the normal form:

$$v^2dt-tdv = 0$$

whose inverse image under ρ is:

$$(xdx+ydy) + \frac{u}{2}(xdy-ydx) = 0$$

For these normal forms, we use the flat first integrals:

$$K_+(v,t) = t^{-1}e^{-1/v} \quad (\mathrm{Re}\, v>0) \qquad K_-(v,t) = te^{1/v} \quad (\mathrm{Re}\, v<0)$$
$$H_+(x,y) = \eta^{-1}u^{1/2}e^{1/iu} \quad (\mathrm{Im} u>0) \qquad H_-(x,y) = \eta u^{-1/2}e^{-1/iu} \quad (\mathrm{Im} u<0)$$

with the relations:

$$\frac{1}{2i}H_+^2 = K_+\circ\rho \qquad 2iH_-^2 = K_-\circ\rho$$

Denote by $G = (G_+,G_-)$ the *characteristic cocycle of the Riccati equation* $\omega'=0$, and set $K\circ G = g\circ K$. The results of ([16],p.154-57) show that $g = (g_+,g_-)$ is given by:

$$g_+(z) = z/(1+\mu z) \qquad g_-(z) = z/(1+\nu z)$$

with:

$$\mu = 4ie^{-i\theta}\sin\pi\sqrt{a^2+b^2} \qquad \nu = ie^{i\theta}\sin\pi\sqrt{a^2+b^2}$$

The characteristic cocycle F of $\omega = 0$ is obtained by lifting G through ρ; a straightforward computation gives the result∎

These foci are defined too by the following algebraic vector field of degree three:

$$\left|\begin{array}{l} \dot{x} = -y -xu/2 - ay(x^2-y^2) - 2bxy^2 \\ \dot{y} = x -yu/2 + ax(x^2-y^2) + 2bx^2y \end{array}\right.$$

(notice that $x\dot{x} + y\dot{y} = -u^2/2$)

Bryuno has given in [5] examples of analytic weak foci with a *divergent normalizing transformation*; they are constructed, too, from the study of Riccati equations. He obtains, for each order k and invariant λ, a *discrete family* of examples, which correspond probably (for k=1) to characteristic diffeomorphisms:

$$f_+(z) = z + z^s +\dots \qquad (s \geqslant 2)$$

But the simplest numerical example he gives is a vector field of degree five.

4.4. REMARKS ON THE GENERAL CASE ($k \geqslant 2$)

The covering $\mathcal{U}$ of $\mathbb{C}(u)$ relative to the normal forms of order k consists of 2k sectors with aperture $2\pi/k$, and bissectors the Stokes lines $\mathrm{Re}\, iu^k = 0$.

The "reality condition" for a cocycle F is still:

$$\sigma F\sigma = F^{-1}$$

But now, in contrast with the case of saddles, the involution σ in $\mathbb{C}(u)$ is the reflexion with respect to a *Stokes line*; therefore, no sector $U_i \cap U_{i+1}$ is preserved under σ, and the reality condition means that F consists of k pairs of related diffeomorphisms. Thus, we may conclude:

The cocycles F which classify, up to analytic equivalence, the *weak foci of order k (and invariant $\lambda \in \mathbb{R}$ fixed) are defined by the choice of k local analytic diffeomorphisms of $(\mathbb{C},0)$, tangent to the identity, and otherwise arbitrary.*

As for the (complex) leaf space of a weak focus of order k, it consists of a necklace of 2k Riemann spheres, equipped with an antiholomorphic involution σ; the fixed points of σ are the equatorial circles of 2 of these spheres.

4.5. C^∞ NORMALIZATION

Takens has shown [22] that every weak focus may be transformed, through a C^∞ diffeomorphism of $(\mathbb{R}^2,0)$, into its formal normal form. For analytic foci, this result is an obvious consequence of our theory (more than in the saddle case); the point is that, as we have already noticed, the real plane now lies over a *Stokes line* of $\mathbb{C}(u)$; this means that the normalizing transformation $\hat{\Phi} \in \hat{D}_\pi$ of a weak focus has a *real sum on* $\mathbb{R}^2$ (see 2.5); this sum defines obviously a C^∞ diffeomorphism of $(\mathbb{R}^2,0)$ which normalizes the focus.

Moreover, this normalizing transformation is canonical, analytic outside 0, and belongs to the Gevrey class of order $s = 1+1/k$.

4.6. WEAK FOCI AND THEIR POINCARE MAP.

Consider an analytic weak focus of order k, and a piece of analytic line L going through the origin $0 \in \mathbb{R}^2$, parametrized by a coordinate r (r=0 at the origin). Take any point $m \in L$ (close to 0), and follow the integral curve through m (in counterclockwise direction for instance) until you hit again L, at a point $m' = P(m)$; the mapping P is an analytic germ, and:

$$P(r) = -r + ar^{2k+1} + \dots \qquad (a \neq 0)$$

The square of P (under composition) is the Poincaré map of the focus, but it is more convenient to use P for what we have in mind. The map P is defined up to analytic conjugacy, in the group of analytic diffeomorphisms of $(\mathbb{R},0)$; its conjugacy class is an invariant of the focus.

As a final remark, we are going to deduce from our theory the following:

THEOREM 4.7. *The analytic equivalence classes of weak foci of order k are in one-to-one correspondance with the conjugacy classes of analytic real germs* $P(r) = -r + ar^{2k+1} + \dots$

We emphasize the main fact: each P is the "Poincaré map" of a weak focus.

Sketch of proof. Here, the key word is "holonomy". We have shown in [17] that resonant differential equations have the same moduli space as the resonant germs of diffeomorphisms of $(\mathbb{C},0)$, the relation between the two problems being defined by the holonomy of the separatrices.

Now, the mapping P relative to a weak focus $\mathcal{F}$ is a holonomy mapping. To see this, transform the complex foliation $\mathcal{F}$ into $\tilde{\mathcal{F}}$, through a Hopf blowing up of $\mathbb{C}^2$ at 0; the foliation $\tilde{\mathcal{F}}$ has two singular points $0,\infty$ on the singular divisor $P_1(\mathbb{C})$, and *three separatrices*: the "original"ones S,S', and the divisor $P_1(\mathbb{C})$ (taking out the singular points). It is clear that P is the holonomy map of the leaf $P_1(\mathbb{C})\setminus\{0,\infty\}$. The singular points $0,\infty$ are resonant with eigenvalues $p=1$ (on S,S') and $q=2$ (on $P_1(\mathbb{C})$).

Let Σ be the leaf space of $\mathcal{F}$: it is a necklace of 2k spheres, endowed with an involution σ. Let Σ' be the "orbit space" of the map P: it is also a necklace of 2k spheres, endowed with an involution σ'. Using the above description, we get a *natural analytic isomorphism* between the two necklaces Σ and Σ', compatible with σ and σ'. One checks easily that, when Σ describes the space of all necklaces representing the moduli space of weak foci, Σ' describes the *whole* set of necklaces corresponding to the moduli space of real analytic diffeomorphisms with linear part $r \to -r$. (The two moduli spaces are a priori isomorphic.) ∎

REFERENCES

[1] A. Andreotti, P. Holm: Quasianalytic and Parametric Spaces, in Real and Complex Singularities, Oslo 1976, Sijthoff and Noordhoff Intern. Publishers.

[2] V.V. Basov: Divergence of Transformations of Real systems to the normal form in the case of nonrough focus (in russian), deposited in the All-Union Institute of Scient. and Technical information at n° 1206-78.

[3] A.D. Bryuno: On local invariants of differential equations, Mat. Zametki,14,4(1973),499-507= Math.Notes, 844-48.

[4] A.D. Bryuno: The normal form of real differential equations, Mat. Zametki,18,2(1975),227-41=Math.Notes,722-731.

[5] A.D. Bryuno: Divergence of a real normalizing transformation, Mat. Zametki,31,3(1982),403-410=Math.Notes, 207-211.

[6] A.D. Bryuno: Analytical form of differential equations, Trudy Moskov.Mat. Obsc.25(1971),120-262= Trans. Moscow Math. Soc. 25 (1971),131-288 (1972).

[7] D. Cerveau, J.F. Mattei: Formes intégrables holomorphes singulières, Astérisque 97 (1982).

[8] H. Dulac: Sur les cycles limites, Bull. Soc. Math. France,51 (1923), 45-188.

[9] J. Ecalle: Les champs de vecteurs locaux résonnants de $\mathbb{C}^{\nu}$: classification analytique, to appear in: Publication R.C.P 25, n° 32 , Strasbourg (1984).

[10] L.M. Markhashov: The analytic equivalence and stability of systems of second order under the resonance 1:1 (Russian), Preprint n° 14,Institute of Problems in Mech.,Moscow (1972).

[11] L.M. Markhashov: Analytic equivalence of second order systems with arbitrary resonance, Prikl. Mat. Mekh. 36,6(1972),1030-42.

[12] L.M. Markhashov:On analytic equivalence of systems of ordinary differential equations with resonances, Preprint n°36, Inst. of problems in Mech., Moscow (1974).

[13] L.M. Markhashov: Invariants of multidimensional systems with a resonance relation, Prikl.Mat.Mekh.38,2(1974),233-239.

[14] L.M. Markhashov: On the analytic equivalence of systems of ordinary differential equations under resonance, in: Problems of analytical mech., Theory of stability and control (Russian), Nauka, Moscow (1975), 189-195.

[15] L.M. Markhashov: The method of invariants in problems about equivalence of ordinary equations, in: Cybernetics and Computational techniques (Russian),n° 39, Naukova Dumka, Kiev (1978), 45-53.

[16] J. Martinet, J.P. Ramis: Problèmes de modules pour des équations différentielles non linéaires du premier ordre, Publ. Math. I.H.E.S 55 (1982), 63-164.

[17] J. Martinet, J.P. Ramis: Classification analytique des équations différentielles non linéaires résonnantes du premier ordre, to appear in Ann. Ec. Norm. Sup. (1984).

[18] J.F. Mattei, R. Moussu: Holonomie et Intégrales premières, Ann. Sc. Ec. Norm. Sup. 13 (1980), 469-523.

[19] J.K. Moser, S.M. Webster: Normal forms for real surfaces in $\mathbb{C}^2$ near complex tangents and hyperbolic surface transformations, Acta Math. 150:3-4 (1983), 255-296.

[20] J.P. Ramis: Les séries k-sommables et leurs applications, Springer Lect. Notes in Physics 126 (1980).

[21] R. Roussarie: Modèles locaux de champs et de formes, Astérisque 30 (1975).

[22] F. Takens: Normal forms for certain singularities of vector fields, Ann. Inst. Fourier, XXIII,2(1973),113-195.

Jean Martinet Jean-Pierre Ramis

Institut de Mathématiques
7, Rue René Descartes
67084 Strasbourg Cedex
France

Singularities & Dynamical Systems
S.N. Pnevmatikos (editor)

SINGULARITIES OF HOLOMORPHIC DIFFERENTIAL EQUATIONS

César Camacho

I.M.P.A.

Brasil

Let $U \subset \mathbb{C}^n$ be an open neighborhood of $0 \in \mathbb{C}^n$ and $Z: U \to \mathbb{C}^n$, $Z(0) = 0$, a holomorphic map on U (vector field with a singularity at $0 \in \mathbb{C}^n$). The integral curves of Z are complex curves i.e. Riemann surfaces parametrized locally as the solutions of the differential equation

$$(*) \qquad \frac{dz}{dT} = Z(z), \quad z \in U, \quad T \in \mathbb{C}.$$

These curves define a complex one dimensional foliation $\mathfrak{F} = \mathfrak{F}_Z$ with singularity at $0 \in U$. These differential equations were studied as far back as 1856 by Briot-Bouquet and later by Poincaré, Painlevé, Dulac and others. Our purpose is to relate several results concerning the study of that foliation, its analytic and topological invariants.

I. TOPOLOGICAL CLASSIFICATION. GENERIC THEORY.

In many (generic) situations the foliation $\mathfrak{F}$ is determined by the linear differential equation

$$(**) \qquad \frac{dz}{dT} = A \cdot z, \qquad A = dZ(0)$$

or first approximation of (*).

In fact, let $\Lambda = \{\lambda_1, \ldots, \lambda_n\}$ be the set of eigenvalues of A and assume that $0 \notin \Lambda$. A <u>resonance</u> between the elements of Λ is any relation of the form

$$\lambda_i = \sum_{j=1}^{n} m_j \lambda_j, \quad m_j \geq 0 \text{ integer}, \quad \sum_{j=1}^{n} m_j \geq 2.$$

It is well known (see [10] for instance) that the nonexistence of resonances implies in formal linearization of (*), that is, there is a power series change of coordinates transforming the differential equation (*) into the form (**).

When the convex hull $\mathcal{H}(A)$ of the eigenvalues of A does not contain $0 \in \mathbb{C}$, this formal change of coordinates converges in a neighborhood of $0 \in \mathbb{C}^n$. This is a theorem due to Poincaré [33]. When $0 \in \mathcal{H}(A)$, a formal linearization (if any exists) is analytic provided that

$$(S) \begin{cases} \text{There exists } c > 0, \ \nu > 0 \text{ such that} \\ |\lambda_i - (m_1\lambda_1 + \dots + m_n\lambda_n)| \geq \dfrac{c}{|m_1 + \dots + m_n|^\nu} \\ \text{for any } (m_1, \dots, m_n) \text{ and } i \text{ with } \lambda_i - (m_1\lambda_1 + \dots + m_n\lambda_n) \neq 0. \end{cases}$$

This is Siegel's theorem [37]. This theorem was improved by Brjuno [6] substituting (S) by the following condition:

$$(B) \begin{cases} \text{If } \omega_k = \min\{|\lambda_i - (m_1, \lambda_1 + \dots + m_n\lambda_n)| \neq 0; \ \sum_{i=1}^{n} m_i \leq 2^{k+1}\} \\ \text{then the series } \sum_k 2^{-k} \cdot \log \dfrac{1}{\omega_k} \text{ converges} \end{cases}$$

which clearly implies (S).

Though, in the presence of resonances the linear part of (*) not always suffices to determine the analytic behavior of $\mathcal{F}$, however in very general situations it determines the <u>topological type</u> of $\mathcal{F}$.

DEFINITION. The singularity $0 \in C^n$ of Z is called <u>hyperbolic</u> if $\lambda_i \notin \lambda_j\mathbb{R}$ for any $i \neq j$.

Under this condition it is easy to see that the holonomy of the eigenspace of λ_j for the foliation $\mathcal{F}_A$ is generated by a linear diffeomorphism f_j, $f_j(z_1, \dots, \hat{z}_j, \dots, z_n) = (\exp(2\pi i \frac{\lambda_1}{\lambda_j}) \cdot z_1, \dots, \exp 2\pi i \frac{\lambda_n}{\lambda_j} \cdot z_n)$, i.e. with eigenvalues $\neq 1$ in absolute value.

THEOREM [19]. Suppose $0 \notin \mathcal{H}(A)$ and $\lambda_i \notin \lambda_j\mathbb{R}$, $i \neq j$. Then $\mathcal{F}_Z$ is topologically equivalent to $\mathcal{F}_A$; i.e. there is a homeomorphism $h: U \to V$, $h(0) = 0$, between neighborhoods of $0 \in \mathbb{C}^n$ sending leaves of $\mathcal{F}_Z/U$ onto leaves of $\mathcal{F}_A/V$.

The condition $0 \notin \mathcal{H}(A)$ implies that all leaves of $\mathcal{F}_A$ are transverse to the spheres $S_r = \{z \in \mathbb{C}^n; |z_1|^2 + \dots + |z_n|^2 = r^2\}$. Moreover $\mathcal{F}_A$ is topologically equivalent to the foliation whose typical leaf is a union of all rays of $\mathbb{R}^{2n}$ passing through a leaf of $\mathfrak{F}_A/S_r$. The same is true for $\mathfrak{F}_Z$ if $r > 0$ is small enough. Therefore the topology of $\mathfrak{F}_Z$ is characterized by the topology of $\mathfrak{F}_Z/S_r$. The hyperbolicity condition implies that $\mathfrak{F}_A/S_r$ is a Morse-Smale 1-dimensional foliation and thus $\mathfrak{F}_Z/S_r$ is equivalent to $\mathfrak{F}_A/S_r$.

The case $0 \in \mathcal{H}(A)$ is less simple and its proof requires more elaborate techniques

THEOREM (Chaperon [13,14]). Suppose that $0 \in \mathcal{H}(A)$ and $\lambda_i \notin \lambda_j\mathbb{R}$ if $i \neq j$. Then $\mathfrak{F}_Z$ and $\mathfrak{F}_A$ are topologically equivalent.

In fact, this theorem still holds under more relaxed conditions on the λ's and for other Lie group actions.

The topological classification of hyperbolic singularities is then achieved as follows.

THEOREM ([19,10,21]). Given a linear hyperbolic vector field A with spectrum

$$\Lambda = \text{spectrum of } A = \{\lambda_1,\ldots,\lambda_n\}.$$

If $0 \notin \mathcal{H}(A)$ then A has no topological invariants. If $0 \in \mathcal{H}(A)$ and if $\hat{\Lambda} = \{2\pi i\,\lambda_1^{-1},\ldots,2\pi i\,\lambda_n^{-1}\}$ then the equivalence class of $\hat{\Lambda}$ under the action of $G\ell(2,R)$ is the sole topological invariant of A, i.e., $A' \sim A \Leftrightarrow$ there is $g \in G\ell(2,\mathbb{R})$ such that after reordering the indices we can get $\hat{\lambda}'_j = g\,\hat{\lambda}_j \ \forall\ j = 1,\ldots,n.$

Further Results

In dimension two, resonances are of two types:
Either $0 \notin \mathcal{H}(\lambda_1,\lambda_2)$ and so there is an integer n such that

$$\lambda_1 = n\,\lambda_2, \qquad n \geq 2$$

or $0 \in \mathcal{H}(\lambda_1,\lambda_2)$ and there are relatively prime positive integers m_1, m_2 such that

$$m_1\lambda_1 + m_2\lambda_2 = 0, \qquad m_1 + m_2 \geq 1.$$

In the first case it is well known since Dulac [15] that after an analytic change of coordinates the differential equation is transformed into,

$$(x)\quad \begin{cases} \dfrac{dz_1}{dT} = \lambda_1 z_1 + a_o z_2^n & \lambda_1 = n\lambda_2, \quad a_o \in \mathbb{C}. \\[2ex] \dfrac{dz_2}{dT} = \lambda_2 z_2 \end{cases}$$

which can be studied by direct integration.

In the second case we can use a <u>formal</u> change of coordinates in order to write the equation as:

$$(xx)\quad \begin{cases} \dfrac{dz_1}{dT} = \lambda_1 z_1 + \sum_{j=1}^{\infty} a_j z_1^{jm_1+1} z_2^{jm_2} \\[2ex] \dfrac{dz_2}{dT} = \lambda_2 z_2 + \sum_{j=1}^{\infty} b_j z_1^{jm_1} z_2^{jm_2+1} \end{cases}$$

THEOREM ([8]). Let $\mathfrak{F}$ be the foliation induced by the equation with resonance

$$m_1\lambda_1 + m_2\lambda_2 = 0 \qquad m_1 + m_2 \geq 1.$$

Then $\mathfrak{F}$ is either holomorphically equivalent near $0 \in \mathbb{C}^2$ to the foliation induced by

$$\frac{dz_1}{dT} = \lambda_1 z_1$$

$$\frac{dz_2}{dT} = \lambda_2 z_2$$

or it is topologically equivalent near $0 \in \mathbb{C}^2$ to the foliation associated to

$$\frac{dz_1}{dT} = \lambda_1 z_1 + a_k z_1^{km_1+1} z_2^{km_2}$$

$$\frac{dz_2}{dT} = \lambda_2 z_2 + b_k z_1^{km_1} z_2^{km_2+1}$$

where k is the first integer j for which $m_1 a_j + m_2 b_j \neq 0$ in the normal form (xx) above. Moreover k and (m_1, m_2) are its sole topological invariants.

This theorem was reproved in [29] and [25].

PROBLEM. Characterize the topology of vector fields in dimension n with a resonance between two eigenvalues λ_1, λ_2, of the form $m_1\lambda_1 + m_2\lambda_2 = 0$.

Still in dimension two, when the eigenvalues λ_1, λ_2 of the differential equation are not in resonance and $\lambda_1/\lambda_2 \in \mathbb{R}$, Siegel and Brjuno have given many examples of nonlinearizable differential equations exhibiting countably many invariant manifolds diffeomorphic to $S^1 \times R$ approaching the singular point with increasing periods; i.e. the number of times these manifolds cover the coordinates axis increases as they approach $0 \in \mathbb{C}^2$. In [35] it is shown that the set of numbers $\exp(2\pi i\, \lambda_1/\lambda_2) \in S^1$ for which this phenomenon happens is dense in S^1. Observe that by Siegel theorem this set has measure zero. Along this direction we also have,

THEOREM [31]. The quotient of eigenvalues λ_1/λ_2 (or λ_2/λ_1) is a topological invariant of the differential equation, provided it is real.

A conformal diffeomorphism f with a fixed point $0 \in \mathbb{C}$ has a power series development

$$f(z) = \lambda_z + a_2 z^2 + \dots$$

A <u>resonance</u> for f is a relation of the kind: $\lambda^n = 1$, $n \in \mathbb{N}$. The Theorem of Siegel [36] allows to linearize f by an analytic change of coordinates provided that $\exists\, c, \nu > 0$ such that $|\lambda^n - 1| > \dfrac{c}{n^\nu}$ for any n. On the other hand, if resonances are permitted the analytic classification is achieved by the work of Ecalle [17, 18] (see also [39]).

The topological classification of $\mathbb{C}$-diffeomorphisms in resonance is given in [7] (see also [29,25]).

PROBLEM. Find a topological interpretation for the nonlinearization of a conformal mapping as above. Is it true that any non linearizable f has infinitely many periodic points approaching $0 \in \mathbb{C}$? What is the structure of the invariant sets of f near $0 \in \mathbb{C}$?

<u>Bifurcations through resonances</u>.

Bifurcations of holomorphic differential equations passing through resonances were studied by Arnold [1,2], Pjartly [32], Il'iašenko-Pjartly [23,24] and in [8]. Arnold observed that when a family Z_t, $0 \notin \mathcal{H}(A_t)$, passes through a resonance $\lambda_1(0) = n\,\lambda_2(0)$, a topological bifurcation takes place in the foliation $\mathcal{F}_{Z_t}$. This is in constrast with the real domain where such bifurcation produces no change in the topology of the real solutions.

More precisely, since $0 \notin \mathcal{H}(A_t)$, we can describe the bifurcation in a small 3-sphere S_r centered at $0 \in \mathbb{C}^2$ and transverse to $\mathcal{F}_{Z_t}$. At a resonance $\lambda_1(0) = n\,\lambda_2(0)$ we see from the normal form of Dulac (x) that generically $\mathcal{F}_{Z_0}/S_r$ is a foliation with only one closed leaf; all other leaves accumulate on this in the past and in the future. Then when passing through this resonance another closed leaf branches off on S_r going around the first one n times with linking number one.

In the Siegel domain Pjartly showed that when passing through a resonance $m_1\lambda_1 + m_2\lambda_2 = 0$ in $\mathbb{C}^2$, generically from a separatrix through $0 \in \mathbb{C}^2$ an invariant manifold branches off whose equation has in first approximation the form $z_1^{m_1} z_2^{m_2} = \varepsilon$ where ε is the deviation from the resonance. The global topological behavior of these bifurcations was studied in [8]. It is shown there that any neighborhood of a differential equation in resonance of the kind $m_1\lambda_1 + m_2\lambda_2 = 0$ contains a subset with nonempty interior whose elements have foliations with simple topological behavior in the sense that their limit sets are unions of the singularity with a finite number of real two dimensional cylinders with hyperbolic holonomy. The generalization of this to n-dimensions is still an open problem.

The situation described by Pjartly was generalized to any dimension by Il'iašchenko and Pjartly [23,24]. They proved also that when the linear part of the equation at a singular point is not a resonance but is close to a countable set of Poincaré resonances (for instance the resonance between three numbers whose convex hull contains $0 \in \mathbb{C}$ in its interior), then in each neighborhood of zero there is a countable set of invariant manifolds of the equation that can be seen as the obstruction to the convergence of the normalizing series that linearizes the equation.

II. DESINGULARIZATION OF VECTOR FIELDS [4,34].

The blow up of a point p in a complex manifold M is a process leading to the creation of a new complex manifold obtained by replacing the point p by a complex projective space, considered as the set of limit asymptotic directions of M at p. More precisely, we consider a local coordinate chart $(\mathcal{U},z)$ of M around the point p, i.e. $z(p) = 0$. For simplicity we assume $n = 2$. Define the projective space $\mathbb{C}P(1)$ as the quotient space of $\mathbb{C}^2/(0)$ under the equivalence $z \sim w \Leftrightarrow z = \lambda w$; i.e. $CP(1)$ is the set of complex lines passing through $0 \in \mathbb{C}^2$. We write $\mathbb{C}^2/(0) \xrightarrow{[\ ]} \mathbb{C}P(1)$ the quotient map and introduce in $\mathbb{C}P(1)$ the quotient topology. The projective space is covered by two open subsets:

$$U_1 = \{[\ell_1,\ell_2)]; \ \ell_1 \neq 0\} \qquad U_2 = \{[\ell_1,\ell_2)]; \ \ell_2 \neq 0\}.$$

This allows us to define a structure of complex manifold in $\mathbb{C}P(1)$ with local charts (U_1,ϕ_1), (U_2,ϕ_2) where

$$\phi_1 : U_1 \to \mathbb{C} \quad \text{is given by} \quad \phi_1([\ell_1,\ell_2]) = \frac{\ell_2}{\ell_1}$$

and

$$\phi_2 : U_2 \to \mathbb{C} \quad \text{is given by} \quad \phi_2([\ell_1,\ell_2]) = \frac{\ell_1}{\ell_2}.$$

Consider the set $\tilde{\mathcal{U}} = \{(z,\ell) \in \mathcal{U} \times \mathbb{C}P(1); \ z \in \ell\}$ with natural projection $\pi: \tilde{\mathcal{U}} \to \mathcal{U}$, $\pi(z,\ell) = z$. Then we can endow $\tilde{\mathcal{U}}$ with the topology of $\mathcal{U} \times \mathbb{C}P(1)$ and introduce in $\tilde{\mathcal{U}}$ the structure of a complex manifold of dimension one by taking local coordinates $(\tilde{U}_1,\tilde{\phi}_1)$, $(\tilde{U}_2,\tilde{\phi}_2)$ as follows:

$$\tilde{U}_1 = \pi^{-1}(U_1) = \{(z,\ell); \ \ell_1 \neq 0\} \quad \text{and} \quad \tilde{\phi}_1 : \tilde{U}_1 \to \mathbb{C}, \quad \tilde{\phi}_1(z,\ell) = (z_1, \frac{z_2}{z_1})$$

$$\tilde{U}_2 = \pi^{-1}(U_2) \quad \text{and} \quad \tilde{\phi}_2 : \tilde{U}_2 \to \mathbb{C}, \quad \tilde{\phi}_2(z,\ell) = (z_2, \frac{z_1}{z_2}).$$

The change of coordinates is clearly holomorphic and the projection in these local charts is expressed by:

$$\pi : \tilde{U}_1 \to U_1 : \pi(z_1,t) = (z_1,z_2) \quad \text{and} \quad z_2 = t\, z_1$$

$$\text{and} \quad \pi : \tilde{U}_2 \to U_2 : \pi(z_2,t) = (z_1,z_2) \quad \text{and} \quad z_1 = u\, z_2 .$$

Clearly $\pi: \tilde{U} \to U$ is surjective, $\pi^{-1}(0) = 0 \times \mathbb{C}P(1)$ and $\pi: \tilde{U}-\mathbb{C}P(1) \to U-\{0\}$ is a complex diffeomorphism. The glueing of $M-\{p\}$ with $\tilde{U}$ via π yields another complex manifold $\tilde{M}$ called the blow up of M at p.

Desingularization of a Vector Field.

We consider the differential equation

$$\dot{x} = a(x,y)$$
$$\dot{y} = b(x,y)$$

defined in U with a singular point at $0 \in \mathbb{C}^2$, i.e., $a(0,0) = b(0,0) = 0$. Let ν be the algebraic multiplicity of this equation at $0 \in \mathbb{C}^2$, i.e., its least order at $0 \in \mathbb{C}^2$:

$$\dot{x} = a_\nu(x,y) + a_{\nu+1}(x,y) + \dots$$
$$\dot{y} = b_\nu(x,y) + b_{\nu+1}(x,y) + \dots$$

This equation induces, via π, a differential equation in the open set $\tilde{U}_1$ of $\tilde{U}$, as follows,

$$\begin{cases} \dot{x} = a_\nu(x,tx) + a_{\nu+1}(x,tx) + \dots \\ \dot{t} = \dfrac{\dot{y}-t\dot{x}}{x} = \dfrac{(b_\nu(x,tx)+b_{\nu+1}(x,tx)+\dots) - t(a_\nu(x,tx)+\dots)}{x} \end{cases}$$

or

$$\begin{cases} \dot{x} = x^\nu(a_\nu(1,t) + x\,A(x,t)) \\ \dot{t} = x^{\nu-1}(b_\nu(1,t) - ta_\nu(1,t) + xB(x,t)) \\ \quad = x^{\nu-1}(P_{\nu+1}(1,t) + xB(x,t)) \end{cases}$$

where $P_{\nu+1}(x,y) = xb_\nu(x,y) - ya_\nu(x,y)$ and A and B are holomorphic functions vanishing at $0 \in \mathbb{C}^2$.

We distinguish two cases:

Dicritical Singularities: $(P_{\nu+1}(x,y) \equiv 0)$. In this case the equation above can be written:

$$\dot{x} = x^\nu(a_\nu(1,t) + xA(x,t))$$
$$\dot{t} = x^\nu\, B(x,t).$$

Dividing by the factor x^ν we obtain the differential equation

$$\dot{x} = a_\nu(1,t) + x\,A(x,t)$$
$$\dot{t} = B(x,t)$$

whose integrals are the same outside $(x=0)$, modulo reparametrization, thus inducing a foliation with <u>isolated singularities</u> (i.e. the solutions of $B(0,t) = a_\nu(1,t) = 0$). The foliation on $\tilde{U}_1$ obtained in this way is transversal to the projective line $(x=0)$ at all points where $a_\nu(1,t) \neq 0$.
Combining this with the correspondent expression in the other coordinate system after dividing by y^ν, we obtain equations of a foliation $\mathfrak{F}_Z^{(1)}$ which coincides with $\pi^*\mathfrak{F}_Z$ outside $\pi^{-1}(0)$ except at a finite number of points (the roots of $a_\nu(1,t) = 0$) which may or may not be singularities.

<u>Nondicritical Singularities</u>: $(P_{\nu+1}(x,y) \not\equiv 0)$. Then the equation above is only divisible by $x^{\nu+1}$ yielding

$$\dot{x} = x(a_\nu(1,t) + xA(x,t))$$
$$\dot{t} = P_{\nu+1}(1,t) + xB(x,t).$$

Again this defines a foliation on $\tilde{U}_1$ with singularities at the points in $(x=0)$ given by $P_{\nu+1}(1,t) = 0$. The expression found in the other coordinate system in $\tilde{U}_2$ is similar and fits well with the foliation in $\tilde{U}_1$ to define a foliation $\mathfrak{F}_Z^{(1)}$ in $\tilde{\mathcal{U}}$ having $\pi^{-1}(0)$ as an invariant set. More precisely $\pi^{-1}(0)$ minus a certain number of singularities given by the roots of $b_\nu(1,t) - ta_\nu(1,t) = 0$ is a leaf of $\mathfrak{F}_Z^{(1)}$. Notice that $\mathfrak{F}_Z^{(1)}$ and $\pi^*\mathfrak{F}_Z$ coincide outside $\pi^{-1}(0)$.

In both cases the foliation $\mathfrak{F}_Z^{(1)}$ is locally given by analytic expressions. Therefore we can repeat the process at any of the singularities of $\mathfrak{F}_Z^{(1)}$. A new foliation $\mathfrak{F}_Z^{(2)}$ is found in a neighborhood of a union of projective lines having normal crossings and again exhibiting a finite number of singularities. The process can then be repeated and after k blow ups we have a foliation $\mathfrak{F}_Z^{(k)}$ defined in a neighborhood $\mathcal{U}_Z^{(k)}$ of a union $\mathcal{P}_Z^{(k)}$ of projective lines with normal crossings and a proper analytic projection $\pi_Z^{(k)}: \mathcal{U}_Z^{(k)} - \mathcal{P}_Z^{(k)} \to \mathcal{U}/\{0\}$ is an isomorphism between the foliations $\mathfrak{F}_Z^{(k)}$ and $\mathfrak{F}_Z$. We will write $(\mathcal{U}_Z^{(k)}, \pi_Z^{(k)}, \mathcal{P}_Z^{(k)}, \mathfrak{F}_Z^{(k)})$ to denote a k-th blow up of Z at $0 \in \mathbb{C}^2$; $\pi^{(k)}$ will be called its <u>projection</u> and $\mathcal{P}_Z^{(k)}$ <u>its divisor</u>. The divisor is a union of embedded projective lines intersecting transversally at points called corners each corner being the intersection of two projective lines.

THEOREM [4,34]. Finitely many blow ups are sufficient in order to obtain only simple singularites. A singularity p of Z is simple if the eigenvalues λ_1, λ_2 of $dZ(p)$ are either

i) $\lambda_1 \neq 0 \neq \lambda_2$ and $\lambda_1/\lambda_2 \notin \mathbb{Q}_+$

or

ii) $\lambda_1 \neq 0 = \lambda_2$ or $\lambda_2 \neq 0 = \lambda_1$ (complex saddle-node)

A nice proof of this theorem due to Ven den Essen can be found in [38,30].

PROBLEM. Find a desingularization theorem in higher dimensions.

III. ANALYTIC CLASSIFICATION OF SIMPLE SINGULARITIES

Analytic Classification of Complex Saddle-Nodes. The Theorem of Martinet-Ramis.

We consider germs at $0 \in \mathbb{C}^2$ of differential equations induced by vector fields Z with an isolated singularity at $0 \in \mathbb{C}^2$ and whose linear part has eigenvalues $\lambda_1 = 0$, $\lambda_2 \neq 0$. The foliation $\mathcal{F}_Z$ can be given also by the equation $\omega = 0$ where ω is a 1-form which by Dulac theorem [15] can be written

$$(1) \qquad \omega = x^{p+1}dy - A(x,y)dx, \qquad A(0,y) = \lambda y, \qquad \lambda \neq 0,$$

where p is the multiplicity of $0 \in \mathbb{C}^2$ as zero of ω. Let E_p be the set of equations as in (1). For any $\omega \in E_p$ there is a unique formal power series (formal in x):

$$(2) \qquad \varphi(x,y) = (h(x),\ y + \sum_{n\geq 1} a_n(y)x^n)$$

all $a_n(y)$ holomorphic in a disc $0 \in D \subset \mathbb{C}$, such that $\varphi^*\omega \wedge \omega_{p,\lambda} = 0$, where

$$(3) \qquad \omega_{p,\lambda} = x^{p+1}dy - y(1+\lambda x^p)dx \qquad \lambda \in \mathbb{C}.$$

Let G_o be the group of germs of holomorphic diffeomorphisms of the form $(x,y) \mapsto (x, y + \sum_{n\geq 1} a_n(y)x^n)$ and $\hat{G}_o$ the group of formal (in x) transformations of this kind.

Let $\mathcal{E}_p \subset E_p$ be the set of analytic differential equations that can be reduced to the form (3) via an element of $\hat{G}_o$. The group G_o leaves $\mathcal{E}_p$ invariant and one can check that the group G obtained from G_o adjoining linear maps $(x,y) \mapsto (\alpha x, \beta y)$, $\alpha^p = 1$, $\beta \neq 0$ leaves also $\mathcal{E}_p$ invariant. The classification of the elements of E_p is then reduced to the study of $\mathcal{E}_p/G_o$ and the action of G/G_o on $\mathcal{E}_p/G_o$. This group being isomorphic to the group of linear maps $(x,y) \mapsto (\alpha x, \beta y)$, $\alpha^p = 1$, $\beta \neq 0$.

Given $\omega \in \mathcal{E}_p$ there exists a unique $\varphi \in \hat{G}_o$ such that $\varphi^*\omega_{p,\lambda} \wedge \omega = 0$ for some $\lambda \in \mathbb{C}$. This normalization converges in some sectors in the following sense,

THEOREM [20]. For any sector $U = U_{\theta,\delta} = \{x \in \mathbb{C};\ |\arg x - \theta| < \pi/p,\ |x| < \delta\}$ with $\delta > 0$ small enough, there exists a holomorphic and bounded mapping $\varphi_U\colon U\times\Omega \to U\times\mathbb{C}$, $(x,y) \mapsto (x, \varphi_U(x,y))$ where Ω is a neighborhood of $0 \in \mathbb{C}$, such that:

1) $\varphi_U^*\omega_{p,\lambda} \wedge \omega = 0$

2) φ_U has φ as asymptotic development, i.e. for any $k \geq 2$, $|\varphi_U(x,y) - (y + \sum_{n=1}^{k} a_n(y)x^n)| \leq C_k|x|^{k+1}$.

<u>The Sheaf</u> $\Lambda_{p,\lambda}$. Let Λ^o be the sheaf over S^1 defined as follows: for any $e^{i\theta_o} \in S^1$ an element of $\Lambda^o_{\theta_o}$ is defined by a function $g(x,y) = (x, g(x,y))$ holomorphic in the region $|\arg x - \theta_o| < \delta$, $|x| < \delta$, $|y| < \delta$ such that (i) $g(0,y) = y$ and (ii) On any region: $|\arg x - \theta_o| \leq \delta' < \delta$ $|y| < \delta'$, $g(x,y)$ is C^∞ in the sense of Whitney. Clearly the asymptotic development of such a map belongs to $\hat{G}_o$.

The subsheaf $\Lambda \subset \Lambda^o$ is defined by transformations defined in sectors as above with asymptotic development equal to the identity. Finally $\Lambda_{p,\lambda} \subset \Lambda$ is defined by the mappings which leave invariant the differential equation $\omega_{p,\lambda} = 0$.

The main result of Martinet and Ramis is:

THEOREM. There is a canonical bijection between $\mathcal{E}_p/G_o$ and $H^1(S^1;\Lambda_{p,\lambda})$.

The mapping $\mathcal{E}_p/G_o \to H^1(S^1;\Lambda_{p,\lambda})$ is constructed as follows. Let U be a collection of open sectors U_i $i = 0,1,\ldots,2p-1$, with argument $\frac{2\pi}{p}$ and bisected by rays of argument $\frac{(2i+1)\pi}{2p}$, $i = 0,1,\ldots,2p-1$, (i.e. $\mathrm{Re}\, x^p = 0$). By the theorem of Hukuara-Kimura-Matuda the mapping $g_{i,i+1} = \varphi_{U_i} \circ (\varphi_{U_{i+1}})^{-1}$ is asymptotic to the identity i.e. $|g_{i,i+1}(x,y)-y| \leq c_k|x|^{k+1}$ for any $k \geq 1$.
Let $V_i^+ = U_{2i} \cap U_{2i+1}$ and $V_i^- = U_{2i+1} \cap U_{2i+2}$. Call $g_i^+ = g_{2i,2i+1}$ and $g_i^- = g_{2i+1,2i+2}$. Since $g_i^{\pm *}\omega_{p,\lambda} = \omega_{p,\lambda}$ we obtain

(i) $g_i^+(x,y) = y + a_o[x^\lambda \exp(-\frac{1}{px^p})]$, for $x \in V_i^+$ and

(ii) $g_i^-(x,y) = y + \sum_{n\geq 2} a_n[x^{-\lambda} \exp(\frac{1}{px^p})]^{n-1} y^n$, for $x \in V_i^-$.

Thus, $g = (g_{i,i+1}) \in H^1(S^1,\Lambda_{p,\lambda})$.
We observe that in $V_i^+ \times \Omega$ the solutions of $\omega_{p,\lambda} = 0$ behave like a <u>node</u>, i.e. except $(x=0)$ all other solutions tend to $0 \in \mathbb{C}^2$ along the direction $y = 0$. In $V_i^- \times \Omega$ the sole solutions of $\omega_{p,\lambda} = 0$ tending to $0 \in \mathbb{C}^2$ are $(x=0)$ and $(y=0)$ i.e. they behave like in a <u>saddle</u> point.
In the other direction, let $U = (U_i)_{i=0}^m$ be a <u>good</u> covering of S^1, i.e. $U_i \cap U_{i+1} \neq \emptyset$, $U_{m+1} = U_o$, and any three elements of U have empty intersection. Let $g = (g_{i,i+1}) \in Z^1(U,\Lambda_{p,\lambda})$. We proceed to associate to g a differential equation defined in a manifold M_g isomorphic to $C^2,0$. The definition of M_g goes as follows. Consider the disjoint union $\amalg U_i \times \Omega$, Ω a neighborhood of $0 \in \mathbb{C}$; for $x \in U_i \cap U_{i+1}$ we identify $(x,y) \in U_i \times \Omega$ and $(x,z) \in U_{i+1}\times\Omega$ provided that $(x,y) = g_{i,i+1}(x,z)$. The germ at 0 of this topological manifold is denoted M_g. Clearly the open set $\dot{M}_g = M_g\backslash\{(0,y);\ y \in \mathbb{C}\}$ is a complex manifold and the projection $\pi\colon M_g \to \mathbb{C}$ given by $\pi(x,y) = x$ is continuous and its restriction $\pi/\dot{M}_g$ to $\dot{M}_g$ is holomorphic. Moreover there is on M_g a unique structure of differentiable manifold such that the natural injections $U_i\times\Omega \to M_g$ are C^∞ in the sense of Whitney. This C^∞ structure on $\dot{M}_g$ coincides with the one induced by its holomorphic structure. The integrability theorem of Newlander-Nirenberg is then used in order to show that M_g admits a unique structure of complex manifold

extending that of $\dot{M}_g$. The projection $\pi: M_g \to \mathbb{C}$ becomes holomorphic of maximum rank. Moreover one can construct a holomorphic embeding:

$$H: (M_g,0) \to (\mathbb{C}^2,0), \qquad H(x,[y]) = (x,Y(x,[y])).$$

In coordinates $(x,y) \in U_i \times \Omega$ $Y = \eta_i(x,y)$ and if $(x,z) \in U_{i+1} \times \Omega$ $Y = \eta_{i+1}(x,z)$. For $x \in U_i \cap U_{i+1}$ $\eta_{i+1} = \eta_i \circ g_{i,i+1}$. Let $(x,y) = \varphi_{i+1}(x,Y)$ be such that $(x,y) = \varphi_i(x,\eta_i(x,y))$ for $x \in U_i$ and $(x,z) = \varphi_{i+1}(x,Y)$ such that $(x,z) = \varphi_{i+1}(x,\eta_{i+1}(x,z))$, $x \in U_{i+1}$. From this it follows that $g_{i,i+1} = \varphi_i^{-1} \circ \varphi_{i+1}$. Therefore

$$\varphi_i^*(\omega_{p,\lambda}) = \frac{\partial \varphi_i}{\partial Y} x^{p+1} dY + [x^{p+1} \frac{\partial \varphi_i}{\partial x} - \varphi_i(x,Y)(1+\lambda x^p)]\, dx$$

$$\varphi_{i+1}^*(\omega_{p,\lambda}) \frac{\partial \varphi_{i+1}}{\partial Y} x^{p+1} dY + [x^{p+1} \frac{\partial \varphi_{i+1}}{\partial x} - \varphi_{i+1}(x,y)(1+\lambda x^p)]\, dx.$$

Thus the 1-forms $\omega_i = \left(\frac{\partial \varphi_i}{\partial Y}\right)^{-1} \cdot \varphi_i^*(\omega_{p,\lambda})$ and $\omega_{i+1} = \left(\frac{\partial \varphi_{i+1}}{\partial Y}\right)^{-1} \varphi_{i+1}^*(\omega_{p,\lambda})$ define the same foliation and have the same coefficient in dY, therefore $\omega_i = \omega_{i+1}$ in $\varphi_i^{-1}(U_i \cap U_{i+1} \times \Omega)$. This defines $\omega \in \mathcal{E}_p$ by putting $\omega = \omega_i$ in $\varphi_i^{-1}(U_i \times \Omega)$ for any $i = 0,\ldots,m$. ■

Remarks. The mappings (i) and (ii) above represent, in the space of leaves of $\omega_{p,\lambda}/V_i^+$ and $\omega_{p,\lambda}/V_i^-$, respectively a translation $y \mapsto y+a_0$ and the map $y \mapsto y + \sum_{n\geq 2} a_n y^n$. If we call $\mathcal{H}$ the group of germs of transformations of this kind (with linear part the identity map) we obtain

$$\mathcal{E}_p/G_0 \approx \mathbb{C} \times \mathbb{C}^p \times \mathcal{H}^p.$$

In fact, fixing $\lambda \in \mathbb{C}$ the identification in $V_i^+ \times \mathbb{C}$ is given by $y \mapsto y+a_0$ and in $V_i^- \times \mathbb{C}$ by $y \mapsto y + \sum_{n\geq 2} a_n y^n \in \mathcal{H}$ where $i=1,\ldots,p$.

Martinet and Ramis also proved that the invariants of a saddle-node can be read in the holonomy of the invariant manifold $x = 0$ in the form $\omega = x^{p+1}dy + A(x,y)dx$. This means that two differential equations of this kind are analytically equivalent if and only if the holonomies of their invariant manifolds are equivalent. They carried out the same program for the analytic classification of differential equations in resonance [29] proving that any conformal diffeomorphism of $(\mathbb{C},0)$, in resonance, can be realized as the holonomy of an invariant manifold of a differential equation in resonance (see also [25]).

PROBLEM. Classify under C^0 equivalence the complex saddle-node singularities. Notice that C^1 equivalence for saddle nodes and resonant singularities implies analytic equivalence [29].

IV. EXISTENCE OF FIRST INTEGRALS

The Theorem of Mattei-Moussu.

The simplest foliations with singularities are those given by the level curves of a function $f: (\mathbb{C}^2,0) \to (\mathbb{C},0)$. It is therefore natural to characterize the vector fields Z with a singular point at $0 \in \mathbb{C}^2$ whose induced foliation $\mathfrak{F}_Z$ is of this kind. Clearly if Z is such a vector field we have $df(Z) \equiv 0$ and we say that f is a first integral of Z. It is not difficult to derive the following topological properties of Z:

i) All leaves of $\mathfrak{F}_Z$ are closed in $\mathbb{C}^2-\{0\}$.

ii) Only a finite number of leaves of $\mathfrak{F}_Z$ accumulate in $0 \in \mathbb{C}^2$.

iii) The holonomy of any leaf is finite.

Conversely, we have the following

THEOREM [30]. Let Z be a holomorphic vector field with an isolated singularity at $0 \in \mathbb{C}^2$ such that:

i) All leaves of $\mathfrak{F}_Z$ are closed in $\mathbb{C}^2-\{0\}$.

ii) Only a finite number of leaves accumulate in $0 \in \mathbb{C}^2$.

Then Z admits a holomorphic first integral. Thus the existence of a holomorphic first integral is a topological invariant property of the vector field.

The proof of this theorem uses nicely the desingularization theorem of Bendixson-Seidenberg. It proceeds by induction on the number $N(Z)$ of blow ups necessary to desingularize Z. For $N(Z) = 0$, i.e. when the singularity is simple the theorem follows easily. The case $N(Z) > 0$ requires analytic continuation of local first integrals in order to obtain a first integral in the neighborhood of the divisor.

This theorem extends naturally to completely integrable 1-forms ω in $(\mathbb{C}^n,0)$ (i.e. $\omega \wedge d\omega = 0$) with a singularity at $0 \in \mathbb{C}^n$. J.F. Mattei remarked in his thesis (see also [12] p.74) that the existence of meromorphic first integral is not a topological invariant property. However, it is possible to find a topological criterion for the existence of multiform first integrals of the form $f = f_1^{\lambda_1} \dots f_p^{\lambda_p}$ where $\lambda_j \in \mathbb{R}_+$. This and further results by Cerveau and Mattei concerning the existence of multiform first integrals can be found in [12].

V. THE SEPARATRIX THEOREM

The problem of finding solutions of differential equations

$$\frac{dz}{dT} = Z(z), \qquad Z(0) = 0$$

tending to $0 \in \mathbb{C}^n$, is an old one. This problem was posed and studied by C.A. Briot and J.C. Bouquet [5]. They succeeded in finding these solutions in many particular cases. H. Dulac also considered this problem from another point of view: Given the differential equation whose linear part has nonzero eigenvalues he wished to determine the set of solutions which tend asimptotically to $0 \in \mathbb{C}^2$, [14].

THEOREM ([9]). Consider the differential equation

$$\frac{dz}{dT} = Z(z), \qquad z \in \mathbb{C}^2,$$

with an isolated singularity $0 \in \mathbb{C}^2$. There exists an invariant complex analytic subvariety, of dimension one, passing through $0 \in \mathbb{C}^2$.

This means that there exists an integral complex curve V of this differential equation such that $\bar{V} = V \cup \{0\}$ and $\bar{V}$ is an analytic subvariety.

The proof of this theorem consists of finding, in the desingularization of Z, a singular point which exhibits an invariant manifold transverse to the divisor. This manifold will be projected then to an invariant subvariety of Z through $0 \in \mathbb{C}^2$.

Notice that not all the simple singularities appearing in the desingularization of Z admit an invariant manifold transverse to the divisor. To detect such a singularity an index is introduced as follows:

Consider an isolated singularity $p \in M$ of a foliation $\mathfrak{F}$ near p induced by a holomorphic vector field η. Suppose that S is an invariant submanifold of η passing through p and let (x,y) be a system of coordinates around p, such that $(x(p),y(p)) = (0,0)$ and $S = (y=0)$. Then

$$\eta = A(x,y)\frac{\partial}{\partial x} + B(x,y)\frac{\partial}{\partial y}$$

with $A(0,0) = B(0,0) = 0$ and $B(x,0) = 0$.

DEFINITION. The index of $\mathfrak{F}$ relative to S at p is

$$i_p(\mathfrak{F},S) = \operatorname*{Res}_{x=0} \frac{\partial}{\partial y}\left(\frac{B}{A}\right)(x,0)\,dx.$$

It is easy to see that this index is invariant by change of coordinates.

EXAMPLES: Suppose p is simple singular point with eigenvalues λ_1, λ_2. If $\lambda_1 \neq 0 \neq \lambda_2$ there are invariant manifolds S_1, S_2 tangent to the eigenspaces of λ_1, λ_2. Then $i_p(\mathfrak{F},S_1) = \lambda_2/\lambda_1$, $i_p(\mathfrak{F},S_2) = \lambda_1/\lambda_2$. If $\lambda_1 \neq 0$ and $\lambda_2 = 0$ then $i_p(\mathfrak{F},S_1) = 0$.

The following two propositions are easy to check.

Suppose that $(\mathcal{U}^{(1)},\pi^{(1)},\rho^{(1)},\mathfrak{F}^{(1)})$ is the blow up of $\mathfrak{F}$ at p. Denote $\pi^{(1)-1}(S-\{p\})$ by S again and let $q = S \cap \rho^{(1)}$.

PROPOSITION 1. $i_q(\mathfrak{F}^{(1)},S) = i_p(\mathfrak{F},S) - 1$.

PROPOSITION 2. $\sum_{j=1}^{r} i_{q_j}(\mathfrak{F}^{(1)},\rho^{(1)}) = -1$, $\{q_1,\dots,q_r\}$ = Singular set of $\mathfrak{F}^{(1)}$.

The idea is to consider the desingularization $(\mathcal{U}^{(n)},\pi^{(n)},\rho^{(n)},\mathfrak{F}^{(n)})$ of Z at $0 \in C^2$ and find in $\rho^{(n)}$ a projective line P containing a singularity p of $\mathfrak{F}^{(n)}$, which is not a corner, such that $i_p(\mathfrak{F}^{(n)},P) \neq 0$. If this is the case there exists an $\mathfrak{F}^{(n)}$-invariant manifold S_n transverse to $\rho^{(n)}$ at $p \in P$ and by the Proper Mapping Theorem $V = \pi^{(n)}(S_n)$ is an analytic variety through $0 \in \mathbb{C}^2$, invariant by Z. In order to find such a point $p \in P$ the properties of the index given above are used for an appropriate ordering of the desingularization.

Along this line the following is still an unsolved question:

PROBLEM. Does any vector field in $\mathbb{C}^3$ with an isolated singularity at $0 \in \mathbb{C}^3$ admit an invariant singular curve or surface through $0 \in \mathbb{C}^3$?

VI. MORE ON TOPOLOGICAL INVARIANTS

Given a curve $f: \mathbb{C}^2,0 \to \mathbb{C},0$, singular at $0 \in \mathbb{C}^2$, we define its <u>algebraic multiplicity</u> as the degree of the first nonzero jet of f, i.e. $\underline{\nu} = \nu(f)$ where

$$f = f_\nu + f_{\nu+1} + \dots$$

is the Taylor development of f and $f_\nu \not\equiv 0$.
A well known result by Burau [40] and Zariski [41] states that ν <u>is a topological invariant</u>; that is, given $g: \mathbb{C}^2,0 \to \mathbb{C},0$ and a homeomorphism $h: U \to V$ between neighborhoods of $0 \in \mathbb{C}^2$ such that $h(f^{-1}(0) \cap U) = g^{-1}(0) \cap V$ then $\nu(g) = \nu(f)$. More generally they prove that the desingularizations of f and g are in fact isomorphic. In higher dimensions (for mappings $f: \mathbb{C}^n,0 \to \mathbb{C},0$, $n \geq 3$) this problem, known as one of Zariski problems, is still open [42].
Consider now a vector field Z in $\mathbb{C}^2$ with a singularity at $0 \in \mathbb{C}^2$. If

$$Z = Z_\nu + Z_{\nu+1} + \dots$$

we define $\nu = \nu(Z)$ as the <u>algebraic multiplicity of</u> Z. A natural question, posed by J.F. Mattei is: is $\nu(Z)$ a topological invariant of $\mathfrak{F}_Z$? We pose the following

<u>Conjecture</u>. Suppose Z and $\tilde{Z}$ induce topologically equivalent foliations $\mathfrak{F}_Z$ and $\mathfrak{F}_{\tilde{Z}}$ near $0 \in \mathbb{C}^2$. Then $\mathfrak{F}_Z$ and $\mathfrak{F}_{\tilde{Z}}$ admit isomorphic desingularizations.
Having isomorphic desingularizations means that the same sequence of blow ups will lead to the desingularizations of Z and $\tilde{Z}$. As a result there is a correspondence (isomorphism) between the final disposition of projective lines and marked points (singularities) in the divisor and corresponding singularities are of the same kind.
That isomorphic desingularizations imply the same algebraic multiplicity can be seen as follows.
Consider a singular foliation $\mathfrak{F}$ with an invariant manifold P passing through a singular point p. In local coordinates (x,y)

around p, $x(p) = y(p) = 0$, we can write

$$\mathcal{F}: \begin{cases} \dot{x} = x^m A(x) + y B(x,y) \\ \dot{y} = y C(x,y) \end{cases}, \quad A(0) \neq 0.$$

where $P = (y=0)$. The number m above is called the <u>multiplicity of $\mathcal{F}$ along P at p</u> and denoted $m(p;P)$. Suppose now that $\mathcal{F} = \mathcal{F}_Z^{(\ell)}$ is the foliation obtained by blowing up $\mathcal{F}_Z$ ℓ times and P a projective line in the divisor $\wp_Z^{(\ell)}$. Assume further that $\wp_Z^{(\ell)}$ is invariant by $\mathcal{F}_Z^{(\ell)}$. If $p \in P$ is a singular point of $\mathcal{F}$ we define $\varphi(p,P) = m(p,P)$ if p is not a corner of $\wp_Z^{(\ell)}$ and $\varphi(p,P) = m(p,P) - 1$ otherwise.

The weight $\rho(P)$ of a projective line P in $\wp_Z^{(\ell)}$ is:

(i) 1 if the projective line appears immediately after blowing up $0 \in C^2$

(ii) the sum of the weights of the projective lines meeting at the singularity which was blow up to originate P.

The following proposition where $\nu = \nu(Z)$, can be proved easily.

PROPOSITION [11]. Suppose $\wp_Z^{(\ell)}$ is invariant by $\mathcal{F}_Z^{(\ell)}$. Then

$$\nu+1 = \sum_P \rho(P) \sum_{p \in P} \varphi(p,P).$$

If, on the other hand, dicritical components appear in the process of blow up, other formulas hold, still relating ν with the process of desingularization. In any case one obtains the invariance of ν as a consequence of the invariance of the desingularization. In what follows we relate other recent results supporting the conjecture.

DEFINITION. A <u>generalized curve</u> is a vector field whose desingularization does not contain complex saddle-nodes, i.e. all singular points in its desingularization are hyperbolic.

THEOREM [11]. If Z is a generalized curve and W is topologically equivalent to Z then W <u>is also a generalized curve</u> and both Z and W have isomorphic desingularizations.

Here we sketch the proof of this theorem in the case that the divisor of the desingularization is invariant. The proof of the general case is similar.

1st step. A separatrix of the vector field Z is an integral curve V such that $\bar{V} = V \cup (0)$. Let S_j $j = 1,\dots,r$ the set of separatices and $f_j\colon \mathbb{C}^2,0 \to \mathbb{C},0$ be analytic mappings such that $S_j = f_j^{-1}(0)$ $j = 1,\dots,r$. Write $f = f_1 \dots f_r$ and define the vector field $Z_f = -\frac{\partial f}{\partial z_2}\frac{\partial}{\partial z_1} + \frac{\partial f}{\partial z_2}\frac{\partial}{\partial z_2}$ whose integrals are the level curves of f. Then the desingularizations of Z_f and Z are the same: We follow the sequence of blow ups necessary to desingularize Z_f. Obtain smooth curves $\tilde{S}_1,\dots,\tilde{S}_r$ that project by blowing down to $S_1,\dots,S_r$. Suppose $\mathcal{P}_{Z_f}^{(k)}$ is the divisor. Then $\tilde{S}_j \cap \mathcal{P}_{Z_f}^{(k)}$ is a singular point p_j of $\mathfrak{F}_{Z_f}^{(k)}$. Let ν_j be the algebraic multiplicity of p_j as singular point of $\mathfrak{F}_Z^{(k)}$. Through p_j pass exactly two transversal smooth separatrices. By the proposition above $\nu_j = 1$. Further elementary arguments imply that in fact p_j is simple.

2nd step. Let Z be a vector field with an isolated singularity at $0 \in \mathbb{C}^n$. Let $\mathcal{O}$ be the ring of germs of holomorphic functions defined in some neighborhood of $0 \in \mathbb{C}^n$ and $I(Z_1,\dots,Z_n)$ the ideal generated by the germs at $0 \in \mathbb{C}^n$ of the coordinate functions of Z. We define the number

$$\mu(Z,0) = \dim_{\mathbb{C}} \mathcal{O}_n/I(Z_1,\dots,Z_n).$$

Then μ is a topological invariant of Z.

3rd step. Let Z be a vector field in $\mathbb{C}^2$ with an isolated singularity at $0 \in \mathbb{C}^2$. Let $f = 0$ be the equation of the separatrices of Z. Then $\mu(Z;0) \geq \mu(Z_f;0)$ and equality follows if and only if Z is a generalized curve.

Proof. By induction on the number of blow ups necessary to desingularize Z_f. For $n = 0$ it is clear. In general we use the formula relating $\mu(Z;0)$ with the numbers μ of the singularities appearing in the first blow up:

$$\mu(Z_f,0) = \nu_{Z_f}^2 - (\nu_{Z_f}+1) + \sum_{j=1}^{r} \mu(\tilde{\mathfrak{F}}_{Z_f}^{(1)},p_j).$$

If Z is a generalized curve the points $p_1,\dots,p_r$ are all the singularities of $\mathfrak{F}_Z^{(1)}$. By induction hypothesis and the proposition obtain $\mu(Z,0) = \mu(Z_f;0)$.
If Z is not a generalized curve, either $p_1,\dots,p_r$ are all sin-

gular points of $\mathfrak{F}_Z^{(1)}$ so at least one of them is not a generalized curve or $\mathfrak{F}_Z^{(1)}$ has additional singularities $p_{r+1},\dots,p_k$. In any case using the induction hypothesis and the formula above, get $\mu(Z;0) > \mu(Z_f;0)$.

3nd step end of proof. If Z and W are topologically equivalent $\mu(Z;0) = \mu(W;0)$. Since $\mu(Z,0) = \mu(Z_f;0)$ and $\mu(W_f;0) = \mu(W_f;0)$ then $\mu(W,0) = \mu(W_f;0)$ which means that W is a generalized curve. Moreover, the desingularization of W is that of W_f which is isomorphic to Z_f, thus W and Z possess isomorphic desingularizations.

PROBLEM (P. Sad). Let Z be a generalized curve and $(\mu^{(n)},\pi^{(n)},\rho_Z^{(n)},\mathfrak{F}_Z^{(n)})$ its desingularization. Is the holonomy of $\rho_Z^{(n)}$ relative to the foliation $\mathfrak{F}_Z^{(n)}$, a topological invariant of Z ?

REFERENCES

[1] Arnold V.I., Remarks on singularities of finite codimension in complex dynamical systems. Funct. Anal. and Appl. 3 (1969), 1-5.

[2] Arnold V.I., Bifurcations of invariant manifolds of differential equations and normal forms in neighborhoods of elliptic curves. Funct. Anal. and Appl. 10 (1976), 249-259.

[3] Arnold V.I., Chapitres supplementaires de la theorie des équations differentielles ordinaires. Moscou MIR (1978).

[4] Bendixson I., Sur les points singulieres d'une équation differentielle linéaire. Ofv. Kongl. Venteskaps Akademiens Forhandlinger 148 (1895), 81-89.

[5] Briot C.A., Bouquet J.C., Recherches sur les fonctions définies par des équations différentielles, Journal de l'École Polytechnique, XXI (1856) 134-198.

[6] Brjuno A.D., Analytical forms of differential equations. Trans. Moscow Math. Soc. 25 (1971), 83-94.

[7] Camacho C., On the local structure of conformal mappings and holomorphic vector fields in $\mathbb{C}^2$. Asterisque 59-60 (1978), 83-94.

[8] Camacho C., Sad P., Topological classification and bifurcations of holomorphic flows with resonances in $\mathbb{C}^2$. Inventiones Mathematicae 67 (1982), 447-472.

[9] Camacho C., Sad P., Invariant varieties through singularities of holomorphic vector fields. Annals of Mathematics 115 (1982), 579-595.

[10] Camacho C., Kuiper N., Palis J., The topology of holomorphic flows with singularity. Publ. Math. IHES 48 (1978), 5-38.

[11] Camacho C., Lins Neto A., Sad P., Topological invariance and equidesingularization for holomorphic vector fields. Preprint.

[12] Cerveau D., Mattei J.F., Formes intégrables holomorphes singulières. Asterisque 97 (1982).

[13] Chaperon M., Propriétés génériques des germes d'actions differentiables de groupes de Lie commutatifs élémentaires, Thèse, Université de Paris 7 (1980).

[14] Chaperon M., Differential geometry and dynamical systems: two examples. This volume.

[15] Dulac H., Recherches sur les points singuliers des équations différentiables. Journal de l'École Polythecnique 2 Sec 9 (1904), 1-125.

[16] Dulac H., Solutions d'un système d'équations différentielles dans le voisinage des valeurs singulières. Bull. Soc. Math. France 40 (1912), 324-383.

[17] Ecalle J., Theorie iterative. Introduction a la theorie des invariantes holomorphes. J. Math. Pures et Appl., 54 (1975), 183-258.

[18] Ecalle J., Les fonctions résurgentes et leurs applications, Tomes I, II, III, to appear.

[19] Guckenheimer J., Hartman's theorem for complex flows in the Poincaré domain. Compositio Mathematicae 24 (1972), 75-82.

[20] Hukuara H., Kimura T., Matuda T., Équations differentielles ordinaires du premier ordre dans le champ complexe. Publ. Math. Soc. of Japan 1961.

[21] Il'iašenko J., Remarks on the topology of singular points of analytic differential equations in the complex domain and Ladis' theorem. Funct. Anal. and Appl. 11 (1977), 105-113.

[22] Il'iašenko J., Global and local aspects of the theory of complex differential equations. Proceed. of the Intern. Cong. of Math. Helsinki, 1978.

[23] Il'iašenko J., Pjartly A., The materialization of Poincaré resonances and divergence of normalizing series. Trudy Seminar Petrovsky 7 (1981), 3-49 (in Russian).

[24] Il'iašenko J., Pjartly A., The materialization of resonances and the divergence of normalizing series for polynomial differential equations. Trudy Seminar Petrovsky 8 (1982), 111-127 (in Russian).

[25] Elizarov P.M., Il'iašenho J., Remarks on the orbital analytic classification of germs of vector fields. Mat. Sb. 121 (163).

[26] Lefschetz S., On a theorem of Bendixson. Journ. of Diff. equat. 4 (1968), 66-101.

[27] Malgrange B., Travaux d'Ecalle et de Martinet-Ramis sur les systèmes dynamiques. Seminaire Bourbaki, Exposé 582 (Nov. 1981) Asterisque 92-93.

[28] Martinet J., Ramis J.P., Problèmes de modules pour des équations différentielles non linéaires du premier ordre. Publ. Math. IHES 55 (1982), 63-164.

[29] Martinet J., Ramis J.P., Classification analytique des équations différentielles non linéaires résonantes du premier ordre. Public. de IRMA Strasbourg (1983).

[30] Mattei J.F., Moussu R., Holonomie et intégrales premières. Ann. Sc. Ec. Norm. Sup. 13 (1980), 469-523.

[31] Naishul V.A., Topological equivalence of differential equations in $\mathbb{C}^2$ and $\mathbb{CP}^2$. Moscow Univ. Math. Bull. 36 nº 4 (1981). Also: Topological invariants of germs of analytic mappings and mapping preserving area. Funct. Analys. and Applic. vol. 14, nº 1 (1980), 73-74.

[32] Pjartly A., Birth of complex invariant manifolds close to a singular point of parametrically dependent vector field. Funct. Anal. and Appl. 6 (1972), 339-340.

[33] Poincaré H., Sur les propriétés des fonctions définies par les équations aux différences partielles. Thèse, Paris (1879).

[34] Seidenberg A., Reduction of singularities of the differential equation Ady = Bdx. Amer. Journ. of Math. (1968), 248-269.

[35] Sad P., A note on nonlinearizable analytic functions. Bol. Soc. Bras. de Mat. 11 (1980), 31-36.

[36] Siegel C.L., Iteration of analytic functions. Ann. of Math. 43 (1942), 607-616.

[37] Siegel C.L., Über die Normal form analytischer Differentialgleichungen in der Nähe einer Gleichgewichtslösung. Nachr. Akad. Wiss. Göttingen Math. Phys. (1952), 21-30.

[38] Ven den Essen A., Reduction of singularities of the differential equation Adx + Bdy = 0. Springer-Verlag Lect. Not. in Math. 670, 53-58.

[39] Voronin S.M., Analytic classification of germs of conformal mappings $(C,0) \to (C,0)$ with identity linear part. Funct. Anal. and its Applic. 15 (1) 1981.

[40] Burau W., Kennzeichnung der schlauchknoten. Abh. Math. Sem. Ham. Univ. 9 (1932), 125-133.

[41] Zariski O., On the topology of algebroid singularities. Amer. Journ. of Math., 54 (1932), 453-465.

[42] Lê D.T. and Teissier B., Report on the problem session. Proceed. of Symposia in Pure Mathematics, AMS 40 (1982).

César Camacho

Instituto de Matemática Pura e Aplicada (IMPA)

Estrada Dona Castorina 110, Jardim Botânico

CEP 22.460 - Rio de Janeiro, Brasil.

Singularities & Dynamical Systems
S.N. Pnevmatikos (editor)

HOLONOMIE EVENESCENTE DES EQUATIONS DIFFERENTIELLES DEGENEREES TRANSVERSES

Robert Moussu

Université de Dijon

France

RESUME : Un germe en $0 \in \mathbb{C}^2$ d'équation différentielle holomorphe $\omega = 0$ tel que $j^1\omega = y\,dy$ et $j^2\omega$ est "générique" possède une unique séparatrice X_ω d'équation $y^2 + x^3 = 0$. Deux telles équations sont holomorphiquement conjuguées si et seulement si elles ont la même holonomie évanescente. Cet invariant est plus précis que l'holonomie de la séparatrice : il existe des équations $\omega = 0$ dont toutes les variétés intégrales adhèrent à $0 \in \mathbb{C}^2$ et telles que X_ω à une holonomie nulle.

1. INTRODUCTION et RESULTATS.

Soit $\omega = a\,dx + b\,dy = 0$ une équation différentielle holomorphe sur une petite boule $\varepsilon B = \{(x,y) \in \mathbb{C}^2 / x\bar{x} + y\bar{y} = \varepsilon\}$dont 0 est le seul point singulier. Elle définit un feuilletage holomorphe $\mathcal{F}_\omega$ de $\varepsilon B - \{0\}$ sans singularité. C. Camacho et P. Sad ont montré dans [2] que $\mathcal{F}_\omega$ possède au moins *une séparatrice*, c'est à dire une feuille X^0 dont l'adhérence X dans εB est un ensemble analytique de codimension 1 qui contient 0, $X^0 = X - \{0\}$.

Soit f une équation réduite de X sur un voisinage de 0. On peut supposer, ε étant pris assez petit,que εB est une boule de Milnor pour f [11]. Pour comprendre la géométrie de $\mathcal{F}_\omega$ au voisinage de X^0 il est naturel de comparer $\mathcal{F}_\omega$ à la fibration de Milnor de f. Un outil classique pour cette étude est *l'holonomie de* X^0 : au générateur du groupe fondamental de X^0 (homéomorphe à $S^1 \times \mathbb{R}$ d'après [11]) correspond un difféomorphisme d'holonomie de la feuille X^0 , $h_{\omega,X}$, qui est un élément de $\mathrm{Diff}(\mathbb{C},0)$, $\mathrm{Diff}(\mathbb{C}^n,0)$ désignant le *groupe des germes de difféomorphismes* holomorphes qui fixent $0 \in \mathbb{C}^n$. Plus précisément, c'est seulement la *classe de conjugaison* de $h_{\omega,X}$, $[h_{\omega,X}]$, ie sa classe d'équivalence modulo les automorphismes intérieurs de $\mathrm{Diff}(\mathbb{C},0)$ qui est bien définie ; nous l'appelons holonomie de la séparatrice X.

Il est clair que si deux équations $\omega = 0$, $\omega' = 0$ sont *holomorphiquement conjuguées* , i.e

$$\omega \wedge \phi^*(\omega') = 0 \qquad \text{avec} \qquad \phi \in \mathrm{Diff}(\mathbb{C}^2,0) \quad ,$$

leurs séparatrices ont les mêmes holonomies. Réciproquement, si $\omega = 0$ *n'est pas dégénérée*, i.e la matrice représentant le 1-jet du champ $X_\omega = b\frac{\partial}{\partial x} - a\frac{\partial}{\partial y}$ possède au moins une valeur propre non nulle, $\omega = 0$ à une ou deux séparatrices (Poincaré [14], Dulac [7]) dont les holonomies caractérisent "en général" la classe de conjugaison de l'équation $\omega = 0$; c'est toujours le cas si les deux valeurs propres de J^1X_ω sont non nulles (voir par exemple [9],[10],[13]). On pourrait espérer qu'un tel résultat subsiste si $\omega = 0$ n'est pas très dégénérée. Mais nous allons voir que ce n'est pas le cas.

Si l'équation $\omega = 0$ est dégénérée mais son 1-jet non nul, elle possède une forme normale formelle ([16],[12]) du type :

$$d(y^2 + x^n) + F(x)dy = 0 \qquad , \qquad F \in \mathbb{C}[[X]]$$

DEFINITION 1 : $\omega = 0$ *est dite dégénérée - transverse si* $n = 3$, *i.e. il existe des des coordonnées* $(x,y) \in \mathbb{C}^2$ *telles que*

$$j^2\omega = d(y^2 + x^3) + \lambda x^2\,dy \quad , \quad \lambda \in \mathbb{C} .$$

La condition $n = 3$ peut-être interprétée en termes de transversalité dans l'espace des 2-jets d'équation $\omega = 0$, voir par exemple [16].

PROPOSITION 1 : Si $\omega = 0$ *est dégénérée - transverse elle possède une unique séparatrice* X *d'équation* $y^2 + x^3 = 0$ *dans de bonnes coordonnées* (x,y) *et l'holonomie* $[h_{\omega,X}]$ *de* X *est tangente à* $1_{\mathbb{C}}$, *i.e* $h'_{\omega,X}(0) = 1$.

<u>Exemples et conjectures de R. Thom</u> : Rappelons tout d'abord les trois conjectures que R. Thom avait proposé lors de ses séminaires à l'IHES dans les années 74-75. Notons encore $\mathcal{F}_\omega$ l'ensemble des feuilles de feuilletage défini par une équation $\omega = 0$ et soit

$$A = \{L \in \mathcal{F}_\omega \,/\, \bar{L} \ni 0\} \;,\quad B = \{L \in \mathcal{F}_\omega \,/\, L \text{ est analytique}\}$$

1ère conjecture : $B = \mathcal{F}_\omega$ et $A \cap B$ fini $\overset{?}{\Longrightarrow}$ ω possède une intégrale première holomorphe.

2ème conjecture : $A = B = \mathcal{F}_\omega$. $\overset{?}{\Longrightarrow}$ ω possède une intégrale première méromorphe.

3ème conjecture : $A = \mathcal{F}_\omega$ et $A \cap B$ est fini $\overset{?}{\Longrightarrow}$ Les séparatrices portent de l'holonomie.

La première conjecture est vraie [13] , la deuxième est fausse [5] ainsi que la troisième , comme le montrera l'exemple suivant. L'équation

$$\omega_1 = d(y^2 + x^3) + x(2y\,dx - 3x\,dy) = 0$$

est dégénérée-transverse et $X = \{y^2 + x^3 = 0\}$ est son unique séparatrice. Nous verrons dans V que chaque feuille de $\mathcal{F}_{\omega_1}$ contient $0 \in \mathbb{C}^2$ dans son adhérence et que l'holonomie $h_{\omega_1,X} = 1_{\mathbb{C}}$.

L'équation $\omega_0 = d(y^2 + x^3)$ possède, comme ω_1, une unique séparatrice $X = \{y^2 + x^3 = 0\}$ dont l'holonomie est $1_{\mathbb{C}}$ et elle n'est pas conjuguée à ω_0. Ainsi l'holonomie de la séparatrice d'une équation dégénérée transverse n'est pas un invariant qui caractérise cette équation.

Nous montrerons dans le paragraphe III qu'à une forme dégénérée-transverse, ω, est associée une représentation

$$H_\omega : \mathbb{Z}/3\mathbb{Z} * \mathbb{Z}/2\mathbb{Z} \to \mathrm{Diff}(\mathbb{C},0)$$

qui est bien définie à un automorphisme intérieur près de $\mathrm{Diff}(\mathbb{C},0)$. Par définition la classe d'équivalence $[H_\omega]$ de H_ω pour cette relation est appelée *holonomie évanescente* de ω. Le qualificatif "évanescent" est justifié dans IV : on peut considérer que l'holonomie évanescente est portée par les cycles évanescents de $y^2 + x^3 = C^{ste}$, $y^2 + x^3 = 0$ étant l'équation de la séparatrice de ω. Remarquons que $[H_\omega]$ est déterminé par le couple (f,g) avec

$$f = H_\omega(1 * 0) \quad , \quad g = H_\omega(0 * 1).$$

Puisque $f^3 = g^2 = 1_{\mathbb{C}}$ on peut prendre pour représentant de $[H_\omega]$ un couple $(\varphi^{-1} \circ J \circ \varphi, S)$ avec

$$S(x) = -x \quad , \quad J(x) = jx$$

où $j = \exp(2i\pi/3)$ et $\varphi \in \mathrm{Diff}(\mathbb{C},0)$. Un petit calcul montre qu'il existe un unique φ du type $\varphi(x) = x + P(x^2)$ tel que

$$[H(\omega)] = [(f,g)] = [\varphi^{-1} \circ J \circ \varphi, S]$$

THEOREME 1 : *Deux équations dégénérées-transverses $\omega = 0$, $\omega' = 0$ sont conjuguées holomorphiquement si et seulement si $[H_\omega] = [H_{\omega'}]$. D'autre part $[h_{\omega,X}]$ l'holonomie de la séparatrice X de ω est déterminée par $f \circ g$ si $H_\omega = (f,g)$; en particulier $h_{\omega,X} = 1_{\mathbb{C}}$ si et seulement si $(f \circ g)^6 = 1_{\mathbb{C}}$*

Par exemple, pour les équations $\omega_0 = 0$, $\omega_1 = 0$ on a

$$H_{\omega_0} = (J,S) \quad , \quad H_{\omega_1} = (\varphi^{-1} \circ J \circ \varphi, S) \quad , \quad h_{\omega_0,X} = h_{\omega_1,X} = 1_{\mathbb{C}}$$

avec $\varphi(x) = x/1+ax$, $a \in \mathbb{C} - \{0\}$.

THEOREME 2 : Soient $f,g \in \mathrm{Diff}(\mathbb{C},0)$ *tel que* $f^3 = g^2 = 1_{\mathbb{C}}$. *Il existe* ω *dégénérée-transverse telle que* $[H_\omega] = [(f,g)]$.

Ce théorème de synthèse est un cas particulier d'un résultat plus général que j'ai montré avec D. Cerveau. Il sera l'objet d'une publication ultérieure. Ces deux théorèmes posent plus de problèmes qu'il n'en résolvent, par exemple :
1. Existe-t-il des équations dégénérées transverses ω telles que $h_{\omega,X} = 1_{\mathbb{C}}$ qui ne sont pas holomorphiquement conjuguées à ω_0 et ω_1 ? Cette question est équivalente (d'après le théorème 2) à la suivante : existe-t-il $f,g \in \mathrm{Diff}(\mathbb{C},0)$ tels que

$$f^3 = g^2 = (f_\circ g)^6 = 1_{\mathbb{C}} \quad \text{et} \quad [(f,g)] \neq [H_{\omega_0}], [H_{\omega_1}] \quad ? \quad (*)$$

2. D. Cerveau a montré qu'une 1-forme ω dégénérée transverse possède une forme "normale" du type :

$$\omega = a\, d(y^2 + x^3) + (y^2 + x^3)\eta + \lambda x(2x\, dy - 3y\, dx)$$

où a est une fonction holomorphe telle que $a(0) \neq 0$, η est une 1-forme holomorphe et $\lambda \in \mathbb{C}$.
Quelle relation existe-t-il entre cette forme normale et $[H_\omega]$? Comment intervient à travers η et $x(2x\, dy - 3y\, dx)$ la connexion de Gauss-Manin de $y^2 + x^3$ dans le calcul de $h_{\omega,X}$, de H_ω ?

Les techniques de J. Ecalle sur "les fonctions résurgentes permettent peut être de résoudre la première question ; quant à la deuxième elle peut-être abordée en utilisant des arguments de B. Malgrange [8] .

II. DESINGULARISATION ([1],[13],[15]) DES FORMES DEGENEREES-TRANSVERSES.

La désingularisation d'une équation $\omega = 0$ est la même que celle de $J^N\omega = 0$ pour N grand. En utilisant la forme normale de Takens,[16],[12], nous pouvons donc supposer que :

$$\omega = d(y^2 + x^3) + x^2 P(x)\, dy \qquad , P \in \mathbb{R}[x]$$

Nous allons constater que ω est désingularisée (ou réduite au sens de [13]) par les trois éclatements suivants (Figure 1)

<u>1er éclatement</u> : $\pi_1 : M_1 \to \mathbb{C}^2$ est l'éclatement de $0 \in \mathbb{C}^2$. Les points de M_1 sont repérés par les deux cartes :

$(t,x),(t',y)$ avec $tt' = 1$, $y = tx$;

$\pi_1(t,x) = (x,tx)$, $\pi_1(t',y) = (t'y,y)$

On note $P_1 = \pi_1^{-1}(0)$ et $A = (t = 0, x = 0)$.

(*) Récemment J.Ecalle et Y.C.Yoccoz ont montré l'existence d'une infinité de [(f,g)] [(f,g)] distincts.

2ème éclatement : $\pi_2 : M_2 \to M_1$ est l'éclatement de $A \in M_1$. Les points de M_2 sont repérés par (t',y) et les deux cartes :

$$(u,t),(u',x) \quad \text{avec} \quad uu' = 1 \quad , \; x = tu \; ;$$

$$\pi_2(u,t) = (t,tu) \qquad \pi_2(u',x) = (x,u'x).$$

On note $P_2 = \pi_2^{-1}(A)$ et $B = P_1 \cap P_2 = (u = 0,\; t = 0)$.

3ème éclatement : $\pi_3 : M_3 \to M_2$ est l'éclatement de $B \in M_2$. Les points de M_3 sont repérés par (t',y), (u',x) et les deux cartes

$$(v,t),\; (v',u) \quad \text{avec} \quad vv' = 1 \quad , \; u = tv$$

$$\pi_3(v,t) = (vt,t) \quad , \; \pi_3(v',u) = (u,v'u).$$

On note $P_3 = \pi_3^{-1}(B)$ et $C = P_1 \cap P_3$, $C' = P_3 \cap P_2$

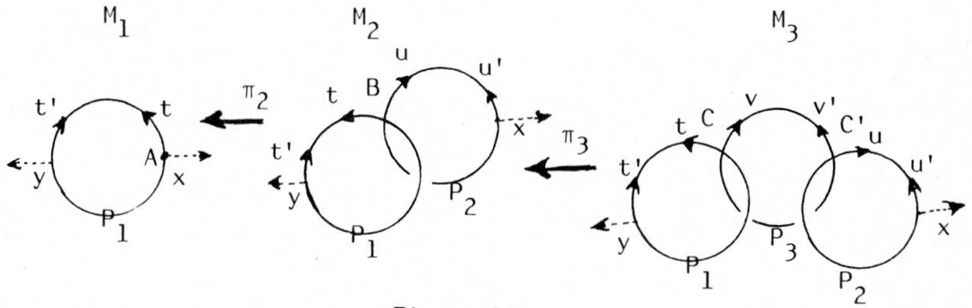

Figure 1.

En résumé les points de M_3 sont repérés par les quatre cartes (t',y), (v,t),(v',u) (u',x) avec $(tt' = 1,\; y = t^3v)$, $(vv' = 1,\; u = vt)$, $(uu' = 1,\; x = uv')$.

Dans ces cartes l'application $\pi = \pi_3 \circ \pi_2 \circ \pi_1$ s'écrit :

$$\pi(t',y) = (t'y,y) \quad , \quad \pi(v,t) = (t^2v,\; t^3v)$$

$$\pi(v',u) = (u^2\, v',u^3v'^2) \quad , \quad \pi(u',x) = (x,u'x).$$

Dans les cartes (t',y), (u',x) les éclatés "divisés" [13] de ω n'ont pas de points singuliers ; en effet

(1) $\quad \pi^*(\omega)/y = U(y,t')dy + y\, dt'\; (\ldots) \qquad$ avec $U(0,t') \neq 0$

(2) $\quad \pi^*(\omega)/x = V(x,u')dx + y\, dx'\; (\ldots) \qquad$ avec $V(0,u') \neq 0$

et ainsi $P_1 - C$ et $P_2 - C'$ sont des feuilles du feuilletage $\tilde{\mathcal{F}}_\omega$ image réciproque de $\mathcal{F}_\omega$ par π. Dans la carte (v,t) l'éclaté divisé de $\tilde{\omega}$

(3) $\quad \tilde{\omega} = \pi^*(\omega)/t^5v = 6v(1+v + t\, A(t,v)\; dt + t(2 + 3v + tB(v,t))dv$

a pour seuls points singuliers $C = (0,0)$ et $C'' = (-1,0)$ et ces singularités sont réduites [13] . Enfin pour étudier le dernier point singulier, C', plaçons-nous dans la carte (v',u) :

(4) $\quad \tilde{\omega}' = \pi^*(\omega)/u^5v' = 2v'(1 + A'(v',u))du + u(1 + B'(v',u))dv'$

avec $A'(0,0) = B'(0,0) = 0$; cette singularité est aussi réduite.

Démonstration de la proposition 1 : L'image réciproque $\widetilde{X}$ par π des, ou plutôt de la séparatrice X de ω, coupe le diviseur exceptionnel en l'unique point singulier de l'éclaté divisé de ω distinct des "coins" C, C', c'est à dire en C''. Dans la carte (v,t), d'après (3) (voir [7],[14]) $\widetilde{X}$ a une équation du type :

$$v + 1 + t^2 f(t) = 0$$

Par un calcul élémentaire on en déduit que l'unique séparatrice X de ω a une équation du type :

$$y^2 + x^3 + \sum_{i+j \geq 3} \alpha_{i,j} x^i y^j = 0 \qquad \text{avec} \quad \alpha_{3,0} = 0$$

et que $h'_{\omega,X}(0) = 1$ d'après 3 (voir [13] -[11]).

III. HOLONOMIE EVANESCENTE :

La construction du morphisme H_ω repose sur le lemme suivant :

LEMME 1 (*existence d'intégrales premières holomorphes dans les coins*). *Les feuilles de* $\widetilde{\mathcal{F}}_\omega = \pi^{-1}(\mathcal{F}_\omega)$ *coupent des voisinages assez petits de* $P_1 - C$, $P_2 - C'$, C, C' *respectivement suivant les courbes de niveau des fonctions holomorphes* F_1, F_2, F, F' *qui s'écrivent dans les cartes correspondantes*

$$F_1 = y\, U_1 \;,\; F_2 = x\, U_2 \;,\; F = t^3 v\, U \;,\; F' = U^2 v'\, U'$$

$$U_1(t',0) = U_2(u',0) = U(0,0) = U'(0,0) = 1.$$

Démonstration : D'après 1.II, 2.II, $P_1 - C$, $P_2 - C'$ sont des feuilles lisses simplement connexes de $\widetilde{\mathcal{F}}_\omega$ d'équations respectives $y = 0$ dans (t',y), $x = 0$ dans (u',x). Il existe donc F_1, F_2 comme dans l'énoncé du lemme.

L'équation $\widetilde{\omega}$, 3. II, qui définit $\widetilde{\mathcal{F}}_\omega$ au voisinage de C a pour 1-jet

$$J^1 \omega = 2(3v\, dt + t\, dv).$$

C'est une équation résonnante dans le domaine de Siegel. Elle possède une intégrale première holomorphe $F = t^3 vU$ avec $U(0) \neq 0$ puisque sa séparatrice P_1 a une holonomie triviale (d'après [7] ,[13],[10]) . Par un argument analogue on montre l'existence de F'.

Description de $\widetilde{\mathcal{F}}_\omega$ au voisinage de P_3. Plaçons-nous dans la carte (v,t) et notons D, D', D'' des petits disques dans $P_3 = \{t = 0\}$ centrés en C, C', C'' et soient

$$Q_3 = P_3 - D \cup D' \cup D'' \;,\; U_3 = \{(v,t) \;/\; v \in Q_3 \;,\; |t| < \varepsilon\}$$

Pour $\varepsilon > 0$ assez petit, $\widetilde{\mathcal{F}}_\omega / U_3$ qui a pour équation 3.II :

$$\widetilde{\omega} = 6v(1 + v + tA)dt + t(2 + 3v + tB)dv = 0$$

est transverse aux fibres $v \times \{|t| > \varepsilon\}$ de U_3. Le germe de ce feuilletage le long de Q_3 est caractérisé par l'holonomie de sa feuille Q_3, c'est cette holonomie que nous appelons *holonomie evanescente* de ω. Pour bien la définir, choisissons les générateurs $\alpha,\beta \in \pi_1(Q_3,C''_o)$, $C''_o \in \partial D''$ de la façon suivante : α (resp. β) est le lacet obtenu en parcourant le segment C''_oC_o (resp. $C''_oC'_o$) où $C_o \in \partial D$ (resp. $C'_o \in \partial D'$), puis le cercle ∂D dans le sens indirect (resp. le cercle $\partial D'$ dans le sens direct) et enfin le segment C_oC'' (resp. le segment $C_oC''_o$). Par construction $\gamma = \beta\cdot\alpha$ est homotope au bord $\partial D''$ et D'' parcouru dans le sens indirect.

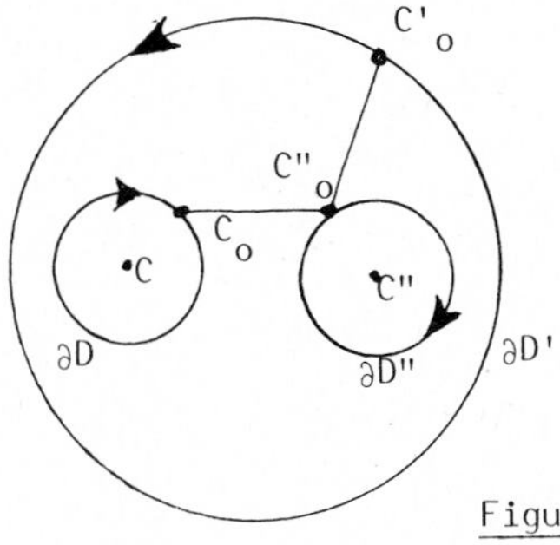

$$\alpha = \overrightarrow{C''_oC_o}\cdot\widehat{\partial D_o}^{-} \cdot \overrightarrow{C_oC''_o}$$
$$\beta = \overrightarrow{C''_oC'_o}\cdot\widehat{\partial D'_o}^{+}\cdot\overrightarrow{C'_oC''_o}$$

Figure 2.

Désignons par f,g les difféomorphismes d'holonomie correspondants à α,β. Puisque le long de ∂D et $\partial D'$, $\tilde{\omega}$ possède les intégrales premières (lemme 1) t^3vU et $u^2v'U'$ on a

$$f^3 = g^2 = 1_{\mathbb{C}}.$$

Ainsi on peut factoriser l'homomorphisme d'holonomie

$$h_\omega : \pi_1(Q_3,C''_o) = \mathbb{Z} * \mathbb{Z} \longrightarrow \mathrm{Diff}(\mathbb{C},0)$$

en $h_\omega = H_\omega \circ q$ où q est la projection canonique de $\mathbb{Z}*\mathbb{Z}$ sur $\mathbb{Z}/3\mathbb{Z} * \mathbb{Z}/2'\mathbb{Z}$ et

$$H_\omega : \mathbb{Z}/3\mathbb{Z} * \mathbb{Z}/2\mathbb{Z} \longrightarrow \mathrm{Diff}(\mathbb{C},0)$$
$$H_\omega = (1 * 0) = f \quad , \quad H_\omega(0 * 1) = g.$$

DEFINITION 2 : La classe d'équivalence de H_ω *,*$[H_\omega]$ *, modulo les automorphismes intérieurs de* Diff$(\mathbb{C},0)$ *est appelée* <u>*holonomie évanescente*</u> *de* ω *ou de* $\mathcal{F}_\omega$.

<u>Démonstration du théorème 1</u> : Il est clair que si ω,ω' sont holomorphiquement conjuguées par $\phi \in \mathrm{Diff}(\mathbb{C}^2,0)$, alors ϕ se relève en un germe de difféomorphisme de M^3 le long du diviseur exceptionnel $P_1 \cap P_2 \cap P_3$ qui applique $\widetilde{\mathcal{F}}_\omega$ sur $\widetilde{\mathcal{F}}_{\omega'}$. Les holonomies de Q_3 considérée comme feuille de $\widetilde{\mathcal{F}}_\omega$ et $\widetilde{\mathcal{F}}_{\omega'}$ sont donc conjuguées, i.e $[H_\omega] = [H_{\omega'}]$.

Avant de montrer la réciproque, montrons l'assertion finale du théorème : f o g détermine l'holonomie $H_{\omega,X}$ de la séparatrice X si $H_\omega = (f,g)$. D'après [13], ou plus précisément un théorème de J. Martinet, J.P. Ramis [10] montre que le difféomorphisme d'holonomie $h = g \circ f = H_\omega(\gamma)$ détermine complètement la classe de conjugaison du germe de $\tilde{\omega} = 0$ en C" puisque $\tilde{\omega} = 0$ est une équation résonnante de 1-jet en $C'' = (-1,0)$:

$$J^1_{C''}\omega = 6(v+1)dt + t\,dv.$$

En particulier $h = g \circ f$ détermine l'holonomie de sa deuxième variété invariante qui n'est rien d'autre que l'holonomie $h_{\omega,X}$.

Supposons que deux équations $\omega = 0$, $\omega' = 0$, dégénérées-transverses aient la même holonomie évanescente ou plus précisément que

$$h_{\omega} = h_{\omega'} : \pi_1(Q_3, C''_o) \to \mathrm{Diff}(\mathbb{C},0) \quad , \quad Q_3 = P_3 - D \cap D' \cap D''.$$

Nous allons construire par la méthode classique du relèvement des chemins dans $P_3 - \{C,C',C''\}$ un difféomorphisme holomorphe fibré $\tilde{\phi}$:

$$\tilde{\phi} : (v,t) \to (v, \varphi(v,t))$$

de $\tilde{U} - k$ sur $\tilde{U}' - k$ qui appliquent $\tilde{\mathcal{F}}_{\omega}$ sur $\tilde{\mathcal{F}}_{\omega'}$, où

$$k = P_1 \cup P_2 \cup \{t' = 0\} \cup \{u' = 0\}$$

et $U = \pi(\tilde{U})$, $U' = \pi(\tilde{U}')$ sont des voisinages de $0 \in \mathbb{C}^2$. Nous utiliserons des chemins C_v d'origine v d'extrémité C''_o tels que si v appartient à l'un des disques D,D',D", la restriction de C_v à ce disque soit portée par un rayon.

<u>1ère étape</u> : Construction de $\tilde{\phi}$ au-dessus de Q_3. Si $|t| < \eta$ assez petit, le chemin C_v^{-1} se relève (pour $(v,t) \to v$) dans la feuille de $\tilde{\mathcal{F}}_{\omega}$ (resp. de $\tilde{\mathcal{F}}_{\omega'}$) passant par (C''_o,t) en un chemin d'origine ce point et d'extrémité (v,t_1) (resp. (v,t'_1)). Par construction l'application

$$\tilde{\phi} : (v,t_1) \to (v,t'_1) = \varphi(v,t))$$

est un difféomorphisme holomorphe d'un voisinage de Q_3 qui applique $\tilde{\mathcal{F}}_{\omega}$ sur $\tilde{\mathcal{F}}_{\omega'}$.

<u>2ème étape</u> : Prolongement de $\tilde{\phi}$ au-dessus de D". D'après le théorème de [10] cité plus haut ou plus précisément sa démonstration, si $v \in D'' - C''$, on peut encore définir φ de la même façon au voisinage du point C". Par construction (voir [10] ou[13]) φ est bornée au voisinage de $v+1 = 0$; elle s'étend de façon holomorphe sur un voisinage de C" dans cette droite.

<u>3ème étape</u> : Prolongement de $\tilde{\phi}$ au-dessus de P_1. Le même argument (que dans la deuxième étape) permet de prolonger $\tilde{\phi}$ à un voisinage V de C. Plus précisément les difféomorphismes d'holonomie pour $\tilde{\mathcal{F}}_{\omega}$, $\tilde{\mathcal{F}}_{\omega'}$ correspondants à ∂D étant conjugués à la rotation d'angle $2\pi/3$, il existe un voisinage U_1 de P_1, contenant D, tel que les feuilles $\tilde{\mathcal{F}}_{\omega}/\tilde{U}_1$, $\tilde{\mathcal{F}}_{\omega'}/\tilde{U}_1$ soient les surfaces de niveau des fonctions holomorphes $F,F' : U_1 \to \mathbb{C}$ qui s'écrivent dans la carte (t,v)

$$F(v,t) = t^3 v(1 + O(v,t))$$
$$F'(v,t) = t^3 v(1 + O'(v,t))$$

où $\mathcal{O},\mathcal{O}'$ s'annulent sur P_1 , i.e $\mathcal{O}(0,t) = \mathcal{O}'(0,t) = 0$.

En utilisant la forme normale de D. Cerveau (voir l'introduction), il est possible de choisir la carte (t,v) , i.e les coordonnées $(x,y) \in \mathbb{C}^2$, telle que les droites v = constante soient transverses aux feuilletages $\widetilde{\mathcal{F}}_{\omega}$, $\widetilde{\mathcal{F}}_{\omega'}$ sur $\widetilde{U}_1$. Alors pour $(t,v) \in \widetilde{U}_1$, le chemin radial

$$\rho_v : \lambda \to v + (1 - \lambda)v_1 \qquad , v_1 = v/|v|$$

se relève dans une feuille de $\widetilde{\mathcal{F}}_{\omega}$ en un chemin d'origine ce point et d'extrémité $(v_1,t_1) \in V$. Notons

$$(v_1,t'_1) = \widetilde{\phi}(v_1,t_1) = (v_1, \varphi(v_1,t_1)).$$

Le chemin ρ_v^{-1} se relève dans la feuille de $\widetilde{\mathcal{F}}_{\omega'}$ en un chemin d'origine (v_1,t'_1) et d'extrémité (v,t'). On définit alors $\widetilde{\phi}$ sur $\widetilde{U}_1 - \{t' = 0\} \cup P_1$ par

$$\widetilde{\phi}(v,t) = (v,t') = (v, \varphi(v,t')).$$

En utilisant les intégrales premières F,F' qui, dans la carte (t',y), s'écrivent

$$F(y,t') = y(1 + \mathcal{O}(t'^3v , \frac{1}{t'})$$

$$F'(y,t') = y(1 + \mathcal{O}'(t'^3v, \frac{1}{t'}) ,$$

on montre que φ est borné au voisinage de $P_1 \cup \{t' = 0\}$. On l'étend à nouveau par continuité.

Par le même argument, $\widetilde{\phi}$ se prolonge à un voisinage de P_2 ; finalement nous obtenons un difféomorphisme holomorphe d'un voisinage $\widetilde{U}$ de $P_1 \cup P_2 \cup P_3$ qui applique $\widetilde{\mathcal{F}}_{\omega}$ sur $\widetilde{\mathcal{F}}_{\omega'}$. Soit U l'image de $\widetilde{U}$ par π ; puisque π est un isomorphisme de $\widetilde{U} - \pi^{-1}(0)$ sur U ,

$$\phi = \pi \circ \widetilde{\phi} \circ \pi^{-1} / U - \{0\}$$

est un difféomorphisme holomorphe qui applique $\mathcal{F}_{\omega}/U$ sur $\mathcal{F}_{\omega'}/(U)$ et qui se prolonge par continuité en 0 en un élément de $\mathrm{Diff}(\mathbb{C}^2,0)$.

IV. HOLONOMIE EVANESCENTE ET CYCLES EVANESCENTS :

Nous allons tout d'abord "rechercher" les cycles évanescents de la fibre de Milnor de $y^2 + x^3 = \eta$ en étudiant l'équation $\omega_o = d(y^2+ x^3) = 0$. Soit $\widetilde{L}_{\eta}$ la feuille de $\widetilde{\mathcal{F}}_{\omega_o}$ image réciproque de $y^2 + x^3 = \eta$ pour η petit. La restriction p_{η} de la projection p de U_3 sur Q_3 qui s'écrit

$$p(v,t) = v$$

à $\widetilde{L}_{\eta}$ est le revêtement à 6 feuillets déterminés par

$$t^6 v^2(1+v) = \eta \ , \ Q_3 = P_3 - D \cup D' \cup D'' \ , \ U_3 = \{(v,t)/v \in Q_3 \ ,|t|< \alpha\}$$

que l'on peut symboliquement représenter par la figure 3.

$$\bar{\partial D}_1 \cup \bar{\partial D}_2 = p^{-1}(\partial D)$$

$$\bigcup_{i=1}^{3} \bar{\partial D}'_i = p^{-1}(\partial D')$$

$$\widetilde{\partial L}_\eta = \bar{\partial D}'' = p^{-1}(\partial D'')$$

$$\bar{C}''_o \in p^{-1}(C''_o)$$

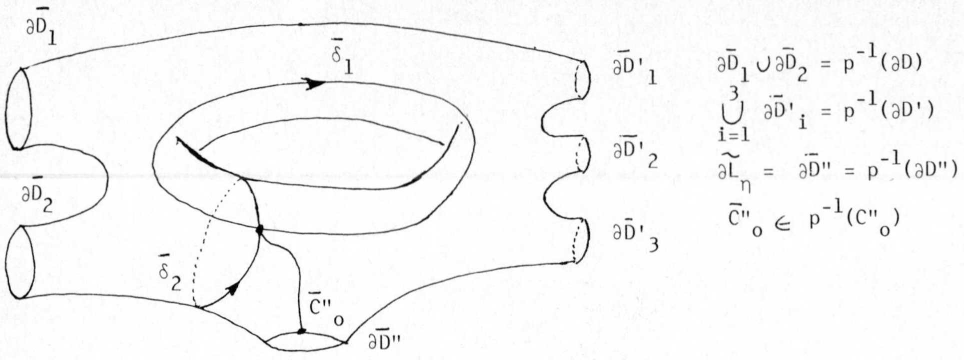

Figure 3.

Par une étude élémentaire (voir par exemple le cours de Valiron d'analyse ou N.A. Campo [3])du revêtement $p_\eta : \widetilde{L}_\eta \cap U_3 \to \mathcal{Q}_3$, on peut montrer le lemme suivant :

LEMME 2 : Le relèvement pour p_η d'origine C''_o du lacet $\delta_i = \beta^{-1}. \alpha^{-i}\beta \ \alpha^i$, avec $i = 1,2$ est un lacet $\bar{\delta}_i$ et les lacets $\bar{\delta}_1, \bar{\delta}_2$ sont des générateurs de $\pi_1(\widetilde{L}_\eta, C''_o)$. D'autre part le relèvement $\bar{\gamma}^6$ de $\gamma^6 = (\beta.\alpha)^6$ pour p_η est librement homotope au bord $\partial D''$ de $\widetilde{L}_\eta$.

Soit maintenant $\omega = 0$ une équation dégénérée-transverse d'holonomie évanescente (f,g). Le difféomorphisme d'holonomie correspondant à δ_i est le commutateur $[f^i,g]$. D'après le lemme, il décrit la façon avec laquelle le cycle évanescent $\bar{\delta}_i$ est "déroulé" en spirale par son relèvement dans les feuilles de $\widetilde{\mathcal{F}}_\omega$. C'est de cette manière que JU.B Il'Jasenko [17],[18] étudie la naissance de cycles limites pour des perturbations de champs hamiltoniens. En utilisant la forme normale de D. Cerveau de ω (de I) et des arguments de B. Malgrange [8] cette interprétation de H_ω peut être précisée.

Le relèvement de γ^6 dans les feuilles de $\widetilde{\mathcal{F}}_\omega$ permet de retrouver l'holonomie de la séparatrice $X = \{y^2 + x^3 = 0\}$. En particulier un calcul élémentaire montre que les trois égalités suivantes sont équivalentes :

$$h_{\omega,X} = [[f,g],[f^2,g]] = (f \bullet g)^6 = 1_{\mathbb{C}}.$$

V. ETUDE DE $\omega_1 = d(y^2+x^3) + x(2x \ dy - 3y \ dx)$

Cette étude a essentiellement pour but de montrer :

PROPOSITION 3 : Il existe des équations $\omega = 0$ *dégénérées transverses telles que toutes les feuilles de* $\mathcal{F}_\omega$ *sont adhérentes à* 0 *et l'holonomie de sa séparatrice* $H_{\omega,X}$ *est* $1_{\mathbb{C}}$.

Si $H_\omega = (f,g)$ est l'holonomie évanescente d'une telle équation alors

$$(f\circ g)^6 = f^2 = g^3 = 1_{\mathbb{C}} \quad \text{et} \quad [f,g] \neq 1_{\mathbb{C}}.$$

En effet si f et g commutent le couple (f,g) est conjugué à (J,S) avec

$$J(x) = jx \quad , \quad S(x) = -x \qquad j = \exp(2i\,\pi/3).$$

D'après le théorème 1, $\omega = 0$ est holomorphiquement conjugué à $\omega_0 = d(y^2+x^3)=0$ puisque $H_{\omega_0} = (J,S)$ et les feuilles de $\mathcal{F}_\omega$ ne sont pas adhérentes à 0.

Calculons l'équation $\tilde{\omega}_1 = 0$ (3.II, 4.II) de $\tilde{\mathcal{F}}_{\omega 1}$ au voisinage de P_3. La carte (v,t) n'est pas très bonne puisqu'elle ne permet pas de voir ce qui se passe en C'. Prenons la carte (s,z) avec

$$v = \frac{2s}{1-s} \quad , \quad t = z(1-s)$$

Dans cette nouvelle carte

$$C = (0,0) \;,\; C' = (1,0) \;,\; C'' = (-1,0)$$
$$\pi(s,z) = 2z^2\, s(1-s) \;,\; 2z^3\, s(1-s)^2) \;,$$
$$\tilde{\omega}_1 = 3s(s+1)(s-1)dz + z\, ds(3s^2+s-1+zs(1-s)).$$

Nous constatons que $\tilde{\omega}_1$ est une équation de Ricatti sur le fibré "canonique" de base $\mathbb{C}P(1)$, de fibre $\mathbb{C}P(1)$ de classe de Chern -1, ou plus exactement, dans la carte (s,z) de ce fibré ; les trois autres cartes étant

$$(s',z') \;,\; (s,w) \;,\; (s',w') \quad \text{avec}$$
$$ss' = 1 \;,\; z' = sz \;,\; w = 1/z \;,\; w' = 1/z' \;.$$

Nous notons $p : E \to \mathbb{C}P(1)$ ce fibré. Le feuilletage $\tilde{\mathcal{F}}_{\omega_1}$ de E possède les propriétés suivantes :

- Sa section nulle $P_0 = \{z = 0\}$ est une feuille qui porte les trois points singuliers C,C',C".
- il est transverse à la fibration en dehors des "séparatrices"

$$\tilde{X} = p^{-1}(C) \;,\; \tilde{X}' = p^{-1}(C') \;,\; \tilde{X}'' = p^{-1}(C'')$$

- La représentation d'holonomie de la feuille $P_0 - \{C,C',C''\} = L_0$ est un homomorphisme

$$h_{\tilde{\omega}_1} : \pi_1(L_o) = Z * Z \longrightarrow \mathcal{K} \subset \mathrm{Diff}(\mathbb{C},0)$$

où $\mathcal{K}$ est le sous-groupe de Klein formé des homographies qui laissent fixe $z = 0$. Ainsi on peut supposer (modulo un changement de coordonnées dans $\mathbb{C}P(1)$) puisque $\tilde{\omega}_1$ possède des intégrales premières en C et C" (lemme 1) :

$$h_{\tilde{\omega}_1}(\beta) = h_{\tilde{\omega}_1}(0 * 1) = H_{\omega_1}(0 * 1) = g = S \quad , \text{ avec } S(x) = -x$$

$$h_{\tilde{\omega}_1}(\alpha) = h_{\tilde{\omega}_1}(1 * 0) = H_{\omega_1}(1 * 0) = f \quad , \quad \text{avec} \quad f(x) = \frac{jx}{1+ax}$$

Un calcul élémentaire au plus élégamment un petit raisonnement géométrique que nous a communiqué D. Sullivan permet de montrer que $(f \circ g)^6 = 1_{\mathbb{C}P(1)}$. Ainsi d'après le théorème 1 la séparatrice $X = \pi(\tilde{X})$ n'a pas d'holonomie.

Pour achever la démonstration de la proposition il suffit de montrer que f et g ne commutent pas, c'est à dire que $a \neq 0$. En effet, on vérifie alors facilement que l'orbite d'un point de $P\mathbb{C}(1) - \{0\}$ sous l'action du groupe engendré par (f,g) contient $0 \in P\mathbb{C}(1)$ dans son adhérence . Ainsi toutes les feuilles de $\tilde{\mathcal{F}}_{\omega_1}$ distinctes du diviseur exceptionnel et de la séparatrice contiennent le diviseur exceptionnel dans leur adhérence et toutes les feuilles de $\mathcal{F}_{\omega_1}$ adhèrent à $0 \in \mathbb{C}^2$.

Supposons que f et g commutent, alors $H_{\omega_1} = H_{\omega_o}$ avec $\omega_o = d(y^2 + x^3)$ et d'après le théorème 1, ω et ω_o sont holomorphiquement conjuguées. D'après le théorème de division de G. de Rham on a

$$\phi^*(d(y^2 + x^3)) = g\,\omega_1$$

où $\phi \in \mathrm{Diff}(\mathbb{C}^2,0)$, $g(0) \neq 0$. Puisque $\phi^*(d(y^2+x^3))$ possède encore $y^2 + x^3$ comme séparatrice il existe une unité U telle que :

$$\phi^*(d(y^2+x^3)) = d(U(y^2+x^3)) = d(y^2+x^3) + x(2y\,dx - 3x\,dy) .$$

Une telle égalité est impossible, son premier membre ne possède pas de termes en x y dx.

REFERENCES

[1] I. Bendixon. Sur les points singuliers des équations différentielles. Ofv. Kangl. Vetenskaps. Akade. Eörhandlinger, Stokholm. 9 (1898),635-658.

[2] C. Camacho & P. Sad. Invariant varieties through singularities of holomorphic vector fields - Ann. of Math. 115(1982), 579-595.

[3] N.A. Campo. Tresses, monodromie et le groupe symplectique - Commentari Math. Helv. (1979)

[4] D. Cerveau & J.F. Mattei. Singularities of codimension one complex foliations, A partial survey. In this volume.

[5] D. Cerveau & J.F. Mattei . Formes intégrables singulières - Astérisque n° 97 - (1982)

[6] G. de Rham. Sur la division des formes et des courants par une forme linéaire Commentari Math. Hel. 28 (1954)

[7] H. Dulac. Recherches sur les points singuliers des équations différentielles- Journal de l'Ecole Polytechnique,2,sec.9 (1904),1-125.

[8] B. Malgrange. Intégrales asymptotiques et monodromie - Ann. Eco. Norm. Sup. Paris 7 (1974), 405-530.

[9] J. Martinet & JP.Ramis. Problèmes de modules pour les équations différentielles non linéaires du premier ordre - Publi. Math. I.H.E.S. n° 55 (1982) 63-164.

[10] J. Martinet & JP. Ramis : Classifications analytiques des équations différentielles résonnantes. A paraître dans Astérisque (1984)

[11] J. Milnor. Singular points of complex hypersurfaces - Ann. of Math. Studies 61 (1968).

[12] R. Moussu. Symétrie et forme normale des centres et foyers dégénérés - Ergod. Th. et Dynam. Sys. (1982) - 2 , 241-251.

[13] J.F. Mattei & R. Moussu : Holonomie et intégrales premières - Ann. Scient. Ec. Norm. Sup. Ser. 4.13 (1980), 469-523.

[14] H. Poincaré. Mémoire sur les courbes définies par une équation différentielle. J. Math. Pures et Appl. 3-7 (1881), 375-422.

[15] A. Seidenberg. Reduction of singularities of the differential equation A dy = B dx - Amer. J. Math. 79 (1968), 248-269.

[16] F. Takens. Normal forms for certain singularities of vector fields - Ann. Inst. Fourier-Grenoble 23 (1973), 163-195.

[17] S. Il'Jasenko. The origin of limit cycles... Mat. Sbornik,78 (1969) N° 3, 353-363

[18] S. Il'Jasenko. The multicity of limit cycles - Amer. Math. Soc. Transl. 2 , vol. 118 (1982).

R. MOUSSU
Université de Dijon
Département de Mathématiques
Laboratoire de Topologie - ERA 07.945
B.P. 134
21004 DIJON CEDEX - FRANCE

Singularities & Dynamical Systems
S.N. Pnevmatikos (editor)

SINGULARITIES OF CODIMENSION ONE COMPLEX FOLIATIONS
A PARTIAL SURVEY

Dominique Cerveau & Jean-François Mattei
Université de Dijon   Université de Toulouse
France

About singularities of foliations: Brasilians and French' works.

In this lecture we consider near the origin $0 \in \mathbb{C}^n$ holomorphic foliations which are given by integrable Pfaffian forms ω :

$\omega = \Sigma\, a_i(x)dx_i$, $\omega \wedge d\omega = 0$, $a_i \in \mathcal{O}_n$, where $\mathcal{O}_n$ denotes the local ring of convergent series in the variable $x = (x_1,\ldots,x_n)$. The singular locus $S(\omega)$ of ω is the set of points $m \in \mathbb{C}^n,0$ such that $\omega(m) = 0$.

We are interested by a qualitative description of the foliation $\mathcal{F}_\omega$ associated to ω:so it is possible to suppose that cod $S(\omega) \geq 2$ (it is possible to change the equation ω such that this fact is verified).

§1. SEPARATRIX

We consider a particular solution of the foliation $\mathcal{F}_\omega$. This is called a separatrix ; that is an holomorphic hypersurface of $\mathbb{C}^n$ given by an equation (f=0), (where f is a reduced element in $\mathcal{O}_n$) satisfying the following :

$\omega \wedge df = f.\eta$, η being germ of 2-form.

This means precisely that the manifold :

$X = \{(f=0) - S(f=0)\}$(where $S(f=0)$ is the singular locus of (f=0)) is a leaf of the regular foliation $\mathcal{F}_\omega|\mathbb{C}^n - S(\omega)$.

An integral curve of the foliation $\mathcal{F}_\omega$ is a non trivial holomorphic path $\gamma : \mathbb{C},0 \to \mathbb{C}^n,0$ such that :

$$\begin{cases} \gamma^*\omega = 0 \\ \gamma(\mathbb{C},0) \not\subset S(\omega) \end{cases}$$

It is clear that for n=2 separatrix and integral curve are the same notions.

In dimension $n \geq 3$ for certain types of singularities, all the integral curves are contained in the separatrices (for example consider a foliation given by the level sets of an holomorphic function) ; but it is not a general fact as it will be shown later.

A very important fact which was conjectured by René Thom is the following :

THEOREM 1. [Camacho-Sad]. Let ω = Adx + Bdy *be an holomorphic one form near the origin* O *in* $\mathbb{C}^2$. *Then* ω *has a separatrix (or an integral curve).*

The proof of this nice result is given by a good understanding of the resolution of Pfaffian form in $\mathbb{C}^2$. Recall that a resolution of ω is a proper morphism

$$\pi : M \to \mathbb{C}^2, O$$

where M is a regular two dimensional surface such that :

1) Outside $\pi^{-1}(O)$, π is an isomorphism. The divisor $\pi^{-1}(O)$ is a curve with normal crossings ; in fact π is obtained by successive blowing up of points and $\pi^{-1}(O)$ is an union of projective lines with transversal intersections.

2) The singularities of the foliation $\pi^*(\mathcal{F}_\omega)$ are reduced.

A germ of 1-form Ω at m is reduced iff the first jet $j^1\Omega$,m is up to equivalence, of one of the two types :

1) $\lambda_1\, y\, dx \qquad , \lambda_1 \neq 0$
2) $\lambda_1\, y\, dx + \lambda_2\, x\, dy$ with $\lambda_1/\lambda_2 \notin \mathbb{Q}_-$.

Using planar sections, Camacho-Sad's theorem proves that any integrable Pfaffian form in $\mathbb{C}^n, O$ $n \geq 3$ has many integral curves. A natural question arising is :

"Does any integrable form possess a separatrix"?

Here the answer is negative :

THEOREME 2 [Jouanolou]. In general, a Pfaffian equation in $\mathbb{P}\mathbb{C}(2)$ *has no algebraic solution.*

Recall that a Pfaffian equation in $\mathbb{P}\mathbb{C}(2)$ is an homogeneous integrable form ω which contains all the lines thought the origin in its kernel; by Chow's theorem any separatrix of ω gives an algebraic solution in $\mathbb{P}\mathbb{C}(2)$,

Nevertheless there exists a generalization of theorem 1 in the following way :

THEOREM 3 [Cerveau-Mattei]. *Let ω be a Pfaffian integrable form near the origin in $\mathbb{C}^n$; then there exists an holomorphic application $F : \mathbb{C}^{n-1},o \to \mathbb{C}^n,o$ with proper critical set such that $F^*\omega = 0$.*

Actually there is no test to decide if a given integrable 1-form possesses or not separatrices. About this, we have the following conjecture which is motivated by the study of several specific examples:

Conjecture : Suppose that for a general imbedding $i : \mathbb{C}^2,o \to \mathbb{C}^n,o$ the Pfaffian form $i^*\omega$ has a finite number af irreductible separatrices, namely $g_1 = o,\ldots,g_p = o$. Then ω has separatrices. More precisely there exists a germ of reduce holomorphic function

$$G : \mathbb{C}^n,o \to \mathbb{C},o$$

such that :

$$\begin{cases} \omega \wedge dG = G.\eta. \\ G \circ i = g_1\ldots g_p. \end{cases}$$

It seems that a good approach to the conjecture is to establish resolution theory for foliations in the same way that for analitic subspaces [Hironaka].

§2. CONSTRUCTION OF EXAMPLES.

The more simple example is given by the classical Frobenius'Theorem : every integrable Pfaffian form ω in $\mathbb{C}^n,0$ without singularities has an ordinary first integral ; more explicitely there exist unit U and submersion x_1 such that $\omega = Udx_1$. By pull back of such a foliation, we construct all the foliations which possess ordinary first integral. A classical remark is the following : the foliation $\mathcal{F}_{Udx_1}$ is given by an action of the commutative Lie group $\mathbb{C}^{n-1}$. This innocent remark suggests to look at codimension one foliation given by actions of classical groups.

- Action of affin group.

Let be $R : \sum_{i=1}^{3} x_i \frac{\partial}{\partial x_i}$ the "radial" vector field in $\mathbb{C}^3$ and $\mathcal{L}_\nu = \sum_{i=1}^{3} A_i \frac{\partial}{\partial x_i}$ an homogeneous vector field of degree ν , i.e the A_i are homogeneous polynomials of degree ν in the variables x_1,x_2,x_3.

We will suppose that the vector fields R and $\mathcal{Z}_\nu$ are not always colinear. By Euler's identity we have :

$$[R, \mathcal{Z}_\nu] = (\nu-1)\, \mathcal{Z}_\nu .$$

So the two vectorfields generate a foliation given by the Pfaffian form

$$\omega_{\nu+1} = i_R\, i_{\mathcal{Z}_\nu}\, dx_1 \wedge dx_2 \wedge dx_3$$

Naturally this foliation is a projective foliation. This type of form is very rigid ; suppose, and that is a generic assumption on $\mathcal{Z}_\nu$, O is an algebraicaly isolated singularity for $d\omega_{\nu+1}$ in $\mathbb{C}^3$; then :

THEOREM 4 [*Camacho-Lins*]. *If* ω *is a germ of integrable holomorphic Pfaffian form at* $O \in \mathbb{C}^p$ *and* $i : \mathbb{C}^3, O \to \mathbb{C}^p, O$ *a germ of immersion such that :*

$$j_0^{\nu+1}\, i^* \omega = \omega_{\nu+1}$$

Then there exists a germ of submersion

$$\mathcal{J} : \mathbb{C}^p, O \to \mathbb{C}^3, O$$

such that $\mathcal{J}^* \omega = \omega_{\nu+1}$.

Roughly speaking this means particulary that any unfolding of $\omega_{\nu+1}$ is trivial.

- Abelian actions of $\mathbb{C}^{n-1}$ on $\mathbb{C}^n$.

Let $A_i = \sum_{j=1}^{n} A_i^j x_j \frac{\partial}{\partial x_j}$ i=1...n-1 be n-1 linear diagonal independant vector fields in $\mathbb{C}^n$. We have $[A_i, A_j] = 0$ and the homogeneous one form :

$$\overline{\omega}_{n-1} = i_{A_1} \dots i_{A_{n-1}}\, dx_1 \wedge \dots \wedge dx_n$$

is trivialy integrable and of the following type :

$$\overline{\omega}_{n-1} = x_1 \dots x_n \sum_{i=1}^{n} \lambda_i \frac{dx_i}{x_i}$$

As an illustration of notions introduced in §1 it is easy to see that the hyperplanes $(x_i = 0)$ are separatrices of $\overline{\omega}_{n-1}$. When the complex numbers λ_i are independant over the positive integers $\mathbb{N}$ all the integral curves are contained in the separatrices $\cup(x_i = 0)$. Morever, with some additional conditions on the λ_i, we obtain an analogous version of theorem 4 (replace $\omega_{\nu+1}$ by $\overline{\omega}_{n-1}$, 3 by n and $\nu+1$ by n-1)

[Cerveau-Lins].

One of the goals of the work in this subject is to prove that a lot of singular foliations in dimension higher than two are pull-back of forms of type $\bar{\omega}_{n-1}$ (see §5); for this foliations there exist a resolution processus.

Problem : Classify general Lie group actions on $\mathbb{C}^n,0$ with codimension 1-leaves.

§3. INTEGRATING FACTOR. INTEGRATION AND EXTENSION.

Let be $f : \mathbb{C}^n,0 \to \mathbb{C},0$ a germ of holomorphic function. We say that f is an integrating factor of the integrable one form ω if :

$$d\left(\frac{\omega}{f}\right) = 0$$

Example : $x_1 \ldots x_n$ is an integrating factor of $\bar{\omega}_{n-1}$. To compute the coefficients λ_i in $\bar{\omega}_{n-1}$ it is suffisiant to calculate the integral

$$\int_{\gamma_j} \frac{\bar{\omega}_{n-1}}{x_1 \ldots x_n}$$

where γ_j is a little cycle around $(x_j = 0)$

This is a general fact :

THEOREM 5 [Cerveau-Mattei] : *let be ω a germ of holomorphic integrable form. Suppose that ω has an integrating factor.*

$$f = f_1^{n_1} \ldots f_p^{n_p} \quad (f_i \text{ irreducble, } n_i \in \mathbb{N})$$

Then

$$(*) \quad \frac{\omega}{f} = \Sigma \lambda_i \frac{df_i}{f_i} + d\left(\frac{\alpha}{f_1^{n_1-1} \ldots f_p^{n_p-1}}\right)$$

In this formula α is a germ of holomorphic function and λ_i are complex numbers given by :

$$\lambda_j = \frac{1}{2i\pi} \int_{\gamma_j} \frac{\omega}{f}$$

where γ_i is a little cycle with index 1 around $(f_j = 0)$.

Remarks: 1) it is very easy to see by $(*)$ that the multiform function $\exp \dfrac{\alpha}{f_1^{n_1-1} \ldots f_p^{n_p-1}} \cdot f_1^{\lambda_1} \ldots f_p^{\lambda_p}$ which is well defined on the universal covering of $\mathbb{C}^n - U(f_i = 0)$ is a first integral of ω. When f is reduced ($n_i = 1$), by changing the f_i we see that ω has first integral of type $f_1^{\lambda_1} \dots f_p^{\lambda_p}$; the foliation appears as pull-back by the morphism $(f_1,\dots,f_p) : \mathbb{C}^n,0 \to \mathbb{C}^p,0$ of the standard form $\bar{\omega}_{p-1} = x_1 \dots x_p \Sigma \lambda_i \frac{dx_i}{x_i}$ (and so of a group action).

2) In general there is at most one integrating factor (up to multiplication by a constant). This is a consequence of the well-known following fact : the quotient of two integrating factors is a meromorphic first integral.

The existence of integrating factor can be shown in low dimensions:

THEOREM 6 [*Cerveau-Moussu*] : *let* ω *be a germ of integrable holomorphic form near* 0 *and* $i : \mathbb{C}^p,0 \to \mathbb{C}^n,0$ $2 \leq p \leq n$ *a generic immersion (ie* cod $S(i^*\omega) \geq 2$). *Then* ω *has an holomorphic integrating factor if and only if this is true for* $i^*\omega$.

This result, which is technically easy, is very useful for the following reason : to construct an integrating factor it is sufficient to work in a plane, and in $\mathbb{C}^2$ we have a resolution σ-process.

§4. INTEGRABLE HOMOGENEOUS FORMS.

The study of these forms is natural because if we take the Taylor expansion of an integrable holomorphic Pfaffian form :

$\omega = \omega_\nu + \omega_{\nu+1} + \dots$

the first significant term, namely ω_ν , is trivialy integrable. We denote by $\mathfrak{I}_\nu(n)$ the algebraic set of integrable homogeneous forms in $\mathbb{C}^n$ of degree ν :

$$\mathfrak{I}_\nu(n) = \{\omega_\nu = \Sigma a_i dx_i,\ \omega_\nu \wedge d\omega_\nu = 0,\ \ a_i(tx) = t^\nu a_i(x)\}.$$

We consider a partition of $\mathfrak{I}_\nu(n)$ in two sets : the set of dicritical forms, ie projective forms, and its complement.

1) Dicritical forms : $\omega_\nu \in \mathfrak{I}_\nu(n)$ is to be said dicritical if $\omega_\nu(R) = 0$ where $R = \Sigma x_i \frac{\partial}{\partial x_i}$ is the "radial" vector field. The set $\mathcal{D}_\nu(n)$ of dicritical form is an algebraic subvariety of $\mathfrak{I}_\nu(n)$ and an element of $\mathcal{D}_\nu(n)$ represent an algebraic foliation on the projective space $\mathbb{P}\mathbb{C}(n-1)$. For n=3, clearly $\mathcal{D}_\nu(3)$ is a vector space, and in fact in this case for $\nu \geq 3$, $\mathcal{D}_\nu(3)$ is a irreductible component of $\mathfrak{I}_\nu(3)$ [Cerveau-Mattei].

2) Non-dicritical forms : Let ω_ν be an element in $\mathfrak{I}_\nu(n)-\mathcal{D}_\nu(n)$; we consider the homogeneous polynomial of degree $\nu+1$:

$$P_{\nu+1} = \omega_\nu(R)$$

The integrability condition $(\omega_\nu \wedge d\omega_\nu = 0)$ in addition to Euler's identity gives :

$P_{\nu+1}$ is an integrating factor of ω_ν. Using the integration theorem 5 we can describe explicitely the algebraic set $\overline{\mathfrak{I}_\nu(n) - \mathcal{D}_\nu(n)}$.

Let $\Sigma^p_{\nu_1 \dots \nu_p}(n)$ be defined as follows :

$\Sigma^p_{\nu_1 \dots \nu_p}(n) = \{\omega_\nu \in \mathfrak{I}_\nu - \mathcal{D}_\nu$ s.t. $P_{\nu+1}$ is a reduced polynomial with p components $P_1,\dots,P_p$ of degree ν_i : $P = P_1 \dots P_p$, $\nu(P_i) = \nu_i\}$.

Clearly an element $\omega_\nu \in \Sigma^p_{\nu_1 \dots \nu_p}(n)$ is of the type :

$$\omega_\nu = P_1 \dots P_p \, \Sigma \, \lambda_i \frac{dP_i}{P_i}$$

Remark : Since $P_{\nu+1} = P_1 \dots P_p$ by Euler's identity we find $\Sigma \lambda_i \nu_i = 1$ (and $\nu+1 = \Sigma \nu_i$).

We have the following statement :

THEOREM 7 [*Cerveau-Mattei*]: *a)* $\overline{\Sigma^p_{\nu_1 \dots \nu_p}(n)}$ *are irreductible algebraic sets and the* $\Sigma^p_{\nu_1 \dots \nu_p}(n)$ *are smooth.*

b) the algebraic decomposition in irreductible components of $\overline{\mathfrak{I}_\nu - \mathcal{D}_\nu}$ *is given, for* $n \geq 3$, *by* :

$$\overline{\mathfrak{I}_\nu - \mathcal{D}_\nu} = \bigcup \overline{\Sigma^p_{\nu_1 \dots \nu_p}(n)}.$$

Remark : In dimension 3, the algebraic decomposition of $\mathfrak{I}_\nu(3)$ is :

$\mathfrak{I}_\nu(3) = \bigcup \overline{\Sigma^p_{\nu_1 \dots \nu_p}(3)}$ for $\nu \leq 2$.

$\mathfrak{I}_\nu(3) = \mathcal{D}_\nu(3) \bigcup \overline{\Sigma^p_{\nu_1 \dots \nu_p}(3)}$ for $\nu > 2$.

§5. VARIOUS FIRST INTEGRALS (Pot pourri)- TOPOLOGICAL CONSIDERATIONS.

5.a. Holomorphic uniform first integral-Malgrange's theorem.

In this paragraph we focalise our attention on the rechearch of an holomorphic uniform first integral of an integrable Pfaffian form ω , that is an holomorphic function $f : \mathbb{C}^n,0 \to \mathbb{C},0$ such that

$$\omega \wedge df = 0$$

In such a situation the leaves of ω are the connected components of the levels $f^{-1}(\varepsilon)$, and $f^{-1}(0)$ is a separatrix for ω. Small singularities implie existence of such a first integral:

THEOREM 8 : Singular Frobenius theorem [*Malgrange*] *.*
Suppose that cod $S(\omega) \geq 3$, *then* ω *has a first integral* f.

In fact Malgrange proves that if there exists formal first integral then there exists another convergent. (A simple algorithmic argument by Martinet-Moussu says that cod $S(\omega) \geq 3$ implies existence of such a formal first integral).

In general it is possible to describe the topology of the foliation $\mathcal{F}_\omega = \mathcal{F}_{df}$ when 0 is an isolated singularity of $f^{-1}(0)$: all the topology of the foliation is contained in the hypersurface $f^{-1}(0)$.

A topological version of Malgrange theorem, conjectured by René Thom was stated by [Mattei-Moussu] : the foliation $\mathcal{F}_\omega$ given by ω is said to be simple if there exists an open neigbourood U of 0 such that :

a) the leaves of $\mathcal{F}_{\omega \mid U-S(\omega)}$ are closed .
b) the set of leaves with 0 in their closure is countable.

THEOREM 9 [*Mattei-Moussu*] *. The following two conditions are equivalent :*
a) $\mathcal{F}_\omega$ *is simple*
b) ω *has a first integral.*

5.b. Meromorphic first integrals.

Here we are interested by foliations $\mathcal{F}_\omega$ with a first integral $f = F/G$, $F(0) = G(0) = 0$ with F and G without common factor. For such a foliation :

✱ ✱ "all the leaves are holomorphic hypersurfaces and all of them contain 0 in their closure".

A natural question, also asked by René Thom, is the following :

"is (✱ ✱) a caracterization of foliations with a meromorphic first integral ?".

Perhaps this question was motivated by the projective version :

Suppose that for an algebraic projective foliation $\mathcal{F}_\omega$ all the leaves are complex hypersurfaces. Then ω has a rationnal first integral [see Jouanolou] .

Unfortunately, following [Susuki], (✱ ✱) is not a good test ; the 1-form η in $\mathbb{C}^2,0$:

$$\eta = (y^3+y^2-xy)dx - (2xy^2+xy-x^2)dy$$

has property (✱ ✱) (see this by blowing up). But a simple argument proves that there is no meromorphic function constant on leaves of $\mathcal{F}_\eta$. Nevertheless it's possible that Thom's idea is topologically true : in [Cerveau-Mattei] it's proved that the foliation $\mathcal{F}_\eta$ is topologically the same that the foliation given by the level sets of the rational function $\frac{y^2+x^3}{x^2}$. Is this a general fact ?

5.c. Multiforme first integral of type $f_1^{\lambda_1}\dots f_p^{\lambda_p}$, $\lambda_i \in \mathbb{C}$.

The first examples of such phenomeneous are given by forms of type $\bar{\omega}_{n-1}$ and homogeneous forms in $\Sigma^p_{\nu_1\dots\nu_p}(n)$. We define a "determinantal" subset $\mathcal{A}_\nu$ in $\mathcal{J}_\nu(n) - \mathcal{D}_\nu(n)$ in the following way ; $\omega_\nu \in \mathcal{A}_\nu$ iff :

1) $P_{\nu+1} = \omega_\nu(R)$ is a reduce polynomial, $\omega_\nu = P_1\dots P_p \, \Sigma\lambda_i \frac{dP_i}{P_i}$, $\Sigma\lambda_i\nu_i = 1$, where all λ_i are $\neq 0$, $-\lambda_i \notin \mathbb{N}$, $-\frac{1}{\lambda_i} \notin \mathbb{N}$ and if $p \geqslant 2$ one of the λ_i is a non real complex number (or an irrationnal number with bad approximations by rationnals).

2) $\pi_1(\mathbb{P}\mathbb{C}(n-1) - (P_{\nu+1}=0))$ is abelian.

Remark : 1) Condition two is verified if singularities of $P_{\nu+1} = 0$ in $\mathbb{P}\mathbb{C}(n-1)$ are normal crossings [Deligne] .

2) $\mathcal{A}_\nu$ contains an open dense subset of $\mathcal{J}_\nu - \mathcal{D}_\nu$.

Now we have the following statement.

THEOREM 10 [Cerveau-Mattei] . Let be ω a germ of holomorphic integrable form in $0 \in \mathbb{C}^n$. We suppose that the first significant jet ω_ν of ω is in $\mathcal{A}_\nu$. Then ω has a first integral of type $f_1^{\lambda_1}\dots f_p^{\lambda_p}$.

It's possible to obtain raffinements of this result for certain forms with dicritic first jet ω_ν lying in an open dense subset of $\overline{\mathcal{J}_\nu - \mathcal{D}_\nu} \cap \mathcal{D}_\nu$ [Cerveau].

A natural question is to ask if the existence of a first integral of type $f_1^{\lambda_1} \ldots f_p^{\lambda_p}$ is a topological fact ; according to §5.a. this is true for λ_i positive integers (in any dimension). It's always true when the λ_i are positive real numbers (or more generally when λ_i are on an half line through O in $\mathbb{C}$) [Cerveau-Mattei] .

One of the more recent result in this area is an unpublished work due to Lins : in dimension two the existence of multiform first integral $f_1^{\lambda_1} \ldots f_p^{\lambda_p}$, with λ_i complex numbers independant over $\mathbb{Q}$, is purely topological.

Actually different works and programs are in development to a better understanding of local complex codimension one foliations with singularities : numerical invariants [Camacho-Lins-Sad];(see the Camacho's exposition in this book) problems of moduli (see for details Martinet-Ramis exposition in this book) Martinet-Ramis Cerveau-Sad - polynomial foliations [Jouanolou] [Palmeira] [Cerveau] - foliations with parameters [Kabila] - Unfoldings [Suwa] [Cerveau-Moussu] - finite determination [Camacho-Lins] [Cerveau-Lins] [Cerveau-Mattei] - Real singularities [Moussu] . In most of these works it appears that the biggest difficulties live in two-dimension.

BIBLIOGRAPHIE

[Camacho-Lins] : The topology of integrable form near a singularity Pub.Math. I.H.E.S. n° 55 (p.5-36) (1982).

[Camacho-Lins-Sad] : Preprint (I.M.P.A. Rio) (1983).

[Camacho-Sad] : Invariant varieties trough singularities of holomorphic vector fields. Ann of Math, 115 (1978), 579-595.

[Cerveau] : Un theorème d'existence d'integrales premières multiformes dans le cas dicritique CRAS 297 (sept 83) série I- 105.

[Cerveau] : Feuilletages algèbriques de $\mathbb{C}^n$ avec singularités (préprint Dijon 1984).

[Cerveau-Lins] : Formes integrables tangentes à des actions commutatives. CRAS t.291 (8 déc.80) série A p. 647-649.

[Cerveau-Mattei] : Formes intégrables holomorphes singulières - Astérisque n° 97.

[Cerveau-Moussu] : Extension de facteurs intégrants et applications

CRAS t. 294 Série I p.17-19.

[Cerveau-Sad]: Rigidité de certains germes de feuilletages holomorphes. (préprint Dijon 1984).

[Deligne]: Le groupe fondamental du complément d'une courbe plane n'ayant que des points doubles ordinaires est abélien. Séminaire Bourbaki vol 79/90 nov. 179.

[Hironaka]: Introduction to the theory of infinitely near singular points. Mem. Mat. del Instituto "Jeorge Juan", 28 Madrid, 1977.

[Jouanolou]: Equations de Pfaff algebriques. Lec. Notes in math. N° 708 . Springer Verlag.

[Kabila]: Thèse de 3ème cycle Dijon 1983.

[Malgrange] : Frobenius avec singularité I : codimension I, Publ. Math I.H.E.S. 46 (1976) p. 163-173.

[Martinet-Ramis] : Problemes de modules pour des equations différentielles non linéaires du premier ordre. Publ. Math I.H.E.S. 55 (1982) p. 63-164. And:classification des equations différentielles holomorphes résonnantes (to appear in Annales de l'E.N.S.).

[Mattei-Moussu]: Holonomie et intégrales premières. Ann. Sc. Ec. Norm. Sup. 13 (1980) p. 469-523.

[Palmeira]: Preprint (P.U.C. Rio).

[Moussu]: Classification C^{∞} des équations de Pfaff complètement intégrables à singularité isolée. Invent. Math. 73, 419-436 (1983).

[Suwa] : A theorem of versality for unfolding of complex analytic foliation singularities (preprint Hokkaido University).

[Suzuki]: Sur les intégrales premières de certains feuilletages analytiques complexes, lect. Notes in Math. , 670, Springer-Verlag, p. 53-58.

D. CERVEAU
Université de Dijon
Département de Mathématiques
Laboratoire de Topologie ERA 07 945
B.P. 138
21004 - Dijon Cedex - France.

J.F. MATTEI
Université des Sciences
Sociales de Toulouse
Place Anatole France
31000 - Toulouse -
France.

Singularities & Dynamical Systems
S.N. Pnevmatikos (editor)

DIFFERENTIAL GEOMETRY AND DYNAMICS : TWO EXAMPLES

Marc Chaperon
Centre de Mathématiques de l'Ecole Polytechnique
France

In recent years, there has been much interplay between dynamical systems and differential geometry. We shall give two rather paradoxical examples of this interaction : in the first, a problem in differential geometry is solved by dynamical methods, whereas, in the second, a seemingly topological -or complex analytic- question in dynamical systems is answered using an intermediate, differential-geometric result.

I - THE ARNOL'D CONJECTURES IN SYMPLECTIC GEOMETRY.

0. Conventions.

All maps are assumed to be smooth. Given a space $\mathcal{F}$ of maps, a *path* in $\mathcal{F}$ is a mapping $t \mapsto f_t$ of $I = [0,1]$ into $\mathcal{F}$, such that $(t,x) \mapsto f_t(x)$ is smooth. When $\mathcal{F}$ is a group with unit e, such a path (f_t) in $\mathcal{F}$ has *compact support* iff $f_t(x) = e(x)$ outside some compact subset.Given a smooth manifold X, we shall denote by $\mathcal{E}(X)$ the space of functions $f : X \to \mathbb{R}$.

1. Statement of the problem.

Let M denote the n-dimensional torus $\mathbb{T}^n = \mathbb{R}^n/\mathbb{Z}^n$. According to classical results of Morse and Lyusternik-Schnirelmann, *every* $f \in \mathcal{E}(M)$ *has at least* n+1 *critical points, and at least* 2^n *if none of them is degenerate.*

Let us first formulate this theorem in a more complicated fashion : recall that the *Liouville form* of the cotangent bundle of T*M is the 1-form λ on T*M such that $\alpha^*\lambda = \alpha$ for every 1-form α on M (considered as a map $\alpha : M \to T^*M$). The *canonical symplectic structure of* T*M is $d\lambda$. An immersion $j : M \to T^*M$ is called *lagrangian* iff $j^*d\lambda = 0$; if, moreover, $j^*\lambda$ is exact, then j is an *exact lagrangian immersion*. When j is an embedding, these are conditions on j(M). We shall denote the space of lagrangian immersions (resp. embeddings) $M \to T^*M$ by $\mathrm{Imm}(M,d\lambda)$ (resp. $\mathrm{Emb}(M,d\lambda)$).

Example. A 1-form α on M is a Lagrangian embedding iff it is closed, and an exact one iff it is exact, hence the

SECOND FORMULATION OF THE MORSE-LYUSTERNIK-SCHNIRELMANN THEOREM :

Let $\pi : T^*M \to M$ *be the canonical projection, and let* $j \in \mathrm{Emb}(M,d\lambda)$ *be exact and such that* $\pi \circ j$ *is a diffeomorphism -which is always the case when* j *is* C^1*-close enough to the zero section* $0_M : M \to T^*M$ *of* π. *Then* j(M) *intersects* $\Sigma = 0_M(M)$ *at least at* n+1 *points, and at least at* 2^n *points if* $j \pitchfork \Sigma$.

In 1965, V.I. Arnol'd conjectured [1] that one could get rid of the hypothesis on $\pi \circ j$. In 1983, I proved [10] that he was right :

THEOREM. Let $j \in \mathrm{Emb}(M,d\lambda)$ *satisfy the following two conditions* :

(*i*) *It is exact.*

(*ii*) *The exists a path* (j_t) *in* Emb(M,dλ) *such that* $j_0(M) = \Sigma$ *and* $j_1 = j$.

Then $j(M) \cap \Sigma$ *contains at least* n+1 *points, and at least* 2^n *if* $j \pitchfork \Sigma$.

Before explaining how this result can be obtained, let us make some

Remarks. a. Its proof was suggested to me by the work [13] of C. Conley and E. Zehnder on another related conjecture of Arnol'd -see 4 below. It is but fair to mention that, once I had the idea of it, much of the proof could be taken from [13] without any modification.

b. One really needs the exactness condition (i), for it is easy to construct closed 1-forms on M which vanish nowhere.

c. If $n = 1$, the theorem is trivial : condition (i) just means that, on the cylinder T*M, the two regions contained between j(M) and Σ on either side of Σ have the same |dλ|-area. THUS, IN THE SEQUEL, WE SHALL ASSUMME $n > 1$.

d. If one replaces Emb(M,dλ) by Imm(M,dλ), the theorem becomes very false ; here is a counterexample with $n = 1$ and $j(M) \cap \Sigma = \emptyset$:

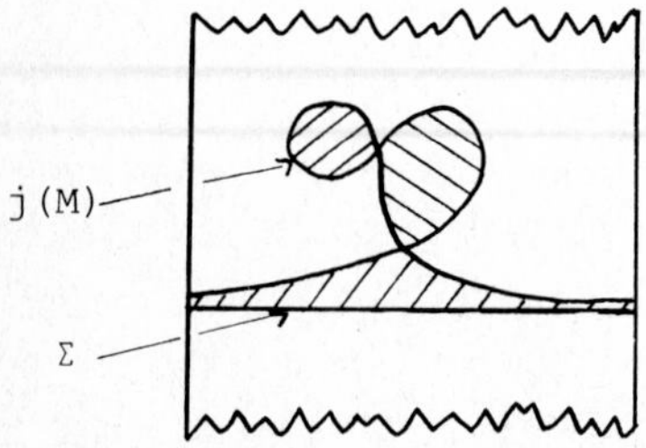

(The cylinder T*M has been cut along a generatrice and then unrolled ; exactness means that the two differently shaded regions have the same |dλ|-area.)

2. Beginning of the proof : how this is transformed into a boundary value problem for a system of ordinary differential equations.

An *isotopy* of $d\lambda$ is a path (φ_t) in the group $\mathrm{Aut}(d\lambda)$ of diffeomorphisms of T^*M which preserve $d\lambda$, such that $\varphi_o = \mathrm{id}$. Given such an isotopy (φ_t), the vector field $X_t := (\frac{d}{dt}\varphi_t)\circ\varphi_t^{-1}$ satisfies $L_{X_t} d\lambda = d(i_{X_t} d\lambda) = 0$ for every $t \in I$. Isotopy (φ_t) is called *hamiltonian* iff there exists a path (h_t) in $\mathcal{E}(T^*M)$ such that $i_{X_t} d\lambda = dh_t$ for every $t \in I$. If (φ_t) is hamiltonian and has compact support, then (h_t) can be so chosen as to have compact support, and this choice is unique : path (h_t) will then be called *the hamiltonian of* (φ_t).

LEMMA 1. *Under the hypotheses of the theorem, there exists a compactly supported hamiltonian isotopy* (φ_t) *of* $d\lambda$ *such that* $j(M) = \varphi_1(\Sigma)$.

This lemma holds for any compact manifold M. It is obtained by first noting ([10], théorème 1.1.3) that (j_t) can be so chosen as to be a path in the space of exact lagrangian embeddings of M into T^*M, and then ([10], théorème 0.4.2) "extending" (j_t) to a compactly supported hamiltonian isotopy (φ_t) of $d\lambda$ -which means that $j_t = \varphi_t \circ j_o$ for every t. □

Given an isotopy (φ_t) as in lemma 1, one has that $y \in j(M) \cap \Sigma$ iff $y = \varphi_1(x)$ with $x \in \Sigma$ and $\varphi_1(x) \in \Sigma$. In other words, if (h_t) denotes the hamiltonian of (φ_t),

$j(M) \cap \Sigma$ *is in bijection with the set of solutions of the following boundary value problem : find* $\gamma : I \to T^*M$, *with both endpoints in* Σ, *such that* $i_{\dot\gamma(t)} d\lambda = dh_t(\gamma(t))$ *for every* $t \in I$.

(This is because $i_{\dot\gamma(t)} d\lambda \equiv dh_t(\gamma(t))$ iff $\gamma(t) \equiv \varphi_t(\gamma(0))$).

3. Why this is a variational problem, and how it is solved.

Let Γ denote the set of those $\gamma = I \to T^*M$ of Sobolev class H^1, with both endpoints in Σ. It is readily checked that

the solutions of our boundary problem are the critical points of $f : \Gamma \to \mathbb{R}$, *given by*

$$f(\gamma) = \int_0^1 (-\gamma^*\lambda - h_t(\gamma(t))dt) \quad .$$

This infinite dimensional variational problem looks much more complicated than the geometric question we started with, but we shall in fact reduce it to a comparatively simple *finite dimensional* variational problem -this is the place where we need M to be a torus (or at least a flat manifold; see [10]), and not just any compact manifold.

First notice that T^*M is canonically isomorphic to $M \times (\mathbb{R}^n)^*$, and thus endowed with an abelian group structure. The covering projection $\tilde{p} : T^*\mathbb{R}^n \to T^*M$ induced by the canonical morphism $p : \mathbb{R}^n \to \mathbb{R}^n/\mathbb{Z}^n = M$ is just the group morphism $\mathbb{R}^n \times (\mathbb{R}^n)^* \to M \times (\mathbb{R}^n)^*$ given by $\tilde{p}(x,\xi) = (p(x),\xi)$, and $\tilde{p}^*d\lambda$ is the canonical symplectic structure $d\tilde{\lambda}$ of $T^*\mathbb{R}^n$.

Let $\mathbb{C}^n$ be equipped with its standard hermitian product $(z,z') \mapsto z.\bar{z}'$, hence with the euclidean product $(z,z') \mapsto (z|z') = \mathcal{R}e(z.\bar{z}')$ and with the symplectic structure σ given by $\sigma(z)(u,v) = \mathcal{I}m(u.\bar{v})$. Let $\psi : \mathbb{R}^n \times (\mathbb{R}^n)^* \to \mathbb{C}^n$ be defined by $\psi(x,\xi) = x + i\xi^\#$, where $\xi^\# \in \mathbb{R}^n$ is such that $(\xi^\# | y) = \xi(y)$ for every $y \in \mathbb{R}^n$. One has that $\psi^*\sigma = d\tilde{\lambda}$, hence (with $q = \tilde{p} \circ \psi^{-1}$) the following

LEMMA 2. *There exists a locally isometric covering projection $q : \mathbb{C}^n \to T^*M$ which is a group morphism satisfying $q^*d\lambda = \sigma$ and $q^{-1}(\Sigma) = \mathbb{R}^n$.* □

Let $H := \{u \in L_2(I,\mathbb{C}^n) : \mathcal{R}e \int_0^1 u(t)\, dt = 0\}$ be endowed with the real Hilbert space structure $(u,v) \mapsto \langle u,v\rangle = \int_0^1 (u(v)|v(t))\, dt$, and let $E \subset H$ be the dense subspace consisting of paths of class H^1 with both endpoints in $\mathbb{R}^n$. Denote the inclusion $M \overset{0_M}{\hookrightarrow} T^*M$ by $\theta \mapsto \theta_o$.

LEMMA 3. *The mapping $(\theta,u) \mapsto \theta_o + q \circ u$ is a bijection of $M \times E$ onto Γ. Using this identification, one has that*

$$f(\theta,u) = \frac{1}{2} \langle -i\dot{u},u\rangle - \int_0^1 h_t(\theta_o + q \circ u(t))\, dt$$

for every $(\theta,u) \in M \times E$. □

Let $A : E \to H$ and $\Phi : M \times H \to \mathbb{R}$ be defined by

$$Au = -i\dot{u} \quad , \quad \Phi(\theta,u) = \int_0^1 h_t(\theta_o + q \circ u(t))\, dt \quad .$$

LEMMA 4. (i) *A is a bijection of E onto H, whose inverse A^{-1} is a compact self-adjoint operator of H, with eigenvalues $1/m\pi$, $m \in \mathbb{Z} \setminus \{0\}$. For each such m, the eigenspace $E_m = \mathrm{Ker}(A^{-1} - id/m\pi)$ is $\mathbb{R}^n u_m$, where $u_m(t) = e^{im\pi t}$ for every $t \in I$.*

(ii) *Φ is C^2 and C^2-bounded on $M \times H$, smooth on $M \times E$.*

(Assertion (ii) is due to the fact (h_t) has compact support and q is a group morphism. To prove that A is a bijection, just exhibit its inverse, which is given by

$$A^{-1}u(t) = \int_0^t iu(s)\,ds - \int_0^1 (\mathcal{R}e \int_0^t iu(s)\,ds)\,dt \quad ,$$

hence compact. The rest of the assertion (i) is easy.) □

By the spectral theorem for self-adjoint compact operators, we have that $H = \oplus E_m$. For each $N \in \mathbb{N}$, let

$$Z_N = \bigoplus_{|m|\leq N} E_m \quad \text{and} \quad Y_N = \bigoplus_{|m|>N} E_m \quad .$$

LEMMA 5. *There exists* $N_o \in \mathbb{N}$ *such that, for every* $N \geq N_o$ *and every* $(\theta,z) \in M \times Z_N$, *if* $E = Z_N \oplus (Y_N \cap E)$ *is identified with* $Z_N \times (Y_N \cap E)$, *then, equation*

(1) $$\frac{\partial f}{\partial y}(\theta,z,y) = 0$$

admits a unique solution $y = v_N(\theta,z) \in Y_N \cap E$. *Moreover, the mapping* $v_N : M \times Z_N \to Y_N \cap E$ *is smooth (for the* H^1 *norm).*

Proof. Define $\nabla_2\Phi : M \times H \to H$ in the following way : for each $(\theta,u) \in M \times H$, $\nabla_2\Phi(\theta,u)$ is the value at u of the gradient of $x \mapsto \Phi(\theta,x)$ with respect to <.,.>. For each $N \in \mathbb{N}$, equation (1) reads

(2) $$y = A^{-1} P_N \nabla_2 \Phi(\theta,z,y) \quad ,$$

where $P_N : H \to Y_N$ is the orthogonal projection. Now, by lemma 4 (ii), $\nabla_2\Phi$ is C^1-bounded. Therefore, by Lemma 4 (i), there exists $N_o \in \mathbb{N}$ such that, for every $N \geq N_o$ and every $(\theta,z) \in M \times Z_N$, the mapping $y \mapsto A^{-1} P_N \nabla_2\Phi(\theta,z,y)$ is a strict contraction of Y_N. This proves that v_n exists and is unique. Its smoothness comes from the implicit function theorem and the fact Φ is smooth on $M \times E$. □

Given $N \geq N_o$, let $Z := Z_N$ and $v := v_N$, and let $g : M \times Z \to \mathbb{R}$ be defined by

(3) $$g(\theta,z) = f(\theta,z+v(\theta,z)) \quad .$$

By the very definition of v, the embedding $(\theta,z)\mapsto(\theta,z+v(\theta,z))$ restricts to a bijection of the critical set of g onto the critical set of f. Now, as Φ is C^1-bounded , the gradient of g is equivalent at infinity to the gradient of $(\theta,z)\mapsto\frac{1}{2}\langle Az,z\rangle$, which implies by more or less standard arguments ([10], [13]) that g (hence f) has the required number of critical points.

4. Some consequences.

Let ω denote the *standard symplectic structure of* $\mathbb{T}^{2n}$ obtained from σ via the canonical projection $\mathbb{C}^n\to\mathbb{T}^{2n}=\mathbb{C}^n/(\mathbb{Z}^n+i\,\mathbb{Z}^n)$, and let G_ω be the component of the identity in the group $\mathrm{Aut}(\omega)$ of those diffeomorphisms of $\mathbb{T}^{2n}$ which preserve ω. The following deep result was proved (for an arbitrary compact symplectic manifold) by Banyaga [4] :

THEOREM. A diffeomorphism φ of $\mathbb{T}^{2n}$ belongs to the commutator subgroup $[G_\omega,G_\omega]$ of G_ω iff there exists a hamiltonian isotopy φ_t of ω with $\varphi_1=\varphi$. □

Let $M=\mathbb{T}^{2n}$, and let $\overset{2}{\omega}$ be the symplectic structure $pr_2^*\omega - pr_1^*\omega$ on $M\times M$, where pr_1, $pr_2 : M\times M\to M$ denote the projections. Clearly, a diffeomorphism φ of M is in $\mathrm{Aut}(\omega)$ iff $\overset{2}{\omega}$ induces the zero form on the graph of φ, a remark which makes the following easy ([10]) result less mysterious :

LEMMA 6. (i) There exists a covering projection $p : T^*M\to M\times M$ such that $p|\Sigma$ is a diffeomorphism onto the diagonal ΔM and $p^*\overset{2}{\omega}$ is the canonical symplectic structure of T^*M.

(ii) For every $\varphi\in[G_\omega,G_\omega]$, there exists $j : M\to T^*M$, satisfying the hypotheses of our theorem (with n replaced by 2n), and such that $p\circ j(x)=(x,\varphi(x))$ for each $x\in M$. □

Under the hypotheses of (ii), $pr_1\circ p|j(M)\cap\Sigma$ clearly is an injection into the fixed point set of φ, hence the following

COROLLARY 1 (Conley-Zehnder [13]). Every $\varphi\in[G_\omega,G_\omega]$ admits at least 2n+1 fixed points, and at least 2^{2n} if all of them are non-degenerate. □

Let us conclude with Arnold's starting point ([1], [2]) : a *distorsion* of the annulus $\mathbb{A}=\mathbb{T}^1\times I$ is a diffeomorphism φ of $\mathbb{A}$ which has a lifting $\tilde{\varphi}$ to the universal covering $\tilde{\mathbb{A}}=\mathbb{R}\in I$ of $\mathbb{A}$ with the following property : for $j=0$, 1 and $x\in\mathbb{R}$, one has that $\tilde{\varphi}(x,j)=(\varphi_j(x),j)$ and $\varphi_0(x)<x<\varphi_1(x)$.

COROLLARY 2 (Poincaré-Birkhoff). Every area-preserving distorsion ψ of $\mathbb{A}$ has at least two fixed points.

The idea of the proof (Arnol'd) is as follows : enlarge $\mathbb{A}$ into wider annulus $\mathbb{A}'$, and extend ψ to an area-preserving distorsion ψ' of $\mathbb{A}'$, in such a way that the following two properties hold :

(i) ψ' has the same fixed points as ψ.

(ii) Let $\mathbb{T}'$ be the 2-torus obtained by gluing together two copies of $\mathbb{A}'$ along their boundaries, and let ω' be the symplectic structure on $\mathbb{T}'$ induced by the volume form of $\mathbb{A}'$; the homeomorphism φ of $\mathbb{T}'$, equal to ψ' in each of the two copies of $\mathbb{A}'$, belongs to $[G_{\omega'}, G_{\omega'}]$.

Now, by a classical theorem of Moser [16], there exists a diffeomorphism $\mathbb{T}' \to \mathbb{T}^2$ sending $k\omega'$ onto ω, where k is a positive constant. Therefore, the Conley-Zehnder theorem implies that φ has at least three fixed points ; since φ has twice as many fixed points as ψ, we are done. □

The reader is referred to [1], [2], [4], [10], [13] and [20] for further information on this very attractive field of mathematics -it should be mentioned that many basic global problems in symplectic geometry (including the rest of the Arnol'd conjectures, see [10]) are still unsolved.

II - NORMAL FORMS, LINEARIZATION AND CENTRALIZERS OF ABELIAN GROUP ACTION-GERMS.

0. Notations and conventions.

Given two manifolds M and N, we shall denote by $J^k(M,N)$ the space of k-jets (i.e. jets of order k) of C^k-mappings $f : M \to N$, and by $j^k f : M \to J^k(M,N)$ the mapping which to each $x \in M$ associates the k-jet of f at x.

A *subspace* of $\mathbb{C}^n$ is a complex sub-vector space of $\mathbb{C}^n$.

1. Introduction : two results on local holomorphic vector fields.

Let $X : (\mathbb{C}^n,0) \to (\mathbb{C}^n,0)$ be a local holomorphic vector field. For each $x \in \mathbb{C}^n$ near 0, let $t \mapsto \exp tX(x)$ denote the germ $(\mathbb{C},0) \to (\mathbb{C}^n,x)$ obtained by integrating X. Another such vectorfield X' is C^k- (resp. *holomorphically*) *isomorphic to* X iff there exists a local C^k- (resp. holomorphic) diffeomorphism $\varphi : (\mathbb{C}^n,0) \to (\mathbb{C}^n,0)$ such that

$\varphi(\exp tX(x)) = \exp tX'(\varphi(x))$ for small enough $x \in \mathbb{C}^n$ and $t \in \mathbb{C}$. For $k \geq 1$, this is equivalent to saying that $\varphi_* X = X'$ *and* $\varphi_*(iX) = iX'$ in some neighborhood of 0.

The *linear part* of X is $dX(0) \in g\ell(n,\mathbb{C})$. An element L of $g\ell(n,\mathbb{C})$ is called *hyperbolic* when its eigenvalues $\lambda_1,\ldots,\lambda_n$ are simple and satisfy $\lambda_j \notin \mathbb{R}\lambda_k$ for $j \neq k$.

In 1980, I proved [7], [8] the following complex analogue of the classical Hartman-Grobman linearization theorem (a weaker form of this result had been conjectured by Camacho, Kuiper and Palis [5]):

THEOREM 1. *Let* $L \in g\ell(n,\mathbb{C})$ *be hyperbolic. Then, every local holomorphic vector field* $X : (\mathbb{C}^n,0) \to (\mathbb{C}^n,0)$ *with linear part* L *is* C^0*-isomorphic to* L.

The proof uses differential geometry *via* the following intermediate result :

THEOREM 2. *For each* $k \in \mathbb{N}^*$, *every local holomorphic vector field* $X : (\mathbb{C}^n,0) \to (\mathbb{C}^n,0)$ *with hyperbolic linear part* L *is* C^k*-isomorphic to a polynomial vector field on the form* $L + N$ *with* $j^1N(0) = 0$ *and* $[L,N] = 0$.

When L is the *Poincaré domain* (i.e. when the convex hull of its spectrum in $\mathbb{C}$ does not contain 0), this is easy : Poincaré and Dulac proved that every X with linear part L was *holomorphically* isomorphic to such a polynomial field (Theorem 3 below). The case when L is in the *Siegel domain* (i.e. the remaining situation) is much more difficult ; if we denote by $\mathcal{S} \subset g\ell(n,\mathbb{C})$ the set of those L's which are in the Siegel domain, and by $\mathcal{H}$ the set of those $L \in \mathcal{S}$ which are hyperbolic, the situation is as follows :

- A celebrated theorem of Siegel [18] states that *there exists* $\mathcal{S}_1 \subset \mathcal{S}$ *such that every local holomorphic vector field* X *with linear part* $L \in \mathcal{S}_1$ *is holomorphically isomorphic to* L.

- Dumortier and Roussarie [14] proved that *there exists* $\mathcal{S}_2 \subset \mathcal{H}$ *such that every local holomorphic vector field* X *with linear part* $L \in \mathcal{S}_2$ *is* C^∞*-isomorphic to* L ($\mathcal{S}_2$ is the set of those $L \in \mathcal{H}$ such that every X with linear part L is *formally* isomorphic to L).

- $\mathcal{S}_1$ *is not included in* $\mathcal{H}$, whereas $\mathcal{S}_1 \cap \mathcal{H}$ *is strictly contained in* $\mathcal{S}_2$. An unpleasant feature of both $\mathcal{S}_1$ and $\mathcal{S}_2$ is that *they have empty interior* -but *both have full measure, and* $\mathcal{S}_2$ *is residual* ; moreover, as both are defined -even locally- by an infinite set of inequalities, neither result is very effective.

- The method of Dumortier and Roussarie yields a result [14] which does not deserve the above criticisms : *for every* $k \in \mathbb{N}^*$, *the set of those* $L \in \mathcal{H}$ *such that every* X *with linear part* L *is* C^k*-isomorphic to* L *contains an open and dense full-measure subset of* $\mathcal{S}$, *locally defined by a finite number of analytic inequalities* (this can be proven exactly as theorem 2 above).

- According to Camacho, Kuiper and Palis [5] , the elements of $\mathcal{H}$ are very unstable with respect to (local) topological equivalence . This shows that Theorem 1 above is essentially different from (and, in a vay, more surprising than) the Hartman-Grobman theorem.

- When one ventures outside $\mathcal{H} \cup \mathcal{S}_1$, the holomorphic -and even C^1-classification of local holomorphic vector fields may exhibit a feature which can be described as extremely interesting or extremely unpleasant according to the taste of the teller : even in the simplest cases, it has been shown by Martinet and Ramis [15] that a given formal isomorphism classe could contain an infinite dimensional space of holomorphic (or C^1) isomorphism classes. Camacho and Sad [6] proved that, in these "simplest" cases, the C^0 classification was better behaved, but the following example shows that the situation is not so good : if x and y denote the standard coordinates on $\mathbb{C}^2$, the Euler vector field $X = x^2 \frac{\partial}{\partial x} + (y - x^2) \frac{\partial}{\partial y}$ is formally isomorphic to $X' = x^2 \frac{\partial}{\partial x} + y \frac{\partial}{\partial y}$, but there does not even exist a germ of a homeomorphism $(\mathbb{C}^2,0) \to (\mathbb{C}^2,0)$ sending the germ at 0 of the holomorphic foliation defined by X' onto the germ at 0 of the foliation defined by X (this is because such a local homeomorphism should send the germs at 0 of $x^{-1}(0)$ and $y^{-1}(0)$ onto germs of *holomorphic* curves through 0 such that X is tangent to both ; now $x^{-1}(0)$ is the only holomorphic curve through 0 whose germ at 0 has this property).

While Siegel's result is a masterpiece in pure analysis, the proof of Theorems 1 and 2 above involves more geometry than analysis -especially since we do not strive for the least possible degree of N in Theorem 2. The interested reader will find in [8] the idea of how Theorem 1 is deduced from Theorem 2. Here, we shall concentrate on the methods used in the proof of Theorem 2 (and of similar results on general abelian group actions as well, see [1], chapitre 3).

2. Main analytical tools for the proof of Theorem 2.

Let M be a compact manifold, let E^+ and E^- be two nontrivial euclidean spaces, and let

$$\begin{cases} V = M \times E^+ \times E^- \quad , \quad W^+ = M \times E^+ \times \{0\} \subset V \quad , \\ W^- = M \times \{0\} \times E^- \subset V \quad , \quad \Sigma = M \times \{0\} \times \{0\} = W^+ \cap W^- \quad . \end{cases}$$

Let $h : (V,\Sigma) \to (V,\Sigma)$ be a local smooth diffeomorphism, *normally hyperbolic at* Σ, *with stable manifold* W^+ *and unstable manifold* W^-, *in the following very special sense* : there exists a positive constant c such that, in some neighbourhood of Σ,

$$|h(x)_+| \leqslant e^{-c}|x_+| \quad \text{and} \quad |h(x)_-| \geqslant e^{c}|x_-| \quad ,$$

where $x \mapsto x_+ \in E^+$ and $x \mapsto x_- \in E^-$ denote the canonical projections and $|.|$ the euclidean norms.

(All the results of this section hold if the trivial vector bundles $V \to M$, $W^+ \to M$ and $W^- \to M$ are replaced by more general (riemannian) vector bundles, but we shall not need this generalization).

Two local diffeomorphisms $g, h : (V,\Sigma) \to (V,\Sigma)$ are C^k-*conjugate* iff there exists a local C^k-diffeomorphism $\varphi : (V,\Sigma) \to (V,\Sigma)$ such that $h = \varphi_* g := \varphi \circ g \circ \varphi^{-1}$ near Σ.

LEMMA 1. *Given* h *as above, there exists a mapping* $s_1 : \mathbb{N}^* \to \mathbb{N}^*$ *with the following property : for each* $k \in \mathbb{N}^*$, *every local* C^{s_1}-*diffeomorphism* $f : (V,\Sigma) \to (V,\Sigma)$ *with the same* $s_1(k)$-*jet as* h *at* Σ *is* C^k-*conjugate to a local* C^k-*diffeomorphism with the same* k-*jet as* h *at* $W^+ \cup W^-$. *This holds for* $k = \infty$ *with* $s_1(\infty) = \infty$.

The proof is as follows : set $(j^k\varphi)|W^+ := \lim_{n\to\infty} ((j^k(f^{-n} \circ h^n))|W^+)$ and $(j^k\varphi)|W^- := \lim_{n\to\infty} ((j^k(f^n \circ h^{-n}))|W^-)$ -these two uniform limits are seen to exist near Σ as soon as f and h have high enough contact at Σ, using an elementary fixed point theorem (see [11], section (4.2) and appendice 6). Notice that this definition yields $(j^k\varphi)|\Sigma = (j^k \mathrm{id})|\Sigma$ because $s_1(k)$ is at least equal to k, and then use (a simple special case of) the Whitney extension theorem to get φ from its k-jet $(j^k\varphi)|\, W^+ \cup W^-$, which clearly satisfies

$$(j^k(\varphi \circ h))|W^+ \cup W^- = (j^k(f \circ \varphi))|W^+ \cup W^- \quad . \quad \square$$

As a by-product of this proof, one gets

LEMMA 2. *Let* h *and* s_1 *be as above, with* $s_1(\infty) = \infty$. *Let* $\varphi : (V,\Sigma) \to (V,\Sigma)$

be a local $C^{s_1(k)}$*-diffeomorphism,* $1 \leq k \leq \infty$*, such that*

$$\begin{cases} (j^{s_1(k)}\varphi)|\Sigma = (j^{s_1(k)}\mathrm{id})|\Sigma \\ (j^k(\varphi \circ h))|W^+ \cup W^- = (j^k(h \circ \varphi))|W^+ \cup W^- \end{cases} .$$

Then, one has that

$$(j^k\varphi)|W^+ \cup W^- = (j^k\mathrm{id})|W^+ \cup W^- \quad \textit{near } \Sigma \quad .$$

Similarly, let X *be a* $C^{s_1(k)}$*-vector field defined in a neighbourhood of* Σ *in* V*, vanishing up to order* $s_1(k)$ *at* Σ*, and such that* h_*X *and* X *have the same* k*-jet at* $W^+ \cup W^-$*. Then,* X *vanishes up to order* k *along* $W^+ \cup W^-$ *in some neighbourhood of* Σ*.* □

Notes. 1. Let φ be as in our "proof" of Lemma 1. Then, in particular, the stable manifold of $\varphi^*f := \varphi^{-1} \circ f \circ \varphi$ at Σ is W^+, and its unstable manifolds is W^-.

2. A rather subtle feature of Lemma 2 is the following : while its hypothesis seems to depend on h itself, and not only on its jet at $W^+ \cup W^-$, its conclusion shows that this is not really the case.

3. Both the hypothesis of Lemma 2 and its conclusion do not involve φ itself, but only its jet at $W^+ \cup W^-$. For $k = \infty$, Lemma 2 leads to the following result : *Let* $\mathcal{D}$ *be the group of all germs at* Σ *of local* C^∞*-diffeomorphisms, and let* Z_0 *denote the centralizer of* $(j^\infty h)|\Sigma$ *in the group* $\{(j^\infty\varphi)|\Sigma : \varphi \in \mathcal{D}\}$*. If*

$$Z = \{(j^\infty\varphi)|W^+ \cup W^- : \varphi \in \mathcal{D} \ \textit{ and } \ (j^\infty(\varphi \circ h))|W^+ \cup W^- = (j^\infty(h \circ \varphi))|W^+ \cup W^-\},$$

then Z *is a group and the canonical (restriction) morphism* $Z \to Z_0$ *is injective (bijective, in fact).*

4. Lemma 1 and Lemma 2 also hold when E^+ or E^- is trivial. In this special case, Lemma 1 is a conjugacy result, and Lemma 2 is a rigidity result for the centralizer Z of (the germ of) h in $\mathcal{D}$ -for example, if Σ is finite, Z is a finite dimensional (algebraic) Lie group. This should be contrasted with Corollary 2 below.

LEMMA 3. *Given* h *as in Lemma 1, there exists a mapping* $s_2 : \mathbb{N}^* \to \mathbb{N}^*$ *such that, for every* $k \in \mathbb{N}^*$ *and every local* $C^{s_2(k)}$*-diffeomorphism*

$f:(V,\Sigma)\to(V,\Sigma)$ *with the same* $s_2(k)$*-jet as* h *at* $W^+\cup W^-$ *near* Σ, *the following holds : if* Ω *is an open neighbourhood of* Σ *in* V *and* $\varphi:\Omega\smallsetminus W^-\to V$ *denotes a* $C^{s_2(k)}$*-mapping with*

$$\begin{cases}(j^{s_2(k)}\varphi)\,|\,W^+\smallsetminus\Sigma = (j^{s_2(k)}\mathrm{id})\,|\,W^+\smallsetminus\Sigma \\ \varphi\circ h = f\circ\varphi \text{ on } (\Omega\smallsetminus W^-)\cap h^{-1}(\Omega\smallsetminus W^-)\end{cases},$$

then φ *can be extended to a* C^k*-mapping* $\Phi:\Omega\to V$, *which has the same* k*-jet as the identity at* W^- *near* Σ *-in particular,* Φ *can be restricted to a local diffeomorphism* $(V,\Sigma)\to(V,\Sigma)$ *(of course,* Φ *is unique and such that* $\Phi\circ h=f\circ\Phi$ *on* $\Omega\cap h^{-1}(\Omega)$*). The same holds for* $k=\infty$ *with* $s_2(\infty)=\infty$).

The proof of this lemma goes as follows (see [11], (4.2.3), Théorème 2, for an easier but less natural argument) : let f, Ω and φ satisfy its hypotheses with $k=s_2(k)=\infty$, and let r be a positive real number such that $W_r^+ := \{x\in W^+ : |x_+|\leqslant r\}$ is contained in Ω, and that f and h have the same $s_2(k)$-jet at W_r^+.
If

$$S_r := \{x\in\Omega : |x_+|<r\} \quad\text{and}\quad D_r := S_r\smallsetminus h(S_r) \quad ,$$

then

$$\Omega_r := (W^-\cap\Omega)\cup\{y\in\Omega : \exists\, m\in\mathbb{N},\ \exists\, x\in D_r : y=h^m(x)\}$$

is an open neighbourhood of $(W^+\cup W^-)\cap\Omega\cap S_r$, and one defines a mapping $m:\Omega_r\smallsetminus W^-\to\mathbb{N}$ by

$$h^{-m(x)}(x)\in D_r \quad .$$

Therefore, if $\varphi_r := \varphi\,|\,D_r$, one has that

$$\varphi(x) = f^{m(x)}(\varphi_r(h^{-m(x)}(x)))$$

for every x close engough to Σ in $V\smallsetminus W^-$.From this formula, direct estimates yield the required result. The case when k is finite is treated in the same way. The reader is referred to [11], Appendice 6 for details. □

Remark. The geometrical meaning of this proof -and that of Corollary 1 below- may appear more clearly once the reader has looked at the corresponding statement for flows, i.e. Corollary 4 below.

3. Some consequences.

Let h, s_1 and s_2 be as in 2., with $s_1(\infty) = s_2(\infty) = \infty$.

COROLLARY 1. *For* $1 \leqslant k \leqslant \infty$, *every local* $C^{s_2(k)}$*-diffeomorphism* $f : (V,\Sigma) \to (V,\Sigma)$ *with the same* $s_2(k)$*-jet as* h *at* $W^+ \cup W^-$ *near* Σ *is* C^k*-conjugate to* h.

Proof. We just have to construct (the germ at W^+ of) a map φ satisfying the hypotheses of Lemma 3. Let U be an open neighbourhood of Σ in V, contained in the domains of definition of f and h, and let $r > 0$ be such that $W_r^+ = \{x \in W^+ = |x_+| \leqslant r\}$ is contained in U, and that f and h have the same $s_2(k)$-jet at W_r^+. By the proof of Lemma 3, the germ of φ at $W^+ \cup W^-$ will be determined by its restriction ψ_r to

$$\Delta_r := \{x \in U \ : \ |x_+| \leqslant r\} \setminus h(\{x \in U \ : \ |x_+| < r\}) \ ,$$

and we can take for ψ_r just any germ at W_+ of a $C^{s_2(k)}$-mapping $g_r : \Delta_r \to V$ with the following two properties :

(i) g_r has the same $s_2(k)$-jet as id along W_+ ;

(ii) its $s_2(k)$-jet at $\partial\Delta_r$ satisfies

$$j^{s_2(k)}(g_r \circ h)(x) = j^{s_2(k)}(f \circ g_r)(x)$$

for every x close enough to W^+ with $|x_+| = r$.
By the Whitney extension theorem, the set of all such choices of germ ψ_r is an infinite dimensional space, hence in particular our Corollary. □

Let $\mathcal{D}$ denote the group of all germs at Σ of local C^∞-diffeomorphisms $(V,\Sigma) \to (V,\Sigma)$. Taking $f = h$ in the above proof, we get

COROLLARY 2. *The centralizer of (the germ at* Σ *of)* h *in* $\mathcal{D}$ *is infinite dimensional.* □

Lemma 1 and Lemma 3 yield

COROLLARY 3. *Let* $s = s_1 \circ s_2$. *For* $1 \leqslant k \leqslant \infty$, *every local* $C^{s(k)}$*-diffeomorphism* $f : (V,\Sigma) \to (V,\Sigma)$ *with the same* $s(k)$*-jet as* h *at* Σ *is* C^k*-conjugate to* h. □

We shall now see that the infinitesimal version of Lemma 3 (and Corollary 1) above can be deduced from Lemma 3 : let X be a C^∞-vector field, defined in some neighbourhood of Σ in V, and such that

$h^t := \exp tX$ satisfies

$$|h^t(x)_+| \leqslant e^{-ct}|x_+| \quad \text{and} \quad |h^t(x)_-| \geqslant e^{ct}|x_-|$$

for $t \geqslant 0$ whenever defined, where c is a positive constant. For each $r > 0$, let $Q_r := \{x \in V : |x_+| = r\}$ and $q_r := Q_r^+ \cap W^+$.

COROLLARY 4. *There exists a mapping* $\sigma : \mathbb{N}^* \to \mathbb{N}^*$ *such that, for every* $k \in \mathbb{N}^*$ *and every local* $C^{\sigma(k)}$*-vector field* Y *with the same* $\sigma(k)$*-jet as* X *along* $W^+ \cup W^-$*, the following holds : if* $r > 0$ *is small enough and if* $j : (Q_r, q_r) \to (V, q_r)$ *is a local* $C^{\sigma(k)}$*-mapping with the same* $\sigma(k)$*-jet at* q_r *as the inclusion* $Q_r \to V$*, then the Cauchy problem*

$$(1) \qquad \begin{cases} \Phi \circ h^t = (\exp tY) \circ \Phi & \textit{when defined} \\ \Phi | Q_r = j \end{cases}$$

admits a local continuous solution $\Phi : (V, \Sigma) \to (V, \Sigma)$*, which is of class* C^k *and has the same* k*-jet as* id *along* W^-*. In particular,* Φ *is a local diffeomorphism, and* $\Phi_* X = Y$ *near* Σ*. The same is true for* $k = \infty$*, with* $\sigma(\infty) = \infty$.

Proof. Let $\sigma : \mathbb{N}^* \cup \{\infty\} \to \mathbb{N}^* \cup \{\infty\}$ be the mapping s_2 associated with $h := h^1$ by Lemma 3. There exists an open neighbourhood U of Σ in V such that the relation

$$h^{-\tau(x)}(x) \in Q_r$$

defines a smooth function $\tau : U \setminus W^- \to \mathbb{R}$. Thus, there is an open neighbourhood Ω of Σ in V such that $\varphi := \Phi | \Omega \setminus W^-$ is well defined by (1), and given by

$$\varphi(x) = (\exp \tau(x) Y)(j(h^{-\tau(x)}(x))) \quad .$$

The hypotheses of Lemma 3 are satisfied by $h := h^1$, $f := \exp Y$ and φ, hence our Corollary. □

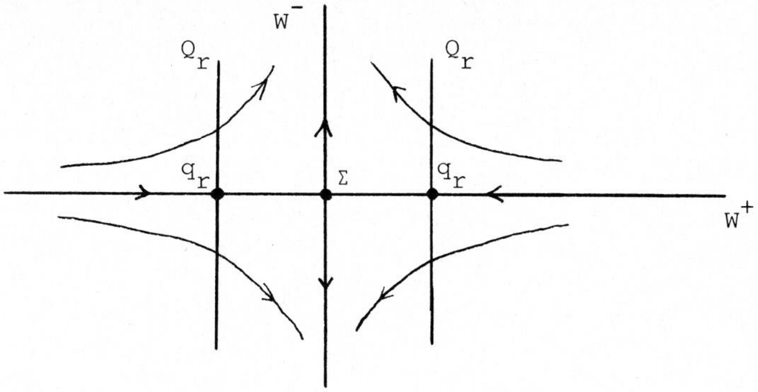

Note. Most important in this proof is the following fact : given a flowline L of (h^t) which passes close enough to Σ, either L is included in W^-, or L intersects Q_r transversally, at exactly one point. In other words, the germ of Q_r at q_r is a realization of the "orbit space" of the germ of $(h^t|V\setminus W^-)$ at Σ.

Let us now prove the infinitesimal version of Corollary 3 :

COROLLARY 5. *Given* X *as in Corollary 4, there exists a mapping* $\sigma' : \mathbb{N}^* \to \mathbb{N}^*$ *such that, for each* $k \in \mathbb{N}^*$, *every local* $C^{\sigma'(k)}$*-vector field* X' *with the same* $\sigma'(k)$*-jet as* X *at* Σ *is* C^k*-isomorphic to* X : *there is a local* C^k*-diffeomorphism* $\varphi : (V,\Sigma) \to (V,\Sigma)$ *such that* $X = \varphi_* X'$ *near* Σ. *The same is true for* $k = \infty$ *with* $\sigma'(\infty) = \infty$.

Proof. First assume that $k = \infty$. Applying Lemma 1 to $h := \exp X$ and $f := \exp X'$, we get a local C^∞-diffeomorphism $\psi : (V,\Sigma) \to (V,\Sigma)$ such that $\psi \circ f \circ \psi^{-1}$ and h have infinite contact at $W^+ \cup W^-$ near Σ. Therefore, $h_* \psi_* X'$ has infinite contact with $(\psi \circ f \circ \psi^{-1})_* \psi_* X' = \psi_* X'$ at $W^+ \cup W^-$ near Σ ; moreover, by the proof of Lemma 1, we may assume that ψ has infinite contact with id at Σ, hence $(j^\infty(\psi_* X' - X))|\Sigma = 0$. Thus, by Lemma 2, applied to $\psi_* X' - X$, the two vector fields $\psi_* X'$ and X have infinite contact at $W^+ \cup W^-$ near Σ, and we can apply Corollary 4 to $Y := \psi_* X'$. The same proof works for $k < \infty$. □

Note. Lemma 1 and Lemma 3 above were suggested to me by the work of Sternberg [19] and Nelson [17]. The only distinctive feature of Lemma 1 is that Σ is an arbitrary compact manifold, and not just a point -this generalization is not quite trivial, since a (riemannian) connection is needed to define higher order derivatives of mappings $V \to V$, which are used in the proof (see [11]). Lemma 3 is more original , for it comes from a geometric reading of Nelson's analytical proof of the Sternberg theorem -Sternberg's more geometric argument uses a different "Cauchy problem". Lemma 3 is a very flexible and powerful tool : the above five corollaries -and especially the nice proof of Corollary 4- are examples of this ; even more striking is the use of Lemma 3 in the study of conjugacy problem in the group of germs of local diffeomorphisms $(V,\Sigma) \to (V,\Sigma)$ which preserve a symplectic or contact structure (see [9], theorem 6), and in the study of (germs of) abelian group actions, to which we shall now turn in the particular case of holomorphic flows.

4. How theorem 2 can be proved.

The general idea is the same as in the proof of Corollary 5 above. In order to explain what this means, let us first describe the analogue of $W^+ \cup W^-$ for our local holomorphic vector fields : given $L \in gl(n,\mathbb{C})$, a *strongly invariant manifold* (s.i.m.) of L is a subspace W of $\mathbb{C}^n$ which is the unstable manifold of $\exp \alpha L$ at 0 for some $\alpha \in \mathbb{C}$. As each such W contains LW, *there is but a finite number of them*. Moreover, by the local uniqueness of the unstable manifold, *every local holomorphic vector field* $N : (\mathbb{C}^n,0) \to (\mathbb{C}^n,0)$ *such that* $[L,N] = 0$, *is tangent to all the s.i.m.'s of* L. Clearly, L *is in the Poincaré domain iff* $\mathbb{C}^n$ *itself is one of its s.i.m.'s* ; therefore, the Poincaré-Dulac theorem which was mentioned after Theorem 2 above is a particular case of

THEOREM 3. *For each* $k \in \mathbb{N}^*$, *every local holomorphic vector field* $X : (\mathbb{C}^n,0) \to (\mathbb{C}^n,0)$ *with linear part* L *is holomorphically isomorphic to a local holomorphic vector field* Y *with the following property : there exists a polynomial vector field* N *on* $\mathbb{C}^n$ *with* $j^1N(0) = 0$ *and* $[L,N] = 0$, *such that* Y *and* $L + N$ *have the same* k*-jet at every s.i.m.* W *of* L *near* 0 *(in particular,* Y *is tangent to* W*).*

Proof. For each $s \in \mathbb{N}^*$, *there exists a polynomial vector field* $L + N_s$ *of degree* s *and a local holomorhic diffeomorphism* $\varphi_s : (\mathbb{C}^n,0) \to (\mathbb{C}^n,0)$ -which may be chosen polynomial of degree s- *such that* $X_s := \varphi_s^* X$ *and* $L + N_s$ *have the same* s*-jet at* 0 (this is the standard theory of formal normal norms ; see for instance [3]).

LEMMA 4. *Let* F *be an intersection of s.i.m.'s of* L. *For each* $k \in \mathbb{N}$, *there is an integer* $\ell(F,k)$ *-determined by* L, F *and* k *-such that, for every large enough* s, *there exists a local holomorphic diffeomorphism* $\varphi_{F,s,k} : (\mathbb{C}^n,0) \to (\mathbb{C}^n,0)$ *with the following properties :*

(i) $\varphi^*_{F,s,k} X_s$ *and* $L + N_s$ *have the same* 2k*-jet at* F *near* 0.

(ii) $z_{F,s,k} := (j^{2k}\varphi_{F,s,k})|F$ *and* $(j^{2k}\mathrm{id})|F$ *have the same* $\ell(F,k)$*-jet at* 0.

(iii) The germ at 0 *of* $z_{F,s,k}$ *is determined by (i) and (ii) in a unique fashion.*
Moreover, the order of contact between $z_{F,s,k}$ *and* $(j^{2k}\mathrm{id})|F$ *at* 0 *is an increasing function of* s.

This lemma can be proved as follows : let $\alpha \in \mathbb{C}$ be such that F is contained in the unstable subspace of $\exp \alpha L$, and, for each $s \in \mathbb{N}^*$, let $h_s := \exp \alpha(L + N_s)$ and $f_s := \exp \alpha X_s$. Then, as in the proof of Lemma 1, there exists an increasing integer-valued function $s \mapsto m(F,s)$, determined by L and F, such that, for some choice of α, the following holds : for $0 \leqslant m \leqslant m(F,s)$, the sequence $((j^m(f_s^n \circ h_s^{-n}))|F)_{n\in\mathbb{N}}$ converges uniformly in some neighbourhood of 0; since a uniform limit of holomorphic maps is holomorphic, the limit of this sequence is the m-jet at F of a holomorphic function $\varphi'_{F,s,m}$ which clearly satisfies

$$(i)'_m \qquad (j^m(\varphi'_{F,s,m} \circ h_s - f_s \circ \varphi'_{F,s,m}))|F = 0 \quad \text{near } 0 \quad .$$

In order to show that $((j^m(f_s^n \circ h_s^{-n}))|F)_{n\in\mathbb{N}}$ converges uniformly near 0, one in fact proves that it converges in a space $\mathcal{F}_m$ of local holomorphic maps $F \to J^m(\mathbb{C}^n,\mathbb{C}^n)$ which have a contact of order $\ell'(F,m)$ with $(j^m\mathrm{id})|F$ at 0, where $m \mapsto \ell'(F,m)$ is an increasing function, determined by L and F. In particular,

$$(ii)_m \qquad z'_{F,s,m} := (j^m \varphi'_{F,s,m})|F \quad \text{and} \quad (j^m\mathrm{id})|F \quad \text{have the}$$

same $\ell'(F,m)$-jet at 0.

Even more precisely, $z'_{F,s,m}$ is obtained as the unique fixed point of a map $\mathcal{F}_m \to \mathcal{F}_m$; therefore, its germ at 0 is determined by $(i)'_m$ and $(ii)_m$ in a unique fashion. Moreover, one has that

$$(i)_m \qquad (j^m(\varphi'_{F,s,m} X_s - (L + N_s)))|F = 0 \quad \text{near } 0 \quad ,$$

by a slightly modified version of Lemma 2 above. Thus, Lemma 4 is proved with $\ell(F,k) := \ell'(F,2k)$ and $z_{F,s,k} := z'_{F,s,2k}$ for $m(F,s) \geqslant 2k$. □

End of the proof of Theorem 3. If L is in the Poincaré domain, then, taking $F := \mathbb{C}^n$ in Lemma 4, we are done.

Now, assume that L is in the Siegel domain ; let $(x_1,\dots,x_n)$ be a system of $\mathbb{C}$-linear coordinates on $\mathbb{C}^n$ in which the matrix of L is in Jordan normal form, and let $E_I := \bigcap_{j\in I} x_j^{-1}(0)$ for each $I \subset \{1,\dots,n\}$. Then, there exists $I_1,\dots,I_p \subset \{1,\dots,n\}$ such that $E_{I_1},\dots,E_{I_p}$ are exactly those s.i.m.'s of L which are maximal for inclusion. Given $k \in \mathbb{N}^*$, let $s \in \mathbb{N}^*$ be so large that $z_{F,s,k}$ as in Lemma 4 is well defined for each $F = E_{I_j} \cap E_{I_\ell}$ with $1 \leqslant j \leqslant \ell \leqslant p$. Then, for each j, the germ at 0 of $z_{E_j,s,k}$ can be identified with a unique convergent power series of the form

$$\sum_{\substack{m\in\mathbb{N}^n \\ |m_{I_j}|\leqslant 2k}} a_{j,m}\, x^m \in \mathbb{C}\{x_1,\dots,x_n\} \otimes_{\mathbb{C}} \mathbb{C}^n \tag{2}$$

where, for $I \subset \{1,\dots,n\}$ and $m = (m_1,\dots,m_n) \in \mathbb{N}^n$, we used the notations $x^m := x_1^{m_1} \dots x_n^{m_n}$ and $|m_I| := \sum_{j\in I} m_j$. By the uniqueness of $z_{F,s,k}$ in Lemma 4 and our hypothesis on s, one has that $a_{j,m} = a_{\ell,m}$ for $|m_{I_j \cup I_\ell}| \leqslant 2k$, hence in particular for $|m_{I_j}| \leqslant k$ *and* $|m_{I_\ell}| \leqslant k$. Therefore, the convergent power series

$$\sum_{1\leqslant j\leqslant p} \quad \sum_{\substack{m\in\mathbb{N}^n,\, |m_{I_j}|\leqslant k \\ |m_{I_\ell}|>k \text{ for } 1\leqslant \ell<j}} a_{j,m}\, x^m$$

is the Taylor expansion at 0 of a local holomorphic diffeomorphism $\psi_{s,k} : (\mathbb{C}^n,0) \to (\mathbb{C}^n,0)$, which obviously has the same k-jet at each E_{I_j} as the local holomorphic diffeomorphism whose Taylor expansion at 0 is given by (2). By the definition of (2) and Lemma 4 (i), we have that

$$(j^k(\psi^*_{s,k} X_s))\,|E_{I_j} = (j^k(\psi^*_{s,k}\, \varphi^*_s X_s))\,|E_{I_j} = (j^k(L + N_s))\,|E_{I_j}$$

for $1 \leqslant j \leqslant p$, hence Theorem 3. □

Let us go back to the proof of Theorem 2 : let $L \in g\ell(n,\mathbb{C})$ be hyperbolic and in the Siegel domain, and let $(x_1,\ldots,x_n)$ be a system of $\mathbb{C}$-linear coordinates on $\mathbb{C}^n$ such that

$$x_j \circ L = \lambda_j x_j \quad \text{for} \quad 1 \leqslant j \leqslant n \quad ,$$

where $\lambda_1,\ldots,\lambda_n \in \mathbb{C}$ are the eigenvalues of L. Define $F : \mathbb{C}^n \to \mathbb{C}$ by

$$F(z) = \sum_{1 \leqslant j \leqslant n} \lambda_j \, |x_j|^2 \quad .$$

For each regular value u of F, let $\hat{\mathcal{V}}_u$ denote the union of those s.i.m.'s of L which do not intersect $F^{-1}(u)$. It can easily be proved ([11], Chapitre 3) that, *for each orbit $\mathcal{O}$ of the complex flow* $(\exp tL)_{t\in\mathbb{C}}$ *, either $\mathcal{O}$ is contained in $\hat{\mathcal{V}}_u$, or $\mathcal{O}$ intersects* $F^{-1}(u)$ *transversally, at exactly one point* -in order words, $F^{-1}(u)$ is a realization of the orbit space of $(\exp tL|\mathbb{C}^n \setminus \hat{\mathcal{V}}_u)_{t\in\mathbb{C}}$.

Let $\mathcal{V}$ denote the union of all the s.i.m.'s of L. Given a local holomorphic vector field $L : (\mathbb{C}^n,0) \to (\mathbb{C}^n,0)$ with linear part L, Theorem 3 allows us to assume that, near 0, X has a contact of arbitrarily high order ℓ at $\mathcal{V}$ with a polynomial vector field $L + N_\ell$ such that $j^1 N_\ell(0) = 0$ and $[L,N_\ell] = 0$. For each $k \in \mathbb{N}^*$, if ℓ is large enough, then a C^k-isomorphism between X and $Y := L + N_\ell$ can be obtained by the same device as in Corollary 4 above, replacing real t by complex t, Q_r by $F^{-1}(u)$ for some small regular value u of F, q_r by $F^{-1}(u) \cap \mathcal{V}$ and W^- by $\hat{\mathcal{V}}_u$. The proof, though much more difficult, is roughly the same as for Corollary 4 : the Cauchy problem which corresponds to (1) (where j is just the germ of the inclusion, for example) can be solved in $U \setminus \hat{\mathcal{V}}_u$ for some open neighbourhood U of 0 in $\mathbb{C}^n$, using purely geometric considerations ; let $\varphi : U \setminus \hat{\mathcal{V}}_u \to \mathbb{C}^n$ be the mapping thus defined. In order to extend φ to the whole of U, some geometry and repeated use of Lemma 3 -or a refinement of it- are needed. The reader is referred to [11] and [12] for details.

Notes. 1. There does exist ([7], [8], [12]) a proof of Theorem 2 which uses only Lemma 1-2-3 above as its analytical tools. Although this other proof is more elementary, it have the inconvenient of being longer and less easily grasped at first glance than the above "brute force" agurment.

2. Theorem 3 is not really needed in the above "proof" of Theorem 2, but it has some intrinsic interest and simplifies the exposition : using only Lemma 1 and Lemma 2, we should have to deal

with differentiable $\mathbb{R}^2$-actions instead of holomorphic $\mathbb{C}$-actions in the last part of the proof ; I find it nicer to stay in the holomorphic category as long as possible.

3. Call $L \in gl(n,\mathbb{C})$ *weakly hyperbolic* if the line segment between two of its eigenvelues in $\mathbb{C}$ never contains the origin. Theorem 2 is true when L is only weakly hyperbolic, whereas the C^o-isomorphism classes of local holomorphic vector fields with linear part L are the elements of a finite dimensional "moduli space".

***

REFERENCES

[1] V.I. ARNOL'D, *Sur une propriété topologique des applications globalement canoniques de la mécanique classique*, C. R. Acad. Sc. Paris, t. 261 (Novembre 1965), Groupe 1, 3719-3722.

[2] V.I. ARNOL'D, Méthodes mathématiques de la Mécanique classique, Mir, Moscou, 1976.

[3] V.I. ARNOL'D, Chapitres supplémentaires de la théorie des équations différentielles ordinaires, Mir, Moscou, 1980.

[4] A. BANYAGA, *Sur la structure du groupe des difféomorphismes qui préservent une forme symplectique*, Comment Math. Helvetici 53 (1978), 174-227.

[5] C. CAMACHO, N. KUIPER, J. PALIS, *The topology of holomorphic flows with singularity*, Publ. Math. I.H.E.S. 48 (1978), 5-38.

[6] C. CAMACHO, P. SAD, *Topological classification and bifurcation of holomorphic vector fields with resonances*, Inv. Math. 67 (1982), 447-472.

[7] M. CHAPERON, Propriétés génériques des germes d'actions différentielles de groupes de Lie commutatifs élémentaires, Thèse, Université Paris VII, 1980.

[8] M. CHAPERON, *On the local classification of holomorphic vector fields*, in Geometric dynamics, Rio de Janeiro 1981, J. Palis ed., Springer Lecture Notes in Maths. n° 1007,(1983), 96-103.

[9] M. CHAPERON, *Quelques outils de la théorie des actions différentiables*, Semaine de géométrie de Schnepfenried, 1982, in Astérisque 107-108 (1983), 259-275.

[10] M. CHAPERON, *Quelques questions de géométrie symplectique*

[d'après, entre autres, Poincaré, Arnol'd, Conley et Zehnder], Séminaire Bourbaki, exposé n° 610, Juin 1983, Astérisque 105-106 (1983), 231-249.

[11] M. CHAPERON, *Géométrie différentielle et singularités de systèmes dynamiques*, Lectures given at E.N.S. and UFS Car, to appear in Astérisque.

[12] M. CHAPERON, Sur la classification des germes d'actions différentiables de groupes abéliens, to appear.

[13] C.C. CONLEY, E. ZEHNDER, *The Birkhoff-Lewis fixed point theorem and a conjecture of V.I. Arnol'd*, Inventiones Math. 73 (1983), 33-49.

[14] F. DUMORTIER, R. ROUSSARIE, *Smooth linearization of germs of* $\mathbb{R}^2$*-actions and holomorphic vector fields*, Ann. Inst. Fourier, Grenoble, 30, 1 (1980), 31-64.

[15] J. MARTINET, J.P. RAMIS, *Problèmes de modules pour des équations différentielles non linéaires du premier ordre*, Publ. Math. I.H.E.S., 55 (1982), 63-164.

[16] J. MOSER, *On the volume elements on a manifold*, Trans. A.M.S. 120 (1965), 286-294.

[17] E. NELSON, Topics in dynamics, Part I, Flows, Princeton, 1970.

[18] C.L. SIEGEL, *Über die Normalform analytischer Differentialgleichungen in der Nähe eine Gleichgewichtslösung*, Nachrichten der Akademie der Wissenschaften Göttingen, Math.-Phys. Kl. (1952).

[19] S. STERNBERG, *On the structure of local homeomorphisms of Euclidean* n-*space, II*, Amer. J. Math. 80 (1958), 623-631.

[20] A. WEINSTEIN, C^0 *perturbation theorems for symplectic fixed points and lagrangian intersections*, Preprint University of California, Berkeley, 1983.

* Marc Chaperon

Centre de Mathématiques
Ecole Polytechnique

91128 PALAISEAU Cedex
(France)

L.A. du C.N.R.S. n° 169

Singularities & Dynamical Systems
S.N. Pnevmatikos (editor)

EVOLUTION DYNAMIQUE D'UN SYSTEME MECANIQUE EN PRESENCE DE SINGULARITES GENERIQUES

Spyros N. Pnevmatikos
Université de Crète
Grèce

Dans plusieurs problèmes de Physique, lorsqu'on étudie l'évolution d'un système mécanique dans le temps, on représente chaque état de ce système, à un instant donné t, par un point x(t) d'une variété symplectique (W,ω), i.e. d'une variété différentiable W de dimension paire munie d'une 2-forme différentielle ω fermée dont le champ de noyaux est partout nul. Si H est une fonction réelle différentiable sur W désignant l'hamiltonienne de ce système, alors la forme symplectique ω lui associe son champ hamiltonien X_H, i.e. l'unique solution différentiable de l'équation de Hamilton-Jacobi:

$$X_H \lrcorner\, \omega = -dH .$$

Chaque point de la variété symplectique correspond ainsi à un état possible du système, le temps étant considéré comme paramètre réel, et la succession de ces états x(t) au cours du temps est exprimée par une solution de l'équation différentielle

$$\frac{dx(t)}{dt} = X_H(x(t)) .$$

Les courbes intégrales de cette équation représentent les mouvements du système réalisés dans son espace de phases.

Lorsque, dans la variété symplectique (W,ω), on se donne une sous-variété M, dite des *états permis*, il est clair que les mouvements du système de conditions initiales dans M ne se dérouleront pas en général entièrement dans M: le champ hamiltonien X_H peut ne pas être tangent à M. En d'autres termes, la dynamique définie par l'hamiltonienne H dans l'espace de phases ambiant n'est pas toujours admissible par la variété d'états permis M. La première obstruction peut déjà se lire sur la nature de la structure induite sur M, par la non résolubilité éventuelle de l'équation de Hamilton-Jacobi sur M:

$$X \lrcorner\, (\omega|M) = -d(H|M) .$$

Remarquons néanmoins que la non tangence du champ hamiltonien X_H à M n'empêche pas que cette équation ait une solution sur M, autrement dit, le système pourrait évoluer sur M suivant une dynamique définie par $H|M$ distincte de la dynamique de l'espace ambiant. Prendre, par exemple:

$$W = \mathbb{R}^4 \ , \quad \omega = dp_1 \wedge dq_1 + dp_2 \wedge dq_2 \ , \quad H(q_1, q_2, p_1, p_2) = (p_1^2 + p_2^2)/2$$

et considérer les variétés d'états permis M_1 et M_2 définies respectivement par les plongements suivants de $\mathbb{R}^2$ dans $\mathbb{R}^4$:

$$i_1(q_1', q_2') = (q_1'^2/2 \ , q_1' \ , p_1' \ , 0) \quad \text{et} \quad i_2(q_1', p_1') = (q_1' \ , 0 \ , p_1'^2/2 \ , p_1').$$

Le champ hamiltonien associé à H sur la variété ambiante $\mathbb{R}^4$:

$$X_H = p_1 \frac{\partial}{\partial q_1} + p_2 \frac{\partial}{\partial q_2}$$

n'est tangent à aucune de ces variétés d'états permis. On remarque que:

- dans le premier cas, la structure induite n'est pas symplectique: $\omega|M_1 = q_1' \, dp_1' \wedge dq_1'$, et la restriction $H|M_1 = p_1'^2/2$ ne définit pas une dynamique admissible pour M_1: le champ

$$X_{H|M_1} = (p_1'/q_1') \frac{\partial}{\partial q_1'}$$

n'est pas défini sur toute la variété M_1 ;

- dans le second cas, la structure induite n'est pas non plus symplectique: $\omega|M_2 = p_1' \, dp_1' \wedge dq_1'$, mais par contre la restriction $H|M_2 = p_1'^4/8 + p_1'^2/2$ définit une dynamique admissible pour M_2: le champ

$$X_{H|M_2} = (1 + p_1'^2/2) \frac{\partial}{\partial q_1'}$$

satisfait l'équation de Hamilton-Jacobi sur toute la variété M_2.

On est ainsi amené à se poser deux types de problèmes:

1° Etant donné une variété différentiable M munie d'une 2-forme différentielle ω , quelles sont les fonctions $h : M \to \mathbb{R}$ qui peuvent définir une dynamique sur M, autrement dit, telles que l'équation de Hamilton-Jacobi associée à ω soit résoluble sur M ?

2° Etant donné une variété différentiable M plongée dans une variété symplectique (W, ω), quelles sont les fonctions $h : M \to \mathbb{R}$ qui admettent une extension différentiable $H : W \to \mathbb{R}$ dont le champ hamiltonien X_H est tangent à M ?

Le premier problème a été traité génériquement pour les 2-formes différentielles ω sur une variété différentiable M de dimension paire, cf.[12], la généricité étant exprimée par transversalité du 0-jet de ω à la stratification canonique par le rang du fibré $\overset{2}{\wedge} T^*M$. La conclusion de cette étude est que:

THEOREME 1. *Si ω est une 2-forme générique sur une variété différentiable M de dimension paire, alors les fonctions $h: M \to \mathbb{R}$ pouvant définir une dynamique sur M sont celles dont la différentielle annule le noyau de ω en presque tout point où celui-ci est bidimensionel.*

L'idée de démonstration repose essentiellement sur une reduction du problème à l'étude d'un système d'équations linéaires à coefficients différentiables, dont le déterminant est le carré d'un polynôme affin en chaque variable dans un système de coordonnées approprié; ceci est obtenu par des arguments analogues à ceux employés par V.I. Arnold dans [1] et J.N. Mather dans [9]. La "propriété des zéros" du lieu de dégénérescence du rang des 2-formes génériques,cf.[12], permet une seconde reduction, en assurant la suffisance de la résolution du système aux points les moins dégénérés de ce lieu. Par la suite, en ces points, on se ramène à un modèle d'équations linéaires qu'on peut résoudre explicitement.

En ce qui concerne le second problème, notons que la théorie classique des contraintes, telle qu'elle a été introduite par P.A.M. Dirac en 1950 et qu'elle a été développée par les travaux des plusieurs physiciens et mathématiciens et notamment par A. Lichnerowicz en 1975, est basée sur l'hypothèse: la structure symplectique de l'espace de phases ambiant induit sur la variété des états permis une structure qui elle aussi est symplectique ou au moins présymplectique. Ceci signifie que la restriction de la forme symplectique ω à M a un champ de noyaux

$$\Delta_x(\omega|M) = T_xM \cap T_xM^{\perp}$$

de dimension constante sur M, ce qui intervient fortement dans la construction des dynamiques associées à M et de leurs extensions à l'espace de phases ambiant, cf.[7]. Mais, génériquement sur les variétés d'états permis, il peut apparaître un lieu de dégénérescences successives du rang de $\omega|M$, ce qui restreint la classe des fonctions pouvant définir une dynamique sur M et empêche de reprendre les techniques classiques pour effectuer ces constructions. On se place dans le contexte générique pour les variétés d'états permis M de (W,ω), la généricité étant exprimée par transversalité du 1-jet du plongement de M dans W à une stratification canonique du fibré des 1-jets d'immersions de M dans W réalisée par relèvement de la stratification par le rang du fibré $\overset{2}{\wedge} T^*M$ via une submersion naturelle, cf.[13]. La généricité du plongement de M dans W peut ainsi se lire uniquement sur la restriction de la forme symplectique ω à M, ce qui nous ramène à la théorie générale des singularités des formes différentielles développée ces dernières années notamment par J.Martinet [1970], R.Roussarie [1975], F.Pelletier [1980]. On en déduit la description de la structure des singularités pouvant apparaître sur ces variétés d'états permis: Ces singularités apparaîssent comme strates d'une stratification finie et localement triviale en des sous-variétés régulières

$$\Sigma_c(\omega|M) = \{ x \in M \,/\, \dim \Delta_x(\omega|M) = c \}$$

et plus précisement, si $c(c-1)/2$ est strictement supérieur à la dimension k de M alors $\Sigma_c(\omega|M)$ est vide, et dans le cas contraire $\Sigma_c(\omega|M)$ est, si non vide, une sous-variété régulière de codimension $c(c-1)/2$ dans M. Aussi, pour chaque $\Sigma_c(\omega|M)$ qui apparaît, son adhérence est localement difféomorphe à une sous-variété algébrique de $\mathbb{R}^k$ et possède la propriété de frontières:

$$\overline{\Sigma_c(\omega|M)} = \bigcup_{c' \geq c} \Sigma_{c'}(\omega|M) \quad .$$

De plus, cf.[12], le type local différentiable du lieu singulier de $\omega|M$, $\Sigma(\omega|M) = M - \Sigma_o(\omega|M)$, ne dépend que de la valeur du rang de $\omega|M$ au point considéré. En somme, en dimension paire k=2m, on retrouve la situation symplectique sur l' ouvert dense $\Sigma_o(\omega|M)$ de M et le lieu singulier $\Sigma(\omega|M)$, lorsqu'il n'est pas vide, apparaît comme une hypersurface dont la partie lisse est constituée des points où le noyau de $\omega|M$ est bidimensionel; cette hypersurface est donnée localement, dans un système de coordonnées approprié, par les zéros d'un polynôme homogène affin en chaque variable, cf. [12]. Par contre, en dimension impaire k=2m+1, aucun point régulier ne fait son apparition et les points où le noyau de $\omega|M$ est unidimensionel forment un ouvert dense dans M.

Dans [13], on poursuit cette étude et on résoud le second des problèmes posés plus haut en déterminant les dynamiques admissibles sur la variété d'états permis M qui provienent de dynamiques de l'espace de phases ambiant, et ceci pour le cas générique des variétés d'états permis de dimension paire dans la variété symplectique (W,ω). Voici la conclusion de cette étude:

THEOREME 2. *Si M est une variété d'états permis générique de dimension paire dans une variété symplectique (W,ω) et $h: M \to \mathbb{R}$ une fonction différentiable, alors les assertions suivantes sont équivalentes:*

(1) *La fonction h admet une extension différentiable $H: W \to \mathbb{R}$ dont le champ hamiltonien X_H est tangent à M.*

(2) *La fonction h possède un champ hamiltonien X_h sur M, i.e. un champ de vecteurs différentiable solution unique de l'équation de Hamilton-Jacobi pour la restriction de la forme symplectique ω sur M.*

(3) *La différentielle de h annule le noyau de la restriction de la forme symplectique ω sur M.*

(4) *La différentielle de h annule le noyau de la restriction de la forme symplectique ω en presque tout point où celui-ci est bidimensionel.*

Il est intéressant de noter que si de plus on démande au 1-jet de $\omega|M$ d'être transverse à la stratification du fibré des 1-jets de sections de $\overset{2}{\wedge}T^*M$ telle qu'elle est construite dans [8], alors dans ce cas l' ensemble

$$\Sigma_{2,0}(\omega|M) = \{ x \in \Sigma_2(\omega|M) \;/\; \Delta_x(\omega|M) \pitchfork T_x\Sigma_2(\omega|M) \}$$

est ouvert dense dans l'ensemble des points où le noyau de $\omega|M$ est bidimensionel. D'après un théorème classique de J.Martinet donné dans [8], la forme $\omega|M$ peut s'écrire au voisinage d'un point de $\Sigma_{20}(\omega|M)$ dans un système de coordonnées approprié $(q_1,\dots,q_m,p_1,\dots,p_m)$:

$$\omega|M = p_1\, dp_1 \wedge dq_1 + \sum_{i=2}^{m} dp_i \wedge dq_i \quad .$$

Ce modèle est analytiquement stable pour la conjugaison des germes de 2-formes différentielles fermées, cf. [8]; son lieu singulier est défini par l'équation $p_1 = 0$ et son noyau, en ces points, est visiblement engendré par les vecteurs $\partial/\partial q_1 (x)$, $\partial/\partial p_1 (x)$. En explicitant la condition de l'assertion (4), le lecteur peut déterminer l'expression locale des hamiltoniennes admissibles au voisinage de ces points:

$$h(q_1,\dots,q_m,p_1,\dots,p_m) = p_1^2\, A(q_1,\dots,q_m,p_1,\dots,p_m) + B(q_2,\dots,q_m,p_2,\dots,p_m)$$

où A,B sont de fonctions différentiables quelconques, ainsi que l' expression des champs hamiltoniens correspondants:

$$X_h = \frac{1}{p_1}\left(\frac{\partial h}{\partial p_1}\frac{\partial}{\partial q_1} - \frac{\partial h}{\partial q_1}\frac{\partial}{\partial p_1}\right) + \sum_{i=2}^{m}\left(\frac{\partial h}{\partial p_i}\frac{\partial}{\partial q_i} - \frac{\partial h}{\partial q_i}\frac{\partial}{\partial p_i}\right) \quad .$$

Pour les points plus dégénérés de la partie lisse de $\Sigma(\omega|M)$, on peut déterminer l'expression locale des hamiltoniennes admissibles et de leurs champs hamiltoniens à l'aide du modèle correspondant des 2-formes fermées de R.Roussarie donné dans [17]. Pour les strates plus dégénérées de $\Sigma(\omega|M)$, on peut obtenir d'exemples génériques en employant les résultats récents sur les 2-formes fermées de F.Pelletier, cf. [10].

Le lecteur peut vérifier facilement que l'ensemble des hamiltoniennes admissibles $h:M \to \mathbb{R}$ pour une variété d'états permis générique M de dimension paire, possède une structure naturelle d'algèbre de Lie définie par le *crochet générique de Poisson*:

$$\{h,h'\} = (\omega|M)(X_h, X_{h'})$$

qui visiblement est idéntifié à la notion du crochet de Poisson classique sur le lieu régulier et la généralise sur le lieu singulier de M.

Le problème de la détermination des dynamiques admissibles pour les variétés d'états permis et de la construction de leurs extensions à l'espace de phases ambiant, se pose aussi naturellement en Géométrie de Contact: Si (W,α) est une variété de contact, i.e. une variété différentiable W de dimension impaire munie d'une forme de Pfaff α dont l'espace caractéristique est partout nul, on sait qu'à chaque fonction différentiable $F: W \to \mathbb{R}$ est associé un champ de contact X_F, i.e. un champ de vecteurs différentiable unique satisfaisant les conditions

$$\begin{cases} \alpha(X_F) = F \\ L_{X_F}\alpha \wedge \alpha = 0 . \end{cases}$$

Etant donné une variété différentiable M de dimension k plongée dans W, on cherche à déterminer les fonctions $f : M \to \mathbb{R}$ qui admettent une extension différentiable $F : W \to \mathbb{R}$ dont le champ de contact X_F est tangent à M.

La difficulté de la construction de ces extensions, lorsque celui-ci est possible, est due ici à la non constance éventuelle de la dimension de l'espace caractéristique $K_x(\alpha|M)$ de la restriction de la forme de contact α à M ; plus précisement, le noyau de $\alpha|M$ étant de dimension constante k-1 sur M, il faudra ici commencer par connaître non seulement le comportement de la dimension du noyau de $d(\alpha|M)$ mais aussi la façon avec laquelle celui-ci penètre dans le noyau de $\alpha|M$ sur les points de M. Pour préciser le contexte générique dans lequel nous

nous plaçons, on considère le fibré $T^*M \times \overset{2}{\wedge} T^*M$ muni de sa stratification canonique par le rang construite dans [8], et les morphismes de fibrés sur M:

$$\begin{array}{ccc} J^1(T^*M) & \xrightarrow{\Phi_1} & T^*M \times \overset{2}{\wedge} T^*M \\ j^1_x\alpha' & \longrightarrow & (j^0_x\alpha', j^0_x d\alpha') \end{array} \quad \text{et} \quad \begin{array}{ccc} T^*M \times \overset{2}{\wedge} T^*M & \xleftarrow{\Phi_2} & J^1(M, W) \\ (i^*\alpha(x), i^*d\alpha(x)) & \longleftarrow & j^1_x i \end{array}$$

où $J^1(T^*M)$ désigne le fibré des 1-jets de sections de T^*M et $J^1(M,W)$ le fibré des 1-jets d'applications différentiables de M dans W. Comme on peut le prouver facilement, Φ_1 est une submersion, ainsi que Φ_2 en chaque 1-jet d'immersion, et par conséquent la stratification canonique de $T^*M \times \overset{2}{\wedge} T^*M$ se relève par image réciproque en une stratification de $J^1(T^*M)$ et une stratification de l'ouvert des 1-jets d'immersions de $J^1(M,W)$. D'après le théorème de transversalité de R.Thom, les formes de Pfaff, resp. les plongements, dont le 1-jet est transverse à cette stratification, constituent un ouvert dense de l'espace $\Lambda^1(M)$ des formes de Pfaff sur M, resp. de l'espace $P(M,W)$ des plongements de M dans W, muni de la C^∞-topologie de Whitney. Les formes de Pfaff, resp. les plongements, qui satisfont ces conditions de transversalité seront appelés ici *génériques*. En considérant le diagramme commutatif

$$\begin{array}{ccccc} J^1(T^*M) & \xrightarrow{\Phi_1} & T^*M \times \overset{2}{\wedge} T^*M & \xleftarrow{\Phi_2} & J^1(M,W) \\ & \nwarrow^{j^1(i^*\alpha)} & \uparrow & \nearrow^{j^1 i} & \\ & & M & & \end{array}$$

on peut voir qu'un plongement i est générique dans $P(M,W)$, si et seulement si, la forme induite $i^*\alpha$ est générique dans $\Lambda^1(M)$. Ainsi, en se ramenant à la théorie des formes différentielles génériques, on en déduit que les ensembles

$$\Xi_\nu(\alpha|M) = \{ x \in M \ / \ \dim K_x(\alpha|M) = \nu \quad \& \quad (\alpha|M)(x) \neq 0 \}$$

$\nu = 0,1,2,\ldots$ sont des sous-variétés régulières, de codimension $\nu(\nu-1)/2$ lorsqu'ils apparaîssent, et la forme $\alpha|M$ s'annule qu'en des points isolés sur lesquels sa différentielle est de rang maximum. La stratification de M ainsi réalisée est finie et localement triviale; aussi l'adhérence de chaque strate est localement difféomorphe à une sous-variété algébrique de $\mathbb{R}^k$ et possède la propriété de frontières:

$$\overline{\Xi_\nu(\alpha|M)} = \bigcup_{\nu' \geq \nu} \Xi_{\nu'}(\alpha|M)$$

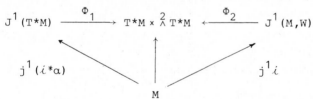

sauf pour $\Xi_0(\alpha|M)$ en dimension paire, et $\Xi_0(\alpha|M)$ et $\Xi_1(\alpha|M)$ en dimension impaire, dont les adhérences peuvent contenir les zéros de $\alpha|M$. Indépendament de la parité ou non de la dimension de M, l'espace caractéristique $K_x(\alpha|M)$ est nul sur l'ouvert dense $\Xi_0(\alpha\ M)$ de M et cesse de l'être sur les points du lieu singulier $\Xi(\alpha|M) = M - \Xi_0(\alpha|M)$. Ce lieu singulier, lorsqu'il apparaît, est une hypersurface dont la partie lisse est constituée des points où l'espace caractéristique est unidimensionel. Son type différentiable local ne dépend que de la dimension de l'espace caractéristique au point considéré; les modèles d'équations algébriques locales en sont donnés dans [15].

Dans [16], on poursuit cette étude et on prouve:

THEOREME 3. *Si M est une variété d'états permis générique de dimension impaire dans une variété de contact (W,α) et $f : M \to \mathbb{R}$ une fonction différentiable, alors les assertions suivantes sont équivalentes:*

(1) *La fonction f admet une extension différentiable $F : W \to \mathbb{R}$ dont le champ de contact X_F est tangent à M.*

(2) *La fonction f possède un champ de contact X_f sur M, i.e. un champ de vecteurs différentiable solution unique du système d'équations*

$$\begin{cases} (\alpha|M)(X_f) = f \\ L_{X_f}(\alpha|M) \wedge (\alpha|M) = 0 \end{cases} .$$

(3) *Les conditions suivantes sont satisfaites en presque tout point où l'espace caractéristique de $\alpha|M$ est unidimensionel:*

(A) $$d(f.\alpha|M) \wedge (d(\alpha|M))^{m-1} = 0$$

et

(B) $$df \wedge (d(\alpha|M))^{m} = 0 .$$

Si on impose une condition de transversalité supplémentaire: le 2-jet de $\alpha|M$ d'être transverse à la stratification du fibré des 2-jets de sections de T^*M construite dans [8], alors

$$\Xi_{1,0}(\alpha|M) = \{ x \in \Xi_1(\alpha|M) \ / \ K_x(\alpha|M) \pitchfork T_x\Xi_1(\alpha|M) \}$$

est ouvert dense dans l'ensemble $\Xi_1(\alpha|M)$ des points où l'espace caractéristique de $\alpha|M$ est unidimensionel. En chaque point de cet ouvert, d'après un théorème classique de J.Martinet cf.[8], la forme $\alpha|M$ peut s'exprimer localement dans un système de coordonnées approprié $(q_1,\ldots,q_m, p_1,\ldots,p_m, z)$ comme:

$$\alpha|M = dq_1 + \sum_{i=1}^{m} p_i\, dq_i + z\, dz \quad .$$

Grâce à ce modèle on prouve que sur les points de $\Xi_{1,0}(\alpha|M)$ la satisfaction de la condition (A) implique celle de (B), et donc sur ces points la satisfaction de la condition (A) équivaut aux assertions (1) et (2) du théorème. Cette condition s'éxprime aussi, à l'aide du champ ξ défini sur $\Xi_{1,0}(\alpha|M)$ par la rélation: $\xi \lrcorner (d\tilde{\alpha})^m = \tilde{\alpha} \wedge (d\tilde{\alpha})^{m-1}$ où $\tilde{\alpha}$ désigne la restriction de $\alpha|M$ sur $\Xi_{1,0}(\alpha|M)$, par:

(4) $$\xi(f) + f = 0 \quad .$$

Notons que l'ensemble des fonctions f qui vérifient les conditions du théorème est stable par le *crochet générique de Lagrange* défini, comme dans le cas classique, par:

$$[\![f , f']\!] = (\alpha|M)([X_f , X_{f'}])$$

et il réçoit ainsi une structure naturelle d'algèbre de Lie. Notons enfin que les trajectoires du champ X_f de conditions initiales dans une strate $\Xi_\nu(\alpha|M)$ ne sortent jamais de cette strate.

REFERENCES

[1] V.I. Arnold, On matrices depending on parameters. Russian Math. Surveys 26(1971),p.29-43.

[2] V.I. Arnold, Lagrangian manifolds with singularities,asymptotic rays, and the open swallowtail. Funct.Anal.and Appl. 15,4(1981),p. 325-246.

[3] V.I. Arnold, Singularities of ray systems. Intern. Cong. Math., Warsaw 1983.

[4] E. Cartan, Les systèmes différentiels extérieurs et leurs applications en Géométrie. Hermann, Paris 1945.

[5] P.A.M. Dirac, Generalized hamiltonian dynamics. Canad. J. Math. 2(1950), p. 129-148.

[6] A. Lichnerowicz, Variétés symplectiques et dynamiques associées à une sous-variété. C.R.Acad.Sciences Paris 280A(1975),p.523-527.

[7] A. Lichnerowicz, Les variétés de Poisson et leurs algèbres de Lie associées. J.Diff.Geometry 12(1977),p.253-300.

[8] J. Martinet, Sur les singularités des formes différentielles. Ann. Inst. Fourier 20,1(1970),p.95-178.

[9] J.N. Mather, Solution of generic linear equations. Dynamical systems, p.185-193, Acad.Press, New-York 1973.

[10] F. Pelletier, Singularités d'ordre supérieur de formes différentielles. Thèse, Université Dijon, Dijon 1980.

[11] S.N. Pnevmatikos, Structures hamiltoniennes en présence de contraintes. C.R.Acad.Sciences Paris, 289A(1979),p.799-802.

[12] S.N. Pnevmatikos, Structures symplectiques singulières génériques. Ann. Inst. Fourier 34,3(1984),p.11-28.

[13] S.N. Pnevmatikos, Contraintes génériques dans les espaces de phases, (à paraître).

[14] S.N. Pnevmatikos, Sur les formes de Pfaff et les structures de contact singulières génériques. C.R. Acad. Sciences Paris, 295A(1982),p.695-698.

[15] S.N. Pnevmatikos, Singularités génériques en Géométrie de Contact. Ann. Inst. H. Poincaré, 1984.

[16] S.N. Pnevmatikos, Contraintes génériques en Géométrie de Contact. Ann. Inst. H. Poincaré, (à paraître).

[17] R. Roussarie, Modèles locaux de champs et de formes. Asterisque 30, Paris 1975.

[18] J.M. Souriau, Structures des Systèmes Dynamiques. Dunod Univ. Paris 1969.

[19] R. Thom, Stabilité Structurelle et Morphogénèse. Benjamin, New-York 1972.

[20] J.C. Tougeron, Idéaux de fonctions différentiables. Springer Verlag, Berlin 1972.

* Spyros N. PNEVMATIKOS

45-49 rue Elie Zervou

11144 Athènes

GRECE

Singularities & Dynamical Systems
S.N. Pnevmatikos (editor)

PSEUDO METRIQUES GENERIQUES ET THEOREME DE GAUSS BONNET EN DIMENSION 2

Fernand Pelletier

Université de Dijon
France

§0. INTRODUCTION.

Le théorème classique de Gauss-Bonnet pour une surface compacte orientable M muni d'une métrique riemannienne g établit que :

$$\frac{1}{2\pi}\int_M \Omega = \chi(M)$$

où Ω est la courbure de la connexion de Levi-Cevita de g et $\chi(M)$ la caractéristique d'Euler Poincaré de M. L'objet de ce travail est d'établir un résultat analogue pour certaines pseudo-métriques génériques.

Une *pseudo-métrique* g sur une variété M est un champ C^∞ de formes bilinéaires symétriques. Lorsque g est définie positive, g est *une métrique riemannienne*. Si g est non dégénérée en tout point de M, g est une *métrique pseudo-riemannienne*. En général, pour une pseudo-métrique g, l'ensemble $\Sigma(g)$ des points x de M où g_x se dégénère est non vide : $\Sigma(g)$ est appelé le *lieu singulier* de g. Le *noyau de g* en $x \in M$ est le sous espace $K_g(x) = \{u \in T_xM / g(u,v) = 0$ pour tout $v \in T_xM\}$.

Lorsque la variété M est une surface, pour une pseudo-métrique générique g, $\Sigma(g)$ est une sous variété régulière de dimension 1, éventuellement vide ; le noyau de g est un champ de droites sur $\Sigma(g)$. De plus, ce champ de droites est transverse à $\Sigma(g)$ sur un ouvert dense de $\Sigma(g)$ et il est tangent à $\Sigma(g)$ en un ensemble de points isolés (Pour plus de détails, voir le paragraphe 1).

Soit g une pseudo métrique générique sur une surface compacte M. Sur la variété ouverte $M' = M-\Sigma(g)$, g induit une métrique pseudo-riemannienne g'. A g' est associée sa connexion de Levi-Cevita ∇'. Le lieu singulier $\Sigma(g)$ étant une variété, g induit sur $\Sigma(g)$ une pseudo-métrique $\bar{g}$. Sur la variété $\Sigma'(g) = \Sigma(g) - \Sigma(\bar{g})$ on a la connexion de Levi-

Cevita $\bar{\nabla}'$ de $\bar{g}$. On peut se demander :

1°) A quelle condition la connexion de Levi-Cevita ∇' sur $M-\Sigma(g)$ se prolonge sur $\Sigma(g)$ par la connexion de Levi-Cevita $\bar{\nabla}'$ de $\bar{g}$?

2°) Lorsque ∇' se prolonge sur $\Sigma(g)$, la forme courbure Ω de ∇' est elle intégrable et quelle relation existe-t-il entre cette intégrale et la caractéristique d'Euler Poincaré $\chi(M)$ de M?

3°) Réciproquement si la 2-forme Ω est intégrable, la connexion ∇' se prolonge-t-elle sur $\Sigma(g)$?

La réponse à toutes ces questions est donnée par les résultats suivants :

THEOREME : *Soit* g *une pseudo métrique générique sur une surface compacte dont la connexion de Levi-Cevita* ∇ *sur* $M-\Sigma(g)$ *se prolonge sur* (g) *(voir déf.2.1.). La forme courbure* Ω *de* ∇' *possède alors les propriétés suivantes :*

1) *Pour toute composante connexe* M_i' *de* $M-\Sigma(g)$ *l'intégrale* $\int_{M_i'}\Omega$ *existe*

2) *On a* $\frac{1}{2\pi}\int_{M_i'}\Omega = \varepsilon_i\chi(M_i')$,

où $\varepsilon_i = 1$ *(resp.* -1*) si la signature de* g *sur* M_i' *est* (2,0) *ou* (1,1) *(resp.* (0,2)*).*

De plus, $\chi(M_i') = 0$ *si la signature de* g *sur* M'_i *est* (1,1).

3) *Soit* Δ *la* 2*-forme égale à* $\frac{\varepsilon_i}{2\pi}\Omega|_{M_i'}$ *sur chaque composante* M_i' *de* $M-\Sigma(g)$.

On a alors :

$$\int_M \Delta = \chi(M).$$

4) *Sur* M, *il existe des pseudo-métriques génériques vérifiant éventuellement (1) et (2), et pour lesquelles* Ω *est intégrable sur* M, *mais la connexion de Levi-Cevita de* g *ne se prolonge pas sur* $\Sigma(g)$.

Ce travail est décomposé en quatre paragraphes. On commence d'abord par rappeler les résultats sur le comportement générique du rang d'une pseudo-métrique (paragraphe 1). Pour les pseudo-métriques génériques, on établit ensuite, dans le paragraphe 2, des conditions nécessaires et suffisantes pour que la connexion de Levi-Cevita se prolonge sur le lieu singulier. On construit également des exemples de pseudo-métriques génériques ayant cette propriété et dont le lieu singulier est non vide. Le paragraphe 3 est consacré à la démonstration des propriétés (1) (2) et (3) du théorème. Dans le paragraphe 4, on construit des exemples de pseudo-métriques génériques vérifiant la propriété (4) du théorème.

§1. COMPORTEMENT GENERIQUE D'UNE PSEUDO-METRIQUE SUR UNE VARIETE [6].

Soit g une pseudo-métrique sur une variété M. Si l'ensemble singulier $\Sigma(g)$ est non vide, $M' = M-\Sigma(g)$ est une réunion d'ouverts connexes M'_i, $i \in I$, et sur chaque M'_i la signature de g est constante. Génériquement, l'ensemble $\Sigma_c(g) = \{x/\dim K_g(x) = c\}$, s'il est non vide, est une sous variété régulière de M de codimension $\frac{c(c+1)}{2}$. Sur chaque $\Sigma_c(g)$ le noyau de g est un champ de c-plans. De manière évidente, si q(x) est la dimension de $K_g(x) \cap T_x\Sigma_c(g)$, le rang en x de la restriction $\bar{g}$ de g à $\Sigma_c(g)$ est inférieur ou égal à $\dim \Sigma_c(g) - q(x)$. En d'autres termes le noyau $K_{\bar{g}}(x)$ de $\bar{g}$ en $x \in \Sigma_c(g)$ contient $K_g(x) \cap T_x\Sigma_c(g)$. Plus généralement, si N est une sous variété de M, et g_N est la restriction de g à M, on a l'inclusion :

$$K_g(x) \cap T_xN \underset{\neq}{\subset} K_{g_N}(x)$$

l'inclusion pouvant être stricte comme le prouve l'exemple suivant : Soit g_o la métrique hyperbolique du $\otimes$ dv sur $\mathbb{R}^2$ et N la sous variété totalement isotrope d'équation u = O ; la restriction g_N de g à N est nulle et on a l'inclusion stricte pour tout $x \in N$:

$$K_{g_N}(x) = T_xN \underset{\neq}{\supset} K_g(x) \cap T_xN = \{O\} .$$

Comme pour les formes différentielles, ([5]) le comportement du rang de g sur $\Sigma_c(g)$ sera décrit par l'étude des espaces

$$K_g(x), K_g(x) \cap T_x\Sigma_c(g) \text{ et } K_{\bar{g}}(x) \text{ pour } x \in \Sigma_c(g),$$

où $\bar{g}$ est la restriction de g à $\Sigma_c(g)$.
On va définir les ensembles :

$$\Sigma_{cc_1 c'_1}(g) = \{x \in \Sigma_c(g) / \dim[K_g(x) \cap T_x\Sigma_c(g)] = c_1 \text{ et } \dim K_{\bar{g}} = c'_1\}.$$

De même, lorsque $\Sigma_{cc_1 c'_1}(g)$ est une sous variété régulière de M non vide, le comportement du rang de g sur $\Sigma_{\substack{cc_1\\ c'_1}}(g)$ sera décrit par l'étude des espaces :

- $K_g(x)$
- $K_g(x) \cap T_x\Sigma_c(g)$, $K_{\bar{g}}(x)$
- $K_g(x) \cap T_x\Sigma_{\substack{cc_1\\ c'_1}}(g)$, $K_{\bar{g}}(x) \cap T_x\Sigma_{\substack{cc_1\\ c'_1}}(g)$, $K_{\bar{\bar{g}}}(x)$

où $\bar{\bar{g}}$ est la restriction de g à $\Sigma_{\substack{cc_1\\ c'_1}}(g)$.

On étudiera l'ensemble $\Sigma_{\substack{cc_1c_2\\ c'_1c'_2\\ c''_2}}(g)$ des $x \in \Sigma_{\substack{cc'_1\\ c_1}}(g)$ tels que :

$$\dim\left[K_g(x) \cap T_x\Sigma_{\substack{cc_1\\ c'_1}}(g)\right] = c_2 \quad ; \quad \dim K_{\bar{g}}(x) \cap T_x\Sigma_{\substack{cc_1\\ c'_1}}(g) = c'_2 \; ;$$

$$\dim K_{\bar{\bar{g}}}(x) = c''_2.$$

Avant d'énoncer le comportement générique du rang d'une pseudo-métrique, nous allons introduire quelques notations. Soient $M_o = M \supset M_1 \supset \ldots \supset M_\ell \supset \ldots \supset M_k$ des sous variétés régulières de M. On note g_ℓ la pseudo-métrique sur M_ℓ induite par g. Pour $R = (r_o,\ldots,r_k)$ et $S = (s_o,\ldots,s_k)$ appartenant à $\mathbb{N}^{k+1}$ on définit l'ensemble $\Sigma_{RS}(M_o,\ldots,M_k;g)$ des x appartenant à M_k tels que :

1°) $\dim K_{g_i}(x) \cap T_xM_k = r_i$

2°) La signature de $g_i(x)$ est $(s_i, \dim M_i - r_i - s_i)$ pour tout $i = 0,\ldots,k$.

Le symbole de g *en* $x \in M_k$ sur $(M_o,\ldots,M_k)$ est l'élément $(R_o,\ldots,R_k,S_k)$ de $\mathbb{N}\times\ldots\times\mathbb{N}^{k+1} \times \mathbb{N}^{k+1}$ défini par :

$R_j = (r_o^j,\ldots,r_i^j,\ldots,r_j^j)$ si $r_i^j = \operatorname{codim}\left[K_{g_i}(x) \cap T_xM_j\right]$.

$S_k = (s_o,\ldots,s_i,\ldots,s_k)$ si la signature de $g_i(x)$ est $(s_i, \dim M_i - s_i - r_i^i)$.

Le symbole de g en x se présente sous la forme d'un tableau triangulaire :

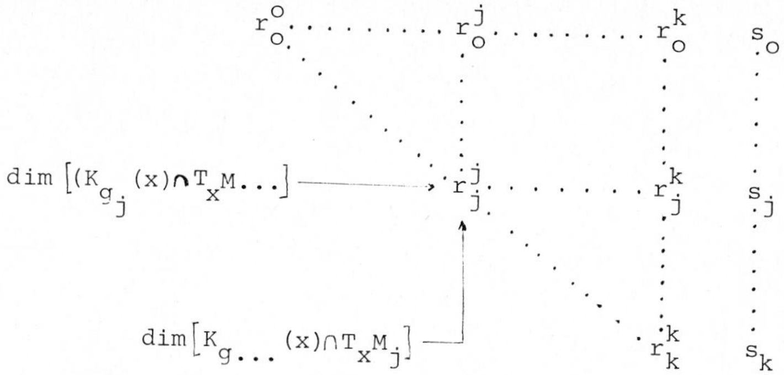

On a alors le résultat suivant :

THEOREME 1.1. [6] *dans l'espace* $\mathcal{P}$(M) *des pseudo-métriques sur* M, *les propriétés suivantes sont génériques :*
Pour tout entier k *et tout* $(R_o,\dots,R_k,S_k)\in \mathbb{N}\times\dots\times\mathbb{N}^{k+1}\times\mathbb{N}^{k+1}$, *si* n *est assez grand, les sous ensembles* $\Sigma_{R_oS_o}(g),\dots,\Sigma_{R_o\dots R_kS_k}(g)$ *sont des sous variétés régulières de* M, *s'ils sont non vides,* $\Sigma_{R_o\dots R_jS_j}(g)$ *étant défini par récurrence de la manière suivante :*

$$\Sigma_{R_o\dots R_jS_j} = \Sigma_{r_jS_j}[M,\Sigma_{R_oS_o}(g),\dots,\Sigma_{R_o\dots R_{j-1}S_{j-1}}(g);g]$$

où $S_j = (s_o,\dots,s_j)$ *si* $S_k = (s_o,\dots,s_j,\dots,s_k)$.
Le symbole d'un élément générique g *est* $(R_o,\dots,R_k,S_k)$ *en tout point de* $\Sigma_{R_o\dots R_kS_k}(g)$. *Lorsque* $\Sigma_{R_o\dots R_kS_k}(g)$ *est non vide sa codimension* $\nu_{R_o\dots R_k}$ *ne depend que de* $(R_o,\dots,R_k)$ *et sa valeur est donnée par la formule de récurrence suivante :*

$$\nu_{R_o} = \frac{r_o^o(r_o^o+1)}{2}$$

$$\nu_{R_o\dots R_k} = \nu_{R_o\dots R_{k-1}} + \frac{(r_k^k-r_{k-1}^k)(r_k^k-r_{k-1}^k+1)}{2} +$$

$$+ r_{k-1}^k(\nu_{R_o\dots R_{k-1}} - \nu_{R_o\dots R_{k-2}} - r_{k-1}^{k-1}+r_{k-1}^k) + \sum_{i=1}^{k} r_{i-1}^k(r_i^{k-1}-r_i^k-r_k^k-r_{k-1}^{k-1}+r_i^k)$$

avec la convention $\nu_{R_o\dots R_{k-2}} = 0$ pour k=1.

Puisque la *codimension de* $\Sigma_{R_o\dots R_kS_k}(g)$ *ne depend par de* S_k *on note*

$\Sigma_{R_o\ldots R_k}(g)$ la réunion des variétés $\Sigma_{R_o\ldots R_k S_k}(g)$ pour $S_k \in \mathbb{N}^{k+1}$. Ce théorème se démontre comme le résultat analogue pour les 2-formes [5] Pour plus de détails voir [6] . Nous allons terminer ce paragraphe en explicitant la description générique que l'on obtient en dimension 2:

(O) le lieu singulier de g est une sous variété régulière de M de dimension 1 : l'ensemble $\Sigma_1(g)$ des points où le noyau de g est de dimension 1. La signature de g est (1,0) ou (0,1) en chaque point de $\Sigma_1(g)$. Avec les notations du théorème 1.1., $\Sigma_1(g)$ est la réunion des variétés :

$\Sigma_{1;1}(g)$ ensemble des points de $\Sigma_{1,0}(g)$ où la signature est (1,0).
$\Sigma_{1;0}(g)$ ensemble des points de $\Sigma_1(g)$ où la signature est (0,1).

(1) La variété $\Sigma_1(g)$ est la réunion des deux variétés suvantes :
$\Sigma_{\substack{10\\0}}(g)$ ensemble des points où le noyau de g est transverse à $\Sigma_1(g)$
$\Sigma_{\substack{11\\1}}(g)$ ensemble des points où le noyau de g est tangent à $\Sigma_1(g)$.

La variété $\Sigma_{\substack{10\\0}}(g)$ *que l'on notera dans la suite* $\Sigma_{10}(g)$, est ouverte dans $\Sigma_1(g)$ et $\Sigma_{\substack{11\\1}}(g)$ *que l'on montrera dans la suite* $\Sigma_{11}(g)$ est un ensemble de points isolés.

Avec les notations du théorèmes 1.1., $\Sigma_{10}(g)$ est la réunion des variétés

$$\left.\begin{cases}\Sigma_{\substack{10;1\\0\ 1}}(g)\\ \Sigma_{\substack{10,1\\0\ 0}}(g)\\ \Sigma_{\substack{10,0\\0\ 0}}(g)\end{cases}\right\}$$ ensemble des points de $\Sigma_{10}(g)$ où la signature de g (resp de la restriction $\bar{g}$ de g à $\Sigma_1(g)$) est : $\begin{cases}(1,0) \text{ resp. } (1,0)\\ (1,0) \text{ resp. } (1,0)\\ (0,1) \text{ resp. } (0,1)\end{cases}$. De même

$\Sigma_{11}(g)$ est la réunion des variétés

$$\left.\begin{cases}\Sigma_{\substack{11;1\\1\ 0}}(g)\\ \Sigma_{\substack{11;0\\1\ 0}}(g)\end{cases}\right\}$$ ensembles des points de $\Sigma_{11}(g)$ où la signature de g (resp. de la restriction $\bar{g}$ de g à $\Sigma_1(g)$) est : $\begin{cases}(1,0) \text{ resp. } (0,0)\\ (0,1) \text{ resp. } (0,0)\end{cases}$

Nous allons illustrer cette description *par un exemple*.
Sur $\mathbb{R}^2$, considérons la pseudo-métrique :

$$g = dx \otimes dx + 2y dx \otimes dy + (x^2-a^2+y^2) dy \otimes dy$$

Le lieu singulier $\Sigma(g) = \Sigma_1(g)$ est la réunion des droites D d'équation x = a et D' d'équation x = -a. L'ensemble $\Sigma_{10}(g)$ est la réunion des demi-droites ouvertes de D et D' obtenue en retirant les points A et A' de coordonnées (a,0) et (-a,0) respectivement. Enfin, $\Sigma_{11}(g)$ est formé des points A et A' (voir la figure ci-dessous) :

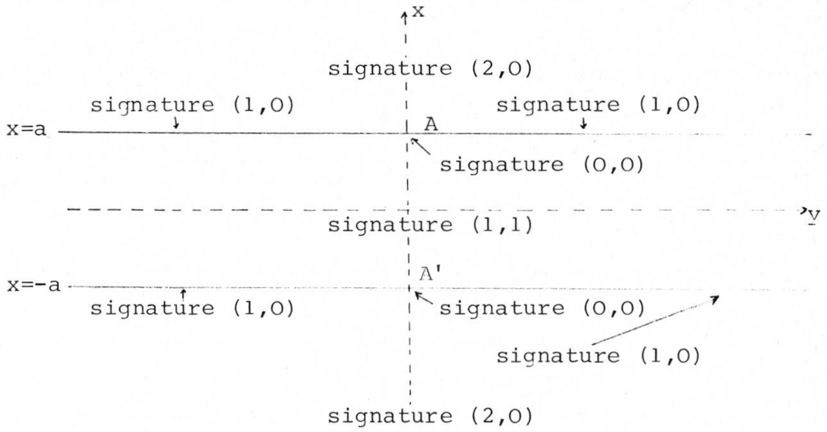

§2. PROLONGEMENT SUR LE LIEU SINGULIER DE LA CONNEXION DE LEVI-CEVITA

Dans tout ce paragraphe, M est une variété de dimension 2 et g une pseudo-métrique générique sur M (au sens du théorème 1.1.). On va d'abord définir la notion de "prolongement" sur $\Sigma(g)$ de la connexion de Levi-Cevita ∇' de $g|M' = M - \Sigma(g)$.
On donnera ensuite deux types de conditions nécessaires et suffisantes pour g pour que ∇' se prolongent sur $\Sigma(g)$: l'une portera sur la dérivée de Lie de g (proposition 2.2.).L'autre sur le comportement de la courbure géodésique des courbes voisines de $\Sigma(g)$ (proposition 2.4.). Le paragraphe s'achèvera par la construction de pseudo-métriques ayant la propriété de prolongement.

Soit $\Gamma(M)$ l'algèbre de Lie des champs de vecteurs sur M. Rapplons qu'une connexion linéaire sur M est une application $\mathbb{R}$-bilinéaire.

$$\nabla : \Gamma(M) \times \Gamma(M) \to \Gamma(M)$$

telle que pour toute fonction f et tout champ de vecteurs X et Y on ait :

(1) $\nabla_{fX}Y = f\nabla_X Y$

(2) $\nabla_X fY = X(f).Y + f\nabla_X Y$.

De plus, si h est une métrique riemannienne (ou pseudo-riemannienne), la connexion ∇ est compatible avec h si elle vérifie :

(3) $X\{h(Y,Z)\}=h(\nabla_X Y,Z) + h(Y,\nabla_X Z)$ pour tout champ X,Y,Z de $\Gamma(M)$.

La connexion de Levi-Cevita de h est l'unique connexion linéaire compatible avec h à torsion nulle c'est-à-dire $\nabla_X Y-\nabla_Y X = X,Y$. Considérons maintenant une pseudo-métrique générique g sur M. On note M' la variété ouverte $M-\Sigma(g)$. Remarquons d'abord que la connexion de Levi-Cevita ∇' de $g_{|M'}$ sur M' ne peut pas se prolonger dans la direction transverse à $\Sigma(g)$, si $\Sigma(g) \neq \emptyset$. En effet, pour tout $x \in \Sigma_{10}(g)$, il existe un voisinage U de x et un champ de vecteur K sans singularités sur U, tel que K engendre le noyau de g sur $\Sigma(g)$. Sur $U-\Sigma(g)$, on aura d'après (3)

$$K\{g(K,K)\} = 2g(\nabla'_K K,K)$$

Si $\nabla'_K K$ est défini sur $\Sigma_{10}(g) \cap U$, K étant le noyau de g sur $\Sigma_{10}(g) \cap U$, on a :

$$g((\nabla'_K K,K) = 0$$

D'autre part, compte tenu de la généricité de g, $g(K,K) = 0$ est une équation de $\Sigma_{10}(g)$ sur U. Comme K est transverse à $\Sigma_{10}(g)$ sur U, $K\{g(K,K)\}$ est non nul sur $\Sigma_{10}(g)$. Soit $\Gamma(g)$ l'algèbre de Lie des champs de vecteurs sur M tangent à $\Sigma(g)$:

Définition 2.1. : On dit que la connexion de Levi-Cevita ∇' de g se prolonge sur $\Sigma(g)$ s'il existe une application $\mathbb{R}$-bilinéaire

$\nabla : \Gamma(M) \times \Gamma(g) \to \Gamma(M)$ ayant les propriétés suivantes :

(1) $\nabla_{fX} Y = f\nabla_X Y$

(2) $\nabla_X fY = X(f).Y+f\nabla_X Y$ $\Big\{$ pour toute fonction f et champ X de $\Gamma(M)$ et Y de $\Gamma(g)$.

(3) $X\{g(Y,Z)\} = g(\nabla_X Y,Z) + g(Y,\nabla_X Z)$ pour tout $X\in\Gamma(M)$ et $Y,Z \in \Gamma(g)$.

(4) $\nabla_X Y =\nabla'_X Y$ pour tout champ de vecteurs X et Y sur M'.

Il est clair que si ∇' se prolonge sur $\Sigma(g)$, l'application bilinéaire

$$\nabla : \Gamma(M) \times \Gamma(g) \to \Gamma(M)$$

qui prolonge ∇' est unique. ∇ induit sur $\Sigma_{10}(g)$ une connexion linéaire compatible avec la métrique $\bar{g}$ definie par g sur $\Sigma_{10}(g)$: c'est la connexion de Levi-Cevita de $\bar{g}$. En d'autres termes, si la connexion de Levi-Cevita ∇' de $g_{|M'}$ se prolonge sur $\Sigma(g)$ alors ∇' se prolonge par la connexion de Levi-Cevita de la restriction de g à $\Sigma_{10}(g)$.

Lorsque $\Sigma(g)$ est vide, g est alors non dégénéré : c'est une métrique pseudo-riemannienne sur M et la connexion de Levi-Cevita ∇ de g est définie sur toute la variété M. En fait, ∇ est caractérisé par l'équation :

$$(5) : 2g(\nabla_X Y,Z)=X\{g(Y,Z)\}+Y\{g(X,Z)\}-Z\{g(X,Y)\} +g([X,Y],Z)-g([X,Z],Y)-g([Y,Z],X).$$

Par contre lorsque $\Sigma(g)$ n'est pas vide, on verra qu'en général la connexion de Levi-Cevita ∇' ne se prolonge pas. Supposons que ∇' se prolonge sur $\Sigma(g)$ par une application $\nabla : \Gamma(M) \times \Gamma(g) \to \Gamma(M)$. Cette application devra vérifier l'équation (5), pour tout champ X et Z de $\Gamma(M)$ et Y de $\Gamma(g)$. En particulier, si K est un champ de vecteurs qui engendre le noyau de g sur $\Sigma(g)$, on aura :

$$g(\nabla_X Y,K) = 0 \text{ sur } \Sigma(g)$$

Par suite, pour que ∇' se prolonge sur $\Sigma(g)$, il faut que le second membre de (5) soit nul sur $\Sigma(g)$, pour tout $X \in \Gamma(M)$, $Y \in \Gamma(g)$, et $Z \in \Gamma(M)$ avec $Z(x) \in K_g(x)$ sur $\Sigma(g)$. Pour $X \in \Gamma(M)$, $Y \in \Gamma(g)$ désignons par α_{XY} la 1-forme sur M définie par : $\alpha_{XY}(Z)$ est égale au second membre de (5) pour tout $Z \in \Gamma(M)$. On a alors :

Proposition 2.2. : Soit $\mathcal{K}(g)$ *l'espace des champs de vecteurs* K *sur* M *tel que :*
$K(x)$ *appartient au noyau* $K_g(x)$ *pour tout* x *de* $\Sigma(g)$. *La connexion de Levi-Cevita* ∇' *de* g *se prolonge si et seulement si :*
(*) *pour tout* X *et* Y *de* $\Gamma(g)$ *et* K *de* $\mathcal{K}(g)$ *la dérivée de Lie* $L_K g(X,Y)$ *est nulle en tout point de* $\Sigma_{10}(g)$.

La démonstration de cette proposition s'appuie essentiellement sur le résultat suivant qui peut être démontré en adaptant le théorème V p. 57 de [7] ou le théorème d'équivalence locale des matrices génériques de [4] :

Lemme 2.3. : Soit g une pseudo-métrique générique sur une variété V. Etant donnée une 1-forme α sur V l'équation $g(X,\ ) = \alpha$ où $X \in \Gamma(M)$ a une solution unique si et seulement si en tout point d'un ensemble dense de $\Sigma_1(g)$ la condition suivante est réalisée : $\operatorname{Ker} \alpha \supset K_g$.

Démonstration de la proposition 2.2. (résumée).

C'est un problème local. Il existe un ouvert U de M tel que $U \cap \Sigma(g) = U \cap \Sigma_{10}(g)$. Soit $\Gamma(U)$ les champs de vecteurs sur U. Compte tenu du lemme 2.3., il faut montrer que pour tout $X \in \Gamma(U)$,

$Y \in \Gamma(g)_{|U}$, $K \in \mathcal{K}(g)_{|U}$, $\alpha_{XY}(K)$ est nul sur $U \cap \Sigma(g)$ si et seulement si (✱) est satisfaite. Comme $\Sigma(g)_{|U} = \Sigma_{10}(g) \cap U$, le noyau de g est transverse à $\Sigma(g)$ sur U. Tout champ X se décompose donc en somme $X_0 + X_1$ où $X_0 \in \Gamma(g)_{|U}$ et $X_1 \in \mathcal{K}(g)_{|U}$. On a alors :

$$\alpha_{XY}(K) = \alpha_{X_0Y}(K) + \alpha_{X_1Y}(K)$$

A partir de (5), on montre que sur $\Sigma(g) \cap U$ on a :

$$\alpha_{X_1Y}(K) = O$$

$$\alpha_{X_0Y}(K) = K\{g(X_0,Y)\} + g([X_0,K],Y) + g([Y,K],X_0) = L_Kg(X_0,Y)$$

C.Q.F.D.

Dans l'espace $\mathcal{P}(M)$ des pseudo-métriques sur M, on désigne par $\mathcal{G}(M)$ l'ensemble des pseudo-métriques génériques g (au sens de 1.1.) et vérifiant :

(✱) $L_Kg(X,Y) = O$ sur $\Sigma(g)$ pour tout $K \in \mathcal{K}(g)$ et X et $Y \in \Gamma(g)$.

L'ensemble $\mathcal{G}(M)$ contient, en particulier, toutes les métriques riemanniennes et pseudo-riemanniennes sur M. A la fin de ce paragraphe, on construira des pseudo-métriques de $\mathcal{G}(M)$ dont le lieu singulier est non vide. Pour les éléments de $\mathcal{G}(M)$, on démontre :

Proposition 2.4. : Pour tout $g \in \mathcal{G}(M)$, *si* $\Sigma(g)$ *est non vide, le noyau de* g *est transverse à* $\Sigma(g)$ *en tout point c'est-à-dire* $\Sigma(g) = \Sigma_{10}(g)$ *et* $\Sigma_{11}(g)$ *est vide.*

Nous allons voir maintenant que la condition (✱) qui caractérise les éléments de $\mathcal{G}(M)$ se traduit par une condition géométrique simple sur le comportement de la courbure géodésique des courbes voisines de $\Sigma(g)$. Pour simplifier, on suppose la variété M orientable, dans le cas contraire, il suffit de considérer le revêtement à deux feuillets de M. Tout voisinage tubulaire de $\Sigma_{10}(g)$ est alors trivial. Il existe donc un plongement $\varphi : \Sigma_{10}(g) \times]-\varepsilon,\varepsilon[\to M$ tel que $V_\varepsilon = \varphi(\Sigma_{10}(g) \times]-\varepsilon,\varepsilon[)$ soit un voisinage tubulaire de $\Sigma_{10}(g)$. Notons $\mathcal{F}_\varepsilon$ le feuilletage $\{\varphi(\Sigma_{10}(g) \times \{t\}) ; -\varepsilon<t<\varepsilon\}$ et F le fibré tangent) $\mathcal{F}_\varepsilon$. Comme g est non dégénérée sur $\Sigma_{10}(g)$, quitte à restreindre V_ε, on peut supposer que $g_{|F}$ est non dégénérée ; on a alors la décomoposition :

$$TM_{|V_\varepsilon} = F \oplus F^\perp$$

où F est l'orthogonal de F relativement g. Chaque feuille L_t de $\mathcal{F}_\varepsilon$ est une variété de dimension 1. Il existe donc une section non nulle

X de F telle que $|g(X,X)| = 1$. Soit K une section non nulle de $F^{\perp}$. la courbure géodésique de la feuille L_t, $t \neq 0$, est alors la fonction: $\rho_t : L_t \to \mathbb{R}$ définie par :

$$\rho_t = \frac{g(\nabla_X X, K)}{\sqrt{|g(K,K)|}} = \frac{-L_K g(X,X)}{\sqrt{|g(K,K)|}} \quad \text{(d'après 5).}$$

On en déduit :

Proposition 2.5. : Soit g une pseudo-métrique générique sur M. La connexion de Levi-Cevita de g se prolonge sur $\Sigma(g)$ si et seulement si la courbure géodésique de L_t tend vers 0 quand t tend vers zéro.

Nous allons terminer ce paragraphe par la construction de pseudo-métriques sur M dont le lieu singulier est non vide. On montrera dans le paragraphe suivant que lorsque M est compacte orientable, toute pseudo-métrique générique de $\mathcal{G}(M)$ est obtenue de cette manière.

Soit C la variété $\mathbb{R}$ ou S^1. On désigne par $\mathcal{M}(C \times]-1,1[)$ l'ensemble des pseudo-métriques de $\mathcal{G}(C \times]-1,1[)$ dont le lieu singulier est $C \times \{-1/2\} \cup C \times \{1/2\}$. On considère des plongements disjoints $\varphi_i : C \to M$ $i=1,\dots,p$. Chaque composante connexe M'_ℓ $\ell=1,\dots,q$, de $M - \bigcup_{i=1}^{p} \varphi_i(C)$ est muni d'une métrique pseudo-riemannienne g_ℓ de signature (2,0) ou (0,2). On note g' la métrique pseudo-riemannienne sur $M' = M - \bigcup_{i=1}^{p} \varphi_i(C)$ ainsi obtenue.

Supposons la variété M orientable. Il existe alors des voisinages tubulaires :

$\psi_i : C \times]-2,2[\to M$ $i=1,\dots,p$ tels que :

- $\psi_i(C \times]-2,2[) \cap \psi_j(C \times]-2,2[) = \emptyset$ pour $i \neq j$
- $\psi_i(C \times \{0\}) = \varphi_i(C)$ $i=1,\dots,p$.

Pour obtenir une pseudo-métrique g de $\mathcal{G}(M)$, on modifie la pseudo métrique g' sur chaque voisinage tubulaire $\psi_i(C_i \times]-2,2[)$, à l'aide de pseudo-métrique $(\psi_i^{-1})^* h_i$, $h_i \in \mathcal{M}(C \times]-1,1[)$ de la manière suivante :

- on choisit dans $\mathcal{M}(C \times]-1,1[)$ une pseudo-métrique h_i ayant même signature sur $C \times]-1,-\frac{1}{2}[\cup C \times]\frac{1}{2},1[$ que g' sur $\psi_i\{C \times]-1,-1/2[\cup C \times]1/2,1[\}$.

- on recolle les pseudo-métriques $(\psi_i^{-1})^* h_i$, $i=1,\dots,p$, et g' par une partition de l'unité.

Le lieu singulier de la pseudo-métrique g ainsi obtenue est :

$$\Sigma(g) = \bigcup_{i=1}^{p} \left[\psi_i(C\times\{-1/2\} \cup \psi_i(C\times\{\tfrac{1}{2}\})\right]$$

Sur $\psi_i\ \{C \times]-\frac{1}{2}, \frac{1}{2}[\}$ la signature de g est (1,1).

Sur les autres composantes connexes de $M-\Sigma(g)$ la signature de g est (2,0) ou (0,2). Dans le cas non orientable, il suffit de remplacer $S^1\times\ ]-1,1[$ par la bande de Moëbus.

§3. THEOREME DE GAUSS BONNET POUR LES METRIQUES GENERIQUES DE $\mathcal{G}(M)$.

Dans tout ce paragraphe, M est une surface compacte orientable et g une pseudo métrique de $\mathcal{G}(M)$. L'objet de ce paragraphe est d'établir le résultat suivant (énoncé dans l'introduction).

THEOREME A : *Soit g une pseudo-métrique de $\mathcal{G}(M)$ et Ω la forme courbure de la connexion de Levi-Cevita de g sur $M-\Sigma(g)$.*

(1) Pour toute composante connexe M'_i de $M-\Sigma(g)$ l'intégrale $\int_{M'_i}\Omega$ est convergente.

(2) On a :

$$\frac{1}{2\pi}\int_{M'_i}\Omega = \varepsilon_i\chi(M'_i)$$

où $\varepsilon_i = 1$, (resp. -1) *si la signature de g sur M'_i est* (2,0),(1,1) (resp. (0,2)). *De plus* $\chi(M'_i) = 0$ *si la signature de g sur M'_i est* (1,1).

(3) Soit Δ la 2-forme égale à $\frac{\varepsilon_i}{2\pi}\Omega_{/M'_i}$ sur chaque composante M'_i de $M-\Sigma(g)$. Alors :

$$\int_M \Delta = \chi(M)$$

Il résulte de la propriété (2) du théorème A que toute composante M'_i sur laquelle la signature de g est (1,1) est difféomorphe à $S^1\times]-1,1[$. On en déduit :

THEOREME B : *Soit g une pseudo métrique générique sur une surface compacte orientable M appartenant à $\mathcal{G}(M)$. Si $\Sigma(g) \neq \emptyset$ alors il existe un nombre fini de plongements $\psi_i : S^1\times\ ]-2,2[\ \to M$ $i=1,\ldots,q$ ayant les propriétés suivantes :*

a) Sur $S^1\times\ ]-2,2[$, $\psi_i^* g$ est une pseudo-métrique générique dont le lieu singulier est $S^1\times\{-1/2\} \cup S^1\times\{\frac{1}{2}\}$.

b) Sur chaque composante connexe de $M - \bigcup_{i=1}^{q}(\psi_i(S^1\times\ ]-2,2[)$ *g ou −g est riemannienne.*

En d'autres termes, toute pseudo métrique g à lieu singulier non vide de $\mathcal{G}(M)$, est obtenue par la construction décrite à la fin du paragraphe 2. La démonstration de la propriété (1) du théorème A va résulter du lemme suivant :

Lemme 3.1. : Soit g une pseudo métrique de $\mathcal{G}(M)$ et $\varphi : \Sigma(g)\times]-\varepsilon,\varepsilon[\to M$ un voisinage tubulaire de $\Sigma(g)$. Sur $V_\varepsilon - \Sigma(g)$, $(V_\varepsilon = \varphi(\Sigma(g)\times]-\varepsilon,\varepsilon[)$, il existe une 1-forme ω' ayant les propriétés suivantes :

a) $d\omega'$ est la forme courbure de la connexion de Levi-Cevita de g sur $V_\varepsilon - \Sigma(g)$.

b) La 1-forme ω' se prolonge sur V_ε en une 1-forme ω de classe C^0 sur $\Sigma(g)$ et ω est nulle sur $\Sigma(g)$.

Démonstration du lemme 3.1. : Comme g est non dégénéré en restriction à $\Sigma(g)$, quitte à restreindre V_ε on peut toujours supposer que la restriction de g à $L_t = \varphi(\Sigma(g)\times\{t\})$ est non dégénérée pour tout $-\varepsilon<t<\varepsilon$. Sur V_ε il existe alors des champs de vecteurs K et X sans singularités tels que : K est orthogonal à L_t pour tout t, X est tangent à L_t pour tout t et $|g(X,X)| = 1$.
Le couple $(K_1 = \dfrac{K}{\sqrt{|g(K,K)|}}, X)$ est un champ de repères orthonormés sur $V_\varepsilon - \Sigma(g)$. Si ∇' est la connexion de Levi-Cevita de g sur $M-\Sigma(g)$, il existe une 1-forme ω' sur $V_\varepsilon - \Sigma(g)$ telle que :

$$\nabla'_Z X = g(K_1,K_1)\omega'(Z)K_1$$

$$\nabla'_Z K_1 = -g(X,X)\omega'(Z)X.$$

La 2-forme $d\omega'$ est la forme courbure de ∇' sur $V_\varepsilon - \Sigma(g)$. Il reste à établir la propriété (2). On a :

$$\omega'(Z) = g(\nabla'_Z X, K_1) = \frac{1}{\sqrt{|g(K,K)|}}\, g(\nabla'_Z X, K)$$

$$\sqrt{g(K,K)}\,\omega'(Z) = g(\nabla'_Z X, K).$$

Si ∇ désigne le prolongement de ∇' à $\Gamma(M)\times\Gamma(g)\to\Gamma(M)$ (voir déf. 2.1.). La 1-forme $\sqrt{|g(K,K)|}\,\omega'$ possède un prolongement C^∞ sur $\Sigma(g)$ et ce prolongement est nul. Désignons par ω_k la 1-forme ainsi définie sur V_ε. Comme g est générique, $g(K,K) = 0$ est une équation de $\Sigma(g)$ sur V_ε. Il en résulte que la 1-forme $\omega = \dfrac{1}{\sqrt{|g(K,K)|}}\,\omega_K$ est un prolongement de ω' à V_ε de classe C^0 sur $\Sigma(g)$ et ω est nulle sur $\Sigma(g)$. C.Q.F.D.

Démonstration de la propriété (1) du théorème A. : (On reprend les notations de la démonstration du lemme 3.1.).
Soit M_t la variété $M-\varphi(\Sigma(g)\times\ ]-t,t[)$ pour $0\leq t\leq\varepsilon$. Pour $t>0$, on note $M_{i,t}$ la composante connexe de M_t qui est contenue dans la composante connexe M_i' de $M-\Sigma(g)$. On a alors :

$$\int_{M_{i,t}}\Omega=\int_{M_{i,\varepsilon/2}}\Omega+\int_{M_{i,t}-M_{i,\varepsilon/2}}\Omega=\int_{M_{i,\varepsilon/2}}\Omega+\int_{\partial(M_{i,t}-M_{i,\varepsilon/2})}\omega'$$

On désigne par $L_{i,t}$ l'intersection de $L_t=\varphi(\Sigma(g)\times\{t\})$ avec $M_{i,t}$.
On a alors :

$$\int_{M_{i,t}}\Omega=\int_{M_{i,\varepsilon/2}}\Omega+\int_{L_{i,\varepsilon/2}}\omega+\int_{L_{i,-\varepsilon/2}}\omega+\int_{L_{i,t}}\omega+\int_{L_{i,-t}}\omega$$

(ω est le prolongement à V_ε de ω'). Comme $L_{i,0}=\partial\overline{M}_i'\subset\Sigma(g)$ et que ω est nulle sur $\Sigma(g)$ (lemme 3.1.(2)), on aura :

$$\int_{M_i'}\Omega=\lim_{t\to 0}\int_{M_{i,t}'}\Omega=\int_{M_{i,\varepsilon/2}}\Omega+\int_{L_{i,\varepsilon/2}}\omega+\int_{L_{i,-\varepsilon/2}}\omega$$

C.Q.F.D.

Soit M_i l'adhérence dans M de la composante connexe M_i' de $M-\Sigma(g)$; M_i est une surface compacte à bord et $g_i=g_{|M_i}$ est une pseudo-métrique sur M_i dont le lieu singulier est ∂M_i et dont le noyau est transverse à M_i. Les propriétés (2) et (3) du théorème A sont des conséquences du lemme suivant :

Lemme 3.2. : Soit W une surface compacte connexe orientable à bord non vide. On considère sur W une pseudo-métrique g dont le lieu singulier est ∂W et dont la restriction à ∂W est non dégénérée. Soit $\varphi:\partial W\times\ ]0,\varepsilon[\ \to W$ un voisinage collier de ∂W et $\mathcal{F}_\varepsilon$ le feuilletage $\{L_t=\varphi(\partial W\times\{t\})\quad 0\leq t\leq\varepsilon\}$. On désigne par ρ_t la courbure géodésique de la feuille L_t relativement à la connexion de Levi-Cevita de $g|W-\partial W$. Si ρ_t tend vers 0 quand t tend vers 0 et si la forme courbure Ω de la connexion de Levi-Cevita de g sur $W-\partial W$ est intégrable alors on a :

$$\int_W\Omega=\varepsilon\,.2\pi.\chi(W)$$

avec $\varepsilon=1$ (resp. -1) si la signature de g est (2,0) ou (1,1) (resp. (0,2) sur $W-\partial W$; $\chi(W)$ est la caractéristique d'Euler-Poincaré de W. De plus si la signature de g sur $W-\partial W$ est (1,1) alors $\chi(W)=0$.

Démonstration du lemme 3.2. : Posons $W_t = W - \varphi(\partial W \times [0,t[)$ pour $0 \leq t \leq \varepsilon$. Supposons d'abord que g est riemannienne sur $W - \partial W$. On a alors :

$$\int_{W_t} \Omega = 2\pi\chi(W_t) - \int_{\partial W_t} \rho_t(s)ds$$

(voir par exemple [2]). Comme $\lim_{t\to 0} \rho_t = 0$, on en déduit que le lemme 3.2. est vrai lorsque g est une métrique riemannienne sur $W-\partial W$. Si la signature de g est (0,2) sur $W-\partial W$, $-g$ est une métrique riemannienne sur $W-\partial W$. La courbure de la connexion de Levi-Cevita de $-g$ est $-\Omega$. On aura donc $\int_W \Omega = -2\pi\chi(W)$. Il reste donc à établir le lemme 3.2. pour les métriques g de signature (1,1) sur $W-\partial W$. Dans cette hypothèse, on va construire une famille g_s, $s \in \mathbb{R}-\{0\}$, de pseudo-métriques sur W ayant les mêmes propriétés que g et telles que :

g_s est riemannienne sur $W - \partial W$ pour $s > 0$

g_s a pour signature (1,1) sur $W - \partial W$ pour $s < 0$ et $g_{-1} = g$.

Soit Ω_s est la forme courbure de la connexion de Levi-Cevita de g_s $s \neq 0$. pour $s > 0$, l'intégrale $\int_W \Omega_s$ est égale à $2\pi\chi(W)$. En calculant l'intégrale $\int_W \Omega_s$ on en déduira comme dans [1] , que la fonction ainsi obtenue est identiquement nulle.

Construisons d'abord g_s. Il existe une décomposition en somme de Whitney $TW = E_0 \oplus E_1$ du fibré tangent à W ayant les propriétés suivantes :

. g est définie positive en restriction à E_0 au dessus de $W - \partial W$

. g est définie négative en restriction à E_1 au dessous de $W - \partial W$

. si V_ℓ est une composante connexe de ∂W, le fibré tangent à V_ℓ coïncide avec $E_{0|V_\ell}$ ou $E_{1|V_\ell}$ selon que g est positive ou négative sur TV_ℓ. Sur W on définit alors la pseudo-métrique g_s, $s \in \mathbb{R}-\{0\}$ de la manière suivante :

$$g_s(u,v) = g(u_0,v_0) - sg(u_1,v_1) \text{ pour tout } u \text{ et } v \in T_xW$$

où u_0 et v_0 (resp. u_1 et v_1) sont les composantes de u et v sur E_0 (resp. E_1). Pour $s > 0$, g est une métrique riemannienne sur $W - \partial W$. Pour $s < 0$, la signature de g_s est (1,1) sur $W - \partial W$ et $g_{-1} = g$. Nous allons d'abord montrer que la forme courbure Ω_s de la connexion de Levi-Cevita de $g_{s|W-\partial W}$ est intégrable. On peut toujours choisir le voisinage collier $\varphi : \partial W \times [0,\varepsilon[\to M$ et la décomposition $TW = E_0 \oplus E_1$ pour que chaque composante connexe de $\varphi(\partial W \times [0,\varepsilon[)$ soit un voisinage collier d'une composante V_ℓ de ∂W et que si $E_{i|_V}$ est tangent à V_ℓ

alors E_i coïncide avec le fibré tangent au feuilletage $\{\varphi(V_\ell \times \{t\})\ 0 \leq t < \varepsilon\}$. Dans ces conditions, soient K et X des champs sans singularités sur $V = \varphi(\partial W \times [0,\varepsilon[)$ tels que K soit orthogonal à $L_t = \varphi(\partial W \times \{t\})$ (relativement à g), X tangent à L_t et $|g(X,X)| = 1$ pour $0 \leq t < \varepsilon$. Comme dans le lemme 3.1., on démontre alors qu'il existe des 1-formes ω et α sur V_ε, de classe C^0 et nulles sur ∂W, telles que :

. $\Omega_s = d\omega_s$ si $\omega_s = \frac{1}{\sqrt{|s|}}\omega + (1+s)\alpha$ sur $V_\varepsilon - \partial W$, (en particulier $\Omega = \Omega_{-1} = d\omega$).

. la courbure géodésique ρ_t^s de L_t relativement à la connexion de Levi-Cevita de g_s est égale à $\omega_s(X/\sqrt{|g_s(X,X)|})$. En particulier, la courbure géodésique ρ_t^s tend vers 0 quand t tend vers 0).

Rappelons que l'on a posé $W_t = W - \varphi(\partial W \times [0,t[)$. On a alors, pour $s \neq 0$

$$\int_W \Omega_s = \lim_{t\to 0} \int_{W_t} \Omega_s = \int_{W_{\varepsilon/2}} \Omega_s + \int_{L_{\varepsilon/2}} \omega_s + \frac{1}{\sqrt{s}} \lim_{t\to 0} \int_{L_t} \omega + (1+s) \lim_{t\to 0} \int_{L_t} \alpha$$

$$= \int_{W_{\varepsilon/2}} \Omega_s + \int_{L_{\varepsilon/2}} \omega_s, \text{ où } \omega_s = \frac{1}{\sqrt{|s|}}\omega + (1+s)\alpha.$$

Ainsi la 2-forme Ω_s est intégrable sur W pour $s \neq 0$. Par un calcul classique, (voir par exemple [1]) on établit que :

$$\int_W \Omega_s = \frac{1}{\sqrt{s}} \frac{P(s)}{s^3}$$

où P est un polynôme en s. Pour $s > 0$, g_s est une métrique riemannienne sur $W - \partial W$, la forme courbure Ω_s est intégrale et la courbure géodésique ρ_t^s de L_t tend vers zéro quand t tend vers 0. D'après la première partie de la démonstration on a donc :

$$\frac{1}{\sqrt{s}} \frac{P(s)}{s^3} = 2\pi . \chi(W)$$

d'où $P(s) = 2\pi.\chi(W).\sqrt{s}.s^3$. On en déduit que $P(s) \equiv 0$ et $\chi(W) = 0$

C.Q.F.D.

Démonstration des propriétés (2) et (3) du théorème A.: La restriction g_i de g à M_i, adhérence dans M de la composante connexe M_i' de $M-\Sigma(g)$, possède les propriétés du lemme 3.1. En effet, d'après la propriété (1) du théorème A, l'intégrale $\int_{M_i} \Omega$ est convergente. D'autre part, $V_\varepsilon \cap M_i$ est un voisinage collier de ∂M_i et d'après la proposition 2.5., si $g \in \mathcal{G}(M)$ la courbure géodésique de $\varphi(\partial M_i \times \{t\})$ tend vers 0 quand t tend vers 0. En appliquant le lemme 3.1. à (M_i, g_i) on obtient la propriété (2) du théorème A. Soit Δ la 2-forme sur $M-\Sigma(g)$ définie par $\frac{\varepsilon_i}{2\pi}\Omega|_{M_i'}$ où $\varepsilon_i=1$ (resp. -1) si la signature de g_i est (2,0) ou (1,1) (resp. (0,2) sur M_i'.

On a alors :

$$\int_M \Delta = \sum_{i=1}^{p} \int_{M_i} \Delta = \sum_{i=1}^{p} \chi(M_i)$$

On a d'autre part :

$$\chi(M) = \sum_{i=1}^{p} \chi(M_i) - \sum_{1\leq i<j\leq p} \chi(M_i \cap M_j)$$

L'intersection $M_i \cap M_j$ est une composante connexe de $\Sigma(g)$ pour $i \neq j$ c'est-à-dire un cercle. On a donc $\chi(M_i \cap M_j) = 0$. On en déduit $\int_M \Delta = \chi(M)$

C.Q.F.D.

§4. PROLONGEMENT DE LA CONNEXION DE LEVI-CEVITA ET INTEGRABILITE DE LA FORME COURBURE POUR LES PSEUDO-METRIQUES GENERIQUES.

Dans le paragraphe précédent, nous avons établi que, sur une surface compacte M, la forme courbure Ω de la connexion de Levi-Cevita de g est intégrable, si g appartient à $\mathcal{G}(M)$. Nous allons montrer que, pour une pseudo métrique générique, l'appartenance de g à $\mathcal{G}(M)$ n'est pas nécessaire pour que l'intégrale $\int_M \Omega$ soit convergente :

PROPOSITION 4.1. : Sur toute surface compacte orientable M, il existe des pseudo métriques génériques $g^{(\ell)}$ $\ell=1,2$, ayant les propriétés suivantes :

i) $\Sigma(g^{(\ell)}) \neq \emptyset$ et la restriction $g^{(\ell)}$ à $\Sigma(g^{(\ell)})$ est non dégénérée

ii) $g^{(\ell)}$ n'appartient pas à $\mathcal{G}(M)$

iii) l'intégrale $\int_M \Omega^{(\ell)}$ est convergente où $\Omega^{(\ell)}$ est la forme courbure de la connexion de Levi-Cevita de $g^{(\ell)}$.

iv) Pour toute composante connexe $M_i^{(1)}$ de $M-\Sigma(g^{(1)})$ on a

$$\int_{M^{(1)}} \Omega^{(1)} = \chi(M_i^{(1)}).$$

Pour toute composante connexe $M_i^{(2)}$ de $M-\Sigma(g^{(2)})$ l'intégrale $\int_{M_i^{(2)}} \Omega^{(2)}$ n'est pas convergente.

On va d'abord construire des pseudo-métriques génériques h_λ sur

$S^1 \times]-2,2[$, λ fonction sur S^1, ayant les propriétés i), ii), iii) de 4.1. et la propriété iv)' suivante :

iv)' $$\int_{S^1 \times \{t\}} \omega'_\lambda = \frac{1}{\sqrt{|t-1|\;|t+1|}} \int_{S_1} \lambda(\theta)d\theta$$

où ω'_λ est une 1-forme sur $S^1 \times]-2,2[$ ayant les propriétés du lemme 3.1. relativement à h_λ .

Considérant des plongements $\psi_i : S^1 \times]-2,2[$ i=1,2 dans M, les pseudo-métriques $g^{(\ell)}$, $\ell=1,2$, seront obtenues en recollant des pseudo-métriques $(\psi_1^{-1})^* h_\lambda$ sur $\psi_i(S^1 \times]-2,2[)$ et une métrique riemannienne sur $M - \bigcup_{i=1}^{q} \psi_i(S^1 \times]-3/2,3/2[)$ par une partition de l'unité.

Construction des pseudo métriques h_λ sur $S^1 \times]-2,2[$.

Considérons une pseudo-métrique h sur $S^1 \times]-2,2[$ ayant une écriture du type

$$a d\theta \otimes d\theta + (t-1)\cdot(t+1)\, dt \otimes dt$$

où a est une fonction sur $S^1 \times]-2,2[$ strictement positive. La pseudo-métrique h sur $S^1 \times]-2,2[$. Le lieu singulier $\Sigma(h)$ de h est $S^1 \times \{1\} \cup S^1 \times \{-1\}$. La restriction de h à $\Sigma(h)$ est non dégénérée.

La signature de g est :

(2,0) sur $S^1 \times]-2,1[$ et $S^1 \times]1,2[$.

(1,1) sur $S^1 \times]-1,1[$.

Le champ de vecteur $K = \frac{\partial}{\partial t}$ est orthogonal (relativement à h) à $S^1 \times \{t\}$ et il engendre le noyau de h sur $\Sigma(h)$. Le champ de vecteurs $X = \frac{1}{\sqrt{a}} \frac{\partial}{\partial\theta}$ est un champ unitaire (relativement à h). Sur $S^1 \times]-2,2[-\Sigma(h)$, soit ω' la 1-forme caractérisée par :

$$\omega'(Z) = h(\nabla'_Z X, K_1) \quad K_1 = \frac{K}{\sqrt{|h(K,K)|}} = \frac{1}{\sqrt{|t-1|.|t+1|}} \frac{\partial}{\partial t}$$

où ∇' est la connexion de Levi-Cevita de h sur $S^1 \times]-2,2[-\Sigma(h)$. (voir la démonstration du lemme 3.1.).

La 1-forme ω' est égale à :

$$\frac{-1}{2\sqrt{|(t-1)(t+1)|}} L_K\, h(X,\frac{\partial}{\partial\theta}) d\theta .$$

Calculons $L_K h(X,\frac{\partial}{\partial\theta})$. On a :

$$L_K h(X,\frac{\partial}{\partial\theta}) = K\{h(X,\frac{\partial}{\partial\theta})\} + h([X,K],\frac{\partial}{\partial\theta}) + h([\frac{\partial}{\partial\theta},K],X) = \frac{\partial\sqrt{a}}{\mu t}$$

En particulier, h appartient à $\mathcal{G}(S^1 \times]-2,2[)$ si et seulement si $\frac{\partial\sqrt{a}}{\partial t} = 0$ sur $\Sigma(h)$.

On a alors :

$$\int_{S^1 \times \{t\}} \omega' = \frac{-1}{2\sqrt{|t-1||t+1|}} \int_{S^1 \times \{t\}} \frac{\partial\sqrt{a}}{\partial t} d\theta .$$

Soit λ une fonction sur S^1 et $m_\lambda \in \mathbb{R}$, $m_\lambda > \sup_{S^1} |\lambda|$. On pose :

$$h_\lambda = (4m_\lambda - 2t\lambda)^2 \, d\theta \otimes d\theta + (t-1)(t+1)dt \otimes dt$$

Avec les notations précédentes, la 1-forme ω'_λ associée à h_λ est telle que :

$$\int_{S^1 \times \{t\}} \omega'_\lambda = \frac{1}{\sqrt{|t-1||t+1|}} \int_{S^1} \lambda(\theta)d\theta .$$

De plus, h_λ appartient à $\mathcal{G}(S^1 \times]-2,2[)$ si et seulement si $\lambda \equiv 0$.

Construction des pseudo-métriques $g^{(\ell)}$ $\ell=1,2$ sur M

Soient $\psi_i : S^1 \times]-2,2[\to M$ $i=1,2$ des plongements de $S^1 \times]-2,2[$ dans M. On considère des fonctions λ et λ' sur S^1 non identiquement nulles et telles que $\int_{S^1} \lambda(\theta)d\theta = 0$ et $\int_{S^1} \lambda'(\theta)d\theta \neq 0$ respectivement.

Etant donnée une métrique riemannienne g' sur $M - \bigcup_{i=1}^{2} \psi_i(S^1 \times]-\frac{3}{2},\frac{3}{2}[)$, on définit la pseudo métrique $g^{(\ell)}$ $\ell=1,2$ de la manière suivante :

$g^{(1)}$ est obtenue en recollant g', $(\psi_1^{-1})^* h$ et $(\psi_2^{-1})^* h$ par une partition de l'unité.

$g^{(2)}$ est obtenue en recollant g', $(\psi_1^{-1})^* h_\lambda$, et $(\psi_2^{-1})^* h_{-\lambda}$, par une partition de l'unité. On vérifie aisément que $g^{(\ell)}$ vérifient les propriétés (i) à (iv) de la proposition 4.1.

BIBLIOGRAPHIE

[1] A.AVEZ : Formule de Gauss-Bonnet en métrique de signature quelconque. C.R.A.S. 255 (1962) p. 2049-2051.

[2] H. CARTAN : Formes différentielles. Hermann 1967.

[3] J. MARTINET : Sur les singularités de formes différentielles. Ann. Inst. Fourier XX n°1 p. 95-178 1970.

[4] J. MATHER : Solutions of generic linear equations - Salvador Symp. on Dynamical systems. p. 185-193 IMPA Rio 1973.

[5] F. PELLETIER : Singularités d'ordre supérieur de formes différentielles. Thèse Université Dijon 1980.

[6] F. PELLETIER : Singularités génériques de pseudo-métriques sur une variété. C.R.A.S. 296 (1983) p. 219-221.

[7] S. PNEVMATIKOS : Etude géométrique des contraintes génériques dans les espaces de phases. Thèse Université Amsterdam 1981.

* Fernand PELLETIER
Laboratoire de Toplogie
ERA 07 945
Département de Mathématiques
Université de Dijon
B.P. 138
21 004 - Dijon cedex.
FRANCE

* * *

Singularities & Dynamical Systems
S.N. Pnevmatikos (editor)

THE INVARIANCE OF MILNOR'S NUMBER IMPLIES THE INVARIANCE OF THE TOPOLOGICAL TYPE IN DIMENSION THREE

Bernard Perron

Université de Dijon

France

§ 0) STATEMENT OF RESULTS : we begin to recall some results of Milnor [7] about the topology of complex hypersurfaces.

Let $f : U \subset \mathbb{C}^n \to \mathbb{C}$ be an analytic fonction defined on a neigborhood U of $0 \in \mathbb{C}^n$ such that $f(0) = 0$. Set $B_\varepsilon = \{z \in \mathbb{C}^n ; \|z\| \leq \varepsilon\}$, $S_\varepsilon = \partial B_\varepsilon$, $D_\eta = \{u \in \mathbb{C} ; |u| \leq \eta\}$.

LEMME M_1 [7] : Let f as above such that $0 \in \mathbb{C}^n$ is an isolated singularity of f. Then there is $\varepsilon_0 > 0$ such that for each ε, $0 < \varepsilon \leqslant \varepsilon_0$, $f^{-1}(0)$ is transversal to S_ε. For each $\varepsilon \in]0,\varepsilon_0]$, there exists $\eta_0 > 0$ such that for each $\xi \in D_{\eta_0}$, then $f^{-1}(\eta)$ is transversal to S_ε.

THEOREM M_2 [7] : With the same hypotheses as above, for $\varepsilon > 0$ small enough, the map $h_\varepsilon : S_\varepsilon - f^{-1}(0) \to S^1$ defined by $h_\varepsilon(z) = f(z)/|f(z)|$ is a C^∞-fibration, C^∞-equivalent to the fibration $k_{\varepsilon,\eta} : B_\varepsilon \cap f^{-1}(\partial D_\eta) \longrightarrow S^1$ defined by $k_{\varepsilon,\eta}(z) = f(z)/|f(z)|$. More over $f^{-1}(D_\eta) \cap B_\varepsilon$ is diffeomorphic to a 4-ball.

THEOREM M_3 ([7];[5]) : The fiber of Milnor's fibration is obtained from the ball B^{2n-2} by attaching μ handles of index $n-1$ where $\mu = \dim_{\mathbb{C}} \mathbb{C}\{z\} / \{\frac{\partial f}{\partial z_1}, \ldots, \frac{\partial f}{\partial z_n}\}$. (Here $\mathbb{C}\{z\}$ is the ring of germs at $0 \in \mathbb{C}^n$ of analytic fonctions and $\{\partial f/\partial z_1, \ldots, \partial f/\partial z_n\}$ is the ideal generated by $(\partial f/\partial z_1, \ldots, \partial f/\partial z_n)$.

Définition : A couple $(\xi,\eta) \in R^+ \times \mathbb{C}$ defined in lemma M_1 is said to be adapted to f at 0.

The two fibrations h_ε, $k_{\varepsilon,\eta}$ are called Milnor's fibrations of f at 0.

Recall the following theorem of Lê-Ramanujan [6] :

THEOREM L-R : Let $f(t,z)$ be an analytic fonction of $z = (z_1,\ldots,z_n) \in U \subset \mathbb{C}^n$ and $t \in V \subset \mathbb{C}$ (U and V neigborhoods of 0) such that $f(t,0) = 0$ for all $t \in V$.

Suppose the fonctions $f_t(z) = f(t,z)$ admit an isolated singularity at $0 \in \mathbb{C}^n$ and the Milnor numbers of f_t are independant of t. Then, for $n \neq 3$, the Milnor's *fibrations of f_0 and f_t, for t small, are isomorphic and the topological type of the singularities are the same.*

Remark 1 : The condition $n \neq 3$ comes from the fact that in the proof, use is made of the celebrated h-cobordism theorem of Smale [8] applied to a cobordism of dimension 2n-2. For n = 3 this theorem is unknown.

The main objective of this paper is to prove the following :

THEOREM 0 : Theorem L-R is true in any dimension.

We recall the proof of theorem L-R to illustrate the difficulties in case n = 3.

Let (ε_0, η_0) a couple adapted to f_0 at $0 \in C^n$. For t small enough, the hypersurface $f_t^{-1}(\eta)$ is transverse to S_{ε_0} for all $\eta \in D_{\eta_0}$. The hypothesis μ_t independant of t implies that f_t has no other critical points than 0 in B_{ε_0}.

Let (ε_t, η_t) a couple adapted to f_t at $0 \in \mathbb{C}^n$ which may be supposed such that $0 < \varepsilon_t \leq \varepsilon_0$, $0 < \eta_t \leq \eta_0$.

Then $H_t = f_t^{-1}(\eta_t) \cap B_{\varepsilon_t}$ (resp. $H_0 = f_t^{-1}(\eta_t) \cap B_{\varepsilon_0}$) is diffeomorphic to the Milnor fiber of f_t (resp. f_0). The hypothetis $\mu_t = \mu_0$ implies also that the cobordism $(\overline{H_0 - H_t}, H_0 \cap S_{\varepsilon_0}, H_0 \cap S_{\varepsilon_t})$ is a $\mathbb{Z}$ - homology cobordism, that is, inclusions induce isomorphisms :

$$H_*(H_0 \cap S_{\varepsilon_0}, \mathbb{Z}) \longrightarrow H_*(\overline{H_0 - H_t}, \mathbb{Z})$$

$$H_*(H_0 \cap S_{\varepsilon_t}, \mathbb{Z}) \longrightarrow H_*(\overline{H_0 - H_t}, \mathbb{Z})$$

For n = 2, the classification of surfaces implies that the cobordism is trivial (that is, diffeomorphic to $(H_0 \cap S_{\varepsilon_t}) \times [0,1]$). For n > 3, by [7], $\overline{H_0 - H_t}$, $H_0 \cap S_{\varepsilon_0}$, $H_0 \cap S_{\varepsilon_t}$ are all simply-connected. With the help of the h-cobordism theorem, we can conclude that the cobordism is trivial. Theorem L-R follows easily.

In case n = 3, I cannot show directly that $(\overline{H_0 - H_t}, H_0 \cap S_{\varepsilon_0}, H_0 \cap S_{\varepsilon_t})$ is an h-cobordism (that is, the inclusions of the boundary components are homotopy equivalences). Even so, the h-cobordism theorem is not yet proved in this dimension.

As a by-product of the proof of theorem 0, one can prove a result concerning families of analytic fonctions with isolated singularity at $0 \in C^n$ (without the hypothesis μ constant) : Let $g_t : (C^n,0) \to (C,0)$ such a family with $g_t(0) = 0$. One knows that $\mu_0 \geq \mu_t$ for t small.

Let (ε_o,η_o) a couple adapted to g_o at 0.

The same construction as above shows that the Milnor fiber G_t of g_t is canonically imbedded in G_o. Let $j_t : G_t \dashrightarrow G_o$ this embedding.

On the other hand, by theorem M_2, there exists canonical embeddings :

$$i_o : G_o \longrightarrow S^{2n-1}$$

$$i_t : G_t \longrightarrow S^{2n-1}$$

THEOREM 1 : The embeddings $i_t : G_t \longrightarrow S^{2n-1}$ *and* $i_o \circ j_t : G_t \longrightarrow S^{2n-1}$ *are isotopic.*

We will only give a sketch of the (long) proof of theorem 0. These results have been announced in [9] . Complete proofs are in [10].

There are essentially two main steps in the proof. The first, using ingredients of analytic geometry, gives a comparative handle decomposition of the Milnor fibers of f_o and f_t (proposition 2 below) . The second step, essentially in the field of differential topology, will show that the fibers are diffeomorphic.

§ 1) STEP ONE : HANDLE DECOMPOSITION OF THE MILNOR FIBER.

All the results below are true in any dimension. For simplicity we take $n=3$.

Let (x,y,z) be coordonates in $\mathbb{C}^3$ such that z is generic. This means in particular that the restrictions $f_{t,0} = f_t|\mathbb{C}^2 \times \{0\} \to C$ have an isolated singularity at 0.

Let (ε_o,η_o) (resp. (ε_t,η_t)) a couple adapted to $f_{o,o}$ (resp. $f_{t,o}$). Then $f^{-1}_{t,0}(D_{\eta_t}) \cap B_{\varepsilon_t}$ and $f^{-1}_{t,o}(D_{\eta_o}) \cap B_{\varepsilon_o}$ are diffeomorphic to 4-dimensional balls (théorem M_2) and by genericity $f_{t,0} : \overline{B_{\varepsilon_o} - B_{\varepsilon_t}} \to \mathbb{C}$ has only quadratic non degenerated critical points (that is, locally analytically equivalent to $x^2+y^2+z^2$) whose critical values are distinct and noted $\{b_j\} \in \overline{D_{\eta_o} - D_{\eta_t}}$.

PROPOSITION 2 : There existe $\mu (= \mu_o = \mu_t)$ circles $S_1,\ldots,S_\mu$ contained in distinct fibers $f^{-1}_{t,0}(p_i) \cap B_{\varepsilon_t}$ of $f_{t,0} : f^{-1}_{t,0}(D_{\eta_t}) \cap B_{\varepsilon_t} \to D_{\eta_t}$ such that :

a) $p_i \in \partial D_{\eta_t}$ and $\arg(p_i) \neq \arg(b_j)$ for all i and j.

b) H_t, the Milnor fiber of f_t , is obtained from the ball $f^{-1}_{t,0}(D_{\eta_t}) \cap B_{\varepsilon_t}$ by attaching handles of index 2 along the cercles S_i (compare with [5])

c) Denote by $\{\widetilde{S_i}\}$ the image of the cercles S_i in $f^{-1}_{t,0}(\partial D_{\eta_o}) \cap B_{\varepsilon_o}$ by the gradient vector field of $|f_{t,0}|$. Then H_o, the Milner fiber of f_o , is obtained from the ball $f^{-1}_{t,0}(D_{\eta_o}) \cap B_{\varepsilon_o}$ by attaching handles along the circles $\{\widetilde{S_i}\}$.

The proof is obtained by a careful study of the one parameter family of polar curves and corresponding Cerf diagrams of the family f_t (see [4] for definitions) using the methods of [5] .

§ 2) STEP TWO : THE FIBERS ARE DIFFEOMORPHIC.

By proposition 2, it is suffisant to prove that the links $\{S_i\} \subset \partial[f_{t,0}^{-1}(D_{\eta_t}) \cap B_{\varepsilon_t}] \approx S^3$ and $\{\widetilde{S_i}\} \subset \partial[f_{t,0}^{-1}(D_{\eta_o}) \cap B_{\varepsilon_o}] \approx S^3$ equipped with particular trivialisation of their normal bundle, are equivalent.

In fact we will prove a stronger result. Let $p \in \partial D_{\eta_t}$, $p' \in \partial D_{\eta_o}$ such that $\arg(p) = \arg(p') \neq \arg(b_j)$ (arg for argument) and set

$F_t = f_{t,0}^{-1}(p) \cap B_{\varepsilon_t}$ and $F_o = f_{t,0}^{-1}(p') \cap B_{\varepsilon_o}$.

Let $\widetilde{F_t}$ the submanifold of F_o obtained by pushing F_t along the vector field grad $|f_{t,0}|$

Consider also the manifold $\widehat{F}_t \subset \partial B_{\varepsilon_t}$ (resp. $\widehat{\widehat{F}}_t \subset \partial B_{\varepsilon_o}$) obtained by pushing F_t (resp. $\widetilde{F_t}$) (see figure 4 below), along the trajectories of a field X_t (resp. X_o) having the following properties :

1) X_t (resp. X_o) is defined on $B_{\varepsilon_t} - f_{t,0}^{-1}(0)$ (resp. $B_{\varepsilon_o} - f_{t,0}^{-1}(D_{\eta_o})$) and non-zero everywhere.

2) X_t (resp. X_o) is tangent to the hypersurfaces $\arg f_{t,0} = \Theta$.

3) the scalar product $<\text{grad } |f_{t,0}| , X_t>$ (resp. $<\text{grad } |f_{t,0}| , X_o>$) is stricly positive.

4) the scalar products $<\text{grad } \delta_o , X_t)$ and $<\text{grad } \delta_o, X_o>$ are stricly positive where δ_o is the distance to the origin.

Such a field exists by Milnor ([7], lemma 5.9).

PROPOSITION 3 : The emmbeddings $\widehat{F}_t \to \partial B_{\varepsilon_t}$ and $\widehat{\widehat{F}} \longrightarrow \partial B_{\varepsilon_o}$ are isotopic.

COROLLARY 4 : The links $\{S_i\}$, $\{\widetilde{S_i}\}$ with their trivialisations are isotopic.

We will give an idea of the proof of proposition 3. Recall (theorem M_3 above) that F_t , the Milnor fiber of $f_{t,0}$ is obtained from D^2 by adding μ'_t 1-handles where $\mu'_t = \dim_{\mathbb{C}} \mathbb{C}\{x,y\} / \{\frac{\partial f_{t,0}}{\partial x} , \frac{\partial f_{t,0}}{\partial y}\}$.

Let $F'_t = L \cup \widetilde{\ell_1} \; \ldots \cup \widetilde{\ell_{\mu'_t}}$ a submanifold of codimension 0 of F_t , where L is a small ball of F_t , $\ell_1, \ldots, \ell_{\mu'_t}$ the hearts of the 1-handles and $\widetilde{\ell_i}$ small tubular neigborhoods of ℓ_i , so that $\overline{F_t - F'_t} \simeq \partial F_o \times [0,1]$ (fig. 1)

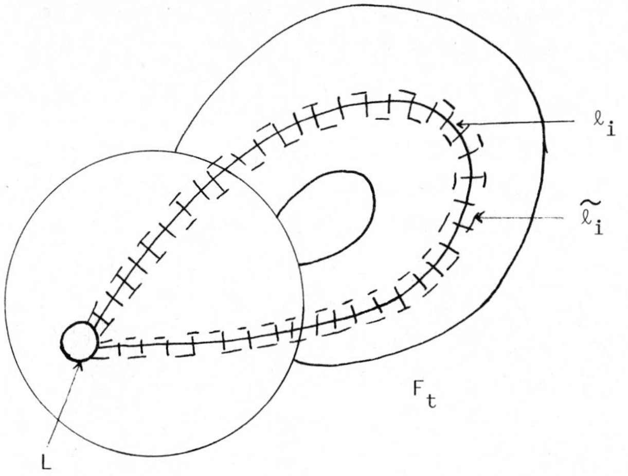

Figure 1.

Let $\widetilde{F'_t} \subset \widetilde{F_t}$ (resp. $\widehat{F'_t} \subset \widehat{F_t}$) the submanifold obtained by pushing F'_t (resp. $\widetilde{F'_t}$) along the vector field grad $|f_{t,0}|$ (resp. χ_t).

By replacing eventually the origin $0 \in \mathbb{C}^2$ by a point nearby, we may suppose that the couple $(|f_{t,0}|\ ,\ \delta_0)$ (δ_0 is the distance to the origine) is generic : this means in particular that $\Sigma^- = \{x \quad \mathbb{C}^2\ ;\ \exists\ \lambda > 0,\ \text{grad}\ |f_{t,0}| = \lambda\ \text{grad}\ \delta_0\}$ is composed of smooth arcs and the points $u \in \Sigma^-$ are of one of the following three types :

a) tangent fold

b) transverse fold

c) transverse cusp

In each of these three cases one has local models for the couple $(|f_{t,0}|^2, \delta_0)$ given by Dufour ([2], [3]) :

$$\text{a)}\quad (|f_{t,0}|^2, \delta_0) \sim (\varepsilon t^2 - x^2 - \varepsilon y^2 + z^2, t) \qquad \varepsilon = \pm 1$$

$$\text{b)}\quad (|f_{t,0}|^2, \delta_0) \sim (-t + \varepsilon_1 x^2 + \varepsilon_2 y^2 + \varepsilon_3 z^2, t)$$

where at most two of the ε_i $(i = 1,2,3)$ are negative.

$$\text{c)}\quad (|f_{t,0}|^2\ , \delta_0) \qquad (-t + \varepsilon_1 tx + \varepsilon_2 x^3 + \varepsilon_3 y^2 + \varepsilon_4 z^2\ ,\ t)$$

where at most one of ε_3, ε_4 is negative.

The equivalence $\sim$ is defined as follows : two C^∞ maps $g_i : (R^n, 0) \to (R^2, 0)$ $(i = 0,1)$ are (locally) equivalent if there exist (local) diffeomorphisms

$h : (R^n,0) \to (R^n,0)$, $k : (R^2,0) \longrightarrow (R^2,0)$ such that $g_1 = k \circ g_0 \circ h$ where k is of the following form : $k(x,y) = (k_1(x),k_2(y))$.

The restrictions about the signs of the ε_i above come from the fact that $f_{t,0}$ is a complex analytic fonction (more precisely from Andreotti - Frankel theorem[1])

Now consider a subdivision $\eta_t = \alpha_0 < \alpha_1 < \alpha_2 < \dots < \alpha_r = \eta_o$ such that the connected components of $\Sigma^- \cap (|f_{t,0}|^{-1}[\alpha_i,\alpha_{i+1}] \cap B_{\varepsilon_0})$ are contained in charts where the local models above are available.

On each piece $T_i = |f_{t,0}|^{-1}[\alpha_i, \alpha_{i+1}]$ we construct a vector field χ_i such that

1) $\chi_i = 0$ exactly on $\Sigma^- \cap T_i$

2) $\chi_i(|f_{t,0}|) > 0$, $\chi_i(\delta_0) > 0$ on $T_i - \Sigma^-$

3) We have an explicit description of the set of points of $|f_{t,0}|^{-1}(\alpha_i)$ which are attracted by $\Sigma^- \cap T_i$: these are submanifolds of dimension ≤ 2 and those of dimension 2 are discs or rings. (For example, for model a) above with $\varepsilon = 1$, the set is a ring). We will denote $\{B_i^j , j \in J_i\}$ these submanifolds of $|f_{t,0}|^{-1}(\alpha_i)$.

Using the vector fields χ_i we can construct an embedding :

$$\varphi : (L \cup \ell_1 \dots \cup \ell_{\mu'_t}) \times [0,1] \to |f_{t,0}|^{-1}[\eta_t, \eta_o]$$

such that :

a) the lines $\varphi(\{x\} \times [0,1]) \cap (|f_{t,0}|^{-1}[\eta_i , \eta_{i+1}])$ for $x \in L$ are trajectories of the field χ_i (L is small).

b) the lines of the cylinder $\mathrm{Im}\,\varphi \cap (B_{\varepsilon_t} - f_{t,0}^{-1}(D_{\eta_t}))$ are the trajectories of the field χ_t.

c) $|f_{t,0}|$ restricted to $\varphi[(\ell_1 \cup \dots \ell_{\mu'}) \times [0,1]]$ has no critical point.

d) δ_0 restricted to $\varphi[(\ell_1 \cup \dots \ell_{\mu'}) \times [0,1]]$ has only Morse critical points, of index 0 and 1.

e) $\mathrm{Im}\,\varphi$ is disjoined from the 2-balls generated by the trajectories of the field grand $|f_{t,0}|$ attracted by the critical points of $f_{t,0}$ other than 0, which are Morse points of index 2.

The idea to construct φ is a follows : suppose we have constructed the embedding $\varphi_i : (L \cup \ell_1 \dots \cup \ell_{\mu'}) \times [0,\tau_i] \to |f_{t,0}|^{-1}[\alpha_0,\alpha_i]$.

If $\varphi_i[(L \cup \ell_1 \dots \cup \ell_{\mu'}) \times \{\tau_i\} \subset |f_{t,0}|^{-1}(\alpha_i)$ does not meet the manifolds $\bigcup_j B_i^j$,then we go to the level $|f_{t,0}|^{-1}(\alpha_{i+1})$ using the trajectories of the vector field χ_i.

Suppose $\varphi_i[(\ell_1 \cup \dots \ell_{\mu'}) \times \{\tau_i\}]$ and $B_i^{j_0}$ intersect transversaly in a finite number of points. We then eliminate these intersection points by introducing critical points of index 0 and 1 for δ_0 by the figure below : we are in a neigborhood of a tangent fold where $(|f_{t,0}|^2, \delta_0) \sim (\varepsilon t^2 - x^2 - \varepsilon y^2 + z^2, t)$ ($\varepsilon = 1$ here). The broken lines represent the trajectories of the field χ_i :

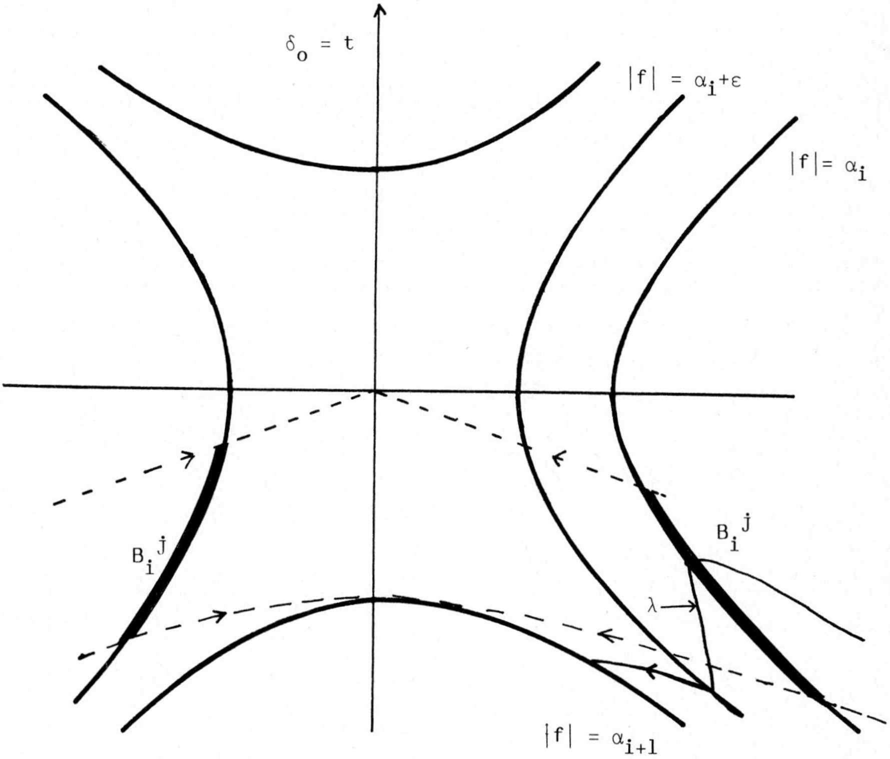

Figure 2.

Let $p \in \varphi_i((\ell_1 \cup .. \ \ell_{\mu'}) \times [0, \tau_i]) \cap B_i^j$.
Take a path λ from p to $p' \in |f_{t,0}|^{-1}(\alpha_i + \varepsilon)$, p' not attracted by $\Sigma^- \cap |f_{t,0}|^{-1}[\alpha_i, \alpha_{i+1}]$ by χ_i, λ being such that δ_o (resp. $|f_{t,0}|$) is decreasing (resp. increasing) along λ . Take a thickening $\tilde{\lambda}$ of λ as in figure 3 an go to $|f_{t,0}|^{-1}(\alpha_{i+1})$ by the vector vector field χ_i from the upper boundary component of the cylinder $\varphi_i((L \cup \ell_1 .. \cup \ell_{\mu'_t}) \times [0, \tau_i]) \cup \tilde{\lambda}$. We then get a cylinder having the desired properties.

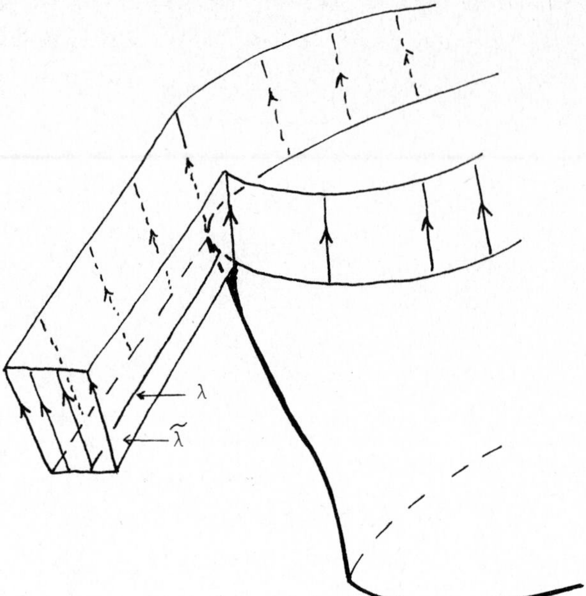

Figure 3

By thichening (with some care) the above situation, we can construct an embedding $\varphi : (F'_t = L \cup \widetilde{\ell}_1 \ldots \cup \widetilde{\ell}_{\mu'_t}) \times [0,1] \to |f_{t,0}|^{-1} [\eta_t, \eta_o]$ such that :

α) $\varphi = \widetilde{\varphi} \,|\, L \cup \ell_1 \ldots \cup \ell_{\mu'}$

β) $|f_{t,0}|$ restricted to $\widetilde{\varphi}(\mathring{F}'_t \times [0,1])$ and $\widetilde{\varphi}(\partial F'_t \times [0,1])$ is without critical point ($(\mathring{\ })$ is the interior).

γ) δ_o restricted to $\widetilde{\varphi}(\mathring{F}'_t \times [0,1])$ is without critical point.

δ) $\delta_o \,|\, \widetilde{\varphi}(\partial F'_t \times [0,1])$ has only critical points of index 0 and 1.

ε) The lines of the cylinder $\mathrm{Im}\, \widetilde{\varphi}$ on $B_{\varepsilon_t} - |f_{t,0}|^{-1}(D_{\eta_t})$ are the trajectories of χ_t.

ν) $\mathrm{Im}\, \widetilde{\varphi}$ does not meet the trajectories of grad $|f_{t,0}|$ attracted by the critical points of $|f_{t,0}|$ other than 0.

The proof of proposition 3 goes now as follows : by condition ε) above and the definition of $\widehat{F'}_t$ (see above) it's clear that $\mathrm{Im}\, \widetilde{\varphi} \cap \partial B_{\varepsilon_t} = F'_t$.

Set $\widetilde{G'} = \mathrm{Im}\, \widetilde{\varphi} \cap [|f_{t,0}|^{-1}(\eta_o)]$ and $\hat{\hat{G'}}$ the submanifold of $\partial B_{\varepsilon_o}$ obtained by pushing $\widetilde{G'}$ along the vector field χ_o.

By condition ν), $\widehat{G'}$ is isotopic (in $|f_{t,0}|^{-1}(\eta_o)$) to $\widetilde{F'}_t$. By the geometric lemma below, $\hat{\hat{F}}_t$ is isotopic to $\hat{\hat{G'}}$.

Proposition 3 then follows from the identification of $f_{t,0}^{-1}(\partial D_{\eta_o}) \cap B_{\varepsilon_o}$ with $\partial B_{\varepsilon_o} - f_{t,0}^{-1}(D_{\eta_o})$ by the vector field χ_o and the isomorphism $\overline{F_t - F'_t} \simeq \partial F_t \times [0,1]$.

GEOMETRIC LEMMA : Let F *be a compact surface with boundary and* $\varphi : F \times I \to S^3 \times I$ *a pseudo-isotopy, (that is an imbedding such that* $\varphi^{-1}(S^3 \times \{i\}) = F \times \{i\}$ *for* $i = 0,1$*). We suppose the projection* $p : S^3 \times I \to I$ *has the following properties :*

a) $p \circ \varphi \,|\mathring{F} \times I$ *is without critical point,*

b) $p \circ \varphi \,|\partial F \times I$ *has only critical points of index* 0 *and* 1 .

Then, modulo $\varphi_0 = \varphi\,|\,F \times \{0\}$, φ *is isotopic to an isotopy, that is, there exists a one parameter family of diffeomorphisms* $\psi_t : S^3 \times I \to S^3 \times I$ *with* $\psi_{t,0} = \psi_t|S^3 \times \{0\} = id_{S^3}$, $\psi_0 = id_{S^3 \times I}$ *such that* $\psi_1 \circ \varphi = \varphi_0 \times id_I$.

The proof is in [10] and uses technics of [11]

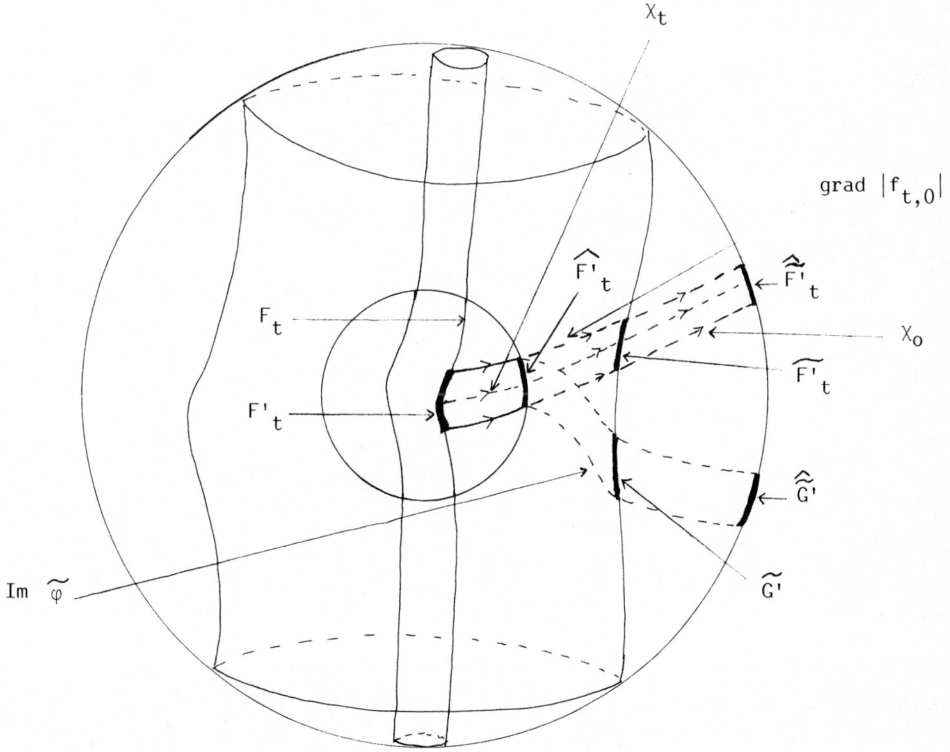

Figure 4

REFERENCES

[1] A. Andreotti & T. Frankel : The Lefschetz theorem on hyperplan sections. Annals of Math. 69 (1959) 713-717

[2] J.P. Dufour : Diagrammes d'applications différentiables. Thèses d'Etat. Montpellier.(1979)

[3] J.P. Dufour : Bi-stabilité des fronces. C.R. Acad. Sc. Paris t. 285 . 26 Sept.(1977) p. 445-448

[4] Lê Dung Trang : Calcul du nombre de cycles évanouissements d'une hypersurface complexe. Ann. de l'Inst. Fourier 23. N° 4 (1973)

[5] Lê Dung Trang & B. Perron : Sur la fibre de Milnor d'une singularité isolée en dimension complexe trois. C.R. Acad. Sciences Paris . t. 289 (9 juillet 1979) p. 115-118

[6] Lê Dung Trang & C.P. Ramanujan : The invariance of Milnor's number implies the invariance of the topological type. Amer. Journal of Math 98 n ° 1 (1976) p. 67-78

[7] J. Milnor : Singular points of complex hypersurfaces. Annals of Math. Studies. Princeton N° 61 (1968)

[8] J. Milnor : Lectures on the h-cobordism theorem. Princeton U. Press (1965)

[9] B. Perron : μ constant implique type topologique constant en dimension complexe trois. Note au C.R. Acad. Sciences. Paris t. 295 (20 déc. 1982) p. 735-738

[10] B. Perron : μ constant implique type topologique constant. Notes multigraphiées - Université de Dijon -(1982)

[11] B. Perron : Pseudo-isotopies de plongements en codimension deux. Bull. Soc. Math. France n° 103 (1975) p. 285-339.

Bernard PERRON
UNIVERSITE DE DIJON
Département de Mathématiques
Laboratoire de Topologie - ERA 07.945
B.P. 138
21004 DIJON Cedex - FRANCE

Singularities & Dynamical Systems
S.N. Pnevmatikos (editor)

REAL CLEMENS STRUCTURES

Claude André Roche
University of Dijon
France

Abstract.

Real Clemens structures are presented in such a way we have a trivialisation, with local models, of the nbhd of a principal divisor with normal crossings This is applied to the desingularisation of an alg. isolated real singularity. Flat forms are found to be preserved by the trivialisation. A generalization of a theorem of Glaeser is also presented.

1. INTRODUCTION

It is well known that if P is a smooth real function on $\mathbb{R}^n$ having an isolated singularity at 0 then there exists an $\varepsilon > 0$ and $\eta > 0$ such that the absolute value $|P|$ restricted to the set $X_{\varepsilon,\eta} = \{x \in \mathbb{R}^n : 0 < |P(x)| < \eta, \ ||x|| < \varepsilon\}$ is the projection map of a trivialisable fiber bundle $|P|_{\varepsilon,\eta} : X_{\varepsilon,\eta} \to (0,\eta)$.

This fiber bundle must be be considered as the real analogue of the Milnor fibration for a complex isolated singularity [8]. It is also known that the geometric study of the germ of P at zero can be done assuming that P is a polynomial function if we are concerned with an algebraically isolated singularity. Now onwards P will be a polynomial function with an algebraically isolated singularity in $0 \in \mathbb{R}^n$.

The purpose of this paper is to give a trivialisation of the bundle $|P|_{\varepsilon,\eta}$.

1980 Mathematics Subject Classification. Primary 58C27, 58A10 ; Secondary 14E15, 58A35.
Key words : Isolated singularity ; Clemens structure ; Composed functions ; stratification.

THEOREM : There exist a trivialisation of the bundle $|P|_{\varepsilon,\eta} : X_{\varepsilon,\eta} \to (0,\eta)$, *say* $T : X_{\varepsilon,\eta} \to |P|^{-1}_{\varepsilon,\eta}(\eta) \times (0,\eta)$, *such that the isomorphism* $T^* : \Lambda^\cdot(|P|^{-1}_{\varepsilon,\eta}(\eta) \times (0,\eta)) \to \Lambda^\cdot(X_{\varepsilon,\eta})$ *gives an isomorphism* $T^* : J^\infty(|P|^{-1}_{\varepsilon,\eta}(\eta) \times \{0\})\Lambda^\cdot(|P|^{-1}_{\varepsilon,\eta}(\eta) \times (0,\eta)) \to$
$\to J^\infty(P^{-1}(0))\Lambda^\cdot(X_{\varepsilon,\eta})$.

The construction of the trivialisation needs the discussion of real Clemens structure for the singular germ P. This consists of a coherent system of germs of tubular neigbourhoods (with additional structure) for the natural stratification of the total transform in an Hironaka's desingularisation of the germ P. This work which is of interest in itself is presented in chapter 2.

These results were used by the author to solve a problem on real relative cohomology proposed by the later Jacques Vey [11]. This was announced in [12] [13] and will appear soon [14] . Some mistakes in these announcements were pointed to me by Ofer Gaber.

Also R. Moussu used the fundamental result of this work to solve the problem of classifying locally integrable pfaffian forms with an algebraically isolated singularity [10] .

In this paper we also present, as an application of the trivialisation theorem a composed function theorem (compare [4]).

THEOREM : Let P be a germ of real smooth function with an algebraically isolated singularity at 0. Let f be a germ of real smooth function constant in the level hypersurface of P. If for every $\varepsilon > 0$ *the set* $\{x \in \mathbb{R}^n : ||x|| < \varepsilon,\ P(x) \neq 0\}$ *has C connected components then there exist C+1 germs of real smooth functions* $\phi_i : (\mathbb{R},0) \to \mathbb{R}$ $i=0,\ldots,C$ *such that :*

each ϕ_i, $i > 0$ *is flat at 0 and* $f = \sum_{i>0} (\phi_i \circ P).I_i + \phi_0 \circ P$.

Where the I_i*'s are the characteristic functions of the connected components of the complement of the singular level.*

One consequence of theorems in chapter 3 is that $(\phi_i \circ P).\ I_i$ is a smooth function.

In this paper smooth function are of class C^∞. We denote $\mathcal{E}(n)$ the ring of germs at the origin of smooth real functions $(\mathbb{R}^n,0) \to \mathbb{R}$, $m(n)$ its maximal ideal, $m^\infty(n)$ the ideal of germs flat at 0. For a real analytic manifold with boundary M, $\mathcal{A}_M$ denotes its structural sheaf,

$\mathcal{E}_M$ the sheaf of germs of smooth real functions on the underlying smooth manifold M, Λ^i_M the sheaf of germs of smooth i-forms of M. The simbol $J^\infty(D)$ for a closed set D of M is for the ideal sheaf of $_M$ of those germs which are flat at D. Let also denote $\varepsilon B(x)$ the open euclidean ball centered at x of radius ε in $\mathbb{R}^n$, $\varepsilon S(x)$ its frontier. We will write only εB or εS if $x = 0$ and $B(x)$ or $S(x)$ if $\varepsilon = 1$.

I wish to thank Ivan Kupka for his guidance in this work and to the members of Dijon's topology Seminar and particularly R. Moussu for enleightening conversations during the preparation of this paper.

2. CLEMENS STRUCTURES.

2.0 Introduction.

Consider as a representant of a finite codimension germ the polynomial function $P : \mathbb{R}^n \to \mathbb{R}$, restricted to the set $X_{\varepsilon,\eta}$ defined below so that O is the sole singularity of P and every $|P|^{-1}(t)$ is transverse to εB for $0 < t < \eta$.

We will use the following theorem of H. Hironaka [5] .

THEOREM 1 : Let P be an analytic function $: \mathbb{R}^n \to \mathbb{R}$ *with an algebraically isolated singularity et* O *in* $\mathbb{R}^n$. *There exists an open nbhd* U *of* O *in* $\mathbb{R}^n$, *a real analytic manifold* M *of dimension* n *and an analytic map* $\theta : M \to U$ *which is surjective proper and algebraic such that*

a) The strict transform of the zeros of P *is an analytic manifold.*

b) The open set $M - \theta^{-1}(O)$ *is dense and* θ *is an isomorphism from* $M - \theta^{-1}(O)$ *to* $U - \{O\}$.

c) The total transform D *of the zeros of* P *presents only normal crossings.*

For an analytic set determined by the ideal sheaf J the total transform F is the analytic set determined by the ideal sheaf $\theta^* J$; the strict transform is the closure of $F - \theta^{-1}(O)$ in M. The analytic set $\theta^{-1}(O)$ is called exeptionnal divisor.

The total transform of the zeros of P is the analytic set determined by the ideal sheaf $(\tilde{P})_M$ where $\tilde{P} = P \circ \theta$.

We can suppose that the nbhd of $O \in \mathbb{R}^n$ in the theorem below contains $\bar{X}_{\varepsilon\eta}$. We have so a diffeomorphism θ of the compact manifold with boundary $\theta^{-1}(\bar{X}_{\varepsilon\eta})$ to $\bar{X}_{\varepsilon\eta}$ outside $\theta^{-1}(O)$. Lets suppose that M coincides with $\theta^{-1}(\bar{X}_{\varepsilon\eta})$ so that the level hypersurfaces of P are all transverse to the boundary of M.

It is clear now that the description of $\tilde{P}$ in the nbhd of D in M is equivalent to the description of the trivialisation of $|P|$. This nbhd will be considered as a reunion of C^{∞}-tubular nbhds of the natural strata of D together with a smooth vector field which gives the trivialisation and has local models : if the stratum is determined by $x_1^{r_1} = 0$, $x_2^{r_2} = 0, \ldots, x_c^{r_c} = 0$ then the field is $\sum_{i=1}^{c} \frac{r_i}{x_i} \frac{\partial}{\partial x_i}$. These local models are important for the results to be stablished in next chapter.

Clemens structure are the real analogue of the construction in [2].

The choice of real vector fields for describing the retraction insted of 1-forms is due to the influence of [16]. The definition presented here differs from that of Arnold [1], [6] in that the role of retractions in tubular nbhds are emplasiced instead of the incidence relations of the boundary of tubes.

2.1. Divisors with normal crossings.

In a real analytic manifold M a locally principal ideal sheaf J_D, of the structure sheaf $\mathcal{A}_M$, determines a divisor $D \subset M$. The divisor is said to present normal crossings if :

i) J_D is locally generated by a monomial function in analytic coordinates i.e. for every point $x \in D$ there exist a local analytic coordinate system (x_i) of M such that over this nbhd J_D is generated by $\Pi x_i^{r_i}$ for certain positive integers r_i.

ii) Global and local analytic irreductible conponents of D coincide i.e. if D_i is an irreductible component of D the trace of D_i in the domain of the coordinate system here above (x_i) is exactly $\{x/x_j = 0\}$ for only one j.

As a consequence of this definition we have that irreductible components of a divisor D with normal crossing are smooth (possibly not reduced) hypersurfaces of M. Let denote this fact by $D = \cup D_i$. Giving an order and a numbering $I \subset \mathbb{N}$ to the component D_i once for all. The natural stratification of D (that obtained by taking succesively regular points) is clearly given by taking connected components of intersections $\bigcap_{i \in J} D_i$ for $J \subseteq I$. Let note this stratification $\mathfrak{D}$ and incidence relation between two strata X and Y by $X \leq Y$ i.e. if $X \subseteq \overline{Y}$.

Definition 1 : The sheaf J determines for every stratum X of $\mathcal{D}$ a numerical information in the form of an integer valued function $r(X) : I \to \mathbb{N}$ by $r(X)(j) = k$ for $j \in I$ if there is a point of X which has a nbhd with an analytic system of coordinates of $M:(U,(x_i)i=1,..,n)$ s.t.

$J_{|U} = \langle x_1^{m_1} \cdot \ldots \cdot x_n^{m_n} \rangle$ (the ideal generated by the monomial in $\mathcal{A}_{M|U}$)

and $D_j \cap U = \{x : x_j = 0\}$ with $m_j = k$. So $r(X)(s) = 0$ if $X \cap D_s = \emptyset$. This is well defined because of connectedness of strata.

We note $r(X)(i)$ by $r_i(X)$ and also r_i if no mistakes are possible. We have $r_i \neq 0$ if and only if $X \subset D_i$ and in that case r_i is the multiplicity of D_i. We find that the codimension of X, noted c(X) or simply c, is the number $\#\{j : r_j(X) \neq 0\}$. The dimension of X is n-c(X). We shall denote $\mathcal{D}$ the category where objects are strata and morphisms are incidence relations plus an identity arrow for each stratum.

2.2. Stratifying fiber bundles.

Let V be a connected embedded analytic submanifold of M.

Definition 1 : An stratifying bundle over V is an analytic manifold U a divisor with normal crossings D, a C^∞ submersion $p : U \to V$ and a C^∞ vector field ξ over U - D non singular and vertical (i.e. $p_*\xi=0$).

Such a bundle is noted (U,V,p,D,ξ) ; (p) for short. We put st. b. for stratifying bundle.

Between such objects, morphisms are those of the category naturally constructed from a graph and a set relations where :

1) Vertices are stratifying bundles over connected embedded analytic submanifolds of M.

2) Arrows and relations are described here below.

Definition 2 : "Isomorphism" arrows. Such an arrow from (U,V,p,D,ξ) to (U',V',p',D',ξ) is determined by a C^∞ diffeomorphism $\psi : U \to U'$ (called underlying map) such that :

a) $(J_{D'} \circ \psi)\, \mathcal{E}_U \cap \mathcal{A}_U = J_D$ (analytic pull back),

b) $\psi_* \xi = \xi'$,

c) $p' \circ \psi = p$.

If ψ happens to be analytic, faithfull flatness of $\mathcal{E}_U$ over $\mathcal{A}_U$

transforms condition a) in a') $\psi^{*}J_{D'} = J_D$.

Definition 3 : "Inclusion" arrows. Such an arrow from (W,X,p_W,S,η) to (U,V,p,D,ξ) is determined by a set inclusion $W \subset U$ (given by the underlying map $i : W \to U$) s.t.

a) W is open in U,

b) $J_S = J_{D|W}$,

c) $\eta = \xi_{|W-S}$ and $p \circ p_W = p$.

Definition 4 : Commutation relations. Two chains in the graph so far determined are equivalent if the composition of underlying maps coincide.

This category of stratifying bundles is denoted SB. Best understanding is obtained from the following proposition.

PROPOSITION 1 : *Every morphism in SB is represented by the composition of an isomorphism and of an inclusion.*

Proof : Clearly a chain of inclusions is equivalent to the inclusion determined by the composition of underlying inclusions of the chain. The same holds for isomorphisms. We prove now that we can transpose the two kinds of arrows.

Let $\phi : (p) \to (p')$ be an inclusion and $\psi : (p') \to (p")$ be an isomorphism arrow. We can build another st. b. (q) included by f in (p") and an isomorphism arrow $g : (p) \to (q)$ s.t. (ψ,ϕ) is equivalent to (f,g) : (q) is $(\psi(U), V, \psi \circ p \circ \psi^{-1}, \psi(D), \psi_{*}\xi)$. The morphism f is the inclusion $\psi(U) \hookrightarrow U"$ and g is $\psi_{|U}$.

Next we deal with locally trivial st. b.

Let $m \geq c \geq 0$ be positive integers and $r \in \mathbb{N}^{*c}$ a multi-index, V a real analytic manifold of dimension m-c supposed connected. Denote (y_i) the natural coordinate system of $\mathbb{R}^c$.

Definition 5 : The (germ of) trivial stratifying bundle of dimension m and multiplicity r over V is the (germ at $\{0\} \times V$ of) canonic projection $\pi : \mathbb{R}^c \times V \to V$ together with the divisor of $\mathbb{R}^c \times V$ determined by (y^r) and the vector field $\sum_{i=1}^{c} \frac{r_i}{y_i} \frac{\partial}{\partial y_i}$.

We naturally define a trivialisable st.b. for m and r as one isomorphic to the trivial st.b. of dimension m and multiplicity r.

(The definition is on germs at the zero section of st.b.). Locally trivial st.b. are those for wich an open covering of the base space exist s.t. by restriction the st.b. is trivialisable for fixed m and r. We can now define for a locally trivial st.b. (p) the multiplicity r(p) as this multi-index.

Consider now locally trivial st.b. with as total space a connected open set of M. This gives a full subcategory of SB wich we denote ℓ.t. SB. We put ℓ.t. st.b. for such bundles.

The divisor underlying a ℓ.t. st. b. has a first stratum for the incidence order relation because this is true for the trivial st. b.. The projection is a diffeomorphism if restricted to this minimal stratum. The germ of ℓ.t. st. b. at this stratum is naturally defined and easily seen to be a fiber bundle. The fiber being $\mathbb{R}^c$ with c the codimension of the minimal stratum.

For the definition of a Clemens structure we shall make an hypothesis on the structural group of a ℓ. t. st. b. .

Let see first what an automorphism of the trivial st. b. looks like : ψ : $(\mathbb{R}^c \times V, V, \Pi, D, \xi)$. We have $y_i \circ \psi = \omega_{\sigma(i)} y_{\sigma(i)}$ where $\omega \neq 0$ as D is preserved. Here σ is a permutation. It's easy to prove that $r_i = r_{\sigma(i)}$. The numbering of irreductible componnents of D is naturally supposed here to be the same as that of coordinates.

The vector field ξ is better descrived by putting $\xi(y_i^2) = 2r_i$ so the axiom $\psi_* \xi = \xi$ gives $\xi(y_i^2 \circ \psi) = 2r_i$ i.e. $\xi(\omega^2_{\sigma(i)}\ y^2_{\sigma(i)}) = 2r_i$. As $r_i = r_{\sigma(i)}$ we have $\xi((\omega^2_{\sigma(i)} - 1)\ y^2_{\sigma(i)}) = 0$.

Let $G_1(r)$ be the group of C^∞ units of V, i.e. of non vanishing C^∞ functions ω of the nbhd L of V in $\mathbb{R}^c \times V$, verifying $(\omega^2-1)y^2$ is first integral of ξ. Let $G(r) = G_1(r) \oplus \ldots \oplus G_1(r)$ (c factors).

Lemma 1 :

a) A ℓ.t. st. b. (p) admit the semi-direct product $G(r(p)) \times_s \mathfrak{S}_c$ as structural group acting on $\mathbb{R}^c$ by compon ent-wise multiplication. (Here $\mathfrak{S}_c$ the symetric group of permutions of c(r(p)) elements).

b) Such structural group can be reduced to the semi-direct product $(\mathbb{Z}/_{2\mathbb{Z}})^c \times_s \mathfrak{S}_c$.

Proof : Assertion a) is clear if we accept to call structural group of a ℓ.t. st. b. the structural group of the underlying locally trivial fiber bundle.

The proof of b) is the same as part of that of lemma 3 §4 of this chapter.

2.3. Real Clemens structures.

Such an structure is a coherent system of tubular nbhds wich are ℓ.t. st. b. with reduced structural groups.

Définition 1 : Let $V \in \mathcal{D}$. An stratifying tube of (M,D) over V is a ℓ.t. st. b. over V with structural group $(\mathbb{Z}/2\mathbb{Z})^{c(V)} \times_s \mathfrak{S}_{c(V)}$ such that the total space is an open set of M with the induced divisor.

Note that such an stratifying tube gives easily a true tubular nbhd of V in M.

Consider now the category $S\tau$ of stratifying tubes of (M,D) as a full subcategory of SB. Put st. t. for short.

Définition 2 : A Clemens structure over D in M is a functor $\mathcal{C}:\mathcal{D} \to S\tau$ such that :

$$r(\mathcal{C}(V)) = r(V).$$

Proposition 1 : Let a Clemens structure $\mathcal{C}$ be given over D in M. There exist a nbhd V of D in M and a vector field ξ on V-D which coincides in a nbhd of each stratum X of D with the field in $\mathcal{C}(X)$.

Remark : For such a field we have a local model. Clearly, if X is in $\bigcap_{i \in J} D_i$ for each point of X there is an analytic system of coordinates $(U,(x_i))$ s.t. $X \cap U = \{x_i = 0,\ i \in J\}$ and C^∞-units α_i giving

$$\xi_{|U} = \sum_{i \in J} \frac{r_i(X)}{\alpha_i x_i} \frac{\partial}{\partial_i x_i \alpha_i}.$$

Le proof is evident as we can give tubular nbhds small enough s.t. two of them doesn't touch if there is no incidence relation between bases.

2.4. Existence of real Clemens structures.

The crucial fact left about real Clemens structures is the existence question.

THEOREM 1 : *Given a compact real analytic manifold with boundary M and a divisor with normal crossings D without singularities on the boundary of* M ; *there exist at least one Clemens structure over* D *in* M.

Proof : It goes by induction on the dimension of strata. The construction is similar to that of Theorem 2.4. of [2] and Proposition 2.4. of [16] . We need three lemmas.

Lemma 1 : Let X be a stratum of D and $(p) = (F,X,p,D,\xi)$ an st. t.. Let $Y \geqq X$ be an incident stratum then there exists an st. t. $\partial_{YX}(p)$ over the open set $B = Y \cap F$. The inclusion $B \subset F$ gives a morphism $\partial_{YX}(p) \to (p)$.

Lemma 2 : Let $Z \geqq Y \geqq X$ be three incident strata, (p) and $\partial_{YX}(p)$ like above. The tube $\partial_{ZY}\,\partial_{YX}(p)$ given by lemma 1 over the open set C of Z coincides with $\partial_{ZX}(p)$ over their common domain.

Lemma 3 : Let X be a stratum of D and L an open nbhd of ∂X in $\overline{X}$. For each st. t. over $X \cap L$: (T) there is an open nbhd $L' \subset L$ of ∂X in $\overline{X}$ and a st. t. (T') which is an extension of $(T)_{|L'}$; i.e. coincides with (T) if both are restricted to $L' \cap X$.

For the theorem we first use Lemma 3 to construct an stratifying tube over minimal strata (which are necessarily closed manifolds).

The final step is to suppose that an st.t. $\mathcal{C}(Y)$ is given for each stratum Y of dimension less than n s. t. for each incident stratum X, we know that in a nbhd of X, $\mathcal{C}(Y)$ coincides with $\partial_{YX}(\mathcal{C}(X))$.

We apply now Lemma 1 for an n+1 dimensional stratum Z to construct $\partial_{Z\Omega}(\mathcal{C}(\Omega))$ for each $\Omega \leqq Z$. All these tubes give, by Lemma 2, an stratifying tube over a nbhd of ∂Z in $\overline{Z}$. This tube we extend by Lemma 3. This proves the existence Theorem.

Proof of Lemma 1.

Let construct $\partial_{ZX}(p)$ first for a trivialising chart of the bundle (p). Say $U \to X \cap U \times \mathbb{R}^{c(X)}$: $x \mapsto (p(x),\ y_i(x),\ i=1,\dots,c(X))$ such that the cocycle with other charts (y_i') is $y_i = s_i\, y'_{\sigma(i)}$ with $s_i \in \mathbb{Z}_2 = \{-1,+1\}$ and $\sigma \in \mathfrak{S}_c$, $r_{\sigma(i)} = r_i$, $\forall\, i$.

In U, $Y \cap U$ is decomposed in connected components : $(Y \cap U)_\lambda$ which are determined by a set I of indices s.t. $(Y \cap U)_\lambda \subset \{y_i = 0,\ i \in I\}$ and a system (ε_i) $(i \in I')$, $\varepsilon_i \in \mathbb{Z}_2$ and I' is $\{1,\dots,c(X)\} - I$: $y_j((Y \cap U)_\lambda)\varepsilon_j > 0$ if $j \in I'$.

For each λ we find a first integral of $\xi_{X|U}$:

$$Y_j = y_j^2 - \frac{r_j}{c(Y)} \sum_I \frac{y_\alpha^2}{r_\alpha}$$

We have that $V_\lambda = \{x \in U : y_j(x)\varepsilon_j > 0\}$ is a nbhd of $(Y \cap U)_\lambda$. In this nbhd we attach a st.t. structure for $\partial_{YX}(p)$ by givng the chart : $V_\lambda \to (U \cap X) \times \mathbb{R}^{I'} \times \mathbb{R}^I : x \mapsto (p(x), \varepsilon_j Y_j^{1/2} (j \in I'), y_i (i \in I))$.
In this coordinates the field can be written $\sum_{i \in I} \frac{r_i}{y_i} \frac{\partial}{\partial y_i}$. The réunion $V_{U,Y}$ of the V_λ is now given the structure of trivial st.t. The reunion $V_Y = \bigcup V_{U,Y}$ for U trivialising (p) is given a ℓ.t.st.b. structure as cocycles of (p) push in cocycles on V_Y : this with the obvious projection, same divisor and field is $\partial_{YX}(p)$.

Proof of Lemma 2.

With the notations just introduced. Let $m' \in Z \cap U$ and let a fixed connex component of $Z \cap U$ be defined by the vanishing of y_i for $i \in J$. Here $J \subset I \subset \{1,\dots,c(X)\}$.

The fiber structures of $\partial_{ZX}(p)$ and $\partial_{ZY}\,\partial_{YX}(p)$ are given by the same coordinates y_j, $j \in J$. Clearly, it suffices to prove that fibers coincide set-theoretically over m'.

Put c(I) = cardinal of I = c(Y) ; $c(J) \leqq c(I)$.

In the case of $\partial_{ZX}(p)$ this fiber is given by equations :

$$\text{(I)} \quad \left\{ \frac{y_i^2}{r_i} - \frac{1}{c(J)} \sum_{\beta \in J} \frac{y_\beta^2}{r_\beta} = A_i \qquad i \notin J \right.$$

for (A_i) a system of constants.

In the other hand, the fiber of $\partial_{ZY}\,\partial_{YX}(p)$ over m' is given by equations :

$$\text{(II)} \quad \left\{ \begin{aligned} &\frac{y_i^2}{r_i} - \frac{1}{c(I)} \sum_{\alpha \in I} \frac{y_\alpha^2}{r_\alpha} = B_i \qquad \text{if } i \notin I \\ &\frac{y_i^2}{r_i} - \frac{1}{c(J)} \sum_{\beta \in J} \frac{y_\beta^2}{r_\beta} = B_i \qquad \text{if } i \in I - J \end{aligned} \right.$$

for (B_i) a system of constants.

Equation II is a consequence of I as :

$$\frac{1}{c(I)} \sum_{\alpha \in I-J} \frac{y_\alpha^2}{r_\alpha} = \frac{1}{c(I)} \sum_{\alpha} (A_\alpha + \frac{1}{c(J)} \sum_{\beta \in J} \frac{y_\beta^2}{r_\beta}) =$$

$$= \sum_{\alpha \in I-J} \frac{A_\alpha}{c(I)} + \frac{c(I)-c(J)}{c(I).c(J)} \sum_{\beta \in J} \frac{y_\beta^2}{r_\beta} .$$

We take now $B_i = A_i + \sum_{\alpha \in I-J} \frac{A_\alpha}{c(I)}$ if $i \notin I$ which is a constant.

For the reciprocal inclution take $A_i = B_i + \sum_{\alpha \in I-J} B_\alpha / c(I)$ for each $i \notin I$.

Proof of Lemma 3.

Let $L' \subset L^+$ be two open nbhds of ∂X in $\bar{X}$ s. t. $\bar{L}' \subset L^+ \subset \bar{L}^+ \subset L$. We can consider a finite open covering of X by coordinate systems of M s.t. D is determined by the right monomial. We will suppose that if any of these coordinate domains meet L^+ then its intersection with X is included in L and that its transverse components are part of a single trivialising atlas of the st. b. (T).

We use the classical extension theorem for tubular nbhds to give on a nbhd T' of X in M a tube projection π which coincides with (T) over L'.

We can now reduce the structural group at (T', π) with the help of a C^∞ partition of the unity associated to a trivialising atlas of π; (ζ_ℓ) : $\Sigma \zeta_\ell = 1$. Suppose this atlas is $(U^\ell, (y_i^\ell))$. The cocycle being $y_i^p = \beta_i^{(p,\ell)} . y_i^\ell$. Putting $y_i'^p = \prod_j (\beta_i^{(j,p)})^{\zeta_j} y_i^j$ we get this reduction.

We can now verify that the field ξ_X given by the local models is well defined all over X.

2.5. The special trivialisation.

In this paragraph we apply the construction of a Clemens structure to the total transform divisor c.f. theorem 1.0. We thus obtain a trivialisation of its nbhd with precise local models.

Consider a real analytic function $\tilde{f} : M \to \mathbb{R}$ s.t. the ideal $(\tilde{f})$ determines a divisor having normal crossings. We suppose that $\tilde{f}$ has no singularity out of D and none lies in the boundary of M. We recall that M is supposed to be compact.

PROPOSITION 1 : Let $\mathcal{C}$ *be a Clemens structure over* D *in* M . *Then there exists a nbhd* $\tilde{U}$ *of* D *in* M *and a vector field* ξ *transverse to* $\tilde{f}$ *on*

$\tilde{U}$-D such that, for each point of D there exists a real analytic coordinate system (V,(Z_i)) and n units (α_i) such that (Z^r) is the ideal $(\tilde{f})_{|V}$ and $\xi = \Sigma \frac{r_i}{x_i}\frac{\partial}{\partial x_i}$ for $x_i = \alpha_i Z_i$.

We can choose $\eta > 0$ and $\tilde{U}$ s.t. ξ and $\tilde{f}$ give a trivialisation $\tilde{U}\text{-D} \to (0,\eta) \times |\tilde{f}|^{-1}(\eta)$.

Proof : Proposition 1.3. give a nbhd $\tilde{U}$ and the field ξ . Local models are those of the Clemens structure.

The function $\tilde{f}$ in the C^∞ coordinates (x_i) can be written as $\tilde{f}(x) = \beta(x)\, x^r$ for a non vanishing C^∞ smooth β.

$$\xi(\tilde{f})\ (x) = \frac{\tilde{f}(x)}{\beta(x)} \sum_i \{\frac{r_i^2\beta(x)}{x_i^2} + \frac{r_i}{x_i}\frac{\partial\beta(x)}{\partial x_i}\} .$$

The term $\frac{r_i^2\beta(x)}{x_i^2} + \frac{r_i}{x_i}\frac{\partial\beta(x)}{\partial x_i} \geqq \frac{r_i}{|x_i|}\{\frac{r_i\beta}{x_i} - |\frac{\partial\beta}{\partial x_i}|\}$ is non vanishing in a nbhd of D (the divisor is here determined by the vanishing of x_i for the i's s.t. $r_i \neq 0$) So $\tilde{f}$ is transverse to ξ.

We consider now a nbhd $\tilde{U}$ of D in M by putting $\tilde{f}^{-1}(-\eta,\eta)$, for η small enought to here $\tilde{f}$ transverse to ξ .

For this nbhd we have the map

$$\psi : \tilde{U} - D \to (0,\eta) \times |\tilde{f}|^{-1}(\eta)$$
$$x \mapsto (|\tilde{f}|(x), \phi(x, t(\varepsilon(x)\eta, x))),$$

for $\phi : \tilde{U} - D \times \mathbb{R} \to \tilde{U} - D$ the flow of ξ, $\varepsilon(x) = \text{sign } f(x)$ and $t(\sigma,x)$ implicitly determined by $f(\phi(x,t(\sigma,x))) = \sigma$.

This map is abviously a C^∞ diffeomorphism.

Here we supposed ξ tangent to the boundary of M inside U. This can be obtained by "shrinking" the manifold M near $D \cap \partial M$.

We finish this chapter by stating a first approximation towards the trivialisation theorem (2.2. of next chapter).

THEOREM 1 : *Let P be a finite codimension germ of polynomial function. There exists a nbhd U of 0, $\eta > 0$ small enought and a map T such that $T : U - P^{-1}(0) \to (0,\eta) \times |P|^{-1}(\eta) \cap U$ is an isomorphism composed of a desingularisation of P like in theorem 1.0. and a diffeomorphism ψ with local models like in proposition 1.*

3. REAL MAPS WITH MODERATE GROUTH IN A DIVISOR D.

3.0. Introduction.

In this important chapter we first introduce a notion : maps with moderate growth and then apply it to the trivialisation with local models obtained in last chapter. We do this to give a description of the behaviour of the trivialisation applied to flat forms.

Our definition of maps with moderate growth is related with Deligne's treatment of growth of functions near an algebraic manifold. In fact, ideas developed in §2 chapter II of [3] underlies the exposition in this chapter. Our scope being less important and our methods more specific than in [3] we will explain the details which apply to smooth functions and forms.

The application to a theorem of composed functions announced is then exposed. We prefer to take the point of view of functions constant over level sets as it generalizes "appartennance biponctuelle" of [4] and is more natural here.

3.1. Maps with moderate growth.

We define here the property verified by the map in the desingularisation theorem (2.0.1.) and the trivialisation map of proposition (2.5.1.). Namely moderate growth at the "infinity". This is enought to stablish the property announced for the trivialisation.

Let B be a smooth reimannian manifold with distance function d and A a compact subset of B.

Definition 1 (Compare VI 3.4. [7]). Let $C \subset B$ be a set, g a C^∞ function defined on B-A. The function g is said to have a zero of infinite order at A over C if for every $p \in \mathbb{N}$ there exist a nbhd L of A and a constant K s.t. for any $x \in C \cap L-A$ $|g(x)| \leq K\, d(x,A)^p$.

Remark 1 : The fact that g is a smooth function on B-A with a zero of infinite order at A doesn't imply g to be flat at A. Example : $\sin(\exp(1/x^2)).\exp(-1/x^2)$ at $0 \in \mathbb{R}$. If g now happens to be smooth in all B, g is flat. (th. VI 1.3.5. [7]).

Remark 2 : If g is a smooth function on B flat at A then every derivative at a point of A has a zero infinite order at A.

Remark 3 : If g is a C^∞ function on B-A s.t. for any coordinate system, every partial derivative (of any order) has a zero of infinite order. Then g can be extended to A in a C^∞ function on B (flat at A): use Lemma IV 4.3. [15] .

Definition 2 : Let g be a C^∞ function on B-A. We say that g has moderate growth in A over C if for every C^∞ differential operator G on B there exist $p \in \mathbb{N}$, $K > 0$ and a nbhd L of A s.t. $|G\, g(x)| \leq Kd(x,A)^{-p}$ for any x in $C \cap L-A$.

These functions are called "multipliers" of the ideal of C^∞ functions on B flat at A. [7] , [15] Proposition (IV 4.2. [15]). If ϕ is a C^∞ function on B-A with moderate growth in A and f is a C^∞ function on B-A such that every derivative has a zero of infinite order at A. Then the function $\phi.f$ extends uniquely to all B. This extension $\phi.f$ is flat at A.

We recall that a function f defined on B-A is said to verify a Łojasiewicz inequality over A if there is a nbhd V of A and two positive constants $C > 0, \alpha \geq 0$ s.t. $|f(x)| \leq Cd(x,A)^\alpha$ for $x \in V-A$. (Compare § 4 Chapter V [15]).

Definition 3 : Let N and B be two C^∞ riemannian manifolds (metric d and d' resp.) ; A and E compact substes of B and N resp. and $g : B-A \to N-E$ a C^∞ map. The map g is said to have moderate growth in A to E if :

1) There exist four positive constants α, β, C_1 and C_2 s.t. $C_1 d(x,A)^\alpha \leq d'(g(x),E) \leq C_2 d(x,A)^\beta$ for x in a nbhd of A. (The left-hand side inequality is a Łojasiewicz inequality for $x \mapsto d'(g(x),E)$ over A).

2) For any C^∞ smooth $f : N \to \mathbb{R}$, f o g has moderate growth in A.

Remark that condition 1) means that values of g approach E at a moderate speed for arguments approaching A.

Functions with moderate growth : $B-A \to \mathbb{R}$ are not of moderate growth if considered as maps $B-A \to S^1 - \{\infty\}$. The right-hand inequality in 1) has been introduced for technical reasons.

A fundamental result is now stablished

PROPOSITION 1 : If $g : B-A \to N-E$ *has moderate growth in* A *to* E *then :*

a) for any $f : N-E \to \mathbb{R}$ *s.t. all its derivatives have a zero of in-*

finite order at E, f o g *and all its derivatives have a zero of infinite order at* A.

b) *for any* f : N-E → ℝ *with moderate growth in* E , f o g *has moderate growth in* A.

Proof : We consider natural normed space structures for the jet spaces J^kB and J^kN over compact nbhds K and K' of A and E s.t. $g(K) \subset K'$.

Assertion a) is stablished by prooving that $\forall N$ there exist a nbhd of A and C > O s.t.

$||j^k(fog)(x)|| \leq Cd(x,A)^N$ in such a nbhd.

Consider the identity $j^k(fog)(x) = (j^kf)(g(x)) \cdot j^kg(x)$ so : $||j^kfog(x)|| \leq ||j^kf(g(x)||.||j^kg(x)||$. Using that $\forall M \; ||j^kf(g(x))|| \leq C_1 \; d'(g(x),E)^M$. (as f and all its derivatives have a zero of infinite order in E). And $||j^kg(x)|| \leq C_3 d(x,A)^{-p}$ for positive constants and a fixed nbhd of A. Point 1) in definition 3 gives

$$||j^kfog(x)|| \leq C_1 \; C_2^M \; C_3 \; d((x,A)^{\beta M-p},$$

take now $M > (N-p)/\beta$.

Assertion b) is now proved using the Łojasiewicz inequality of $d'(g(x),E)$.

3.2. The special trivialisation has moderate growth.

In this paragraph we consider local models of desingularisation map and the trivialisation map to prove they have moderate growth in their "singular loci ".

We first stablish the relation between flat C^∞ forms and mappings with moderate growth in a compact set A.

PROPOSITION 1 : Let g : B-A → N-E *be a mapping with moderate growth in* A *to* E *then if* α *is a smooth differential form of* M *flat at* E, $g^*\alpha$ *extends to* B *in a* C^∞ *form flat at* A.

Proof : A C^∞ k-form α wich is flat at E can be considered as a differentiable function on the fiber bundle $\oplus^k TN$ such that any derivative has a zero of infinite order at $\oplus^k TM_{|E}$.

Now $\oplus^k Tg$ has moderate growth on $\oplus^k TM_{|E}$.

Proposition 1.2. proves that $g^*\alpha$ has as all its "derivatives" a zero of infinite order at A. Then by Remark 3.2. $g^*\alpha$ extends to a flat form in A. The set $\oplus^k TM_{|E}$ is not compact but we can work in a compact nbhd of the null section of $\oplus^k$ TM to prove this result.

Lemma 1 : Let V be an open nbhd of O in $\mathbb{R}^n$, M a real analytic manifold and $\theta : M \to V$ an analytic, proper map. If $E = \theta^{-1}(O)$ and $\theta_{|M-E} : M-E \to V-\{O\}$ is an isomorphism then $\rho = \theta^{-1}_{|M-E}$ is a map with moderate growth in O to E.

Proof : Denote θ' the restriction $\theta_{|M-E}$. Property 2) in defintiion 2.3. is proved by computing in local coordinates the partials of ρ. Let $\nu \in \mathbb{N}^n$, (x_i) natural coordinates of $\mathbb{R}^n$ and $(W,(y_i))$ analytic coordinates in M centered in a point of E. Computation gives :

$$D^\nu \rho(x) = \frac{P_\nu\,(D^\sigma(\theta')^j;\sigma\leqq\nu;j=1,\dots,n)}{(\mathrm{Det\ Jac}\ \theta')^{|\nu|}} \circ \rho(x)$$

where P_ν is a polynomial vector.

We know Det Jac θ to be analytic, and non vanishing outside E. Then we can apply classical Łojasiewicz inequality : $\exists K > 0, \exists N > 0 : \forall s \in V$

$$K|\mathrm{Det\ Jac}\ \theta(s)| \geqq d(s,E)^N.$$

for any $x \in \theta'(V-E)$ we have

(1) $$||D^\nu\rho(x)|| \leqq K' d(\rho(x),E)^{-N|\nu|} \quad \text{for constant } K' > 0.$$

If we write down what the classical Łojasiewicz inequality means for $||\theta||^2 : M \to \mathbb{R}$ we have $\exists C > 0, \exists n > 0$ s.t.

(2) $$C||\theta(s)||^2 \geqq d(s,E)^n.$$

Property (1) together with (2) gives that there exist $K'' > 0$ and $n \geqq 0$ s.t.

(3) $$||D^\nu\rho(x)|| \leqq K''||x||^{-n}$$ for any x in a nbhd of O (E is compact).

For the property 1) in definition 2.3. we use now (2) again. This gives the right-hand side inequality by putting $\rho(x)$ in place of x. The left-hand side inequality follows from differentiability of θ and the fact that $E = \theta^{-1}(O)$.

Corollary 1 : For such a $\theta : M \to U$ we have an isomorphism $\theta^* : m^\infty \Lambda^\cdot(n) \to J(E)\Lambda^\cdot(M)_E$.

We put $J(E) = \Gamma(E, J^\infty(E))$ and $\Lambda'(M)_E = \Gamma(E,\Lambda^\cdot_M)$ to have flat forms and $\Lambda^\cdot(n) = \Lambda^\cdot_{\mathbb{R}^n,O}$

Let $\psi : \tilde{U} - D \to (0,\eta) \times \tilde{P}^{-1}(\{-\eta,\eta\})$ be the trivialisation of theorem 2.5.1. obtained with the help of a Clemens structure over D. The set $\tilde{U}$ is an open nbhd of D in the compact manifold M.

We denote $\Phi = \psi^{-1}$; Y the set $(0,\eta) \times \tilde{P}^{-1}(\{-\eta,\eta\})$ and $Y_o = \{0\} \times \tilde{P}^{-1}(\{-\eta,\eta\})$.

PROPOSITION 2 :

i) The map ψ *has moderate growth in* D *to* Y_o .

ii) The map Φ *has moderate growth in* Y_o *to* D.

Proof : Recall first that the trivialisation ψ is given by the flow $\phi : S \to \tilde{U} - D$ of a field ξ where $S \hookrightarrow \tilde{U} - D \times \mathbb{R}$ is a nbhd of $\tilde{U}- D \times \{0\}$:

$$\psi(y) = (|\tilde{P}(y)|;\phi(y,t(y,\varepsilon(y)\eta))), \quad y \in \tilde{U}\text{-}D$$

where $\varepsilon(y)$ is the sign of $\tilde{P}(y)$ and $t(y,\sigma)$ is given by the implicit relation $\tilde{P}(\phi(y,t(y,\sigma))) = \sigma$.

We will need two lemmas to finish the proof of proposition 1.

Lemma 2 : Let G be a C^∞ differential operator on $\tilde{U}$ and $n \in \mathbb{N}$. There exist a nbhd of D and positive constants C and S s.t. $\left|\frac{\partial^n}{\partial\eta^n} G\, t(y,\eta)\right| \leqq C|\eta|^{-S}$ in the nbhd of $D \times \{0\}$.

Lemma 3 : Let G be a C^∞ differential operator on $\tilde{U}$. There exist a nbhd of D and positive constants K and N s.t. (if G acts over $\tilde{U} \times \mathbb{R}$ in the trivial way)

$$|G\phi^i|(y,t(y,\eta)) \leqq K|\eta|^{-N}$$ in the nbhd of $D \times \{0\}$. (ϕ^i is a component of ϕ in a coordinate system centered in D).

Proof of Proposition 2 :

We first remark that the map Φ is given by $\Phi(\lambda,y) = \phi(y,t(y,\pm\lambda))$ and so Lemma 2 and 3 prove that derivatives of Φ have the right growth bounds.

As $\tilde{P}$ is analytic on $\tilde{U}$, vanishing at D we have a Łojasiewicz inequality :

$\exists \alpha, K > 0$ s.t. in a nbhd of D, $d(y,D) \leqq K|\tilde{P}(y)|^\alpha$ (✱).

We evaluate in $\phi(y,t(y,\pm\lambda))$ to obtain :

$$d(\Phi(\lambda,y),D) \leqq K\,\lambda^{\alpha}.$$

Next we remark that for a compact in a nbhd of $\{x_1 \cdot \ldots \cdot x_c = 0\}$ we have $|x_1^{\alpha_1} \cdot \ldots \cdot x_c^{\alpha_c}| \leqq K|x_i|^{\alpha_i}$ for a constant K and every i. As $\tilde{P}$ is locally a unit times a monomial function ; we can find M and $K' > 0$ s.t. $|\lambda| = |\tilde{P}(\phi(y,t(y,\lambda)))| \leqq K'|\phi^i(y,t(y,\lambda))|^M$.

We get so $K''|\lambda|^{M'} \leqq d(\Phi(\lambda,y),D)$.

The same reasoning goes for ψ exept that for the right-hand side inequality in 1) of def. 2.3. we simply use algebroid local models for $\tilde{P}$ at D.

For the left-hand side we use (*) as $|\tilde{P}(y)|$ is equivalent to $d(\psi(y),Y_0)$. This finishes the proof of the Proposition.

Proof of Lemma 2.

Lets compute $\frac{\partial}{\partial y_i}\tilde{P}(\Phi(y,t(y,\eta)))) = C_i$ where we put $y_o = \eta$ and C_i is 0 or 1 accordingly. We find :

$$\frac{\partial t}{\partial y_i}(y,\eta).\xi(\tilde{P})(\phi(y,t(y,\eta))) = C_i - \Sigma\,\frac{\partial \tilde{P}}{\partial y_\ell} \circ \phi \;.\; \frac{\partial \phi^\ell}{\partial y_j}(y,t(y,\eta)).$$

We prove by induction that for any $\alpha \in \mathbb{N}^n$,

$\xi(\tilde{P})(\phi(y,t(y,\eta))).\frac{\partial^{|\alpha|}}{\partial y^\alpha} t(y,\eta) = P_\alpha$. Where P_α is a polynomial function on : C^∞ functions of y and $t(y,\eta)$; the partials $\frac{\partial^{|\gamma|}\phi^i}{\partial y^\gamma}(y,t(y,\eta))$, $|\gamma| \leqq |\alpha|$; the partials $\frac{\partial^{|\beta|}}{\partial y^\beta}\xi(\ldots\xi(\tilde{P})\ldots)$ evaluated in $\phi(y,t(y,\eta))$ where we have composed q-times ξ for $|\beta| + q \leqq |\alpha|$ and at last the partials of $t(y,\eta)$ with orders strictly smoler than $|\alpha|$.

So, its enough to prove that

$$\left|\frac{\partial^{|\beta|}}{\partial y^\beta}\xi(\ldots\xi(\tilde{P})\ldots)(y,t(y,\eta))\right| \leqq \frac{K}{|\eta|^N} \text{ for K, N > 0 in a nbhd of D.}$$

Remark now that locally $\xi = \frac{\beta}{\tilde{P}}\,.\,\nabla Q$ if $\tilde{P} = \beta Q$, β a non vanishing function, Q a monomial function and $\nabla Q = \Sigma r_i\, y^r/y_i\, \frac{\partial}{\partial y_i}$.

Then $\xi(\xi^{(q-1)}(\tilde{P})) = \frac{\beta}{\tilde{P}}\,\nabla Q(\xi^{(q-1)}(\tilde{P}))$.

This allowes us to make an induction. Case $q = 1$ is evident.

We finish by the remark that :

$|\xi(\tilde{P})\phi(y,t(y,\eta))| \geqq K_1|\eta|$ for a certain $K > 0$ (c.f. the proof of 1.5.2.).

Proof of Lemma 3.

Here we can use an explicit integration of the field ξ which gives that $\frac{\partial^{|\alpha|}\phi^i}{\partial x^\alpha}$ is a polynomial function on partials of the squares of ϕ^i and on $1/\phi^i$. This is bounded near D by $K/|\phi^i|^N$ for K and $N > 0$. We finish by simply evaluating at $(y,t(y,\eta))$.

THEOREM 1 : The map $\psi : \tilde{U}-D \to (0,\eta) \times \tilde{P}^{-1}(\{-\eta,\eta\})$ *gives an isomorphism of* $J^\infty(Y_0)\Lambda^\cdot(Y)$ to $J^\infty(D)\Lambda^\cdot(M)$.

THEOREM 2 : Let P *be a real polynomial with an algebraically isolated singularity at* 0. *Let* $X_{\varepsilon\eta}$ be $\{x \in \mathbb{R}^n : 0 < |P(x)| < \eta, ||x|| < \varepsilon\}$ *for* ε *and* η *small enough and* $X_0 = \overline{X}_{\varepsilon\eta} \cap P^{-1}(0)$. *There exists a trivialisation of* $|P|_{|X_{\varepsilon\eta}} : X_{\varepsilon\eta} \to (0,\eta)$ *say* $T : X_{\varepsilon\eta} \to (0,\eta) \times |P|^{-1}(\eta) \cap \overline{X}_{\varepsilon\eta}$ *which gives an isomorphism of flat forms :*

$$T^* : J^\infty(\{0\} \times |P|^{-1}(\eta))\Lambda^\cdot(0,\eta) \times |P|^{-1}(\eta) \to J^\infty(X_0)\Lambda^\cdot X_{\varepsilon\eta}$$

These theorems are corollaries of Proposition 2 Proposition 1 and Corollary 1.

3.3. The local trivialisation of a singularity. Application.

Here we use the fundamental result of this paper (theorem 2.2.).

We discuss the slight generalization (in the particular case of ours) we obtain of a theorem of Glaeser [4] . This result is better understood in the context of chapter II of [9] .

Prior a discussion with R. Moussu, this result was the particular case of dimension zero of the computation of real relative cohomology spaces. The formal aspect of this computation needs theorem 1. II of [9] in dimension 0 and results of Milnor - Brieskorn - Bloom - Sebastiani in other dimensions.

So consider a germ of a function having at zero an algebraically isolated singularity. We can suppose this a polynomial germ and apply the good trivialisation theorem.

Let recall first the content of theorem 1. II of [9] .

THEOREM : Let g be a germ constant in all level hypersurfaces of f . Then f and g have left-equivalent infinite jets.

Clearly that means that if $dg \wedge df$ vanish identically there exist a power series in one variable ζ s.t. $\zeta \circ \hat{f} = \hat{g}$. Here $\hat{h}$ is the infinite Taylor expansion of a germ h at O.

Using Emile Borel's theorem [7] f.i. we get a germ of C^∞ function ϕ such that $\hat{\phi} = \zeta$ and we find that $\phi \circ f - g$ is a flat function at O.

Lemma 1 : A flat function h at zero, constant on the level hypersurfaces of f is flat at $f^{-1}(0)$.

Proof : The easy way for us is to use desingularisation and a Clemens structure. The behaviour of h at any point of $f^{-1}(0) - \{0\}$ is the same as that $\theta^* h$ at any smooth point of the total transform D (as h is constant in level sets of f).
In particular we can consider a smooth point of the exeptionnal divisor where $\theta^* h$ is flat and we are done.

PROPOSITION 1 : If h is a smooth function constant in the level sets of f and flat at $f^{-1}(0)$ then there exist germs at zero of flat functions in one variable ($C^\infty[0,+\infty)$) : ϕ_i such that

$$h = \Sigma \phi_i \circ f \, . \, I_i .$$

Where I_i is the caracteristic function of a connected component of the germ at zero of the complement of $f^{-1}(0)$; $(\phi_i \circ f).I_i$ is C^∞.

Proof : Use theorem 2.2. to push h to the trivial set $Y - Y_o$. The fact that we push with a trivialisation of $|f|$ a function constant on the level sets of f proves that $T^* h$ is constant on the fibers of the trivial fibration $Y - Y_o \to (0,\eta)$. That is $T^* h$ is a function on $(0,\eta)$ over each connected component of $Y - Y_o$.

Now, connected components of $Y - Y_o$ correspond through T to connected components of the germ of $\mathbb{R}^n - f^{-1}(0)$ at O and the proposition is proved.

As a corollary we have.

THEOREM 1 : Let P be a germ of real smooth function with an algebraically isolated singularity at O. Let f be a germ of real smooth function constant in the level sets of P. If for every $\varepsilon > 0$ the set $\{x \in \mathbb{R}^n : ||x|| < \varepsilon , P(x) \neq 0\}$ has c connected components. Then there exist c+1 germs of real smooth functions $\phi_i : (\mathbb{R},0) \to \mathbb{R}$ i=0,...,c *such that :*

each ϕ_i, $i > 0$ *is flat at* 0 *and*

$$f = \sum_{i>0} (\phi_i \circ P)\, I_i + \phi_0 \circ P.$$

Remark $(\phi \circ P).I$ is a C^∞ germ if I is constant every where exept at a subset of $P^{-1}(0)$ when $\phi \in m^\infty(1)$.

Theorem 2.2. was developed as the principal tool in the computation of real relative cohomology of a finite codimension germ [13] .

We think that others developments of real Clemens structures are possible.

BIBLIOGRAPHY

[1] V.I. Arnol'd : Index of singular points of a vector field, the Petrovskiĭ-Oleĭnik inequalities and mixed Hodge structures. Functional Anal. Appl. 12 (1978).

[2] C.H. Clemens : Picard Lefschetz theorem for families of nonsingular algebraic varieties acquiring ordinary singularities, Trans. Amer. Math. Soc. 136 (1969), 93-108.

[3] P. Deligne : Equations différentielles à points singuliers réguliers, Lectures Notes in Math., vol. 136, Springer-Verlag, Berlin and New York, 1970, pP.

[4] G. Glaeser : Fonctions composées différentiables, Ann. of Math. 77 (1963), 193-209.

[5] H. Hironaka : Resolution of singularities of an algebraic variety, I and II, Ann. of Math. 79 (1964), 109-326.

[6] I. Kupka : L'indice d'un point singulier d'un champ, Astérisque S.M.F. 59-60 (1978), 151-172.

[7] B. Malgrange : Ideals of differentiable functions, Oxford Univ. Press, 1966, pp.

[8] J. Milnor : Singular points of complex hypersurfaces, Ann. of Math. Studies, Vol. 61, Princeton Univ. Press, 1969, pp.

[9] R. Moussu : Sur l'existence d'intégrales premières pour un germe de forme de Pfaff, Ann. Inst. Fourier 26 (1976), 171-220.

[10] R. Moussu : Classification C^{∞} des équations de Pfaff complètement intégrables à singularité isolée, Invent. math. 73 (1983), 419-436.

[11] C.A. Roche : Cohomologie relative dans le domaine réel, Thèse de 3ème cycle. Grenoble (1982).

[12] C.A. Roche : Cohomologie relative, GMEL. Actualités Mathématiques, Gauthier-Villard, Paris, 1982.

[13] C.A. Roche : Real relative cohomology of finite codimension germs, Bull. Amer. Math. Soc. 7 (1982), 596-598.

[14] C.A. Roche : Real relative cohomology, in preparation.

[15] J.C. Tougeron : Ideaux de fonctions différentiables, Ergebnisse der Mathematik und ihrer Grenzgebiete, vol. 71, Springer-Verlag, Berlin and New York, 1972, pp.

[16] V. A. Vasil'ev : Asymptotic behavior of exponential integrals in the complex domain, Functional Anal. Appl. 13 (1979), 239-247.

C.A. ROCHE
Université de Dijon
Laboratoire de Topologie
ERA 07 945 CNRS
B.P. 138
21004 - Dijon Cedex
France.

Singularities & Dynamical Systems
S.N. Pnevmatikos (editor)

RIGIDITY OF WEBS AND FAMILIES OF HYPERSURFACES

Jean-Paul Dufour & Patrick Jean
Université de Montpellier
France

This paper contains two parts : in the first section we study the local smooth classification of (p+1)-webs of codimension n of $\mathbb{R}^{qn}$, with $p \geqslant q$, and we show that it is the same as the topological one ; then we apply our results to the classification of families of hypersurfaces of $\mathbb{R}^n$, to settle that the local topological stability of these families is not generic.

1. LOCAL CLASSIFICATION OF WEBS

We recall ([1]) that a (p+1)-web of codimension n of $\mathbb{R}^{qn}$ is a (p+1)-tuple of p to p transversal foliations of codimension n of $\mathbb{R}^{qn}$. So it is locally equivalent to consider a diagram of p+1 p to p transversal submersions $f_1, f_2, \ldots, f_{p+1}$

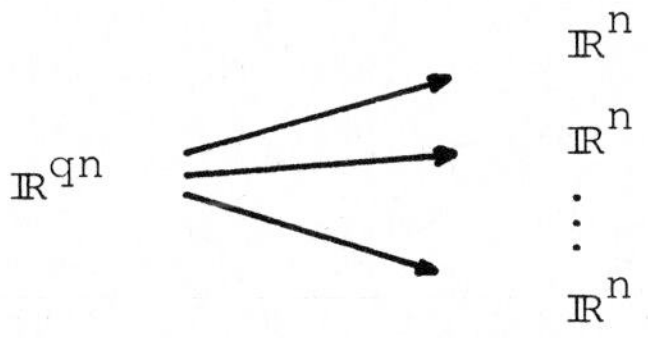

We shall confine our study to a neighbourhood of the origin, in the case $p \geqslant q$, assuming

$$f_i(0) = 0 \quad \text{for all } i , \ 1 \leqslant i \leqslant p+1 .$$

$\mathbb{R}^{qn}$ will be implicitly decomposed into $\mathbb{R}^n \times \mathbb{R}^n \times \ldots \times \mathbb{R}^n$, whose elements will be denoted $(x_1, x_2, \ldots, x_q)$, with x_i in $\mathbb{R}^n$ for all i, $1 \leqslant i \leqslant q$.

Definition 1 : two (p+1)-webs $f = (f_1, f_2, \ldots, f_{p+1})$ and $g = (g_1, g_2, \ldots, g_{p+1})$ are *smoothly equivalent* (respectively *topologically equivalent*) if there exist local diffeomorphisms (respectively homeomorphisms) $H, k_1, k_2, \ldots, k_{p+1}$,

$$H : \mathbb{R}^{qn},0 \longrightarrow \mathbb{R}^{qn},0$$

$$k_i : \mathbb{R}^n,0 \longrightarrow \mathbb{R}^n,0 \quad , \quad 1 \leqslant i \leqslant p+1 ,$$

such that

$$f_i \circ H = k_i \circ g_i \qquad \text{for all } i , \quad 1 \leqslant i \leqslant p+1 .$$

Theorem 1. Any $(q+1)$*-web of codimension* n *of* $\mathbb{R}^{qn}$ *is smoothly equivalent to a web*

$$(x_1,x_2,\ldots,x_q) \begin{array}{l} \nearrow x_1 \\ \nearrow x_2 \\ \quad \vdots \\ \searrow x_q \\ \searrow x_1+x_2+\ldots+x_q+\theta(x_1,x_2,\ldots,x_q) \end{array} ,$$

where $\theta : \mathbb{R}^{qn},0 \longrightarrow \mathbb{R}^n,0$ *is such that*

$$(1) \quad \begin{cases} \theta(x,x,\ldots,x) = 0 \\ \theta(0,\ldots,0,\underset{\substack{\uparrow \\ \text{rank } i}}{x},0,\ldots,0) = 0 \quad \text{for all } i , \quad 1 \leqslant i \leqslant q . \end{cases}$$

More precisely, there are as many classes of equivalence as classes of maps θ *satisfying* (1) *under the relation* $\mathcal{R}$ *defined by* $\theta_1 \,\mathcal{R}\, \theta_2$ *if and only if there exists a linear local homeomorphism* $k : \mathbb{R}^n,0 \longrightarrow \mathbb{R}^n,0$ *such that*

$$\theta_1 \circ (k \times k \times \ldots \times k) = k \circ \theta_2 .$$

Proof. We proceed by successive equivalences from a $(q+1)$-web $f = (f_1,f_2,\ldots,f_q,f_{q+1})$.

First, it is clear, by the hypothesis of transversality, that the map H defined in a neighbourhood of $0_{\mathbb{R}^n}$ by

$$H(x_1,x_2,\ldots,x_q) = (f_1(x_1,x_2,\ldots,x_q),f_2(x_1,x_2,\ldots,x_q),\ldots, f_q(x_1,x_2,\ldots,x_q))$$

is a local diffeomorphism of the origin, which, associated to the diffeomorphisms $k_1 = k_2 = \dots = k_{q+1} = id_{\mathbb{R}^n}$, shows that f is smoothly equivalent to a web f^1

$$(x_1, x_2, \dots, x_q) \longrightarrow \begin{cases} x_1 \\ x_2 \\ \vdots \\ x_q \\ f^1_{q+1}(x_1, x_2, \dots, x_q) \, . \end{cases}$$

By transversality again, for all i, $1 \leq i \leq q$, the map h_i defined in a neighbourhood of $0_{\mathbb{R}^n}$ by

$$h_i(x) = f^1_{q+1}(0, \dots, 0, \underset{\substack{\uparrow \\ \text{rank } i}}{x}, 0, \dots, 0)$$

is a local diffeomorphism of the origin ; then the product map $\tilde{H} = h_1 \times h_2 \times \dots \times h_q$ verifies the same property ; $\tilde{H}$, associated to the local diffeomorphisms $\tilde{k}_i = h_i$ for all i, $1 \leq i \leq q$, and $\tilde{k}_{q+1} = id_{\mathbb{R}^n}$, shows that f^1 is smoothly equivalent to a wef f^2

$$(x_1, x_2, \dots, x_q) \longrightarrow \begin{cases} x_1 \\ x_2 \\ \vdots \\ x_q \\ f^2_{q+1}(x_1, x_2, \dots, x_q) \, , \end{cases}$$

where f^2_{q+1} verifies, in a neighbourhood of $0_{\mathbb{R}^n}$,

$$(2) \qquad f^2_{q+1}(0, \dots, 0, \underset{\substack{\uparrow \\ \text{rank } i}}{x}, 0, \dots, 0) = x \quad \text{for all } i \, , \quad 1 \leq i \leq q \, .$$

At this step of the proof, let us remark that the existence of local diffeomorphisms (respectively homeomorphisms) H, k_i, $1 \leq i \leq q+1$, to obtain a web f^3, smoothly equivalent (respectively topologically equivalent) to f^2,

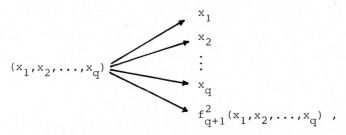

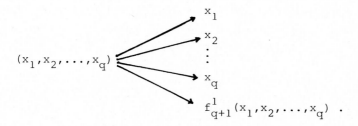

$$(3) \qquad (x_1, x_2, \ldots, x_q) \longrightarrow \begin{cases} x_1 \\ x_2 \\ \vdots \\ x_q \\ f^3_{q+1}(x_1, x_2, \ldots, x_q) \end{cases},$$

with f^3_{q+1} satisfying also (2), implies

$$k_1 = k_2 = \ldots = k_{q+1} = k$$

and

$$H = k \times k \times \ldots \times k \quad ;$$

this comes merely from the property of commutativity of diagrams in any equivalence of webs.

Now let us consider the map F defined in the neighbourhood of $O_{\mathbb{R}^n}$ by

$$F(x) = f^2_{q+1}(x, x, \ldots, x) \quad ;$$

we have

$$(dF)_0 = q \, id_{\mathbb{R}^n} \quad ;$$

then the only eigenvalue of $(dF)_0$ is q, different from 1 ; we deduce from Seidenberg's theorem ([6]) that F is linearizable, that is to say there exists a local diffeomorphism of the origin, k, such that

$$(dF)_0 = k \circ F \circ k^{-1} \quad ;$$

by this diffeomorphism, f^2 is smoothly equivalent to a web f^3 like (3), where f^3_{q+1} satisfies (2) and

$$(4) \qquad f^3_{q+1}(x, x, \ldots, x) = qx \quad \text{in a neighbourhood of } O_{\mathbb{R}^n} .$$

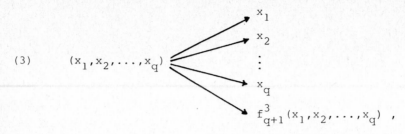

We shall have completed the proof when we have proved that the existence of a smooth equivalence between two webs like (3) satisfying (2) and (4) implies that the diffeomorphism k is linear, at least on some neighbourhood of the origin ; but from (4) , by an obvious induction, we prove that k verifies

$$k(x) = q^p \, k\left(\frac{x}{q^p}\right)$$

for all x close enough to $0_{\mathbb{R}^n}$ and all integers p ; furthermore, as k is differentiable at 0 , we have

$$(5) \qquad k(x) = (dk)_o(x) + \|x\| \, \varepsilon(x) \ ,$$

with

$$\lim_{x \to 0} \varepsilon(x) = 0 \quad ;$$

combining these two relations, we obtain

$$k(x) = (dk)_o(x) + \|x\| \, \varepsilon\left(\frac{x}{q^p}\right)$$

for all integers p ; comparing this expression with (5) , it follows that

$$\varepsilon(x) = 0 \ ,$$

and so

$k = (dk)_o$ on some neighbourhood of the origin, which completes the proof.

Theorem 2. *The smooth and topological equivalences are the same.*

Proof. From theorem 1 , we have only to prove that a homeomorphism k which realizes a topological equivalence between two webs

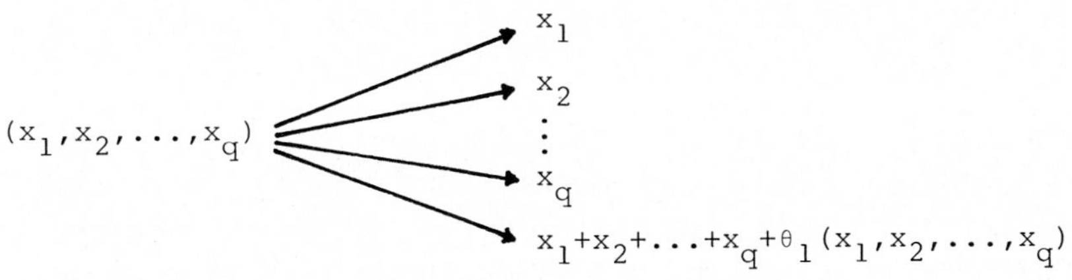

and

$$
(x_1,x_2,\dots,x_q) \longmapsto \begin{cases} x_1 \\ x_2 \\ \vdots \\ x_q \\ x_1+x_2+\dots+x_q+\theta_2(x_1,x_2,\dots,x_q) \ , \end{cases}
$$

where θ_1 and θ_2 verify (1), is a local diffeomorphism of the origin.

First, we show that k is Lipschitz on some neighbourhood of the origin, that is to say there exists a real number M such that

$$\|k(x) - k(y)\| \leqslant M\|x-y\|$$

for every couple (x,y) sufficiently close to $0_{\mathbb{R}^n \times \mathbb{R}^n}$.

So, let us remark, applying the implicit function theorem to the map F defined in a neighbourhood of $0_{\mathbb{R}^{2n} \times \mathbb{R}^n}$ by

$$F((x,y),z) = y + z + \theta_1(y,z,0,0,\dots,0) - x \ ,$$

that for every (x,y) sufficiently close to $0_{\mathbb{R}^n \times \mathbb{R}^n}$, there exists z near $0_{\mathbb{R}^n}$ such that

$$x = y + z + \theta_1(y,z,0,0,\dots,0) \quad ;$$

under these conditions, let us write

$$\text{(6)} \qquad \frac{\|k(x)-k(y)\|}{\|x-y\|} = \frac{\|k(z)\|}{\|z\|} \, \frac{\dfrac{\|k(x)-k(y)\|}{\|k(z)\|}}{\dfrac{\|x-y\|}{\|z\|}}$$

But

$$
\begin{aligned}
\frac{\|x-y\|}{\|z\|} &= \frac{\|z+\theta_1(y,z,0,0,\dots,0)\|}{\|z\|} \\
&\geqslant 1 - \frac{\|\theta_1(y,z,0,0,\dots,0)\|}{\|z\|} \\
&\geqslant \frac{1}{2} \quad \text{for } y \text{ sufficiently close to } 0 \text{, from property (2).}
\end{aligned}
$$

Likewise

$$\frac{\|k(x)-k(y)\|}{\|k(z)\|} = \frac{\|k(z)+\theta_2(k(y),k(z),0,0,\dots,0)\|}{\|k(z)\|}$$

$$\leqslant 1 + \frac{\|\theta_2(k(y),k(z),0,0,\dots,0)\|}{\|k(z)\|}$$

$\leqslant \frac{3}{2}$ for y sufficiently close to 0, because of the continuity of k at 0.

Furthermore, as k is continuous in a neighbourhood of 0, the function G defined by

$$G(z) = \frac{\|k(z)\|}{\|z\|}$$

is bounded on every compact set

$$K_r = \{z \in \mathbb{R}^n / \frac{r}{q} \leqslant \|z\| \leqslant r\}$$

where r is a strictly positive given real number ;
let M_r be an upper bound of G on K_r, and let z be an element of $\mathbb{R}^n$ such that

$$0 < \|z\| \leqslant r \quad ;$$

there exists an integer p such that

$$\frac{\|k(q^p z)\|}{\|q^p z\|} \leqslant M_r \quad ;$$

but, from property (4), we have, for r sufficiently small

$$k(q^p z) = q^p k(z) ,$$

and so, under the same conditions

$$\frac{\|k(z)\|}{\|z\|} \leqslant M_r .$$

Then, from (6), k is Lipschitz on some neighbourhood of the origin ; we deduce from Rademacher's theorem ([4]) k is there almost everywhere differentiable ; in particular, every neighbourhood of the origin contains at least one point at which k is differentiable. It follows that k is differentiable at the origin.

Indeed, by the implicit function theorem, there exists, for all

x sufficiently close to $0_{\mathbb{R}^n}$, y near $0_{\mathbb{R}^n}$, such that

$$x + y + \theta_1(x,y,0,0,\ldots,0) = 0 .$$

Let us choose x such that k is differentiable at x ; for all z close to $0_{\mathbb{R}^n}$, we have, with x and y sufficiently close to the origin

$$(7) \qquad k(x+y+z+\theta_1(x+z,y,0,0,\ldots,0)) =$$
$$(dk)_x(z) + \frac{\partial\theta_2}{\partial x_1}_{(k(x),k(y),0,0,\ldots,0)}((dk)_x(z))$$
$$+ \|z\|\varepsilon_1(z) + \frac{\partial\theta_2}{\partial x_1}_{(k(x),k(y),0,0,\ldots,0)}(\|z\|\varepsilon_1(z))$$
$$+ \|(dk)_x(z) + \|z\|\varepsilon_1(z)\|\varepsilon_2(z) \quad ,$$

with

$$\lim_{z\to 0}\varepsilon_1(z) = \lim_{z\to 0}\varepsilon_2(z) = 0$$

Furthermore, the map H defined by

$$H(z) = x + y + z + \theta_1(x+z,y,0,0,\ldots,0)$$

is a local diffeomorphism of the origin, again with x and y sufficiently close to 0 ; then relation (7) becomes, for all τ sufficiently close to 0

$$k(\tau) = id_{\mathbb{R}^n} + \frac{\partial\theta_2}{\partial x_1}_{(k(x),k(y),0,0,\ldots,0)} \circ (dk)_x \circ (dH^{-1})_o(\tau)$$
$$+ \|\tau\|\varepsilon_3(\tau) \quad ,$$

with

$$\lim_{\tau\to 0}\varepsilon_3(\tau) = 0 \quad ,$$

which means k is differentiable at 0 .

To conclude, we need only repeat the last point of the proof of theorem 1 , where we show then k is necessarily linear on some neighbourhood of the origin.

2. APPLICATION TO THE STUDY OF FAMILIES OF HYPERSURFACES

From our viewpoint, a one parameter family of smooth hypersurfaces of $\mathbb{R}^n$, (σ,f) , is a diagram of smooth maps

$$\mathbb{R} \xleftarrow{\ f\ } U \xrightarrow{\ \sigma\ } \mathbb{R}^n ,$$

where U , the domain of (σ,f) , is an open set of $\mathbb{R}^n$, and where f is a submersion. The hypersurfaces of the family are the sets $\sigma(f^{-1}(\tau))$, where the parameter τ runs over $\mathbb{R}$. Thus, the set of one parameter families of $\mathbb{R}^n$ with domain U , is identified to an open subset Ω_n of the set $C^\infty(U,\mathbb{R}^{n+1})$ of smooth maps from U onto $\mathbb{R}^{n+1}$ endowed with the Whitney smooth topology. Furthermore, the $f^{-1}(\tau)$ define a foliation of U (of codimension 1), so the study of a family of hypersurfaces is the study of the image of a foliated manifold by a smooth map.

Our study will be essentially local.

<u>Definition 2</u> : let (σ,f) and (σ',f') be two one parameter families of hypersurfaces of $\mathbb{R}^n$ with respective domains U and U'; let u be a point in U , and u' a point in U' ; we say that (σ,f) , *in the neighbourhood of* u , *is topologically equivalent* (respectively *strongly smoothly equivalent*) *to* (σ',f') , *in the neighbourhood of* u' , if there exist three local bijections (respectively three local diffeomorphisms) φ, H, K ,

$$\begin{array}{lll}
\varphi & : & \mathbb{R}, f(u) \longrightarrow \mathbb{R}, f'(u') \\
H & : & \mathbb{R}^n, u \longrightarrow \mathbb{R}^n, u' \\
K & : & \mathbb{R}^n, \sigma(u) \longrightarrow \mathbb{R}^n, \sigma'(u')
\end{array}$$

where K is a local homeomorphism , φ and H not necessarily continuous, such that we have a commutative diagram

$$\begin{array}{ccccc}
\mathbb{R} & \xleftarrow{\ f\ } & V & \xrightarrow{\ \sigma\ } & \mathbb{R}^n \\
\varphi \downarrow & & \downarrow H & & \downarrow K \\
\mathbb{R} & \xleftarrow{\ f'\ } & V' & \xrightarrow{\ \sigma'\ } & \mathbb{R}^n
\end{array}$$

where V and V' are respectively neighbourhoods of u in U , and of u' in U' .

<u>Definition 3</u> : let (σ,f) be a one parameter family of hypersurfaces of $\mathbb{R}^n$ with domain U ; let u be a point in U ; we say that (σ,f) is *topologically stable* (respectively *strongly smoothly stable*) *in the neighbourhood of* u , if for every

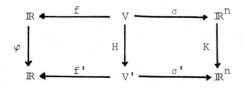

neighbourhood V of u in U there exists a neighbourhood W of (σ,f) in Ω_n such that, for every (σ',f') in W , there exists u' in V with (σ,f) , in the neighbourhood of u , topologically equivalent (respectively strongly smoothly equivalent) to (σ',f') , in the neighbourhood of u' .

Theorem 3. The local topological stability is not a generic property for the one parameter families of smooth curves in the plane.

Proof. Our purpose here is just to show the interest of the theory developed in paragraph 1. For more details, the reader may refer to [3] . So we recall the following results, without proof.

Lemma 1. In Ω_2 , the following property is generic : at every point u of U , we are in one of the 5 cases:

1. (σ,f) is *regular*, that is to say u is a regular point for σ ;
2. (σ,f) is a *transversal fold*, that is to say u is a fold point for σ , (σ,f) is regular at u , and f , in restriction to the singular set of σ , is regular at u ;
3. (σ,f) is a *tangent fold*, that is to say u is a fold point for σ , (σ,f) is regular at u , and f , in restriction to the singular set of σ , is critical non degenerate at u ;
4. (σ,f) is a *transversal umbrella*, that is to say (σ,f) is a Whitney's umbrella at u , the line of double points of which is transversal at u to the direction $\mathbb{R}^2 \times \{0\}$ in $\mathbb{R}^3$;
5. (σ,f) is a *transversal cusp*, that is to say u is a cusp point for σ , and (σ,f) is regular at u .

Lemma 2. Local models for the 5 previous cases are respectively, under strong smooth equivalence :

1. $x \longleftarrow (x,y) \longmapsto (x,y)$;
2. $x+y \longleftarrow (x,y) \longmapsto (x,y^2)$;
3. $x^2+y \longleftarrow (x,y) \longmapsto (x,y^2)$;
4. $x+xy+y^3 \longleftarrow (x,y) \longmapsto (x,y^2)$;
5. $y+g(x,y^3+xy) \longleftarrow (x,y) \longmapsto (x,y^3+xy)$

where g is an arbitrary smooth map (null at 0) .

Lemma 3. If (σ,f) is a transversal cusp at u , it is not strongly smoothly stable in the neighbourhood of u .

By simple geometrical considerations, we see that if (σ,f) , in the neighbourhood of u , is topologically equivalent to (σ',f') , in the neighbourhood of u' , then for both families we are in one of the local situations 1,2,3,4 or 5 . Thus, from lemma 3 , the proof will be completed when we have proved that the notions of local topological and strong smooth equivalences are the same for two transversal cusps. So we admit at first the following result, which is a consequence of the theorem of preparation.

Lemma 4. Let σ be the canonical cusp (x,y) (x,y^3+xy) , the interior of which is denoted Δ (in other words, Δ is the set of the points which have 3 inverse images by σ) ; for any smooth function $g : \mathbb{R}^2,0$ $\mathbb{R},0$, we denote

$$\bar{g}(x,y) = y + g \circ \sigma(x,y) .$$

For each transversal cusp (σ,f) with $f(0) = 0$, there exists a commutative diagram

$$\begin{array}{ccccc} \mathbb{R} & \xleftarrow{f} & \mathbb{R}^2 & \xrightarrow{\sigma} & \mathbb{R}^2 \\ \downarrow{\scriptstyle id} & & \downarrow{\scriptstyle H} & & \downarrow{\scriptstyle K} \\ \mathbb{R} & \xleftarrow{\bar{g}} & \mathbb{R}^2 & \xrightarrow{\sigma} & \mathbb{R}^2 \end{array}$$

where H and K are local diffeomorphisms of the origin, and g is a smooth function : $\mathbb{R}^2,0$ $\mathbb{R},0$. Furthermore, for each couple (g,g') of smooth functions : $\mathbb{R}^2,0$ $\mathbb{R},0$, we have a commutative diagram

$$\begin{array}{ccccc} \mathbb{R} & \xleftarrow{\bar{g}} & \mathbb{R}^2 & \xrightarrow{\sigma} & \mathbb{R}^2 \\ \downarrow{\scriptstyle id} & & \downarrow{\scriptstyle H} & & \downarrow{\scriptstyle K} \\ \mathbb{R} & \xleftarrow{\bar{g}'} & \mathbb{R}^2 & \xrightarrow{\sigma} & \mathbb{R}^2 \end{array}$$

where H and K are as above, if and only if

$g_{|\Delta} = g'_{|\Delta}$ on some neighbourhood of the origin.

Now let us suppose we have a commutative diagram

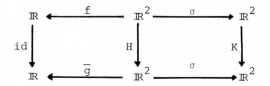

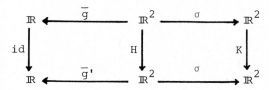

$$\begin{array}{ccccc} \mathbb{R} & \xleftarrow{\ \bar{g}\ } & \mathbb{R}^2 & \xrightarrow{\ \sigma\ } & \mathbb{R}^2 \\ \varphi \big\downarrow & & \big\downarrow H & & \big\downarrow K \\ \mathbb{R} & \xleftarrow{\ \bar{g}'\ } & \mathbb{R}^2 & \xrightarrow{\ \sigma\ } & \mathbb{R}^2 \end{array}$$

where K is a local homeomorphism of the origin, and H and φ are any bijections (preserving 0) .

First, and this is the decisive step, we show that K , in restriction to Δ , is locally smooth. This result is a corollary of theorem 2 , if we note that the curves of the families $(\sigma,\bar{g})$ and $(\sigma,\bar{g}')$ determine in Δ two 3-webs of codimension 1 of $\mathbb{R}^2$ which are exchanged by K .

Thus H , in restriction to $C = \{(x,y)/3y^2 + x < 0\}$, is locally smooth, since σ , in restriction to C , is a diffeomorphism from C onto Δ . It follows then, using the commutativity of the diagram

$$\begin{array}{ccc} \mathbb{R} & \xrightarrow{\ \bar{g}\ } & C \\ \varphi \big\downarrow & & \big\downarrow H \\ \mathbb{R} & \xrightarrow{\ \bar{g}'\ } & C \end{array}$$

and the fact the leaf $\bar{g} = 0$ enters C by transversality, that φ is smooth on some neighbourhood of 0 .

Finally, from the first proposition of lemma 4, we have the following commutative diagram

$$\begin{array}{ccccc} \mathbb{R} & \xleftarrow{\ \bar{g}_1\ } & \mathbb{R}^2 & \xrightarrow{\ \sigma\ } & \mathbb{R}^2 \\ \mathrm{id} \big\downarrow & & \big\downarrow H_1 & & \big\downarrow K_1 \\ \mathbb{R} & \xleftarrow{\ \varphi \circ \bar{g}\ } & \mathbb{R}^2 & \xrightarrow{\ \sigma\ } & \mathbb{R}^2 \end{array}$$

where H_1 and K_1 are local diffeomorphisms, whence the commutative diagram

$$\begin{array}{ccccc} \mathbb{R} & \xleftarrow{\ \bar{g}_1\ } & \mathbb{R}^2 & \xrightarrow{\ \sigma\ } & \mathbb{R}^2 \\ \mathrm{id} \big\downarrow & & \big\downarrow H \circ H_1 & & \big\downarrow K \circ K_1 \\ \mathbb{R} & \xleftarrow{\ \bar{g}'\ } & \mathbb{R}^2 & \xrightarrow{\ \sigma\ } & \mathbb{R}^2 \end{array}$$

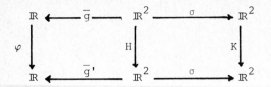

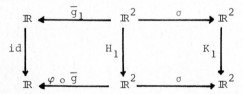

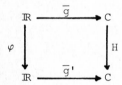

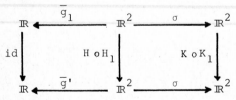

which implies, by elementary calculus,

$$g_{1|\Delta} = g'_{|\Delta} ,$$

which gives us the desired result, by applying the second proposition of lemma 4.

Theorem 3 can be generalized to the case of one parameter families of hypersurfaces of $\mathbb{R}^n$. The proof would be analogous. The characterization given in [2] for the equivalence of two transversal cusps and analogous situations when $n > 2$ may be useful ; in all cases theorem 2 is essential. The case $n = 3$ is entirely studied in [5]. Finally, one may notice, we have proved that the notions of topological and strong smooth equivalences are generically the same for the curves in the plane (using lemma 2) ; we conjecture that this result can be generalized for the one parameter families of hypersurfaces of $\mathbb{R}^n$.

REFERENCES

[1] A. BEAUVILLE, Géométrie des tissus, Séminaire Bourbaki, Lecture Notes in Math., 770, (1979), Springer Verlag.

[2] M.J. DIAS CARNEIRO, On the envelope theory, Thesis, Princeton University, U.S.A. (1980).

[3] J.P. DUFOUR, Familles de courbes planes différentiables, Topology, 22,4 (1983), 449-474.

[4] A. FRIEDMAN, Differential games, Wiley Interscience, New York (1971), chap. 4.

[5] P. JEAN, Familles à un paramètre d'hypersurfaces de $\mathbb{R}^3$, Thèse de 3ème cycle, Université des Sciences de Montpellier, France (1983).

[6] S. STERNBERG, On the structure of local homeomorphisms of euclidean n space II, Am. J. of Math. 80 (1958), 623-631.

Jean-Paul Dufour - Patrick Jean
Université des Sciences et Techniques
Place Eugène Bataillon
34060 Montpellier Cédex
France

Singularities & Dynamical Systems
S.N. Pnevmatikos (editor)

BIFURCATION FROM AN ORBIT OF SYMMETRY

David Chillingworth
University of Southampton
England

We explore two different approaches to investigating what happens to an orbit of critical points of a symmetric function when the continuous symmetry of the function is broken, and outline an application of the second approach to the traction problem in non-linear elastostatics.

Intuition tells us that *symmetric problems have symmetric solutions*[1]. Nevertheless, many examples show that there may be other solutions with less symmetry than the original problem: we can call this *spontaneous symmetry-breaking*. The intuition and its shortcomings have firm foundation in the theory of mappings symmetric with respect to given group actions, and understanding the mathematical mechanisms of symmetry-breaking has become one of the central problems of bifurcation theory: Dancer [5], Golubitsky and Schaeffer [7,8], Michel [13], Sattinger [15], Vanderbauwhede [16].

Variational problems are those whose solutions are stationary or *critical points* of a real-valued function. Thus let M be a Banach manifold, let $f:M \to \mathbb{R}$ be a sufficiently smooth (say C^∞) function, and let G be a compact Lie group acting smoothly on M. Suppose f is G-invariant, i.e. $f(gx) = f(x)$ for all $x \in M$, $g \in G$; then if x is a critical point of f so also is each point of the orbit Gx of x, so Gx forms a *critical manifold* for f. Our interest is in seeing how Gx breaks into smaller critical manifolds (including isolated critical points) as f is perturbed to a less symmetric function.

Write $G_x = \{g \in G: \; gx=x\}$, the *isotropy subgroup* of x. If $x = hy$ then $G_x = hG_yh^{-1}$ and all points on the orbit Gx have isotropy subgroups conjugate to G_x. In suitable local coordinates around x the action of G_x on M is the same as its linearized action (see Palais, [14] Theorem 5.3), from which it follows that the set $F_x = \mathrm{Fix}\, G_x$ of fixed points of G_x forms a smooth submanifold of M passing through x. This submanifold F_x is the equivalence class of x in M under the equivalence relation

$$x \sim y \iff G_x = G_y . \qquad (1)$$

The coarser equivalence relation $x \approx y \iff G_x$ conjugate to G_y (2)

has equivalence classes which are manifolds foliated by the equivalence classes of (1). Those of (2) are usually called the *strata* of the group action, and we shall call those of (1) the *leaves*. There are only finitely many strata, while there may be infinite families of leaves.

EXAMPLES

1. $M = \mathbb{R}^2$, $G = O(2)$ acting by rotations and reflections.

   strata: $\{0\}$ and $\mathbb{R}^2 \setminus \{0\}$.

   leaves: $\{0\}$ and {lines through 0, excluding 0 itself} .

2. $M = \mathbb{R}^2$, $G = Z_2$ generated by reflection $(x_1,x_2) \to (x_1,-x_2)$.

   strata = leaves = x_1-axis and $\{x \mid x_2 \neq 0\}$.

The search for critical points of f is greatly helped by the following property of the leaves:

THEOREM 1. *The point* x *is a critical point for* f *on* M *if and only if it is a critical point for the restriction* $f|\Lambda$, *where* Λ *is the leaf containing* x .

Proof. Suppose x is a critical point for $f|\Lambda$. Then the derivative $df(x)$ annihilates the tangent space $T_x\Lambda$ and is invariant under the linearized action of G_x on T_xM . If $df(x) \neq 0$ there exists $v \in T_xM$ with $df(x).v \neq 0$, and by averaging over G_x we can choose v to be G_x-invariant; then for small $|t| \neq 0$ the vector tv corresponds to a G_x-invariant point of M not lying in Λ . This contradiction shows $df(x) = 0$ after all.

COROLLARY 1.1. *Critical points of* $f|\mathrm{Fix}\,G$ *are critical points of* f .

Since $\mathrm{Fix}\,G$ is a particular stratum (isotropy subgroup $G_x = G$) this is a special case of the Theorem. It is sometimes called the *Principle of Symmetric Criticality*: see Palais [14], Ladyženskaja [9], for discussions of more general conditions and specific circumstances under which it holds.

Note that the Theorem itself implies that it is equivalent to its apparently weaker version with leaf Λ replaced by stratum Σ , the form in which it is more usually seen, e.g. Michel [12]. The advantage in working with strata is that there are only finitely many of them, and when M is compact there are compact strata, namely those corresponding to (conjugacy classes of) isotropy subgroups which are *maximal* among isotropy subgroups.

COROLLARY 1.2. *If* Σ *is compact then* f *has at least one critical point (or two, if* Σ *is other than a single point) on* Σ .

In fact f has to have at least c critical points on Σ , where c is the Liusternik-Šnirel'man category of Σ [4], §2.12. The more complicated the topology of Σ , the larger c has to be.

If y is a point of a critical orbit Gx then the second derivative $D^2f(y)$ must vanish along the tangent space to Gx at y , since f is constant on Gx . Call Gx a *nondegenerate* critical orbit if the kernel of the quadratic form $D^2f(y)$ is precisely this tangent space, so that 'normal' to Gx the function f is nondegenerate; Gx is then an example of a *nondegenerate critical manifold* as defined by Bott [2].

COROLLARY 1.3. *Let* Λ *be a leaf for the action of a compact Lie group* H *on* M , *let* $f:M \to \mathbb{R}$ *be* H*-invariant, and let* N *be any nondegenerate critical manifold for* f . *Suppose* x *is an isolated point of* $N \cap \Lambda$, *and* N, Λ *have no nonzero common tangent at* x . *Then any* H*-invariant function close to* f *has a critical point close to* x .

This follows because nondegeneracy of N implies that x is a nondegenerate critical point of $f|\Lambda$, and therefore persists under perturbation. Stated more precisely, it says that to any C^r family of H-invariant functions f_λ with $f_0 = f$ there corresponds a C^r family of critical points x_λ for f_λ , with $x_0 = x$, at least for λ sufficiently small. A natural setting in which to apply this is where $N = Gx$, and H is a subgroup of G . This is the variational case of the theorems of Dancer [5] and Vanderbauwhede [16], §8. It holds under much weaker assumptions of differentiability of the group action: see Dancer [6].

EXAMPLE. Let $f:\mathbb{R}^2 \to \mathbb{R} : (x_1,x_2) \to 2r^4 - r^2$ where $r^2 = x_1^2 + x_2^2$, and take $G = O(2)$ as in Example 1 and $H = Z_2$ acting as in Example 2. The circle $r = \frac{1}{2}$ is a nondegenerate critical manifold for f , and meets the x_1-axis (an H-leaf) transversely at $(\pm \frac{1}{2},0)$: we see that these two critical points persist under all perturbations of f which are symmetric about the x_1-axis, i.e. H-invariant.

The idea pursued above is that nontrivial isotropy can often be exploited to force the existence of critical points of a function, without further knowledge of the function. Now we turn to cases when there may be no nontrivial isotropy.

As before, let N be a compact nondegenerate critical manifold for $f:M \to \mathbb{R}$, and consider a 1-parameter perturbation f_λ of f. The nondegeneracy and the Implicit Function Theorem show that for sufficiently small λ there is a section s_λ of a tubular neighbourhood of N in M such that for each $y \in N$ the corresponding point $y_\lambda = s_\lambda(y)$ is the unique critical point near y of $f|S_y$, where S_y is the fibre containing y in the tubular neighbourhood. Thus all critical points of f_λ near N must be among the y_λ, and we can in principle locate them using the following routine piece of calculus:

LEMMA. y_λ *is a critical point of* f_λ *precisely when* y *is a critical point of* $\tilde{f}_\lambda = f_\lambda \circ s_\lambda : N \to \mathbb{R}$.

Again we can invoke Liusternik-Šnirel'man category to give a lower bound on the number of critical points of $\tilde{f}_\lambda$ with no further information. If we suppose or can calculate that $\tilde{f}_\lambda$ is a Morse function on N, then the Morse inequalities ([4], §2) give us sharper information. In fact it is not difficult to find sufficient conditions for $\tilde{f}_\lambda$ to be a Morse function, as we now proceed to show.

Write

$$f_\lambda(x) \equiv f(x) + \lambda p_1(x) + \lambda^2 p_2(x) + O(\lambda^3)$$

$$s_\lambda(y) \equiv y + \lambda u(y) + O(\lambda^2)$$

where for the second line to make sense we suppose that the tubular neighbourhood of N in M is parametrized by a neighbourhood of the zero section of the normal bundle of N in M, and $u(y)$ is a vector in the normal space V_y at y. Then for $y \in N$ we have

$$\begin{aligned}\tilde{f}_\lambda(y) &= f_\lambda(s_\lambda(y)) \\ &= f_\lambda(y) + \lambda Df_\lambda(y).u + \tfrac{1}{2}\lambda^2 D^2 f_\lambda(y).u^2 + O(\lambda^3) \\ &= f_\lambda(y) + \lambda^2[Dp_1(y).u + \tfrac{1}{2} D^2 f(y).u^2] + O(\lambda^3)\end{aligned}$$

since $Df(y) = 0$ when $y \in N$; here u means $u(y)$.

Now by construction u satisfies

$$Df_\lambda(y+\lambda u) = O(\lambda^2)$$

as a linear map $V_y \to \mathbb{R}$, so the coefficient of λ gives

$$Dp_1(y) + D^2 f(y).u = 0 \in V_y^* .$$

The nondegeneracy condition on the critical manifold N is that $D^2f(y)$ is invertible, thought of as a linear map $L(y):V_y \to V_y^*$, so writing $Dp_1(y) = \alpha(y) \in V_y^*$ and putting $u = -L(y)^{-1}\alpha(y)$ into the expression for $\tilde{f}_\lambda(y)$ gives

$$\tilde{f}_\lambda(y) = f_\lambda(y) - \tfrac{1}{2}\lambda^2 L(y)^{-1}.\alpha(y)^2 + O(\lambda^3)$$

with $L(y)^{-1}$ regarded as a quadratic form on V_y^*. Since f = constant on N we can suppose $f(y) = 0$ and divide the other terms by λ to yield

THEOREM 2. *The critical points of* $\tilde{f}_\lambda$ *on* N *are those of*

$$f_\lambda^*(y) = p_1(y) + \lambda(p_2(y) - \tfrac{1}{2}B(y)) + O(\lambda^2)$$

where $B(y)$ *denotes the* Betti form $L(y)^{-1}\alpha(y)^2$.

We call $B(y)$ the Betti form because in the application to elasticity described below it happens that $B(y)$ is intimately related to the mechanical principle of Betti reciprocity: see Marsden and Hughes [10] for an account of this.

If p_1 is a Morse function on N then its critical points persist under small perturbations. Therefore Theorem 2 has the immediate consequence:

COROLLARY 2.1. *If* p_1 *is a Morse function on* N *then critical points of* f_λ *close to* N *correspond to those of* p_1 *on* N *for* $|\lambda|$ *sufficiently small.*

If p_1 is not a Morse function on N, but nevertheless its critical set C_1 is a nondegenerate critical manifold (possibly having several components of different dimensions), we can continue by applying Theorem 2 to the function $f_\lambda^*:N \to \mathbb{R}$ as input. This produces:

COROLLARY 2.2. *If* $h_2 = p_2 - \tfrac{1}{2}B$ *is a Morse function on* C_1 *then critical points of* f_λ *close to* N *are close to* C_1 *and correspond to the critical points of* h_2 *on* C_1 *for* $|\lambda|$ *sufficiently small.*

APPLICATION. The traction problem in nonlinear elastostatics.

References for this are Chillingworth, Marsden and Wan [3], Wan and Marsden [18]. See also Marsden and Wan [11], Wan [17] and Marsden and Hughes [10].

Let $\mathcal{B}$ be a compact region in $\mathbb{R}^3$ with smooth boundary: this represents a solid body. Forces (loads) acting on $\mathcal{B}$ are represented by pairs $\ell = (b,\tau)$ where $b:\mathcal{B} \to \mathbb{R}^3$, $\tau:\partial\mathcal{B} \to \mathbb{R}^3$ are *body forces*, *surface tractions* respectively. The general problem is to find equilibrium configurations of $\mathcal{B}$ for a given imposed

ℓ , with the necessary condition that net force and moment of ℓ on $\mathcal{B}$ are zero. A *configuration* is an injective map $\phi : \mathcal{B} \to \mathbb{R}^3$. We suppose that $\mathcal{B}$ is in equilibrium at some *initial configuration* $I_{\mathcal{B}}$ under a load ℓ_0 , and seek equilibrium configurations for ℓ close to ℓ_0 .

Our model for nonlinear elasticity assumes a *stored energy* function W which is a smooth real-valued function defined on the space E of maps ϕ , with a suitable Sobolev topology. Equilibrium configurations corresponding to a load ℓ are taken to be the critical points of the smooth function V_ℓ defined by

$$V_\ell(\phi) = W(\phi) - \langle \ell, \phi \rangle ,$$

where

$$\langle \ell, \phi \rangle = \int_{\mathcal{B}} b.\phi + \int_{\partial \mathcal{B}} \tau.\phi$$

represents work done on the body by the load ℓ moving it to configuration ϕ , giving a nondegenerate bilinear pairing $L \times E \to \mathbb{R}$ where L is the linear space of loads with zero net force and moment, again suitably topologized.

The usual action of $G = SO(3)$ on $\mathbb{R}^3$ induces actions on E and L in an obvious way which preserve the bilinear pairing $= \langle \ell, \phi \rangle = \langle Q\ell, Q\phi \rangle$, $Q \in G$. We suppose that W is G-invariant (the principle of *material frame indifference*), and so V_ℓ is G_ℓ-invariant where G_ℓ is the isotropy subgroup of ℓ . The possibilities for G_ℓ are the two extremes:

(i) $G_\ell = \{id\}$: typical ℓ

(ii) $G_\ell = G$: $\ell = 0$

and one intermediate case:

(iii) $G_\ell \cong S^1$: ℓ is called a *parallel load.*

(Note that in (iii) the *body* is not necessarily symmetric, but the load vectors all point in the same direction.)

If ϕ is a critical point of V_ℓ then the orbit $G_\ell \phi$ is a critical manifold. To apply the general analysis described above we assume:

Hypothesis H. The orbit $N = G_{\ell_0} I_{\mathcal{B}}$ *is a nondegenerate critical manifold for* V_{ℓ_0} .

This entails a usual type of strong ellipticity condition on W , and supposes the problem to be 'typical' in having no fortuitous symmetries of body and load or coincidental degeneracies.

We take ℓ to have the form

$$\ell = \ell_0 + \lambda\ell_1 + \lambda^2\ell_2 + O(\lambda^3) ,$$

so that in our general setting we have

$$f = V_{\ell_0} = W - \ell_0^*$$

$$p_1 = -\ell_1^*$$

$$p_2 = -\ell_2^*$$

where ℓ^* denotes the linear function $\phi \to \langle\ell,\phi\rangle$ on E, and then we consider cases (i)-(iii) separately.

Case (i). Here $I_{\mathcal{B}}$ is a nondegenerate critical point of V_{ℓ_0}, so for small $|\lambda|$ there is a unique equilibrium configuration ϕ_λ close to $I_{\mathcal{B}}$.

Case (ii). In this case we are looking for all equilibria close to all rigid rotations of $I_{\mathcal{B}}$, assuming the load to be of small magnitude. From Theorem 2, the first step is to find critical points of ℓ_1^* on $N = GI_{\mathcal{B}}$. These are points $\phi \in N$ for which ℓ_1^* annihilates the tangent space to N at ϕ, namely the space $\{K\phi : K \in \mathit{skew}\}$, where *skew* denotes the Lie algebra of $SO(3)$ consisting of skew-symmetric 3×3 matrices. Using the so-called *astatic* map $k: L \times E \to M_3$ (= 3×3 matrices) given by

$$k(\ell,\phi) = \int_{\mathcal{B}} b \otimes \phi + \int_{\partial\mathcal{B}} \tau \otimes \phi ,$$

where for $u,v \in \mathbb{R}^3$ we write $u \otimes v = uv^T \in M_3$, (so $\langle\ell,\phi\rangle = \mathrm{trace}\ k(\ell,\phi)$), we see that $\phi = QI_{\mathcal{B}}$ is a critical point for $\ell_1^*|N$ if and only if for all $K \in \mathit{skew}$

$$\langle\ell_1,K\phi\rangle = \langle\ell_1,KQI_{\mathcal{B}}\rangle = 0$$

i.e.

$$\mathrm{trace}\ A_1Q^TK = 0$$

where $A_1 = k(\ell_1,I_{\mathcal{B}})$. This is the condition that Q be a critical point of the linear function

$$M_3 \to \mathbb{R} : X \to \mathrm{trace}\ A_1X^T$$

restricted to $SO(3)$, or, equivalently, that A_1Q^T be symmetric. Thus the infinite-dimensional critical point problem for $\ell_1^*|N$ reduces to a straightforward algebraic or geometric problem about 3×3 matrices, leading to a complete solution as follows:

THEOREM 3. The set C_1 *of critical points for* ℓ_1^* *restricted to* $N \cong SO(3)$ *has of one of five descriptions (according to various properties of the eigenvalues of* A_1*) :*

type name for ℓ_1	description of C_1
0	*four nondegenerate critical points*
1	*two* ndcp's *and one circle*
2	*one* ndcp *and a copy of* $\mathbb{R}P^2$
3	*two circles*
4	*the whole of* N .

If ℓ_1 is of type 0 then $\ell_1^*|N$ is a Morse function and so by Corollary 2.1 for sufficiently small $|\lambda|$ the critical points of V_ℓ are small perturbations of the four for ℓ_1 on N . If ℓ_1 is of type 1-4 then we have the convenient fact:

PROPOSITION. The critical set C_1 *is a nondegenerate critical manifold for* ℓ_1^* *restricted to* N .

This comes from the properties of the right action of $SO(3)$ on M_3 . We wonder how general this result may be. In any case, we are in a position to use Corollary 2.2, which becomes:

COROLLARY. If $h_2 \equiv \ell_2^* + \frac{1}{2} B$ *is a Morse function on* C_1 *then equilibrium solutions to the traction problem for loads* $\ell = \lambda\ell_1 + \lambda^2\ell_2 + O(\lambda^3)$ *correspond to the critical points of* h_2 *on* C_1 *for* $|\lambda|$ *sufficiently small.*

Often ℓ is taken as $\lambda\ell_1$, so that it is the Betti form alone whose critical points on C_1 determine the equilibrium solutions. From explicit parametrizations of C_1 and from the quadratic nature of B the maximum number m of critical points of the Morse function B on C_1 can be calculated according to the following table:

load type :	0	1	2	(3)	4
m :	4	6	14	(8)	40

Type 3 is rather special, and relates to case (iii) below. Note $m \geqslant 4$ in all cases, since the Liusternik-Šnirel'man category of $SO(3)$ is 4 .

Case (iii). Following our general procedure we first look for critical points of ℓ_1^* on $N = G_{\ell_0} I_B$. We find *either* $\ell_1^*|N$ is a Morse function with precisely *two* critical points (maximum and minimum)

or ℓ_1^* is constant on N .

The second case occurs if and only if the direction of ℓ_0 is an eigendirection for A_1 with eigenvalue trace A_1 : see [18]; compare also Bharatha and Levinson [1]. (Warning on notation: our ℓ_0, ℓ_1 here correspond to $\ell_{\mathcal{B}}$, ℓ_0 in [18], §2.) In the first case we apply Corollary 2.1. In the second case, under our hypothesis H , we invoke Corollary 2.2 and study the function $\ell_2^* + \frac{1}{2}B$ on N ; if $\ell_2 = 0$ we find that when B is a Morse function it can have *two* or *four* critical points on N . If, however, the nature of the problem is such that ℓ is restricted to be parallel to ℓ_0 , so that f_λ^* is S^1-invariant, then B necessarily vanishes and a whole circle of solutions close to N persists for $|\lambda|$ small.

Full details of these analyses, with many further observations and extensions on bifurcation diagrams, physical interpretations, perturbation methods and the relation to the classical literature can be found in the references mentioned.

Notes.

1. See Waterhouse [19] for illustrations and interesting historical comments on this principle.

REFERENCES

[1] S. Bharatha and M. Levinson, Signorini's perturbation scheme for a general reference configuration in finite elastostatics. Arch. Rat. Mech. Anal. 67(1978),365-394.

[2] R. Bott, Nondegenerate critical manifolds. Ann. Math. 60(1954),248-261.

[3] D.R.J. Chillingworth, J.E. Marsden and Y.-H. Wan, Symmetry and bifurcation in three-dimensional elasticity. Part I: Arch. Rat. Mech. Anal. 80(1982),295-331; Part II: 83(1983),363-395.

[4] S.-N. Chow and J.K. Hale, Methods of Bifurcation Theory. Springer-Verlag, New York 1982.

[5] E.N. Dancer, On the existence of bifurcating solutions in the presence of symmetries. Proc. Roy. Soc. Edinburgh 85A(1980),321-336.

[6] E.N. Dancer, The G-invariant implicit function theorem in infinite dimensions. Proc. Roy. Soc. Edinburgh 92A(1982),13-30.

[7] M. Golubitsky and D. Schaeffer, Imperfect bifurcation in the presence of symmetry. Commun. Math. Phys. 67(1979),205-232.

[8] M. Golubitsky and D. Schaeffer, A discussion of symmetry and symmetry-breaking. AMS Proc. Symp. Pure Math. 40(1983),499-515.

[9] O. Ladyženskaja, On finding symmetrical solutions of field theories variational problems. Proc. Int. Congress Math. 1982, Warsaw (to appear).

[10] J.E. Marsden and T.J.R. Hughes, The Mathematical Foundations of Elasticity. Prentice-Hall, New Jersey 1983.

[11] J.E. Marsden and Y.-H. Wan, Linearization stability and Signorini series for the traction problem in elastostatics. Proc. Roy. Soc. Edinburgh, 95A(1983),171-180.

[12] L. Michel, Points critiques des fonctions invariantes sur un G-variété, C.R. Acad. Sci. Paris, A272(1971),433-436.

[13] L. Michel, Symmetry defects and broken symmetry. Rev. Mod. Phys. 52(1980),617-751.

[14] R.S. Palais, The principle of symmetric criticality. Commun. Math. Phys. 69(1979),19-30.

[15] D.H. Sattinger, Branching in the Presence of Symmetry. CBMS 40, SIAM, Philadelphia 1983.

[16] A. Vanderbauwhede, Local Bifurcation and Symmetry. Research Notes in Math. 75, Pitman, London 1982.

[17] Y.-H. Wan, The traction problem for incompressible materials (to appear).

[18] Y.-H. Wan and J.E. Marsden, Symmetry and bifurcation in three-dimensional elasticity, Part III. Arch. Rat. Mech. Anal. 84(1983),203-233.

[19] W.C. Waterhouse, Do symmetric problems have symmetric solutions? Amer. Math. Monthly (1983),378-387.

David Chillingworth
Department of Mathematics
University of Southampton
Southampton S09 5NH
England

Singularities & Dynamical Systems
S.N. Pnevmatikos (editor)

A PEDESTRIAN PROOF OF THE HOPF-BIFURCATION THEOREM

Francine and Marc Diener
University of Oran
Algeria

Having shown the short-shadow lemma, we apply this result of non standard analysis to two classical problems : small perturbations of the harmonic oscillator and the Hopf-bifurcation.

One knows that the Hopf-bifurcation consists in the phenomenon that, in a vector field of the plane depending continuously on a parameter a , a limit cycle may disappear by the vanishing of its amplitude : the cycle swallows the focus encircled by it and turns it into a weak focus of reversed stability.

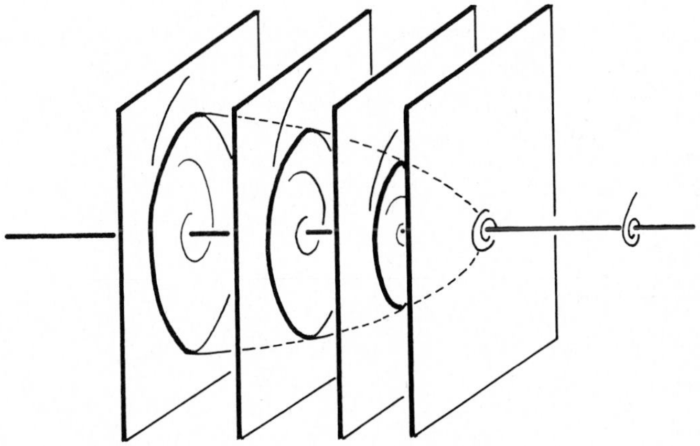

This phenomenon is interesting from several points of view (cf. [5]) and we shall attach our attention to two of them : topologic (existence or not of a cycle) and asymptotic (how fast the amplitude changes near the bifurcation value). The result we give below is not new, at least for standard vector fields. But on the other hand, the proof gives a new way to handle this question. It consists in just observing the solutions, with appropriate instruments, to see if the expected phenomenon does really take place. And, as it is of local nature, no surprise that one uses a magnifying glass adapted with its asymptotic properties.

The main analytical tool is the so-called short-shadow lemma, which will be proved in the first paragraph, and which is of common use in the non standard approach of many questions in ordinary differential equations. For essentially didactic reasons we apply it first, in the second paragraph, to the famous problem of small perturbations of the harmonic oscillator. The approach adopted here allows to weaken the classical assumptions : as no local inverse theorem is called upon, it suffices to assume the perturbation to be locally Lipschitz. For the same reasons, it is also possible to weaken the assumptions in the Hopf-bifurcation theorem. Let us also point out that the method, when applied to an explicitly given equation (say by a polynomial, as for the Van der Pol equation), gives a way to carry the computations explicitely, and according to our taste this is not the least advantage of it.

1. THE SHORT SHADOW LEMMA

Let us give here the key-tool of non standard analysis as applied to small perturbations of vector fields that we shall use in the sequel.

Recall that the *shadow* of a parametrized curve is the unique standard curve which is (infinitely) close to it. It exists provided the curve stays limited and the parametrization is a S-continuous function [2]. One will speak of short-shadow of a curve if one restricts one self to curves $\phi: I \to \mathbb{R}^n$, with I limited, in order that for any t_0 and t in I, $t_0 - t$ is not (infinitely) large. One has the following result :

THE SHORT-SHADOW LEMMA

Let X_0 be a standard, continuous vector field, defined on an open subset $A \subset \mathbb{R}^n$, and let X be a vector field close to X_0 on A.
i) If $\phi: I \to \mathbb{R}$ is a solution of X such that I is limited and any point of $\phi(I)$ is near-standard in A, then the shadow ϕ_0 of ϕ exists and is a solution of X_0 on $I_0 = {}^0I$.
ii) Assume X is locally Lipschitz. If $\phi_0: I_0 \to \mathbb{R}^n$ is a standard solution of X_0 such that $\phi_0(I_0)$ is relatively compact in A, then, for any $t_0 \in I_0$, any solution of X such that $\phi(t_0) \simeq \phi_0(t_0)$ can be extended to the whole interval I_0, and stays (infinitely) close to ϕ_0 on I_0.

Proof. i) By assumption, for any $t \in I$, $\phi(t)$ is near standard, thus $\phi'(t) = X(\phi(t)) \simeq X_0({}^0\phi(t))$ is limited and thus ϕ is S-continuous. So its shadow $\phi_0 : I_0 = {}^0I \to A$ exists and $\phi(t) \simeq \phi_0(t)$ for any $t \in I \cap I_0$. Let t_0 and t be any standard points of $\mathrm{int}(I_0)$. One has

$$\phi_0(t) \simeq \phi(t) = \phi(t_0) + \int_{t_0}^{t} X(\phi(s))\,ds \simeq \phi_0(t) + \int_{t_0}^{t} X(\phi(s))\,ds$$

$$\simeq \phi_0(t_0) + \int_{t_0}^{t} X_0(\phi_0(s))\,ds .$$

As the first and the last terms are standard, they must be equal. This equality holds for any t and t_0 standard in $\mathrm{int}(I_0)$. So it holds for any t and t_0 in $\mathrm{int}(I_0)$, ϕ_0 and X_0 being standard. This result extends to $\mathrm{Fr}(I_0)$ by continuity of ϕ_0.

ii) Let $t_0 \in I_0$, let ϕ be a solution of X such that $\phi(t_0) \simeq \phi_0(t_0)$ and let I be the maximal interval of existence of ϕ. Notice first that it suffices to prove that ϕ_0 is the shadow of ϕ on $J = I \cap I_0$ for, if this "weaker" theorem is true, one sees at once that $J = I_0$. Indeed by assumption, there exist a standard compact neighborhood $K \subset A$ of $\phi_0(I_0)$, and if $\phi(J) \subset \mathrm{hal}\ \phi_0(J) \subset K$, ϕ can be extended at each end of J, and thus $I \supset I_0$.

Now, let $\mathcal{U}$ be the set of all $t \in J$ such that $\phi(\tau) \simeq \phi_0(\tau)$ for any τ between t_0 and t, and let $\mathcal{V}$ be the set of all $t \in J$ such that $\phi(\tau) \in B$ for all τ between t_0 and t. Clearly $\mathcal{U} \subset \mathcal{V}$. But by (i) $\mathcal{U} \subset \mathcal{V}$, thus $\mathcal{U} = \mathcal{V}$. But this implies $\mathcal{U} = J$, otherwise $\mathcal{U}$ would be strictly external and $\mathcal{V}$ being internal they could not be equal [(2)]. QED

2. SMALL PERTURBATION OF THE HARMONIC OSCILLATOR

Let us consider now the following classical equation

(1) $\ddot{x} + x = \varepsilon\, f(x,\dot{x})$

where f is a standard [(3)], locally Lipschitz function, and $\varepsilon > 0$ infinitesimal. The vector field X associated with (1) in the phase plane is, at all limited points (x,y), close to the following standard vector field X_0 :

$$X = \begin{cases} \dot{x} = y \\ \dot{y} = -x + \varepsilon\, f(x,y) \end{cases} \qquad X_0 = \begin{cases} \dot{x} = y \\ \dot{y} = -x \end{cases}$$

It follows from the lemma above that the solution starting at any limited point is defined for all t and its short shadow has equation $x^2 + y^2 = r^2$. Yet this does not allow to decide whether one or the other of those trajectories is indeed closed. In order to answer that question, one has to increase one's hability to distinguish more tiny details, and thus introduce a "magnifying glass" to blow up an infinitesimal neighbourhood of any one of these circles. It is convenient to introduce polar coordinates $(x,y) = [r,\theta] = (r\cos\theta, r\sin\theta)$. Now let

$$R = \frac{r - r_0}{\varepsilon}$$

Equation (1) becomes.

$$\frac{dR}{d\theta} = \frac{-\sin\theta\,\tilde{f}}{1-\varepsilon\cos\theta\,\tilde{f}/(r_0+\varepsilon R)}\text{, where } \tilde{f} =: f((r_0+\varepsilon R)\cos\theta, (r_0+\varepsilon R)\sin\theta))$$

Now this equation is once more a small perturbation of a standard equation, provided r_0 is limited, and

(2) $$\frac{dR}{d\theta} \simeq -\sin\theta\, f(r_0\cos\theta, r_0\sin\theta)$$

Thus, one may apply the short shadow lemma and prove the following result (Poincaré 1893 [6]).

THEOREM

If equation (1) admits a periodic solution (x,y) such that $x^2+y^2 = r_0^2$, then r_0 necessarily satisfies $\bar{f}(r_0) = 0$, with

$$\bar{f}(r) =: \int_0^{2\pi} f(r\cos\theta, r\sin\theta)\sin\theta\, d\theta$$

Conversly, if $\bar{f}(r_0) = 0$ and moreover $\bar{f}'(r_0) \neq 0$, then equation (1) admits such a periodic solution.

Proof. It can essentialy be pictured on figure 2 below.

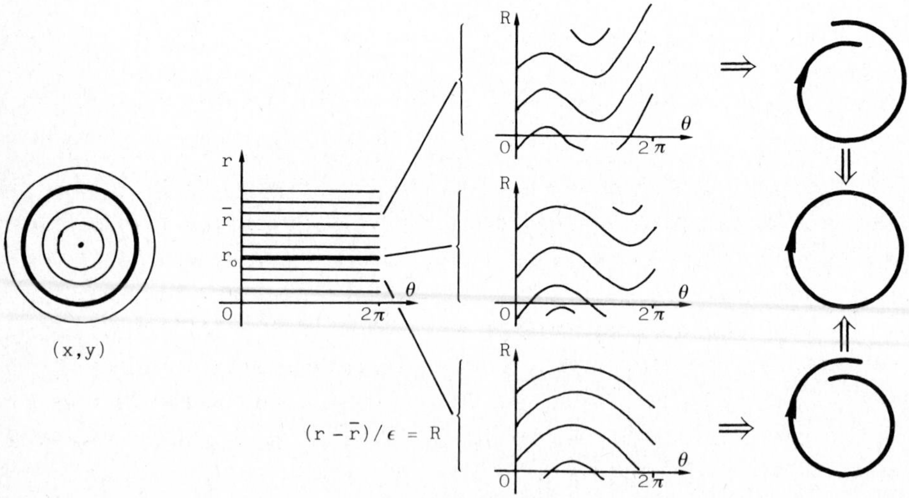

Indeed, one notices that, under the magnifying glass, the behaviour of the shadows of the trajectories depend on $\bar{r}$, that is on the level at which this change of variables has been performed. If $\bar{f}(\bar{r}) = 0$, the solutions in the scope have all their shadow that have the same value for $\theta = 0$ and $\theta = 2\eta$ (that is "after one turn"), and this is of course necessarily the case for a periodic solution. On the other hand, if $\bar{f}(\bar{r}) \neq 0$, the shadows of the solutions get all increased (or decreased) of precisely $\bar{f}(\bar{r})$ between $\theta = 0$ and $\theta = 2\eta$, which shows that, at the original scale, the trajectories spiral, decreasingly or increasingly $(\dot{\theta} = -1)$ according to $\dot{\theta}\bar{f}(\bar{r} \neq 0)$ is positive or negative. So, if $\bar{f}(r_0) = 0$ and $\bar{f}'(r_0) \neq 0$, all trajectories such that $r \simeq r_0$ get trapped between those that, above and below, spiral towards each other (for increasing or decreasing θ). One gets in that way a continuous Poincaré first-return map from a standard interval, transversal to $\{x^2 + y^2 = r_0\}$, into itself ; the fixed point of it necessarily belongs to a closed trajectory.

3. THE HOPF-BIFURCATION THEOREM

Let X be a vector field on $\mathbb{R}^2$, regular enough to ensure local existence and unicity of solutions, and depending continuously on one parameter a, for $a \simeq 0$. One assumes that X admits the following decomposition :

$$X = X_0 + X_1 + \ldots + X_n + \mathrm{Tail}_n$$

where the X_p are one parameter a depending vector fields, the components of which are homogeneous polynomials in $(x,y\,;\,a)$ of degree p, and with $\mathrm{Tail}(\varepsilon x, \varepsilon y, \varepsilon a)/\varepsilon^n \simeq 0$ for all $\varepsilon > 0$, $\varepsilon \simeq 0$ and limited $(x,y\,;\,a)$.

The Hopf-bifurcation phenomenon here under consideration leads to assume that $(0,0)$ is a singular point of X for any a, thus

$$X(0,0\,;\,a) \equiv X_p(0,0\,;\,a) \equiv 0 \quad \text{for all} \quad p \leq n \quad \text{and} \quad a \simeq 0 .$$

Moreover, one assumes that this singular point is, for $a \neq 0$, a non-degenerate focus, that reverses its Lyapunov stability for $a = 0$. Let $\lambda(a) = \alpha(a) \pm i\,\omega(a)$ be the distinct characteristic roots of the linear vector field $X_1(.\,;\,a)$. One assumes that $\alpha(a)$ changes its sign "transversaly" for $a = 0$, that is $\alpha'(0) \neq 0$.

As $X_1(.\,;\,a)$ shows a center for $a = 0$, we assume that

$$X_1(x,y\,;\,0) = \begin{pmatrix} 0 & 1 \\ -1 & 0 \end{pmatrix} \begin{pmatrix} x \\ y \end{pmatrix}$$

which may always be obtained explictly by a linear change of coordinates and a change of variable $t' = t\,\omega(0)$. X_1 being of that form, condition $\alpha'(0) \neq 0$ becomes

$$(-\alpha'(0) =) \int_0^{2\eta} u\cdot(X_2(u;a) - X_2(u;0))/a \, d\theta \neq 0 \quad \text{with} \quad u = (\cos\theta, \sin\theta)$$

Here, at last, is a genericity assumption on X for $a = 0$. For $p = 1,2,3,\ldots$ let

$$d_p = \int_0^{2\eta} u\cdot X_p(u;0)\, d\theta \qquad \text{with} \qquad u = (\cos\theta, \sin\theta) .$$

One assumes that there exist an $\bar{n}$ such that $d_{\bar{n}} \neq 0$, and we shall choose $\bar{n}$ such that $d_p = 0$ for all $p < \bar{n}$. Clearly one has $d_1 = d_{2_m} = 0$; thus $\bar{n}$ has to be odd and at least equal to 3. Notice that, if X is standard, it essentially suffices to assume X to be $C^{\bar{n}}$ (more precisely to belong to a subset of the hypersurface $\{\alpha = 0\}$ of $C^3(0;0)$ containing the open and dense subset $\{\alpha'(0)\cdot d_3 \neq 0\}$). For $p = 2,3,4\ldots$ let

$$c(p) = \alpha'(0) / dp$$

and, for $\varepsilon \simeq 0$, $\varepsilon > 0$, consider the following magnifying glass :

$$p - MG^{\cdot\varepsilon} \qquad \xi = x/\varepsilon \qquad \eta = y/\varepsilon \qquad A = a/\varepsilon^{p-1}$$

THEOREM

The shadows of the solutions of the vector field $\Xi = \bar{n} - MG^{\varepsilon}(X)$ are circles centered at $(0,0)$. The vector field Ξ possesses, for A appreciable (4)*, a cycle having as a shadow a circle of appreciable radius r_0 if and only if*

$$c(\bar{n})\, A > 0 \qquad \text{and} \qquad r_0 = \sqrt[\bar{n}-1]{c(\bar{n})A}$$

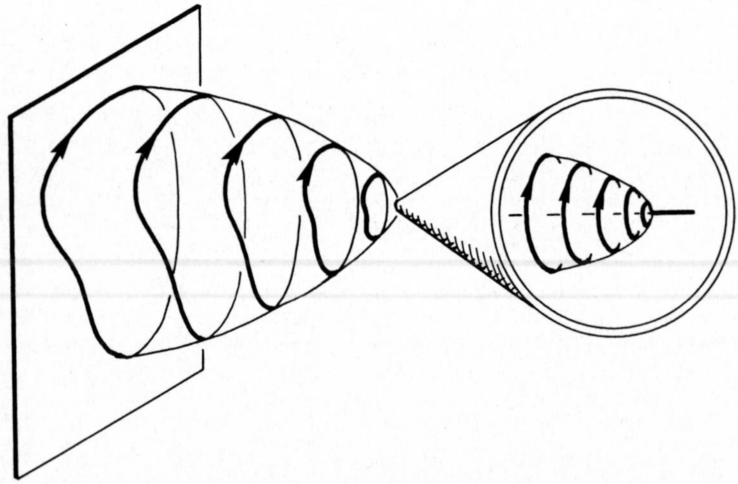

COROLLARY (Hopf-Bifurcation)

Assume for short that $c(\bar{n}) > 0$. *For any* $\delta_0 > 0$, *there exists* $a_0 > 0$ *and a function* $[0,a_0] \ni a \mapsto \gamma_a$, *such that* γ_a *is a periodic solution of* $X(\cdot;a)$, *non stationnary for* $a \neq 0$, *and for any* $t \in \mathbb{R}$ *and* $a \in [0,a_0]$.

$$\| \gamma_a(t) \| = \sqrt[\bar{n}-1]{c(\bar{n})a}\,(1 + \delta_a(t)) \quad , \text{ with } \quad |\delta_a(t)| < \delta_0$$

Proof of the corollary :

This statement being standard, one may assume X and δ standard. Choose $a_0 > 0$ infinitesimal. Let $a \in]0,a_0]$ and $\varepsilon = a^{1/(\bar{n}-1)}$. From the theorem, $\Xi = \bar{n} - MG^{\varepsilon}(X)$ has, for $A = 1$, a non stationnary cycle Γ_1 , with

$$\| \Gamma_1(t) \| = \sqrt[\bar{n}-1]{c(\bar{n})a}\,(1 + \delta) \ , \text{ with } \ \delta \simeq 0$$

Coming back to the initial scale, this ensures the existence of a non stationnary cycle γ_a of $X(\cdot;a)$, and for any $t \in \mathbb{R}$

$$\| \gamma_a(t) \| = \varepsilon\, \sqrt[\bar{n}-1]{c(\bar{n})}\,(1 + \delta) = \sqrt[\bar{n}-1]{c(\bar{n})\,a}\,(1 + \delta) \quad , \text{ with } \ \delta \simeq 0$$

Finally $|\delta| < \delta_0$, for $\delta_0 > 0$ is standard and $\delta \simeq 0$. QED

Proof of the theorem

Through $\bar{n} - MG^{\varepsilon}$, the vector field X becomes the vector field Ξ which, because of the special form of X_1 , is for any limited A, a small perturbation of the vector field associated with the harmonic oscillator in the phase plane. Thus the shadows of the solution of Ξ are circles centered at (0,0) .

Now, one proceeds as in the previous paragraph, introducing polar coordinates $(x,y) = [\rho/\theta]$, which give, under $\bar{n} - MG^{\varepsilon}$ $(\xi,\eta) = [r,\theta]$, with $r = \rho/\varepsilon$, $A = a/\varepsilon^{\bar{n}-1}$, and $dr/d\theta \simeq 0$. As for the harmonic oscillator, in order to determine which circle $r = \bar{r}$ is the shadow of a closed orbit, one performs a second blow-up. But in that case $R = (r - \bar{r})/\varepsilon$ reveals solutions the shadows of which are still periodic [5]. Thus, and in order to be able to conclude, one uses an other, stronger magnifying glass $s = (r - \tilde{r}(\theta))/\varepsilon^{\bar{n}-1}$, with

$$\tilde{r}(\theta) = \int_0^{\theta} u\cdot(r^2 Q_2 + \varepsilon r^3 Q_3 + \ldots + \varepsilon^{\bar{n}-3}\, r^{\bar{n}-1}\, Q_{\bar{n}-1})\, d\theta$$

$$u = (\cos\theta, \sin\theta) \qquad Q_p = X_p\,(u;0) \quad .$$

At this scale, the shadows of the solutions increase (or decrease) when θ changes from 0 to 2η from an amount of

$$s(2\eta) - s(0) \simeq \bar{r}(\bar{r}^{\bar{n}-1} d_{\bar{n}} - A\,\alpha'(0))$$

Reasoning as previously, the result follows. QED

(1) It may be interesting to point out that such a change of coordinates has been used by Hopf in his original paper [4], but has been forsaken after.

(2) This is the most elementary permanence principle of Non Standard Analysis : by definition, a stricktly external set can not be internal.

(3) What follows still holds if f is only near-standard.

(4) Appreciable : neither infinitesimal, nor infinitely large.

(5) This expresses that the Poincaré map is, asymptotically, flat on the identity map.

REFERENCES

[1] V. Arnold, Chapitres supplémentaires de la théorie des équations différentielles ordinaires. Editions MIR, Moscou (1978).

[2] F. Diener, Cours d'analyse non standard. Office des Publications Universitaires, Alger (1983).

[3] F. Diener, Méthodes du plan d'observabilité. Thèse, Strasbourg (1981).

[4] E. Hopf, Abzweigung einer periodischen Lösung von einer stationären Lösung eines Differentialsystems. Ber.Math-Phys. Sachsiche Academie der Wissensdraften, Leipzig 94 (1942) 1-22.

[5] J.E. Marsolen and H. McCracken, The Hopf Bifurcation and its Applications. Springer Verlag, New York (1976).

[6] H. Poincaré, Les méthodes nouvelles de la Mécanique Céleste. Gauthier Villars, Paris (1982).

Francine et Marc DIENER
Institut de Mathématique et Informatique. Université d'Oran.
B.P. 1524 ES SENIA (ORAN)
ALGERIE.

Singularities & Dynamical Systems
S.N. Pnevmatikos (editor)

IN PURSUIT OF BIRKHOFF'S CHAOTIC ATTRACTOR

Ralph H. Abraham
University of California
U. S. A.

A history of the chaotic bagel attractor, in theory and in experiments with forced oscillators, from 1916 to the present, including an account of its occurrence in catastrophic bifurcations.

1. HISTORICAL INTRODUCTION

There is a growing awareness of the gap between theoretical and experimental concepts in chaotic dynamics. As the bagel is unique among chaotic attractors in having a long history in theoretical as well as experimental dynamics, we have chosen to emphasize it here, in hopes of closing this gap.

In 1932, Birkhoff published a remarkable paper on *remarkable curves* [1]. These are curves only in the sense that they are closed subsets of the plane of measure zero, dividing the plane into two components. They originally arose as attractors in twist mappings of a plane annulus, in 1916. Birkhoff showed they are not Jordan curves, so he called them remarkable curves. In fact, they are fractals. The suspension of a twist map provides a flow on a thickened torus, with a remarkable surface, or fractal torus, as attractor. We call this a *Birkhoff bagel*.

Shortly after the appearance of Birkhoff's paper, Levinson conjectured that this bagel might occur in a forced dynamical system of second order [2]. At about the same time, Cartwright and Littlewood [3] guessed that the bagel had already been observed in this context, unknowingly, by Van der Pol and Van der Mark [4] in 1927. In fact, they reported the occurrence of an "irregular noise" in the earphone of a forced neon glow tube device. This device may be regarded as an analog simulator for some forced dynamical system of second order, but probably not the well known Van der Pol system.

In recent years, experimentalists searched in vain for a chaotic attractor in the forced Van der Pol system until 1980, when Shaw announced a sighting of the bagel, at last [5]. Very elusive, this *Van der Pol bagel* is very hard to find, but Shaw discovered a variant forcing of the Van der Pol system in which chaotic bagel attractors abound, which we call *Shaw bagels*. More recently, we have found abundant chaos in a forced Van der Pol system [6].

In this paper, we briefly describe these results, and present some conjectures on catastrophic bifurcations in which a bagel suddenly appears or disappears.

2. LIENARD'S DIFFEOMORPHISM

The Van der Pol system in Cartwright normal form,

$$P \begin{cases} \dot{x} = y \\ \dot{y} = -x + k(1-x^2)y, \ k > 0 \end{cases}$$

is obtained from Rayleigh's model for the clarinet reed,

$$R \begin{cases} \dot{u} = v \\ \dot{v} = -u + k(v-v^3/3)y, \ k > 0 \end{cases}$$

by differentiation. The equivalence of these two systems is conveniently seen by applying Lienard's map,

$$L: \mathbb{R}^2 \to \mathbb{R}^2; \ (x,y) \to (u,v)$$

defined by

$$L \begin{cases} u = -y + k(x - x^3/3), \ k > 0 \\ v = x \end{cases}$$

which is an area-preserving diffeomorphism. That is, $L_* P = R$. This equivalence is useful to obtain the phase portrait of the Van der Pol system, as the Rayleigh system is easier to analyze directly.

For example, the Rayleigh system is analyzed by means of the two characteristic curves,

$$C_1: \dot{u} = 0 \quad (u - \text{axis})$$

$$C_2: \dot{v} = 0 \quad (\text{graph of } u = k(v-v^3/3))$$

from which the limit cycle is found, as shown in Figure 1(a). The inverse image of these two curves, under L, consists of the two axes. However, the analysis of the Van der Pol system directly involves the curves,

$$D_1 : \dot{x} = 0 \qquad (x - \text{axis})$$

$$D_2 : \dot{y} = 0 \qquad (\text{graph of } y = x/k(1 - x^2))$$

as shown in Figure 1(b). Obviously it is easier to argue geometrically with the y-axis than with the three pieces of the curve, D_2.

3. ACCELERATION BIAS

The forced Van der Pol system is conveniently regarded as a serially coupled scheme of two dynamical systems, one of which is the Van der Pol system with an acceleration bias, or constant forcing term,

$$\text{PA} \begin{cases} \dot{x} = y \\ \dot{y} = -x + k(1-x^2)y + b \end{cases}$$

while the forcing system is

$$\dot{\theta} = 2\pi f \qquad (\text{constant})$$

and the coupling function is

$$b = A \sin \theta$$

where $A > 0$, or sometimes, to break the symmetry,

$$b = A \sin \theta + a$$

for some constant, $a > 0$. The factorization of the forced Van der Pol system into these two subsystems allows a geometric intuition on the behavior of the forced system, at least, for slow forcing. For the response diagram of the driven system, PA, may be obtained from the Lienard diffeomorphism. In the Rayleigh coordinates, the pushforward, L_*PA, is the velocity-biased Rayleigh system,

$$\text{RV} \begin{cases} \dot{u} = x - a \\ \dot{x} = -u + k(v - v^3/3) \end{cases}$$

for which the two characteristic curves are as shown in Figure 1(a), except for C_1, which is raised to the horizontal line, $v = a$. Obviously, the limit cycle will disappear for $|a| > 1$, leaving an attractive point. Evidently, there are two Hopf bifurcations in the response diagram of this system, which we call the *red cigar*, shown in Figure 2(a). Verification of this response has been obtained by simulation [6].

Returning to the problem of the elusive bagel, we see that forcing the Van der Pol system will not be likely to produce a chaotic attractor unless the force is large enough to periodically pass at least one of the Hopf bifurcations. Also, it will help if the forcing function is asymmetric, and rapid. In fact, these intuitions have succeeded in producing abundant chaos [6].

4. VELOCITY BIAS

The Shaw variant of the forced Van der Pol system consists of applying the force to the velocity rather than the acceleration. Factoring into a serially coupled scheme of two subsystems, we obtain

$$\mathrm{PV}\begin{cases}\dot{x} = y - c\\ \dot{y} = -x + k(1 - x^2)y\end{cases}$$

As the Lienard diffeomorphism is no help in this case, we may study the direct characteristic curves, as in Figure 1(b), with the horizontal line, D1, raised to level, c. This shows that in general, there are two critical points, a saddle and a repellor. Thus, as the level, c, increases, the limit cycle may interact with the invariant curves of the saddle. In fact, for some critical level, $c = \pm c_0$, the limit cycle vanishes in a blue sky catastrophe, as simulation has shown [6].

Thus the response diagram of this system contains a bounded snake of periodic attractors, the *blue sleeve*, shown in Figure 2(b). This shows that the trajectory grows rapidly when the force exceeds the critical values, $\pm c_0$, and gives some intuitive explanation for the abundance of chaotic bagels found in this system by Shaw.

5. BAGEL CATASTROPHES

The *blue bagel catastrophe* is, roughly, the suspension of the blue sky catastrophe by a periodic forcing system. Previously, we had speculated on the occurrence of this bifurcation in the forced Van der Pol system [7]. In simulation, we did not find it. Instead, we found a *red bagel catastrophe*, in which a periodic orbit catastrophically explodes into a bagel [8]. We end this bagel review with yet another catastrophic scenario, related to the bifurcation of codimension two studied by Chenciner [9].

We consider a flow in $S^1 \times \mathbb{R}^2$, obtained from a forced oscillator with a single control parameter. Before the catastrophic bifurcation, we have an attractive invariant 2-torus (AIT) with a rational flow, and braided periodic attractors and saddles. As the control is increased, the toral flow twists faster, and the inset cylinders of the braided periodic saddles become tangent to the outsets (first on one side, later on the other) and transversely homoclinic. With only one side homoclinic, we have a chaotic limit set of saddle type, as in Smale's original horseshoe. But with both sides homoclinic, the tangle is attractive, and comprises a bagel attractor. Thus, in a bifurcation sequence of codimension one, an AIT *explodes* into an attractive bagel. The rotation number is replaced by a rotation interval, probably containing the original rational in its interior. Many related events, such as the annihilation of two bagels, suggest themselves. It seems increasingly likely that chaotic attractors abound in typical forced oscillators. Some further evidence for this is given in a companion paper [6].

6. ACKNOWLEDGEMENTS

Many valuable discussions took place at the conference. In particular, it is a pleasure to thank Marc Diener for his explanations on the red cigar and the canard which he had independently discovered [10], Alain Chenciner for clarifying the technical details of Birkhoff's attractor, and Jean-Claude Roux for pointing out its occurrence in his experiments [11]. The drawings are by Christopher D. Shaw.

Fig. 1(a). Characteristic curves of the Rayleigh system, R, with k = 1.

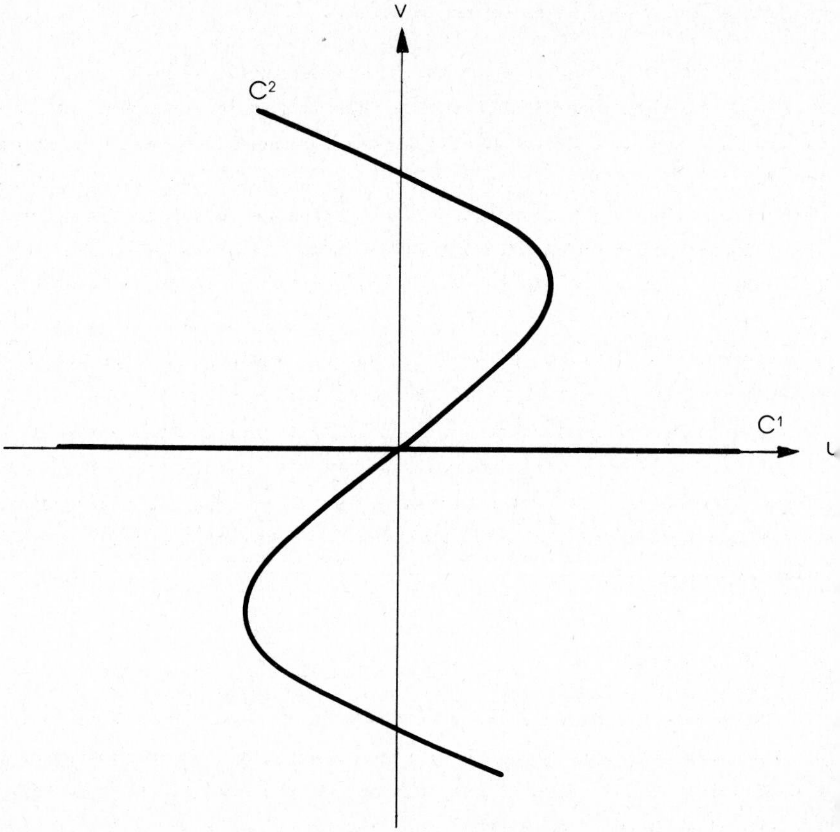

Fig. 1(b). Characteristic curves of the Van der Pol system, P, with k = 1.

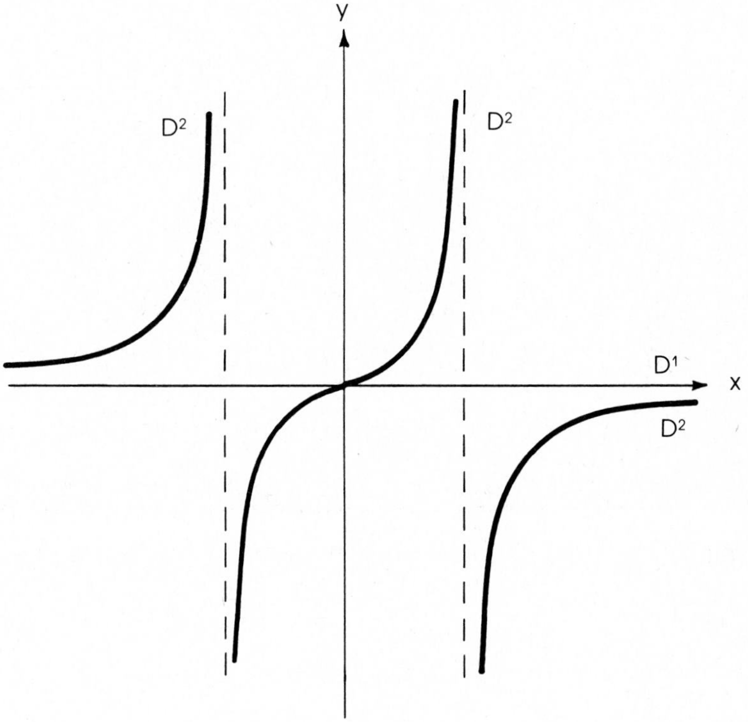

Fig. 2(a). The red cigar, in the response diagram for the acceleration-biased Van der Pol system, PA, or equivalently, the velocity-based Rayleigh system, RV, with k = 1. The control parameter, a, varies from -1 to 1.

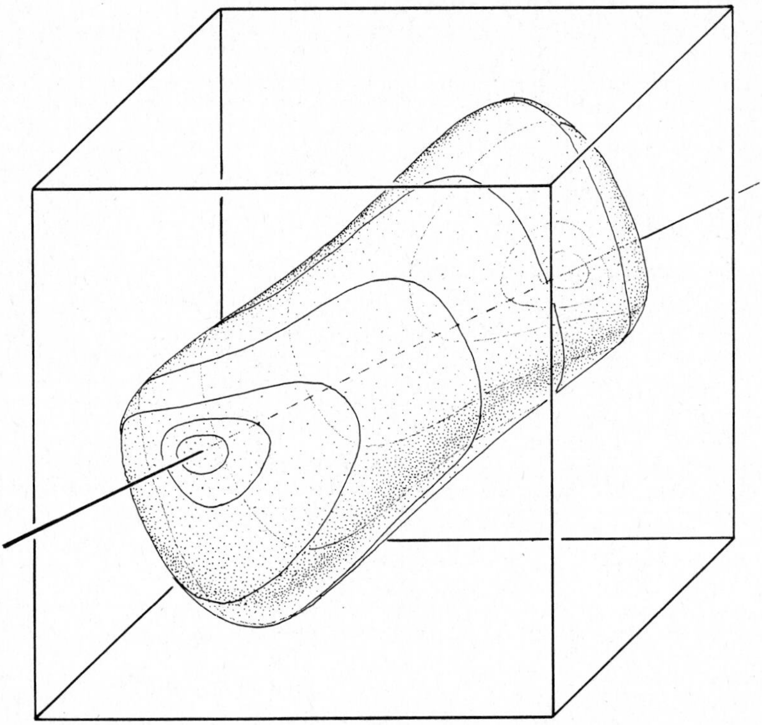

Fig. 2(b) The blue sleeve, in the response diagram for the velocity-biased Van der Pol system, PV, with k = 1. The control parameter, c, varies from -1 to 1.

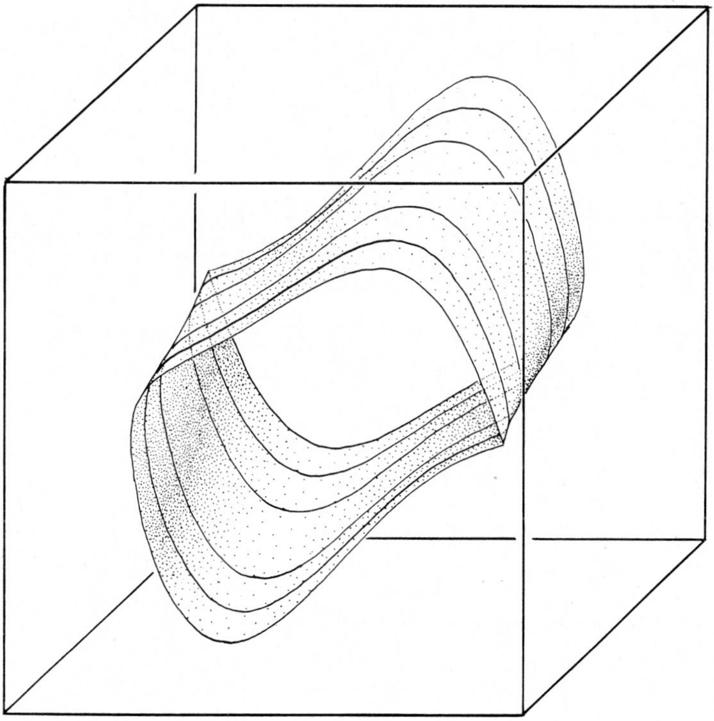

Bibliography

[1] Birkhoff, G. D., Sur quelques courbes fermees remarquables, Bull. Soc. Math. d. France 60 (1932) 1-26.

[2] Levinson, N., Transformation theory of non-linear differential equations of the second order, Ann. Math. 45 (1944) 723-737.

[3] Cartwright, M. L. and Littlewood, J. E., On non-linear differential equations of the second order, J. London Math. Soc. 20 (1945) 180-189.

[4] Van der Pol, B. and Van der Mark, J., Frequency demultiplication, Nature 120 (1927) 363-364.

[5] Shaw, R. S., Strange attractors, chaotic behavior, and information flows, J. Naturforsch. 36a (1981) 80.

[6] Abraham, R. H. and Simo, C., Bifurcations and chaos in forced Van der Pol systems, this volume.

[7] Abraham, R. H., Catastrophes, intermittency, and noise, in Fischer, P. and Smith, W. R. (eds.), Chaos in Dynamics (Dekker, New York, 1984).

[8] Abraham, R. H. and Scott, K. A., Chaostrophes of forced Van der Pol systems, in Fischer, P. and Smith, W. R. (eds.), Chaos in Dynamics (Dekker, New York, 1984).

[9] Chenciner, A., Courbes invariantes, orbites periodiques, ensembles de Cantor invariants dans les bifurcations des points fixes elliptiques non conservatifs, this volume.

[10] Diener, M., Canards, ou comment bifurquent les systemes differentiels lents-rapides, preprint, CERDRO, BP1510 Saint-Charles, Oran.

[11] Roux, J.-C., Bifurcations and complex behavior in physico-chemical systems: recent experimental results, this volume.

Ralph H. ABRAHAM

University of California
Mathematics Board
Santa Cruz, CA 95064
U. S. A.

Singularities & Dynamical Systems
S.N. Pnevmatikos (editor)

BIFURCATIONS AND CHAOS IN FORCED VAN DER POL SYSTEMS

Ralph Abraham & Carles Simó

University of California — Universitat de Barcelona

U. S. A. — SPAIN

The Van der Pol system, modified through the addition of constant force and bias terms and explored through simulation, reveals an unsuspected bifurcation of codimension two. Forcing the system yields an abundance of chaotic attractors and bifurcations.

1. INTRODUCTION

We began our simulations with the intent to illustrate the bifurcation sequences of codimension one, described in the companion paper [1]. These illustrations reveal the detailed structure of the *red cigar* (Sec. 2, Figs. 1-5) and the *blue sleeve* (Sec. 3, Figs. 6-11). Unable to resist the temptation to explore further, we added sinusoidal forcing to the standard shift, and discovered an abundance of chaotic attractors and bifurcations, very similar to the sequence studied by Rossler [2] (Sec. 4, Figs. 12-17). Finally, combining the two bias parameters without forcing, we mapped the full bifurcation set in the control plane, finding the *blue goblet*, an unsuspected bifurcation of codimension two (Sec. 5, Figs. 18-22). This appears to be identical to one discovered by Takens [3; 5, p. 371] and found in a related context by Fitzhugh [11], Guckenheimer and others [4; 5, p. 70].

2. STANDARD BIAS

The equations under consideration are

$$\dot{x} = y,$$

$$\dot{y} = -x + \epsilon(1 - x^2)y + a,$$

$\epsilon > 0$, $a \in \mathbf{R}$. The system has the trivial symmetry $(x,y,a) \longleftrightarrow (-x,-y,-a)$. It has only one critical point, $P = (a,0)$. Figure 1 gives the locus of $\dot{y} = 0$ for $a = 0, 0.9, 1, 1.1, 2$ and $\epsilon = 1$. Note that every line y = constant has, at most, two cuts with the graphs shown in Figure 1.

Figure 2 displays the linear character of P as a function of (ϵ, a). It turns out that this system has exactly one periodic orbit and that it is the only attractor of the system for $a \in (-1,1)$ for any $\epsilon > 0$. For $a \in (-\infty,-1] \cup [1,\infty)$ the only attractor is P.

Figure 3 shows some of these orbits for $\epsilon = 1$ and values of a equal to -.99, -.96, -.84, -.5, 0, .5, .84, .96, .99. For the same value of ϵ in Figure 4 we present the section of the current attractor as a function of a. The family of periodic orbits (PO) can be seen as a *cigar*. In Figure 5 the plot of the period T of the PO against a is shown. The amplitude $(x_{sup} - x_{inf})$ of the PO has a quadratic behavior when $a \to \pm 1$.

Figures 3 bis, 4 bis and 5 bis are obtained when ϵ is set equal to 4. In Figure 3 bis the displayed POs have values of a equal to -0.999, -0.9926, -0.991776, -0.99177072, -0.99177071, -0.9917707, -0.99, -0.5 and 0, respectively. Symmetrical figures are obtained for positive values. The selected values are chosen to display the well known *canard effect* due to the presence of a slow manifold [6]. This effect is hard to see in Figures 4 bis and 5 bis. However it can be checked in Table 1, which offers some numerical data to reproduce Figures 3 bis to 5 bis.

3. VELOCITY BIAS

We merely move the bias to the first equation. The behavior of the system changes in a dramatic way. The equations are

$$\dot{x} = y - c,$$

$$\dot{y} = -x + \epsilon(1 - x^2)y,$$

$\epsilon > 0$, $c \in \mathbf{R}$, with the trivial symmetry $(x,y,c) \longleftrightarrow (-x,-y,-c)$. We always suppose $c \neq 0$. Two critical points appear, $P_\pm = (x_\pm, c)$, where

$$x_\pm = -\frac{1}{2\epsilon c} \pm \left(\frac{1}{4\epsilon^2 c^2} + 1\right)^{1/2}.$$

Note that, letting c go to zero, one of the points goes to the origin and the other escapes to infinity.

The linear behavior of P_+ is shown in Figure 6 as a function of (ϵ, c). It is always a repellor for $\epsilon > 0$. P_- is always a dissipative saddle (i.e., a saddle such that the divergence is negative at it). Figure 7 displays the invariant stable and unstable manifolds of P_- for the following couples of (ϵ, c):

(1,0.5), (1,0.6), (4,0.1), (4,0.2), (0.25,1.8) and (0.25,2.2).

According to our numerical results it seems that for every $\epsilon > 0$ there is exactly one positive c (and, of course, this happens also for $-c$) such that there exists a homoclinic orbit to P_-. Let $c^*(\epsilon)$ be this value. In Figure 8 we present the curve $c^*(\epsilon)$. Table 2 offers some of the related numerical data.

Now we fix $\epsilon = 1$. The attracting POs present for $c = 0$ can be continued in the range $(-c^*(1), c^*(1))$ and it is (according to numerical simulation) the only attractor in this range. The family of POs terminates at each end in the homoclinic orbit (the so-called blue sky catastrophe) and, accordingly, the period T goes to $+\infty$. We shall call this family a . For $c \in (-\infty, -c^*(\epsilon)) \cup (c^*(\epsilon), \infty)$ there is no attractor. Figure 9 shows the PO for c = −.569, −.4, −.2, .4, .569. Figures 10 and 11 are similar to Figures 4 and 5. For $\epsilon = 4$ the results are given at Figures 9 bis (for c = −.159, −.13, −.07, .07, .13, .159), 10 bis and 11 bis. Finally Table 3 offers the related numerical data.

4. STANDARD BIAS WITH FORCING

Here we consider the following equations:

$$\dot{x} = y,$$

$$\dot{y} = -x + \epsilon(1 - x^2)y + a + b\cos\omega t$$

Next we describe the results of a rough exploration by simulation. For $(\epsilon, a, b, \omega) = (4,1,2,3)$, starting at (0,0) for $t = 0$ the orbit becomes attracted by a stable PO. Throughout the exploration we keep $(\epsilon, a, \omega) = (4,1,3)$. Slowly varying b and starting each simulation at the previous attractor one gets a sequence of period-doubling bifurcations. Figure 12 shows, for b = 2, 2.5, 2.84 and 2.87, orbits which we call 1-PO, 2-PO, 4-PO and 8-PO because they are seen as 1, 2, 4, 8 periodic points under the time $-\frac{2\pi}{\omega}$ map (that we design as F).

A slight further increment of b to 2.88 gives what seems to be a 32 piece strange attractor (32-SA). Figure 13 displays the points obtained using F and magnifications. Then we have an inverse cascade of fusions of attractors (see [2, 7, 8, 9]). Figures 14, 15 and 16 show the 4-SA (for b = 2.885), the 2-SA (for b = 2.89) and the orbit for b = 2.89, respectively.

A further increase to 2.9 shows that the SA is destroyed (of course, by heteroclinic tangency, see [7]) and we observe an attracting 7-PO. Figure 17 displays this orbit for b = 2.9, 2.8, 2.6. Now we can go backwards in b. This attracting 7-PO exists till some value near 2.5 for which the orbit disappears through a saddle-node bifurcation. The stable manifold of the corresponding saddle bounds the basin of attraction. It is the transversality of this manifold with the SA, which seems to be the closure of the unstable manifold of a 2-PO of saddle type (the one obtained by continuation of the previously attracting 2-PO), the reason of the destruction of the SA. Then two different attracting POs, or an attracting PO and an SA, can simultaneously coexist. The basins of some of these attractors can be small and difficult to detect if random initial conditions are chosen. Further analysis will be the object of a future paper. Table 4 gives initial conditions (t = 0) for the displayed figures.

5. DOUBLE BIAS

In this section we combine the two previously studied biases. It will be apparent that the symmetry existing before in the termination of the family of attracting POs can be destroyed.

The standing equations are

$$\dot{x} = y - c,$$

$$\dot{y} = -x + \epsilon(1 - x^2)y + a, \quad \epsilon > 0$$

with symmetry $(x,y,a,c) \longleftrightarrow (-x,-y,-a,-c)$. We can suppose $a,c \neq 0$, the other cases being already discussed, and even we can suppose $c > 0$. The critical points are $P_\pm = \left(-\frac{1}{2\epsilon c} \pm \left(\frac{1}{4\epsilon^2 c^2} + 1 + \frac{a}{\epsilon c}\right)^{1/2}, c\right)$. They exist if

$$a > -\epsilon c - \frac{1}{4\epsilon c} \qquad (*)$$

(for positive ϵc). P_+ always has index +1 and P_- is always a saddle. The linear character of $P_\pm$ is given in Figure 18, P_- expansive means div $X(P_-) > 0$. When equality holds in (*), only one critical point appears. In Figure 18 we also show the curve $\rho(a,c) = 0$ (for $\epsilon = 1$) for which we get a homoclinic connection. A similar picture for $\epsilon = 4$ is given in Figure 18 bis. Table 5 offers numerical data for the figures. Table 6 presents values of (ϵ,a) producing homoclinic orbits for increasing values of c.

Let us analyze the neighborhood of the point $a = -1$, $\epsilon c = 1/2$ in the parameter space. First set $a = -1$, $\epsilon c = 1/2$ and change variables through $\xi = x + 1$, $\eta = y - \frac{1}{2\epsilon}$. We get $\dot{\xi} = \eta$, $\dot{\eta} = \xi(2\epsilon\eta - \epsilon\eta\xi - \xi/2)$. We note $\dot{\eta} = 0$ if $\xi = 0$ or $\eta = \frac{\xi/2}{\epsilon(2-\xi)}$. Figure 20 describes the vector field. For a fixed $\eta > 0$ the maximum of $\dot{\eta}$ is obtained for $\xi = \frac{2\epsilon\eta}{1+2\epsilon\eta}$. Then $\dot{\eta} < 2\epsilon\eta$ and the slope in A is less than 2ϵ. A similar result is obtained for $\eta < 0$, $\xi \in (0,2)$, bounding the slope for large enough values of $-\eta$. Therefore, flow entering through the positive η axis escapes through the positive ξ axis, and, after, through the negative η axis. Forward and backward images of the negative ξ axis go to W^u, W^s. Local expressions for $W^{u,s}$ near (0,0) are

$$\eta = \Psi^o_{u,s}(\xi) = a_o(-\xi)^{r_o} + a_1(-\xi)^{r_1} + \ldots,$$

$r_1 > r_o$, $r_o = 3/2$, $a_o = \pm 3^{-1/2}$. For $\xi \to -\infty$ we get

$$W^s: \quad \eta = \Psi^\infty_s(\xi) = b_o(-\xi)^{s_o} + b_1(-\xi)^{s_1} + \ldots,$$

$s_1 < s_o$, $s_o = 3$, $b_o = -\frac{\epsilon}{3}$, and

$$W^u: \quad \eta = \Psi^\infty_u(\xi) = c_o(-\xi)^{t_o} + c_1(-\xi)^{t_1} + \ldots$$

$t_1 < t_o$, obtaining $\Psi^\infty_u(\xi)$ = hyperbola + $\epsilon^{-2}/2\xi^{-4} + \ldots$, where hyperbola means the locus of $\eta = 0$,

$$-\frac{1}{2\epsilon} + \frac{1}{\epsilon}(-\xi)^{-1} - \frac{2}{\epsilon}(-\xi)^{-2} + \frac{4}{\epsilon}(-\xi)^{-3} - \frac{8}{\epsilon}(-\xi)^{-4} + \ldots .$$

A sketch of $W^{u,s}$ is also given in Figure 20.

We next describe the (global) behavior for $(a,\epsilon c)$ near $(-1,1/2)$.

THEOREM: Near $a = -1$, $\epsilon c = 1/2$, the bifurcation diagram given in Figure 21 holds.

Proof: As the only essential modifications to the flow are near the origin, we only need to examine this region. The proof shall also give quantitative information. We set $\epsilon > 0$, $a = -1 - \alpha$, $c = \frac{1+\beta}{2\epsilon}$ with α,β small. The condition $\frac{1}{4\epsilon^2c^2} + 1 + \frac{\alpha}{\epsilon c} \geq 0$ is written as $\beta^2 - 2\alpha(1+\beta) \geq 0$. Hence, curve (1)-(3) is given by $\alpha = \frac{\beta^2}{1+\beta}$. On this curve the field has a double fixed point as at $(\alpha,\beta) = (0,0)$, and the behavior is the same.

From the fact that there are only two fixed points for $\beta^2 - 2\alpha(1+\beta) > 0$, their character already discussed, the only thing to prove is the location of branch (5) for which we get a homoclinic connection. Let $\gamma^2 = \beta^2 - 2\alpha(1+\beta) > 0$, $\gamma > 0$. We introduce the changes

$$\xi = (2\gamma)^{-1}[(1+\beta)x+1+\gamma], \quad \eta = (2\gamma^{3/2})^{-1}[y-(1+\beta)/2\epsilon], \quad ' = \frac{d}{d\tau} = \gamma^{-1/2}\frac{d}{dt}$$

giving

$$\xi' = (1+\beta)\eta$$

$$\eta' = \frac{1}{1+\beta}(\xi-\xi^2) + 4\epsilon\frac{\beta-\gamma}{\gamma^{1/2}}\frac{1+\frac{\beta+\gamma}{2}}{(1+\beta)^2}\eta + 8\epsilon\gamma^{1/2}\frac{1+\gamma}{(1+\beta)^2}\xi\eta - \frac{8\epsilon\gamma^{3/2}}{(1+\beta)^2}\xi^2\eta$$

Let us introduce $k = (\beta-\gamma)/\gamma$. When $\alpha,\beta \to 0$ (and, hence, $\gamma \to 0$) the dominant term and the main perturbations are

$$\xi' = \eta, \quad \eta' = \xi - \xi^2 + \gamma^{1/2}(4\epsilon k\eta + 8\epsilon\xi\eta).$$

For $\gamma = 0$ this is a hamiltonian system, for $H = \frac{1}{2}\eta^2 - \frac{1}{2}\xi^2 + \frac{1}{3}\xi^3$, with a separatrix given by

$$\xi = \frac{3}{2\cosh^2(\tau/2)}, \quad \eta = -\frac{3\sinh(\tau/2)}{2\cosh^3(\tau/2)}.$$

The Melnikov function [5, Ch. 4; 10] is given by

$$4\epsilon\int_R \begin{vmatrix} \eta & 0 \\ \xi-\xi^2 & k\eta+2\xi\eta \end{vmatrix}(\tau)d\tau = 4\epsilon\,[k\int_R \eta^2 + 2\int_R \xi\eta^2]$$

linear in k. The value of k in order to get connection is $k = \frac{-2\int_R \xi\eta^2}{\int_R \eta^2} =$

$$-3\frac{\int_R \sinh^2\cosh^{-8}}{\int_R \sinh^2\cosh^{-6}} = -\frac{12}{7}. \quad \text{Hence } \beta = -\frac{5}{7}\gamma, \text{ or } \alpha = -\frac{12}{25}\beta^2 + \ldots \text{ (for } \beta < 0).$$

This gives (5). Similar checks give the results concerning periodic orbits. □

This result agrees with numerical simulations (see Table 5). Now we look at the behavior of the homoclinic connections for $c \to \infty$. Using the same method of scaling variables and time, displaying a hamiltonian plus perturbations, after some computations to higher order than before, we get $a = 5/7$. This is the asymptote to the curve $\rho(a,c) = 0$ of homoclinic connections for $c \to \infty$ for any ϵ. This agrees with Figures 18 and 18 bis, and with Table 6. The curve $\rho(a,c) = 0$ never reaches the line $a = 1$, and it seems that for $\epsilon \leq 1$, c increases monotonically when $\alpha \to 5/7$, while for $\epsilon > 1$, a reaches a maximum on $\rho(a,c) = 0$ and then, when decreasing to 5/7, c goes monotonically to ∞.

We return to Figure 18. Single arrows mean Hopf bifurcation and double arrows mean creation of a PO by inverse blue sky catastrophe. The conjecture, supported by the previous analytic discussion and several numerical experiments, is that in $S_1 \cup S_2$ there is a punctual attractor; in Q (unbounded region) the attractor is just one PO, and, in $\mathbf{R}^2-(Q \cup S_1 \cup S_2)$ there is no attractor. A path like c_1,c_2,c_3,c_4 in the parameter space (for any $\epsilon > 0$) produces families of POs, as shown in Figure 19: red cigar, blue sleeve, Hopf-connection (direct goblet) and connection-Hopf (inverse goblet), respectively.

6. ACKNOWLEDGEMENTS

The authors express their gratitude to Dr. Gerard Gomez for his valuable cooperation in the program they used for most of the numerical simulations, to Dr. Bruce Stewart and Prof. John Guckenheimer for pointers to the literature on bifurcations of codimension two, and to Christopher Shaw for Fig. 22. The work of the second author has been partially supported by CIRIT, Univ. de Barcelona and the Univ. Autonoma de Barcelona.

7. BIBLIOGRAPHY

[1] Abraham, R. H., In pursuit of Birkhoff's chaotic attractor, this volume.

[2] Rössler, O., An equation for continuous chaos, Phys. Lett. 57A (1976) 397-398.

[3] Takens, F., Forced oscillations and bifurcations, Comm. Math. Inst., Rijkuniversitat Utrecht 3 (1974) 1-59.

[4] Guckenheimer, J., Dynamics of the Van der Pol equation, IEEE Trans. CAS-27 (1980) 983-989.

[5] Guckenheimer, J. and Holmes, P., Nonlinear Oscillations, Dynamical Systems and Bifurcations of Vector Fields (Springer, New York, 1983).

[6] Diener, M., Canards, ou comment bifurquent les systemes differentials lents-rapides, preprint, CERDRO, BP1510 Saint-Charles, Oran.

[7] Simo, C., On the Henon-Pomeau attractor, J. Stat. Phys. 21 (1979) 456-493.

[8] Simo, C. and Garrido, L., Some ideas about strange attractors, in Dynamical Systems and Chaos, L.N. in Phys. 179 (Springer, New York, 1983) 1-23.

[9] Lorenz, E. N., Noisy periodicity and reverse bifurcation, Ann. New York Acad. Sci. 357 (1980) 282-293.

[10] Marsden, J. E., Chaotic orbits by Melnikov's method: a survey of applications, Proc. 22nd IEEE Conf. Decision & Control (Dec. 1983).

[11] Fitzhugh, R., Thresholds and plateaus in the Hodgkin-Huxley nerve equations, J. Gen. Physiol. 43 (1960) 867-896.

a.	x_{inf}	x_{sup}	Period
0	-2.008620	2.008620	6.6633
.1	-1.946181	2.063823	6.6788
.2	-1.874901	2.112698	6.7258
.3	-1.792173	2.155544	6.8051
.4	-1.693699	2.191935	6.9183
.5	-1.571987	2.220220	7.0660
.6	-1.412658	2.235991	7.2454
.7	-1.184127	2.236798	7.4352
.8	- .807333	2.150291	7.5269
.84	- .581432	2.074460	7.4573
.88	- .303718	1.953738	7.2716
.92	- .011883	1.781628	6.9755
.96	- .360095	1.549512	6.6284
.97	- .458713	1.475628	6.5406
.98	- .568854	1.388712	6.4536
.99	- .703418	1.275994	6.3677

Table 1a: ϵ =1

a	x_{inf}	x_{sup}	Period
0	-2.022963	2.022963	10.2035
.2	-1.999139	2.044498	10.3458
.4	-1.972056	2.064318	10.8040
.6	-1.939767	2.082796	11.6967
.8	-1.896882	2.100197	13.3698
.88	-1.872418	2.106901	14.4959
.96	-1.831121	2.113474	16.3747
.99	-1.764276	2.115906	18.2484
.9917	-1.684231	2.116033	19.1733
.99177	-1.514461	2.115715	20.0614
.9917706	-1.405560	2.114634	20.3409
.9917707	-1.176177	2.105610	20.5273
.99177071	- .616834	2.010342	19.1280
.99177072	- .162242	1.841492	16.1990
.99177073	- .090650	1.807267	15.6060
.991771	.120522	1.693909	13.6742
.9918	.406494	1.509359	10.7124
.9926	.625961	1.341057	8.3845
.999	.907310	1.090591	6.4244
.9996	.942492	1.056693	6.3379
.9998	.959613	1.039984	6.3103

Table 1b: ϵ =4

ϵ	$c^*(\epsilon)$
0.1	4.921542
0.2	2.491228
0.3	1.689162
0.4	1.290966
0.5	1.052638
0.7	.779078
1	.569529
2	.307091
3	.210359
4	.159864
5	.128919
7	.092957
10	.065551

Table 2

c	x_{inf}	x_{sup}	Period
.1	-2.024675	1.990147	6.7123
.2	-2.038161	1.969425	6.8719
.3	-2.048942	1.946643	7.1906
.4	-2.056895	1.921988	7.8129
.5	-2.061893	1.895579	9.3739
.55	-2.063237	1.881707	12.1738
.565	-2.063485	1.877454	15.8887
.569	-2.063539	1.876312	21.7364
.5694	-2.063544	1.876198	25.6603
.5695	-2.063546	1.876170	29.8378

Table 3a: ϵ =1

c	x_{inf}	x_{sup}	Period
.04	-2.027075	2.018759	10.4569
.08	-2.031100	2.014461	11.3751
.12	-2.035040	2.010065	13.8633
.14	-2.036979	2.007830	17.2059
.15	-2.037941	2.006703	21.2342
.155	-2.038420	2.006137	25.8253
.157	-2.038611	2.005910	29.5204
.159	-2.038802	2.005683	38.4049
.1596	-2.038859	2.005615	47.5895
.1598	-2.038878	2.005592	58.7766

Table 3b: ϵ =4

b	x	4	Fig
2	.996	-.626	12a
2.5	1.003	-.651	12b
2.84	1.013	-.587	12c
2.87	1.013	-.573	12d
2.88	1.015359	-.579361	13
2.885	1.012503	-.564710	14
2.89	1.017725	-.597039	15,16
2.9	1.101703	-.713349	17a
2.8	1.371443	-.382202	17b
2.6	1.15758	-.59054	17c

Table 4

a	c
-.99999952	.4995
-.99999808	.499
-.99999299	.498
-.9999	.49283590
-.999	.47767821
-.99	.43218198
-.98	.40632647
-.95	.36238802
-.9	.33739337
-.8	.34057534
-.5	.40279066
-.3	.45834335
-.1	.52738296
.1	.61948395
.3	.76258556
.45	.96849067
.5	1.09172511
.55	1.28684964
.6	1.65722509
.65	2.64787700
.7	11.0412399
.71	36.476921
.71272558	100
.714	545.626
.71412984	1000

Table 5a: $\epsilon = 1$

a	c
-.999999	.12481588
-.99999	.12439064
-.9999	.12273488
-.999	.10993464
-.995	.08006557
-.99	.08042535
-.9	.08695508
-.7	.10185668
-.5	.11737645
-.3	.13364175
-.1	.15083604
.1	.16924222
.3	.18934203
.5	.21209498
.7	.24005527
.9	.28910545
.95	.34497425
.9541	.38550468
.95414067	.3893
.9541	.39913685
.95	.47743240
.9	1.01942836
.74291038	10
.71722746	100
.71458062	1000

Table 5b: $\epsilon = 4$

ϵ	a
10	.971481
20	.986726
40	.993566
100	.997471

Table 6a: c = 1

ϵ	a
10	.792950
20	.855072
40	.923795
100	.973105

Table 6b: c = 10

ϵ	a
.1	.689450
.2	.702112
.4	.708495
2	.714842
10	.722878
20	.731684
40	.748630
100	.794949

Table 6c: c = 100

ϵ	a
.1	.711841
.2	.713078
.4	.713709
2	.714341
10	.715151
20	.716052

Table 6d: c = 1000

ϵ	a
.1	.713675
.2	.713984
.4	.714142
1	.714247
2	.714300
4	.714359
10	.714502

Table 6e: c = 4000

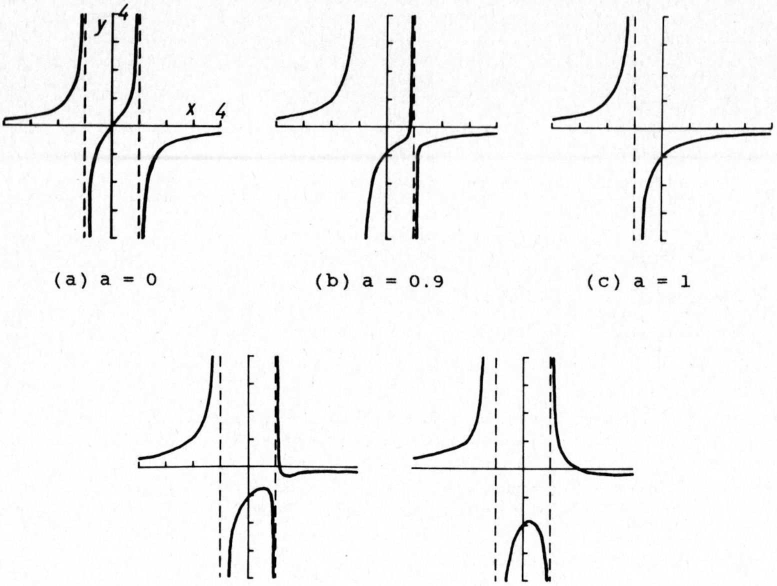

(a) a = 0 (b) a = 0.9 (c) a = 1

(d) a = 1.1 (e) a = 2.

Fig. 1 Characteristic curves with standard bias, $\epsilon = 1$.

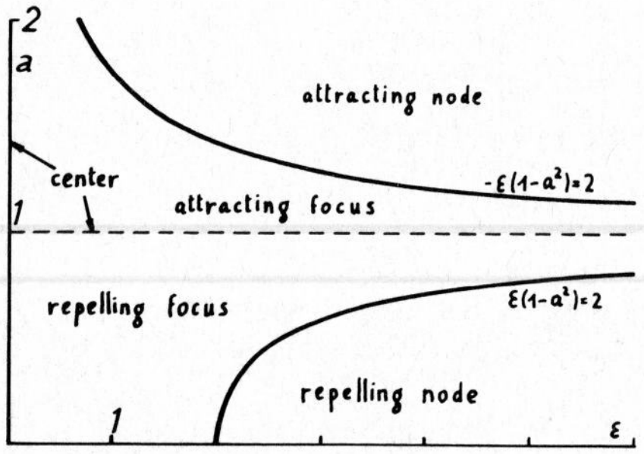

Fig. 2. Qualitative type of the critical point, P, with standard bias.

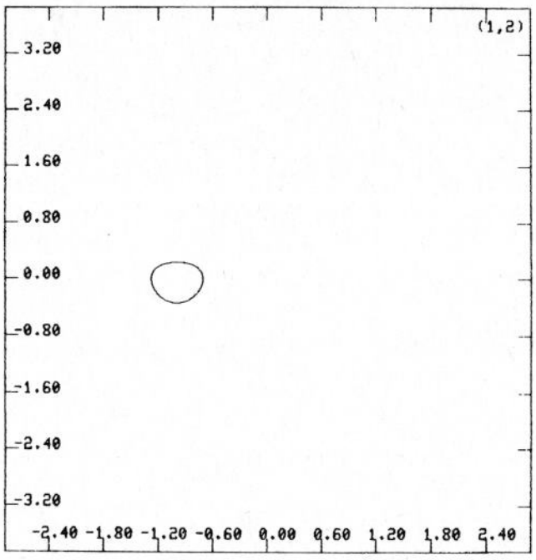

(a) a = −0.99

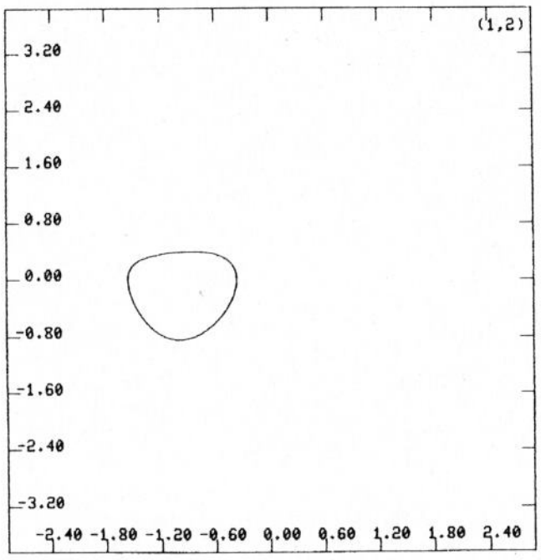

(b) a = −0.96

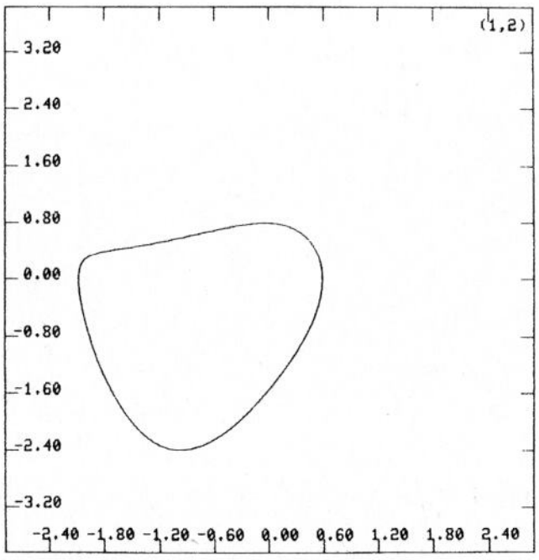

(c) a = −0.84

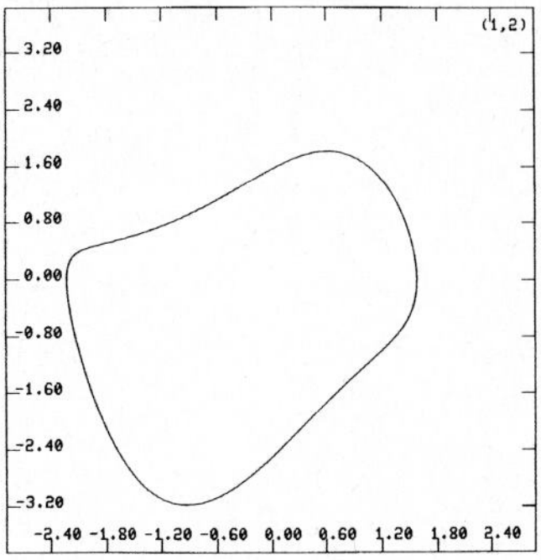

(d) a = −0.50

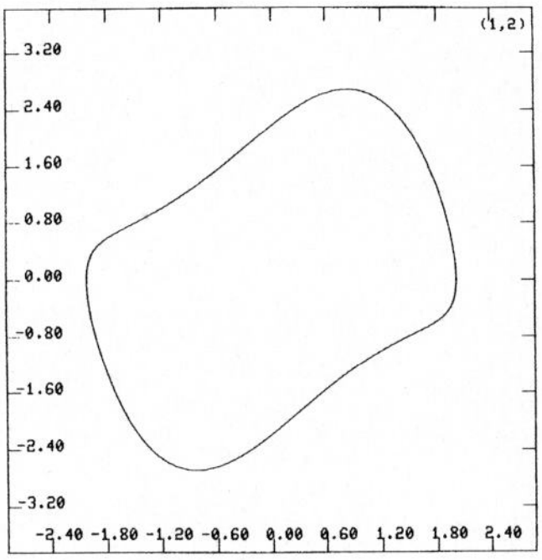

(e) a = 0

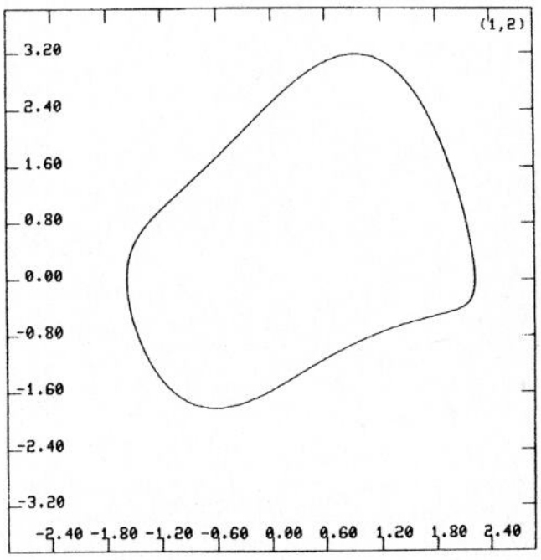

(f) a = 0.50

Fig. 3. Periodic attractors with standard bias, $\epsilon = 1$.

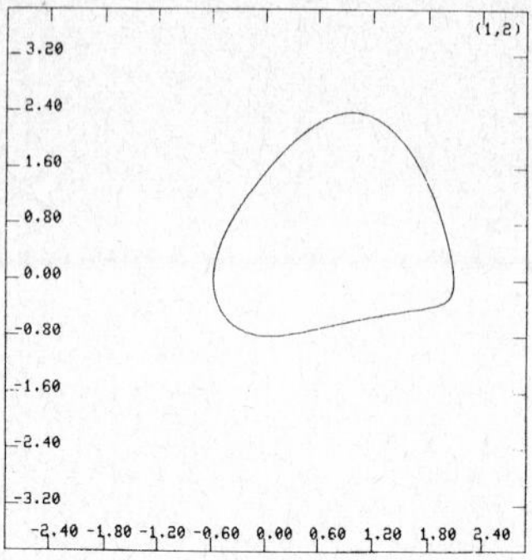

(g) a = 0.84

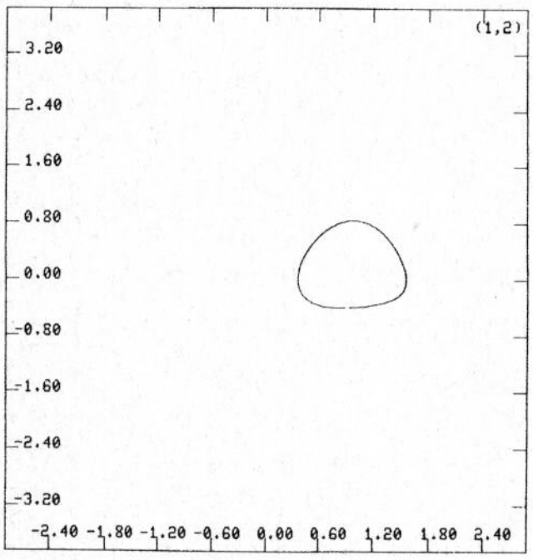

(h) a = 0.96

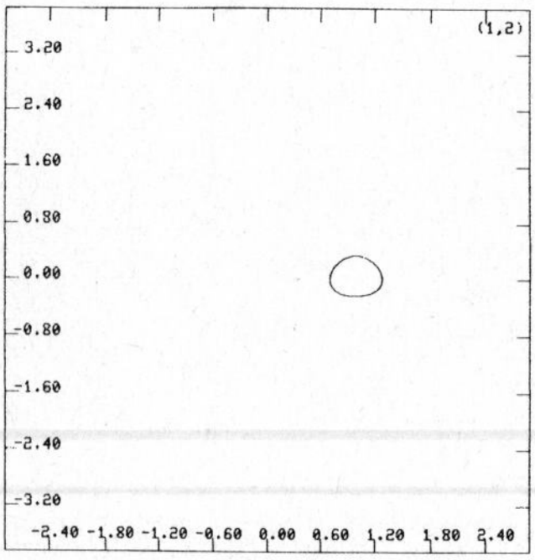

(i) a = 0.99

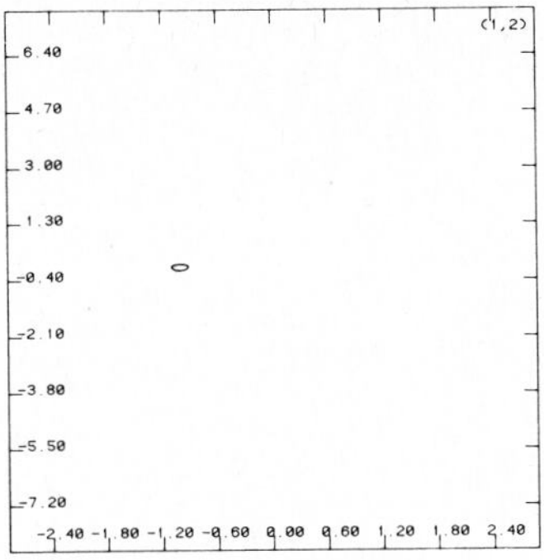

(a) a = -0.0990

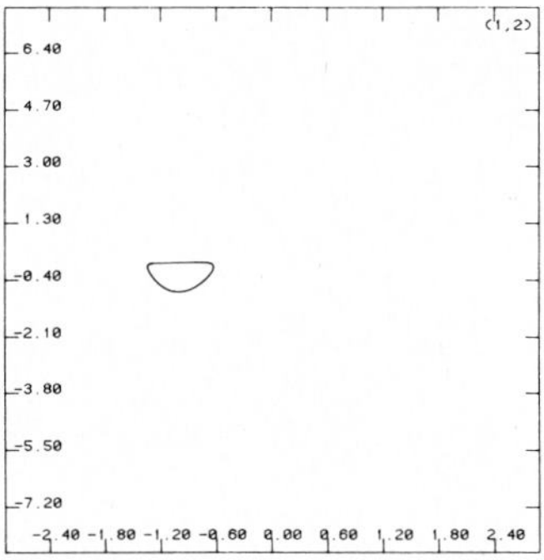

(b) a = -0.9926

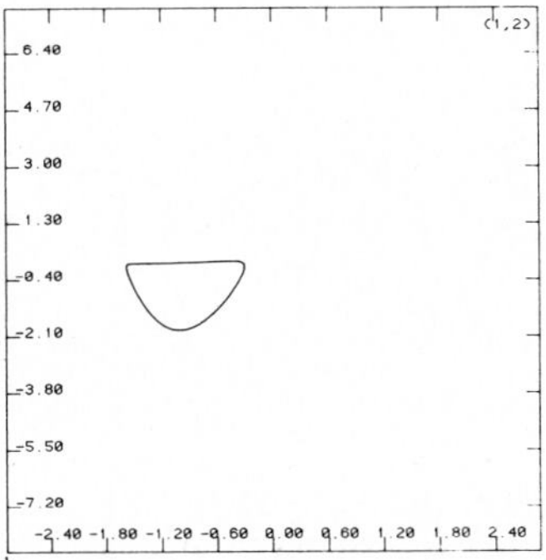

(c) a = -0.991776

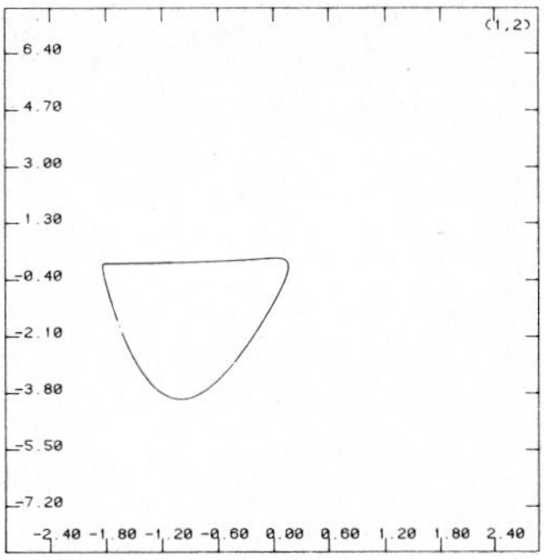

(d) a = -0.99177072

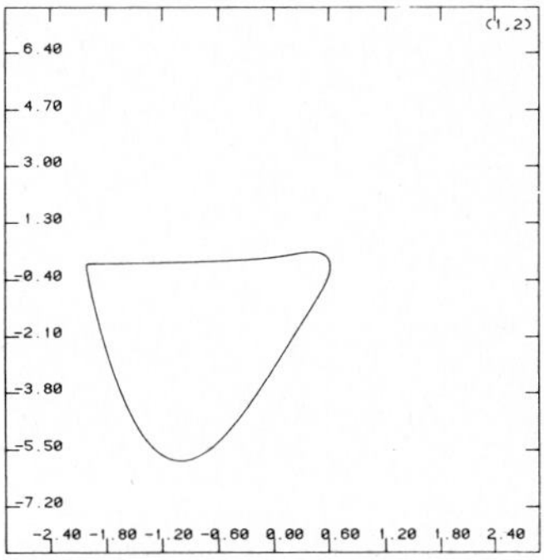

(e) a = -0.99177071

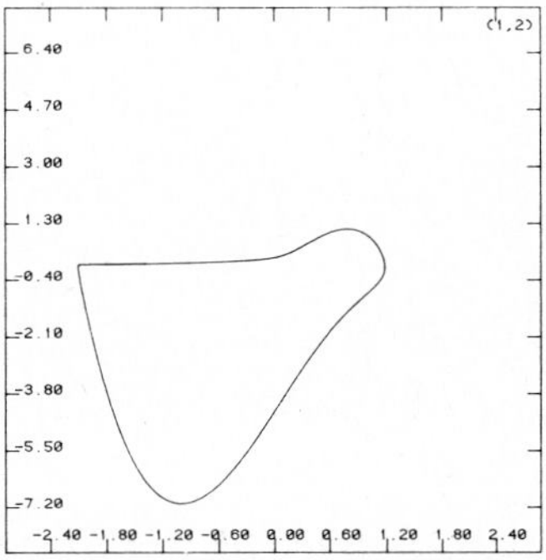

(f) a = -0.99177070

Fig.3 bis. Periodic attractors with standard bias, $\epsilon = 4$, showing the canard.

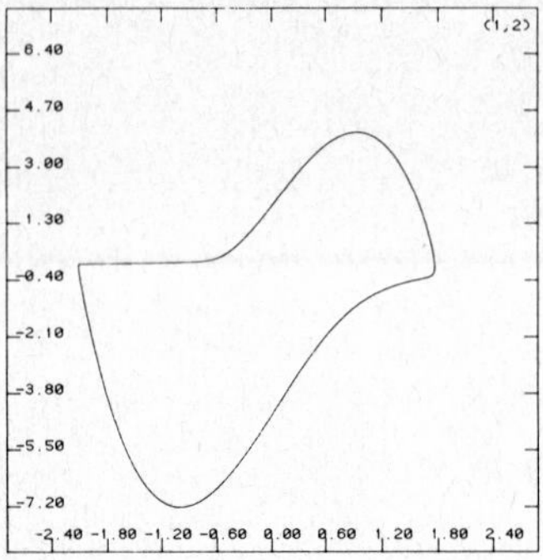

(g) a = -0.990

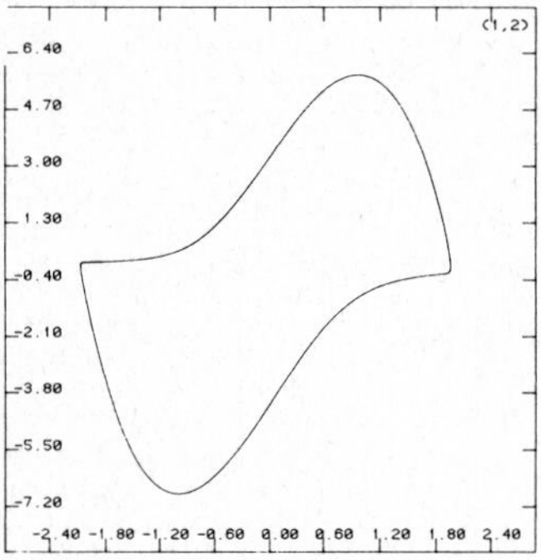

(h) a = -0.5

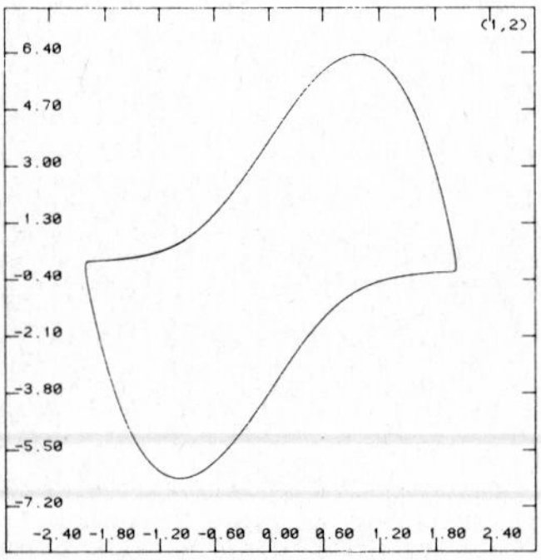

(i) a = 0

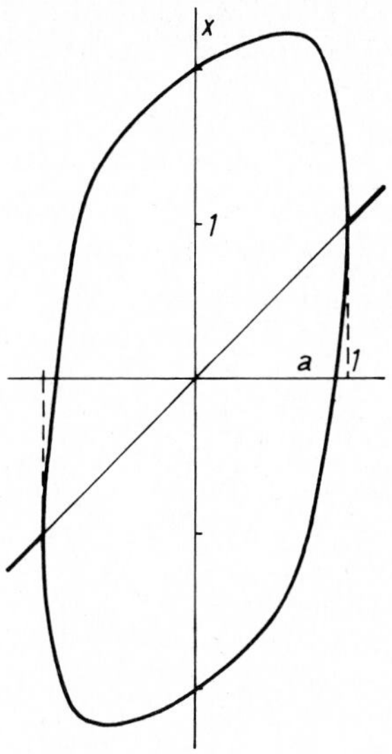

Fig. 4. Cross-section of the red cigar, ϵ =1.

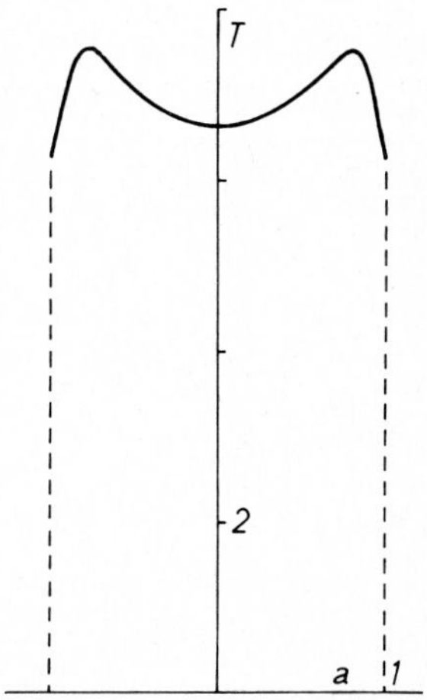

Fig. 5. Periods of the red cigar, ϵ =1.

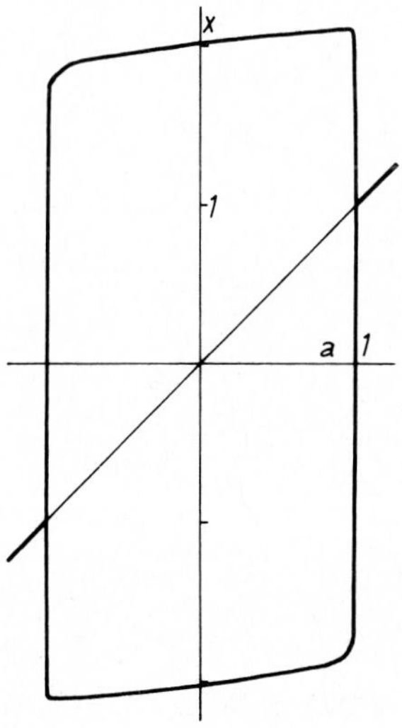

Fig. 4 bis. Cross-section of the red cigar, ϵ =4.

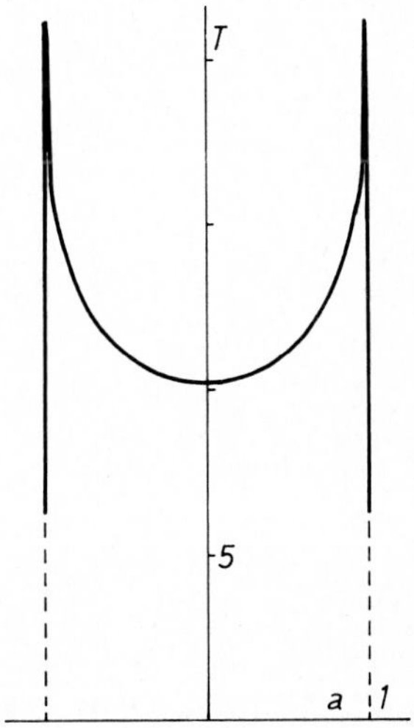

Fig. 5 bis. Periods of the red cigar, ϵ =4.

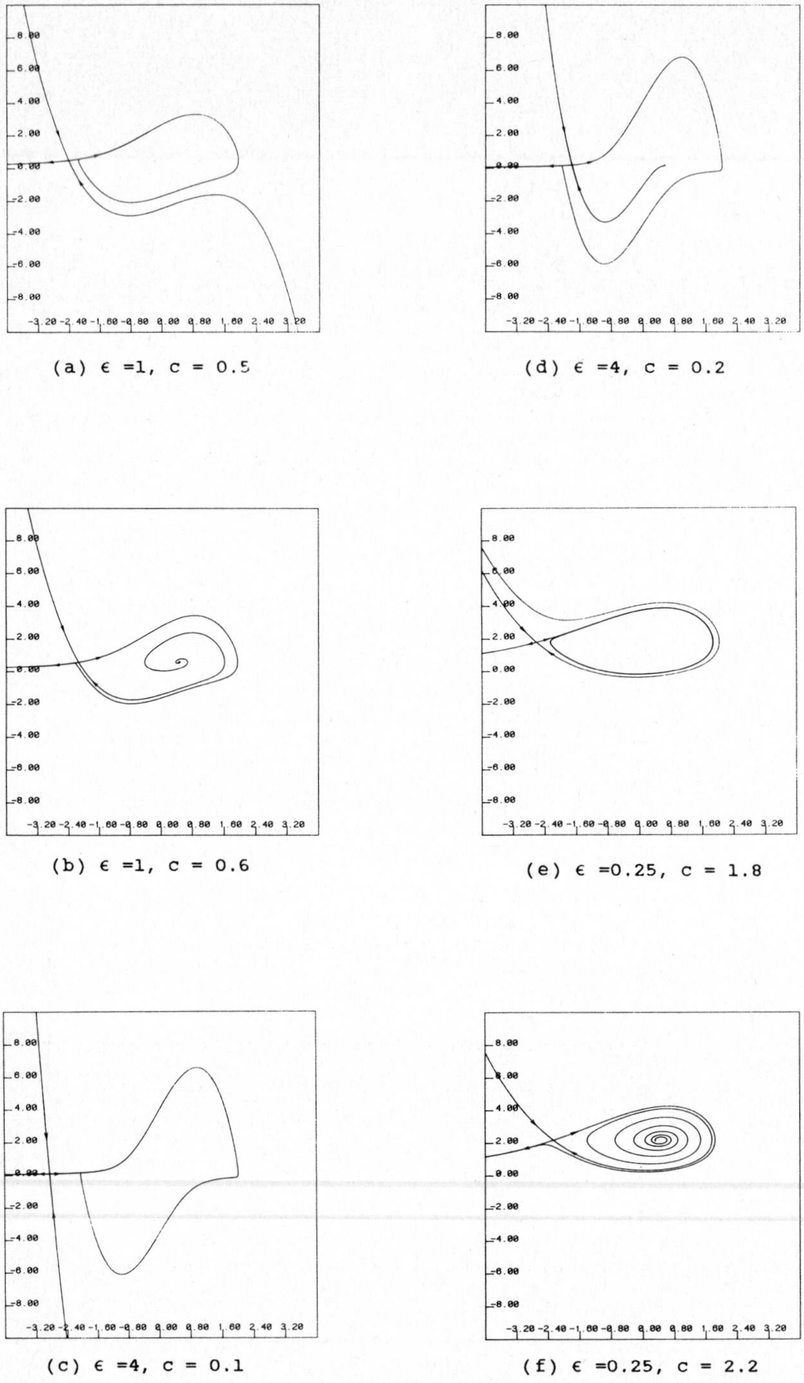

Fig. 7. Invariant curves of the saddle point, P_-, with velocity bias.

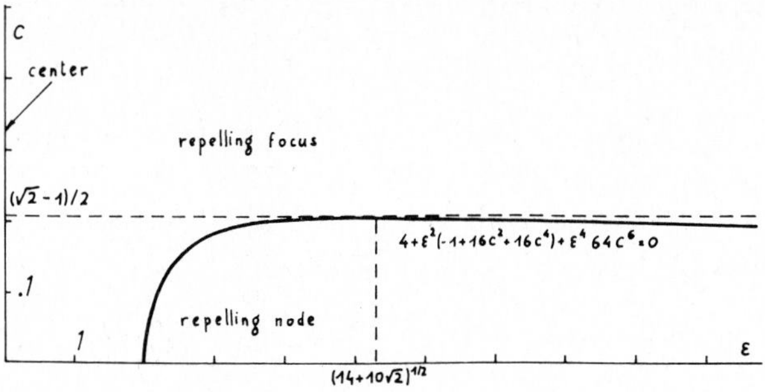

Fig. 6. Qualitative type of the critical point, P_+, with velocity bias.

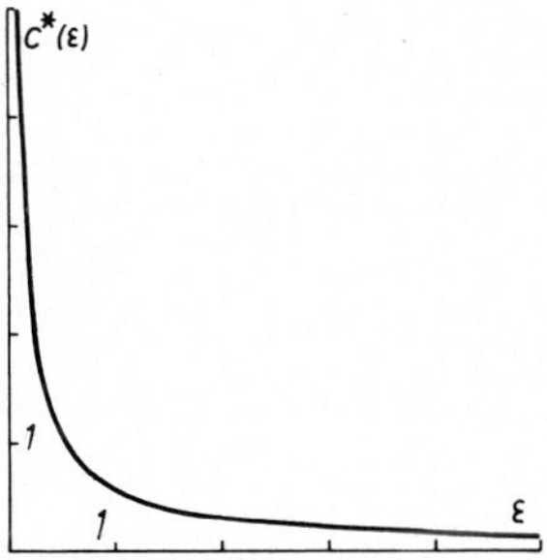

Fig. 8. Velocity bias for the homoclinic saddle connection of P_-.

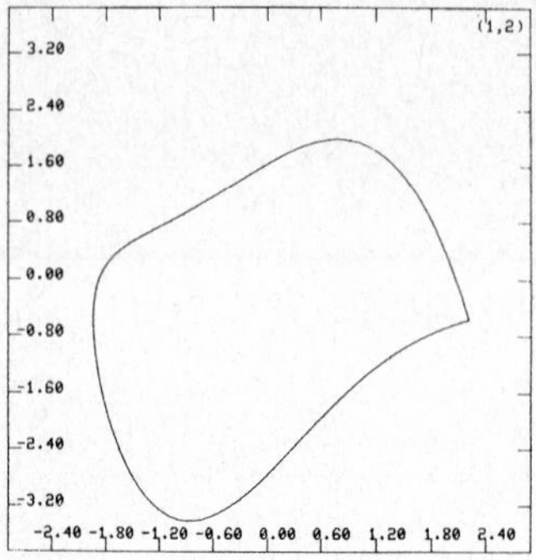

(a) c = −0.569

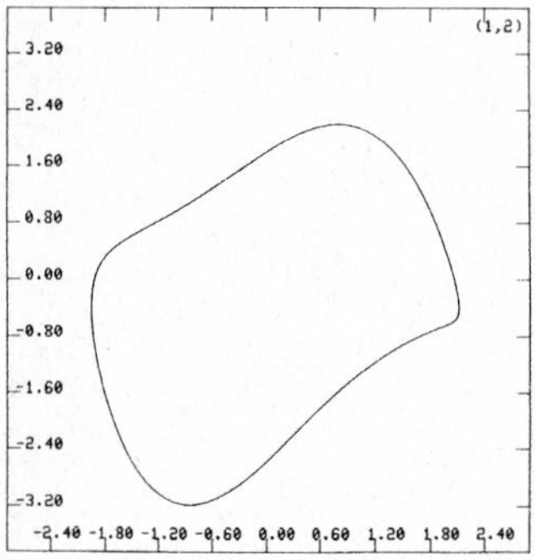

(b) c = −0.4

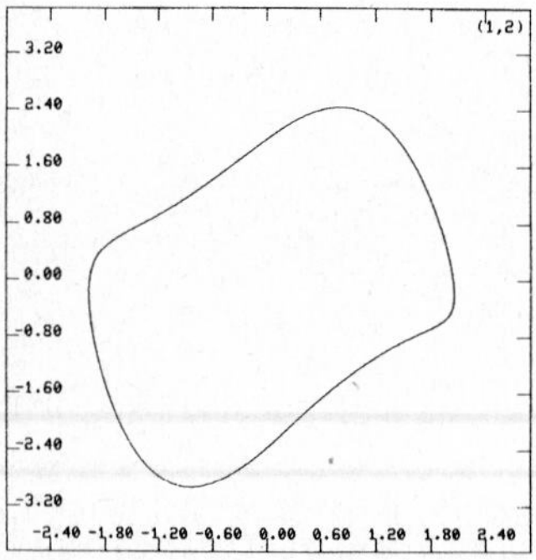

(c) c = −0.2

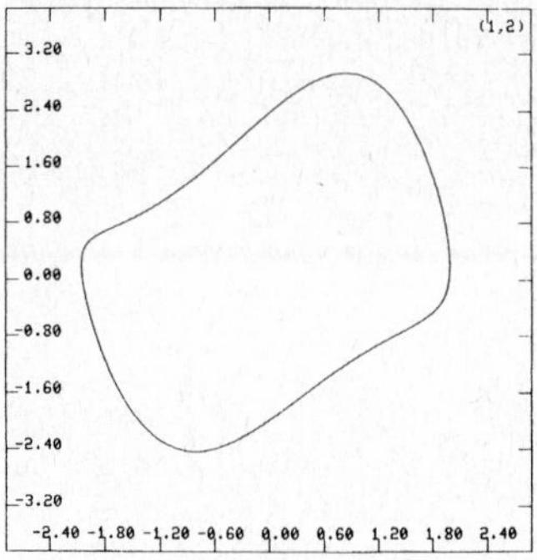

(d) c = 0.2

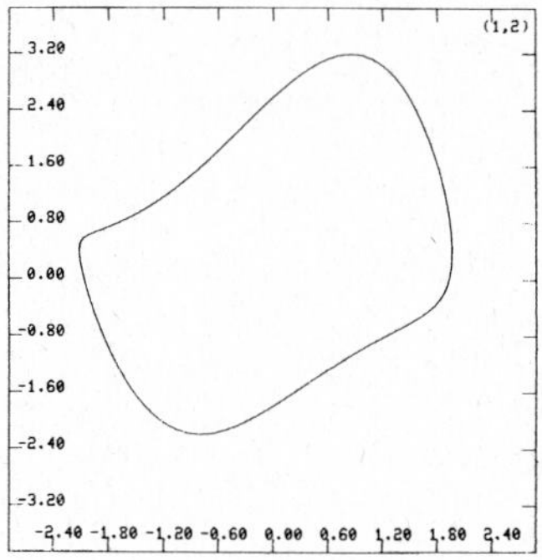

(e) c = 0.4

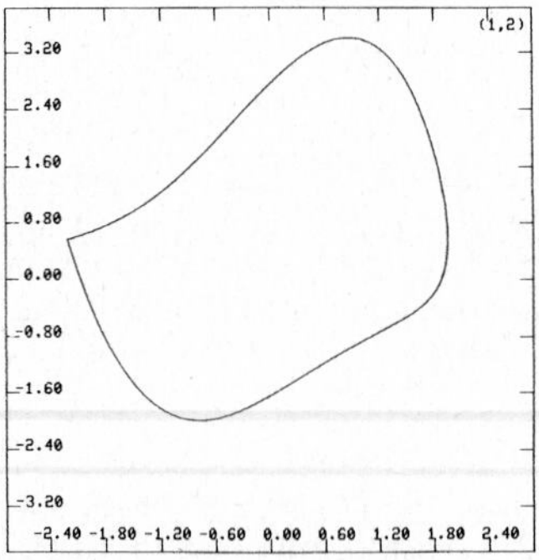

(f) c = 0.569

Fig. 9. Periodic attractors of the blue sleeve, with $\epsilon = 1$ and velocity bias, c.

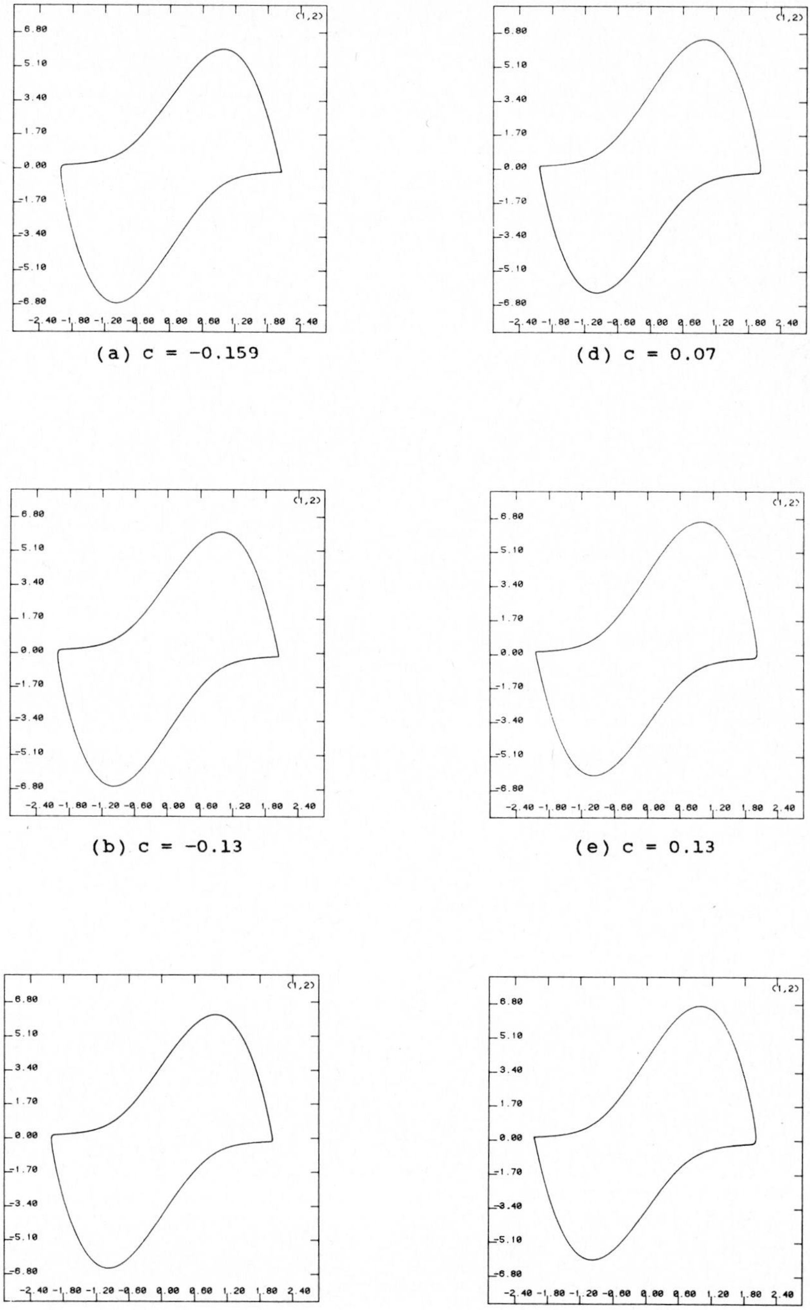

(a) c = -0.159

(d) c = 0.07

(b) c = -0.13

(e) c = 0.13

(c) c = -0.07

(f) c = 0.159

Fig. 9 bis. Periodic attractors of the blue sleeve, with ε =4 and velocity bias, c.

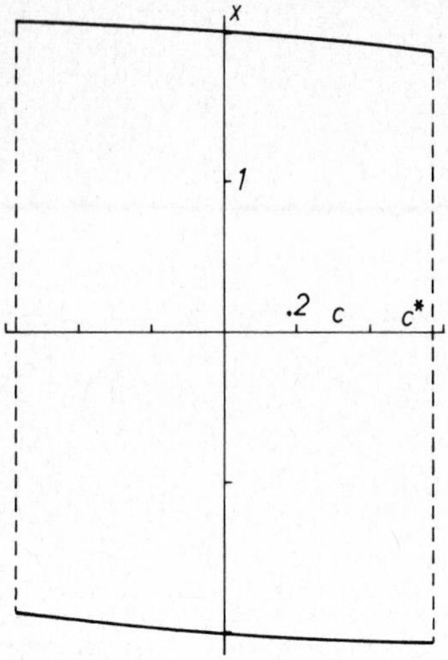

Fig. 10. Cross-section of the blue sleeve, ϵ =1.

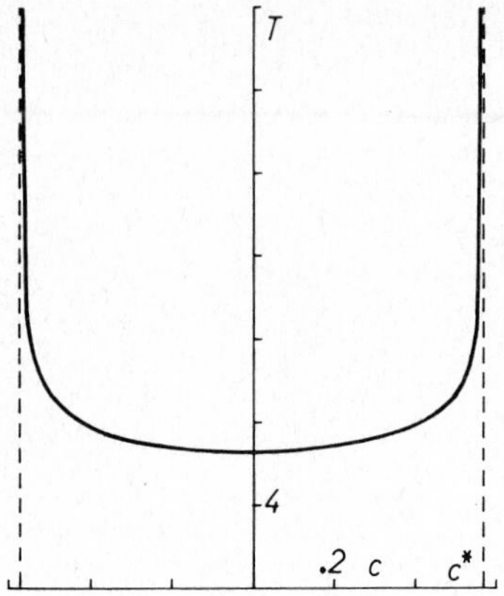

Fig. 11. Periods of the blue sleeve, ϵ =1.

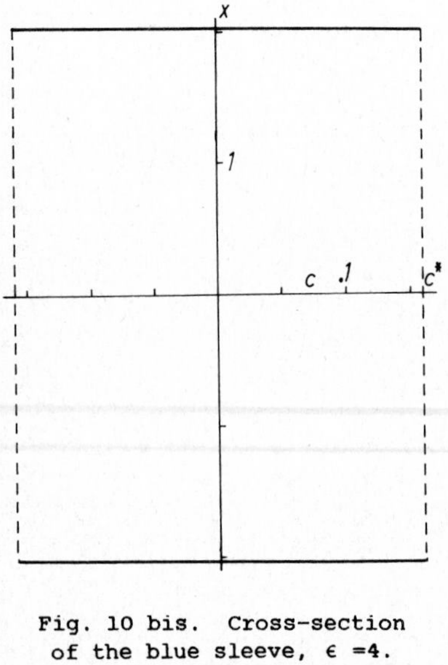

Fig. 10 bis. Cross-section of the blue sleeve, ϵ =4.

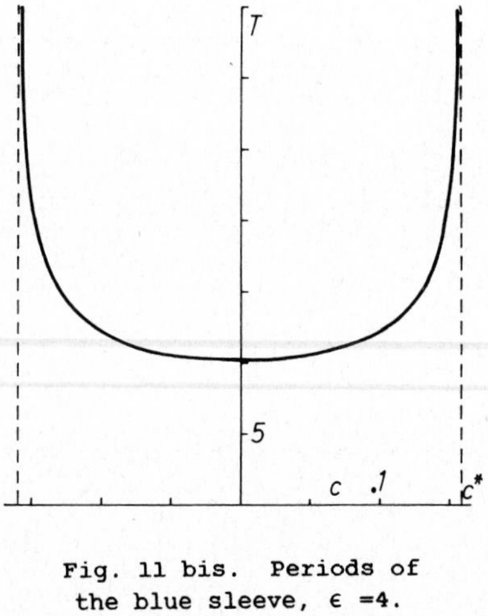

Fig. 11 bis. Periods of the blue sleeve, ϵ =4.

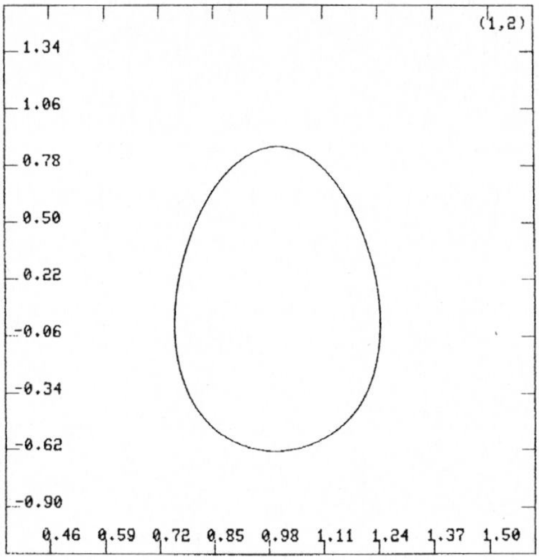

(a) b = 2.000

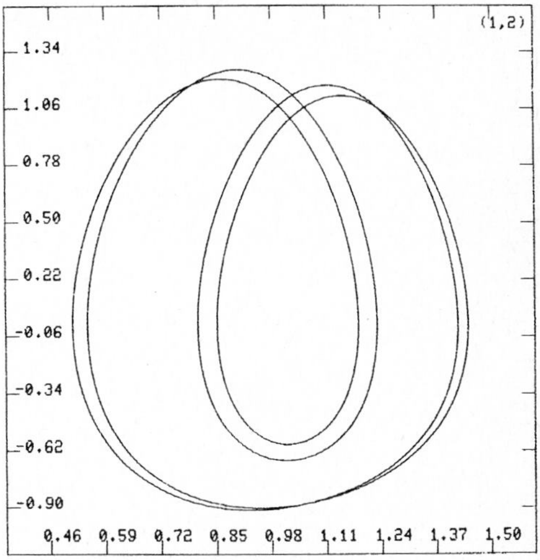

(c) b = 2.840

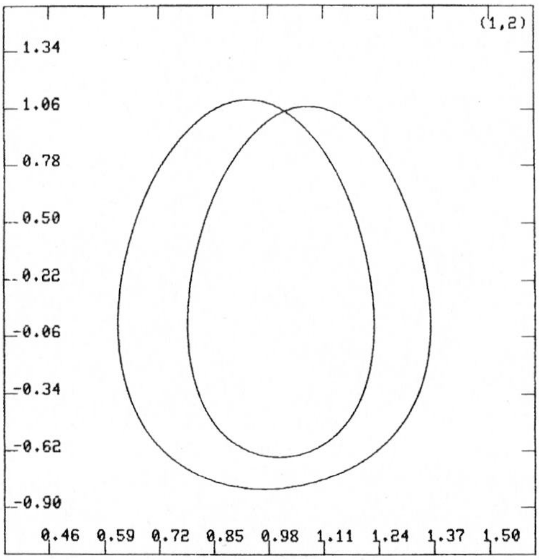

(b) b = 2.500

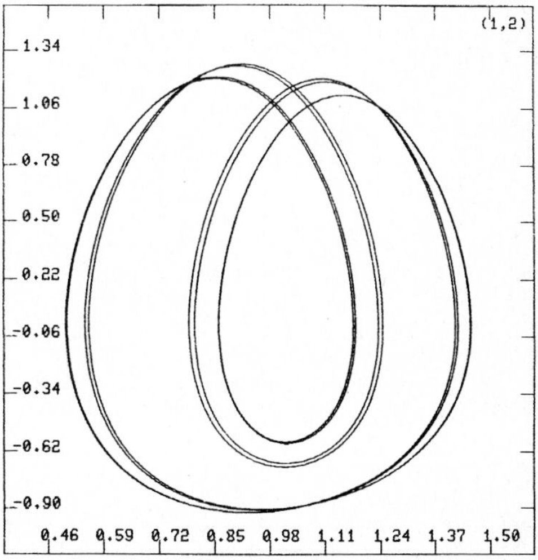

(d) b = 2.870

Fig. 12. Period doubling bifurcation sequence in the Van der Pol system with standard bias and forcing, with $\epsilon = 4$, $a = 1$, $\omega = 3$, and amplitude, b.

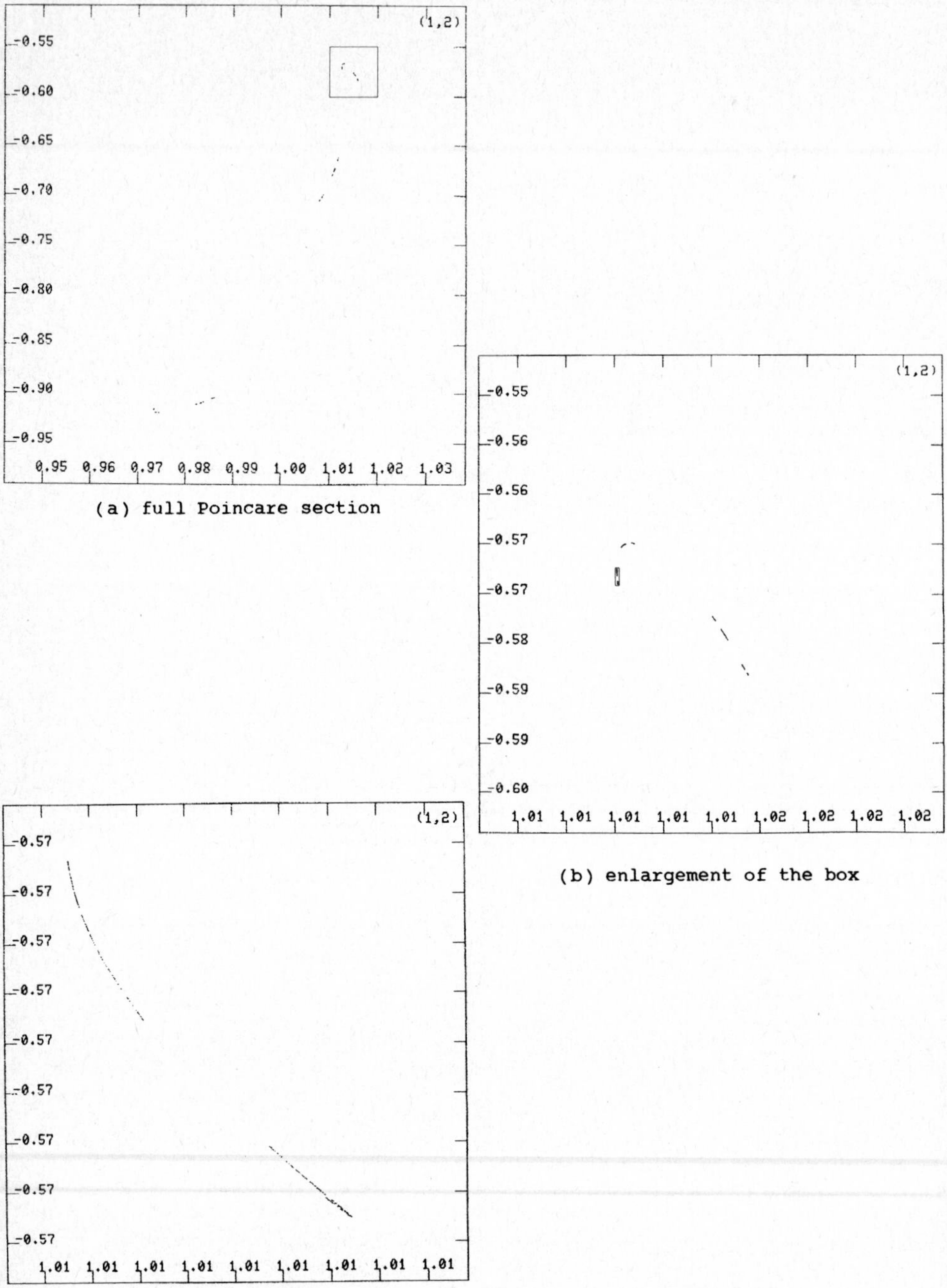

(a) full Poincare section

(b) enlargement of the box

(c) further enlargement

Fig. 13. Sections of the chaotic attractor (32-SA) in the forced Van der Pol system, as in Fig. 12, except b = 2.880.

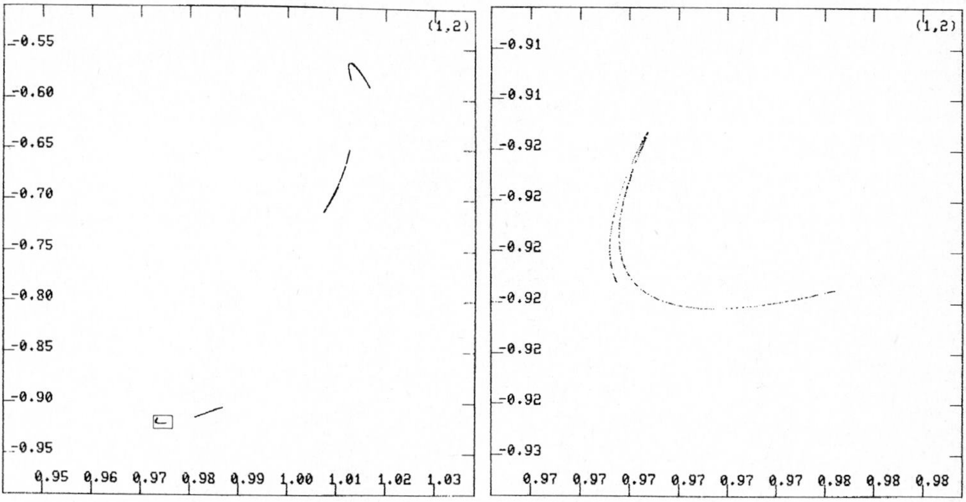

(a) full Poincare system (b) enlargement of the box

Fig. 14. Sections of the chaotic attractor (4-SA) in the forced Van der Pol system, as in Fig. 12, except b = 2.885.

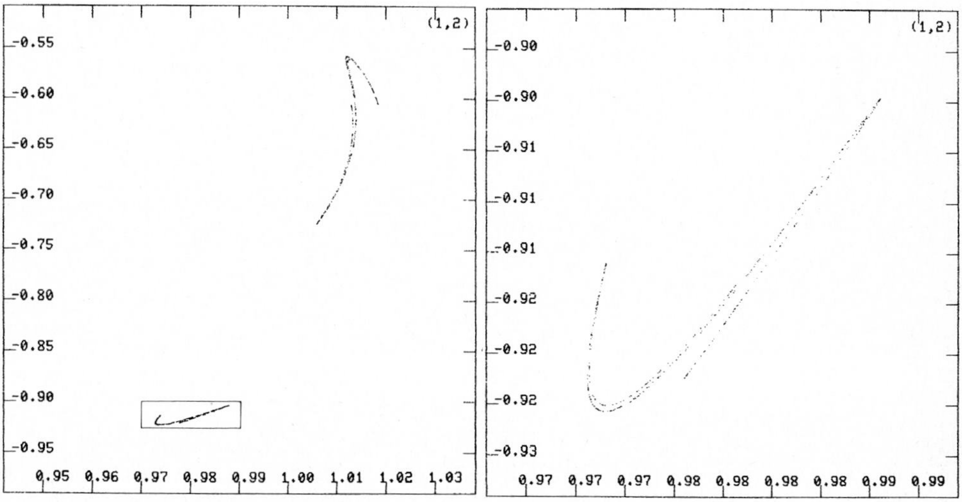

(a) full Poincare system (b) enlargement of the box

Fig. 15. Sections of the chaotic attractor (2-SA) in the forced Van der Pol system, as in Fig. 12, except b = 2.890.

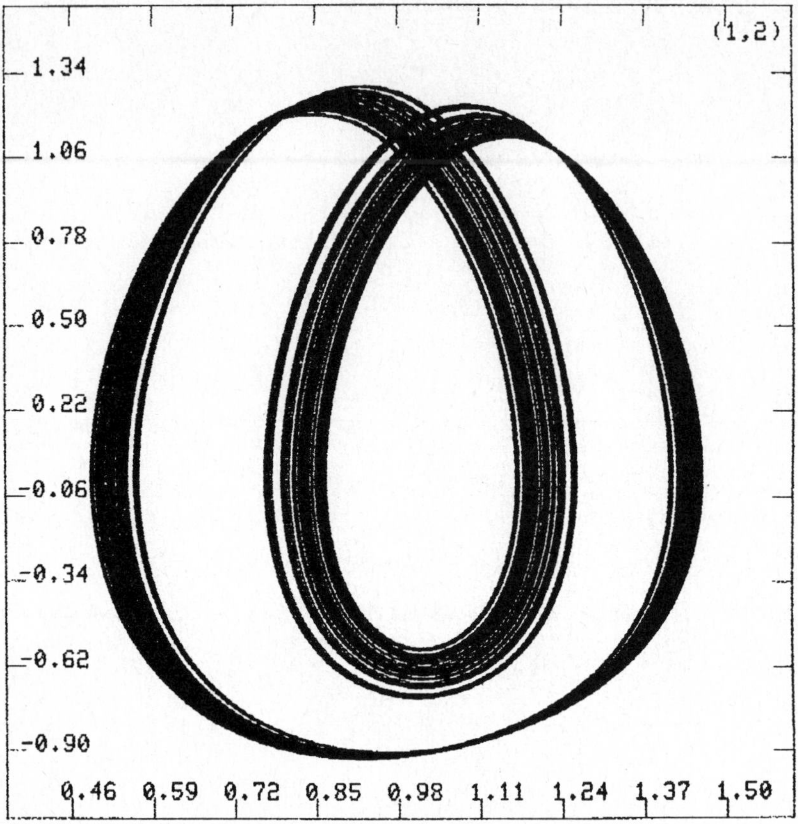

Fig. 16. Chaotic attractor in the forced Van der Pol system, as in Fig. 15.

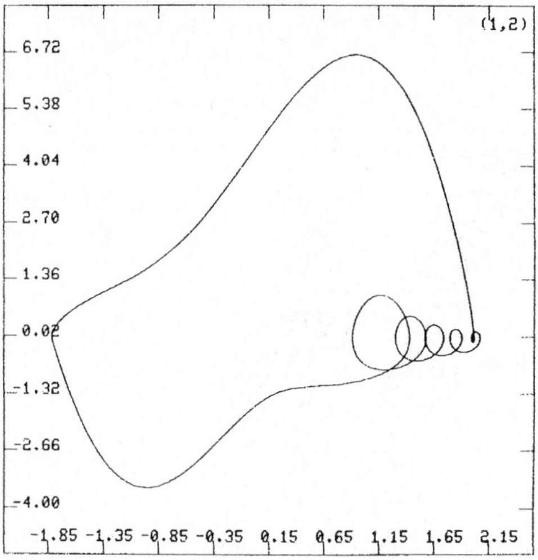

(a) b = 2.9

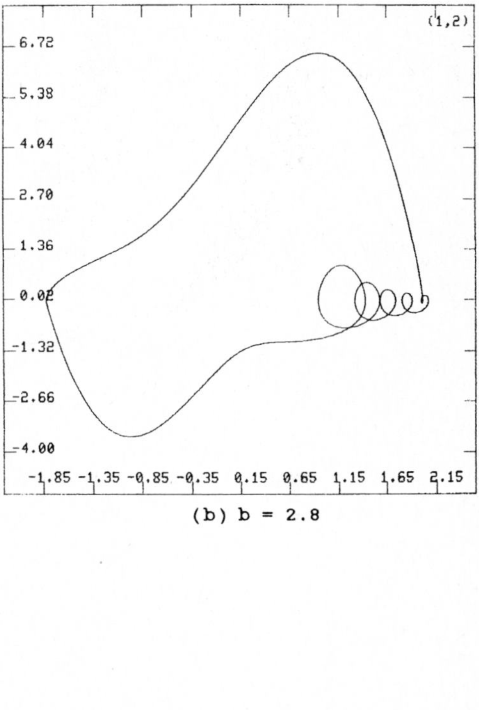

(b) b = 2.8

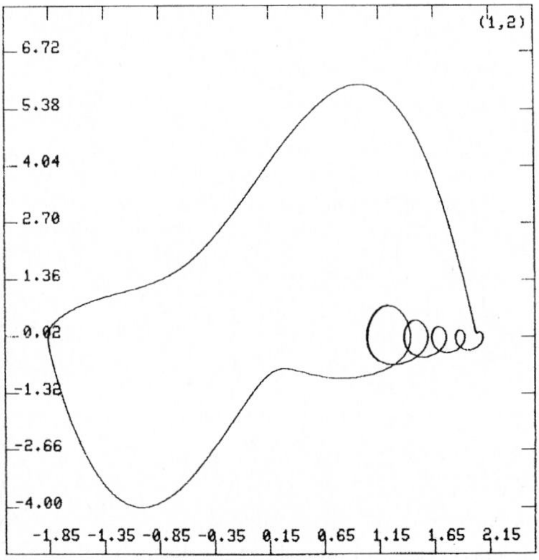

(c) b = 2.6

Fig. 17. Alternative periodic attractor (7-PO) in the forced Van der pol system, as in Fig. 12.

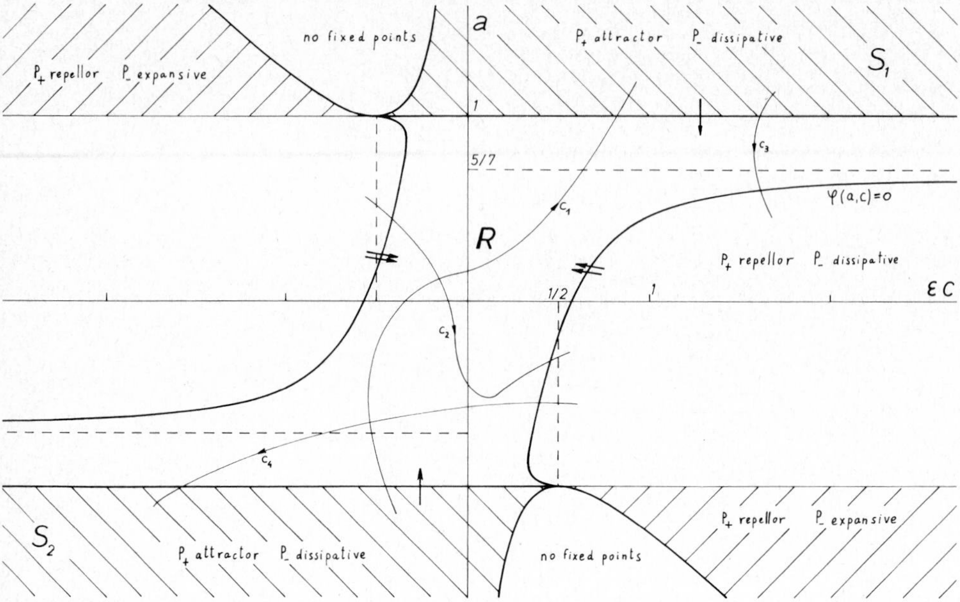

Fig. 18. Bifurcation set of the double biased Van der Pol system, with $\epsilon = 1$.

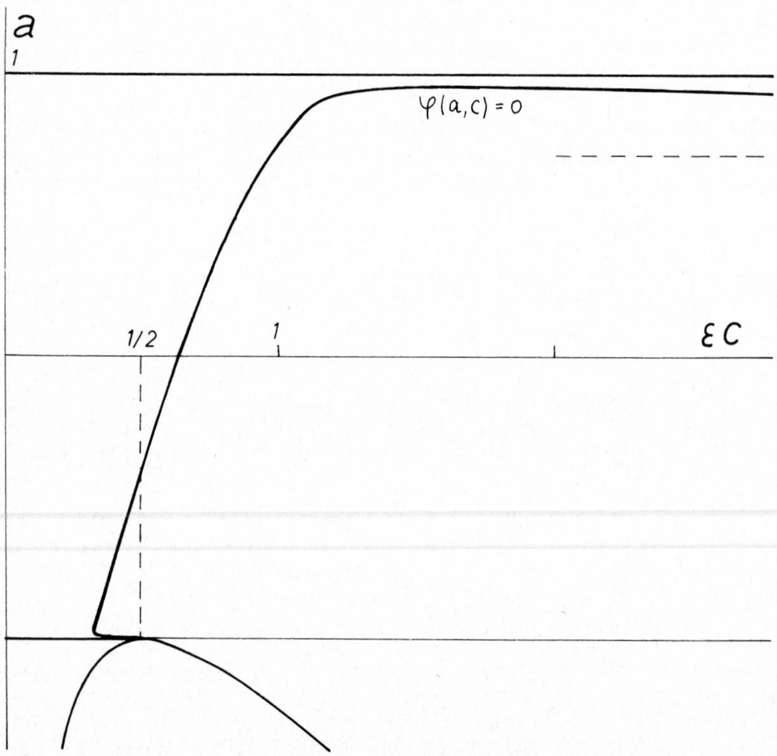

Fig. 18 bis. Bifurcation set of the double biased Van der Pol system, with $\epsilon = 4$.

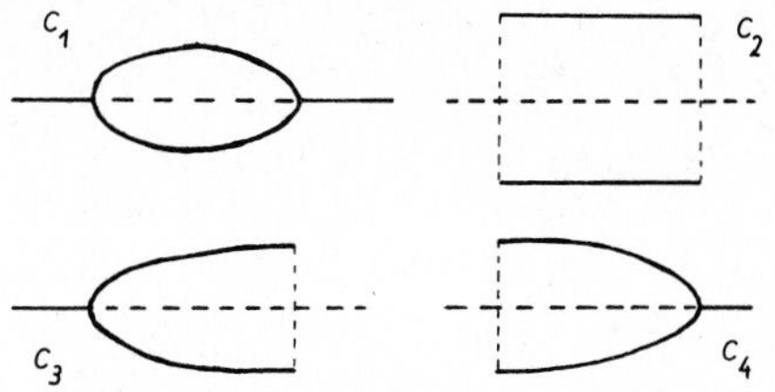

Fig. 19. Bifurcation diagrams for four selected arcs in the control plane, Fig. 18.

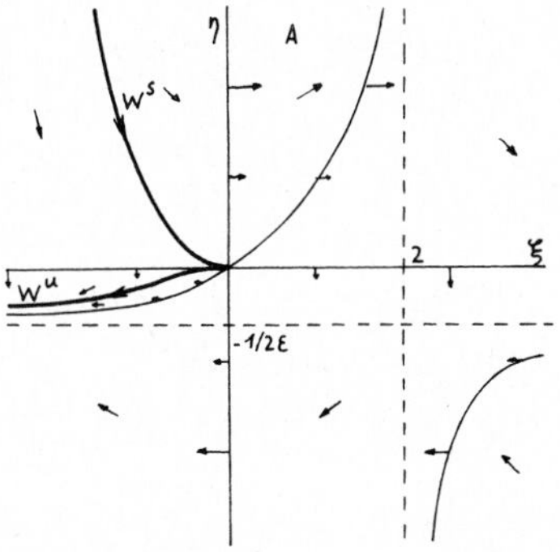

Fig. 20. Vectorfield corresponding to the bifurcation point of codimension two in Figs. 18, 18bis, and 19, showing the stable and unstable manifolds of the doubly degenerate critical point.

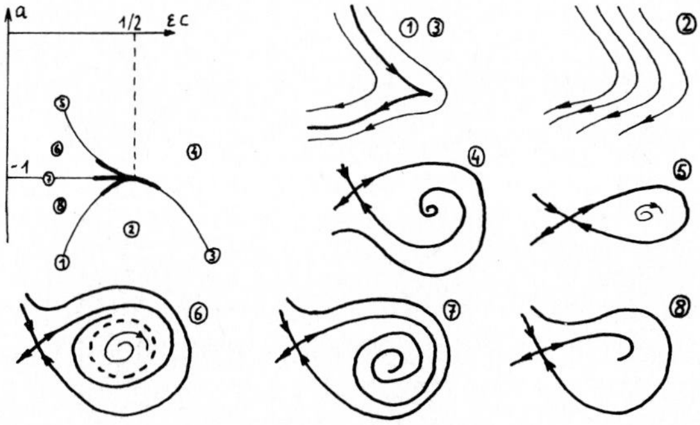

Fig. 21. Bifurcation tableau for the bifurcation of codimension two.

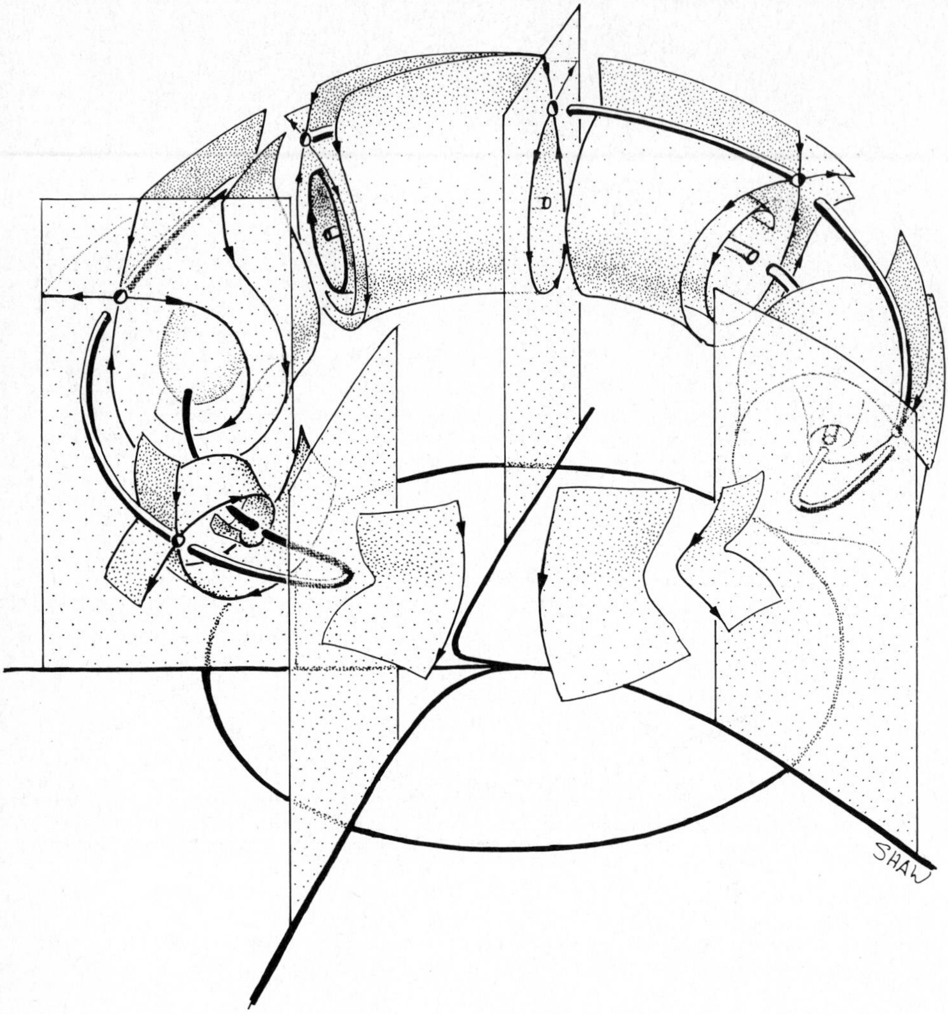

Fig. 22. Response diagram for a cycle around the bifurcation point of codimension two: the blue goblet.

(a) The goblet within the separatrix.

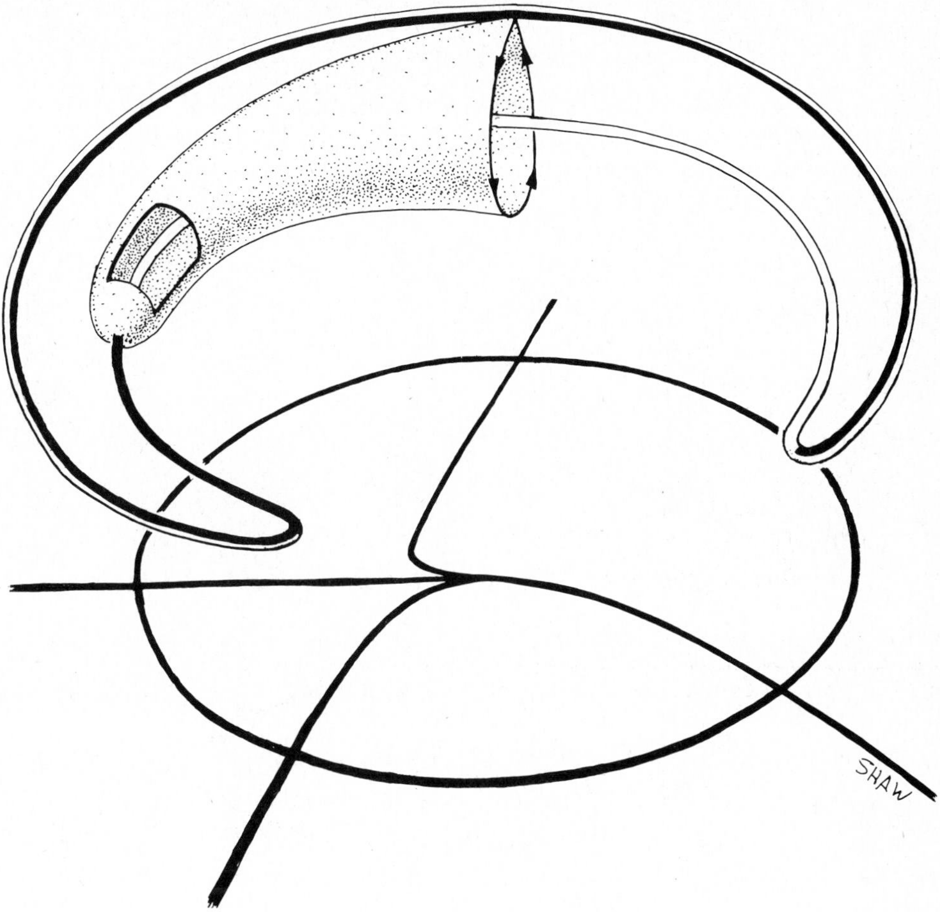

(b) Same, with the separatrix removed for a better view.

Ralph ABRAHAM & Carles SIMO

University of California
Department of Mathematics
Santa Cruz, CA 95064
U.S.A.

Universitat de Barcelona
Facultat de Matematiques
Gran Via 585, Barcelona 7
SPAIN

Singularities & Dynamical Systems
S.N. Pnevmatikos (editor)

CHAOS IN EXPERIMENTAL CHEMICAL SYSTEMS: TWO EXAMPLES

Jean-Claude Roux

Université de Bordeaux

France

In 1973 D. RUELLE (1) suggested that, due to the non-linear character of the differential equations governing the evolution of a chemical system, non periodic dynamics could be observed in realistic chemical kinetics. Such dynamics were later observed by ROSSLER (2),(3) in the simulations of a simple, low dimensional (3 ordinary differential equations) abstract chemical model. The first experimental evidences of non periodic behavior in a real chemical system followed in 1973 by R.A SCHMITZ et al. (4). However in this work and in the following the "evidences" were simply "intuitive" (time series showing some irregularities) even if,as we did in 1979 (5), a Fourier spectra was calculated showing the appearance of a broad band correlated to the occurence of the irregular behavior.

The recent progress in the experimental studies on chemical chaos that I shall devellop in the following were made possible by the development of a new tool for analysing the data : The reconstruction of the trajectories from a single time series (6),(7). Although this method implies some arbitrary choices -the dimensionality of the space and the time delay- it allowed to visualize the first strange attractor from an experimental data (8). In fact, we have shown that the topology of the attractor does not depend heavily on the arbitrary choice of the time delay and that the dimensionality of the attractor can be "reasonably" guessed.(9)

Only a small number of chemical systems exhibits typical non-linear behavior i.e. non monotonous dynamics -oscillating or chaotic-. Among these the BELOUSOV - ZHABOTINSKII (B.Z.) reaction is by far the most studied. It is usually performed in a stirred reactor with continuous feed of reactants and continuous overflow in order to insure the stationarity of the constraints on the chemistry and to allow observations over arbitrary long time scale. The experimentalist can easily act upon all these parameters (temperature, flow rate, concentrations of the chemical in the feeds .....) thus allowing -in theory- to study bifurcations problems of codimension two or more.

In terms of two of these parameters the general behavior of the B.Z. reaction can be schematized as on figure 1. Typical non linear behaviors occur only in a restricted volume of the parameter space and the non-periodic behavior is observed only near some of the boundaries of this volume. The flow rate is usually chosen as control parameter region of parameters which were scanned by the different groups on chemical chaos are represented by the dashed lines figure 1.

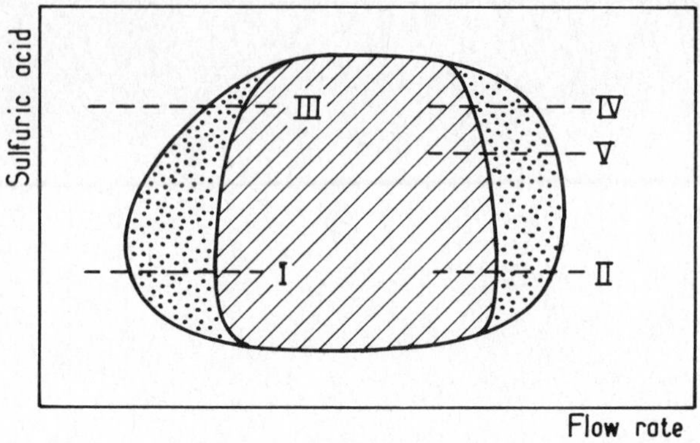

Figure 1: Behavior of the B.Z. reaction as a function of two constraints (schematic). The dashed area represents the periodic regime, the dotted one the "non periodic" dynamics. Horizontal lines I to V represents experimental paths (see text).

In this paper I shall only detail two of these studies: The first concerns the identification of a strange attractor, a work done in Texas with H.L. SWINNEY and the second illustrates a recent different type of chaos obtained in Bordeaux, the wrinkled torus.

Identification of a strange attractor

In this experiment we start with the periodic regime and follow the dashed line I by decreasing the flow rate (9)-(11). After a period doubling cascade (11) we reach the regime depicted figure 2. This time series, although showing some regularity in periodicity of the successive maxima (phase coherence) presents a distribution of amplitude in which no order could be found at first sight. However the trajectories "rebuilt" by the time delay method (7) (figure 3) in a 3D space are not stochastically distributed. Instead they describe a surface (with no apparent thickness). This character is clearly seen when the trajectories are rotated continuously (12). Figure 4a shows an other reconstruction of the trajectories with a shorter time delay. A characteristic Poincare section of this attractor, made by a plane perpendicular to the plane of figure 4a and passing through the dashed line is plotted on figure 4b.

Moreover, on this reconstructed surface the trajectories are not stochastically distributed. The order becomes apparent if we look at the chronology of the succcessive intersections in the Poincare map. Figure 4c shows the corresponding 1D map obtained by plotting the abscissa of an intersection as a function of the

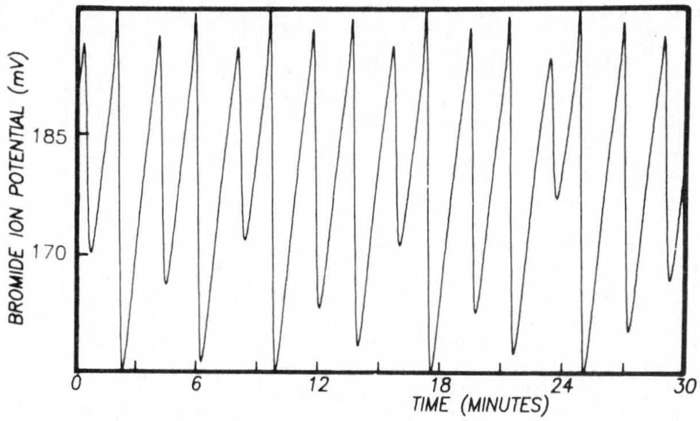

Figure 2: Time series of a chaotic regime.

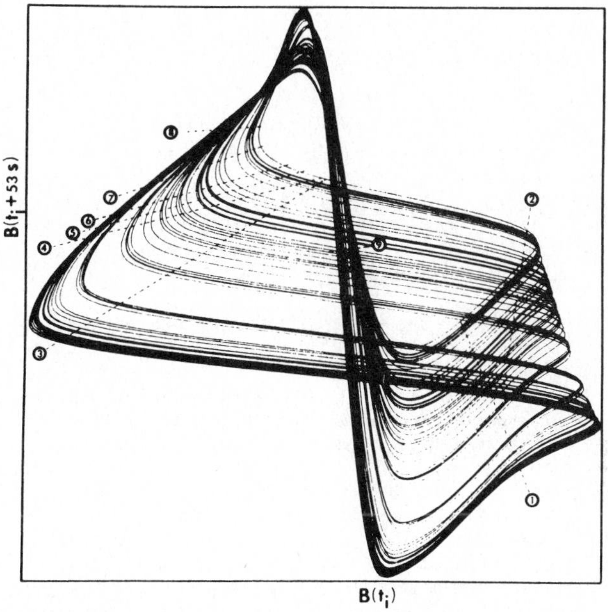

Figure 3: attractor corresponding to the time series figure 2

abscissa of the preceding one: All the data points appear to fall on a single valued curve.

Thus the system is completely deterministic: For any Xn the map determines Xn+1. However it is chaotic: Two trajectories initially chosen arbitrarly closed to each other will separate exponentially fast. This divergence can also be followed on the 1D

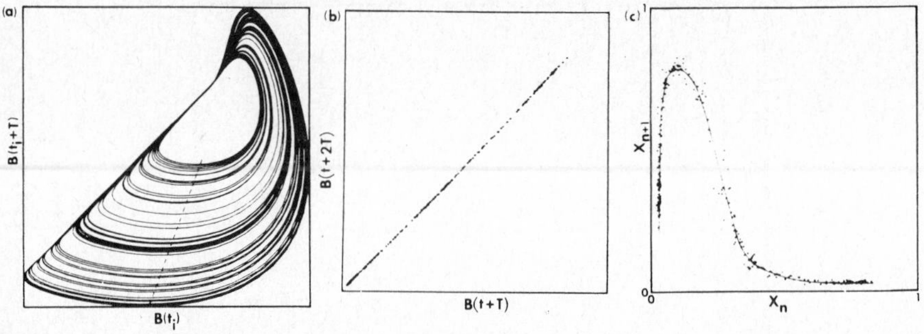

Figure 4: a)The attractor built from the time series of figure 2. b)Poincare section obtained by the intersection of trajectories of fig4a with a plane going through the dashed line and perpendicular to the plane of fig 4a. c)Next return map of the abscissa of figure 4b.

map (fig 4c) where it can be shown (11) that the Lyapunof exponent is:

$$\lambda = 1/n \sum_{i=1}^{i=n} \ln |f'(x_i)|$$

$f'(x_i)$ is the derivative of the map at x_i. The Lyapunof exponent calculated from the data figure 3c was found equal to .3 .1 .Its positive value guarantees the divergence of the trajectories and gives a measure of the sensitive dependance on initial conditions, a land mark in deterministic chaos.

In spite of their exponential divergence trajectories do not go to infinity : They are bounded in a small region of the phase space. To accomodate this contradiction the attractor must present a folding following the stretching. This succession of folding and stretching is evident on the successive Poincare sections (figure 5) along the attractor shown figure 2.

A new type of chaos: wrinkles on a torus.

I shall emphazise in this paragraph on recent results where we observe a direct transition from periodic to chaotic - quasiperiodic regime. When along line V (figure 1) the flow rate is increased the periodic regime gives yield to the regime depicted on figure 6. This regime looks quasi periodic , but a carefull analysis shows that the number of large amplitude oscillations inside each group varies with no apparent order between 24 an 27.

The reconstruction of the attractor is shown figure 7 (two different views). To give an idea of the shape of the attractor intersections of the trajectories with two planes are shown on figure7. It looks like a torus which can be shematized by a sphere with a hole going from pole to pole. This kind of torus had already been predicted by GUCKENHEIMER (13) and observed in simulations by W. LANGFORD (14). It appears as the result of the

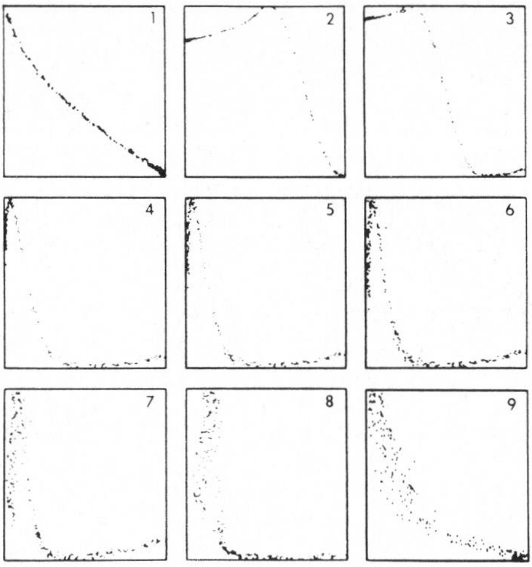

Figure 5: Successive Poincare sections on the attractor fig 2; the intersections on the attractor are the dashed lines on fig 2

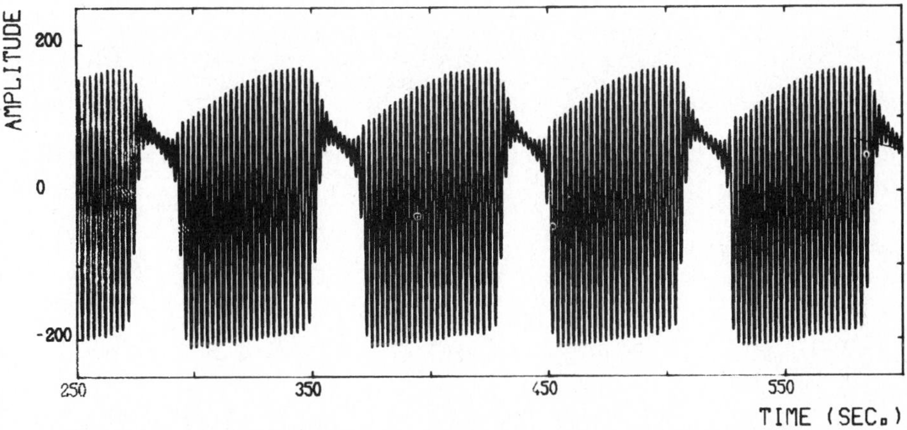

Figure 6:"Pseudo" quasi-periodic regime (only a tenth of the actual time series is shown).

interaction of two instabilities, namely a Hopf bifurcation an a steady state bifurcation.

Poincare sections 1,3 and 2,4 corresponding to the intersection of the torus by the two planes of figure 5 are presented figure 8a and 8b. A stretching in the outer part of the torus is evident between section 1 and 2, and a folding is also clear in sections 2,3 and 4. These stretching and folding are the translation of the chaotic character of the dynamics. The folding becomes more evident if we look at a Poincare map done by a plane

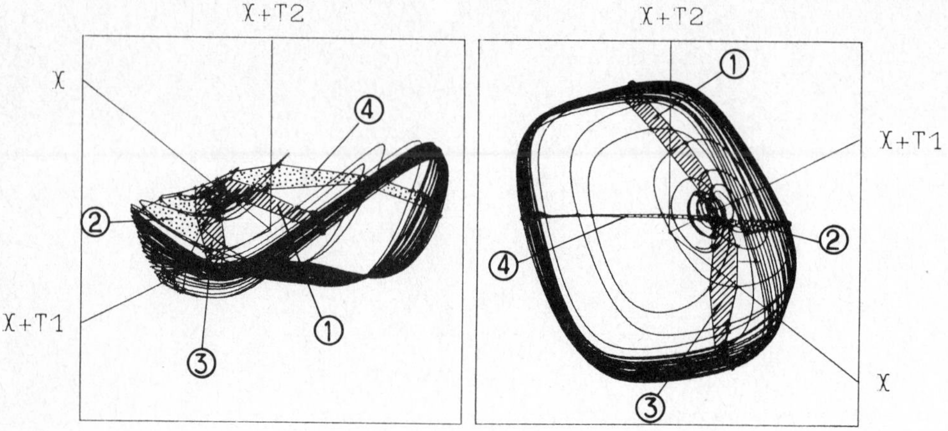

Figure 7: Two views of the attractor corresponding to the time series of figure 6.

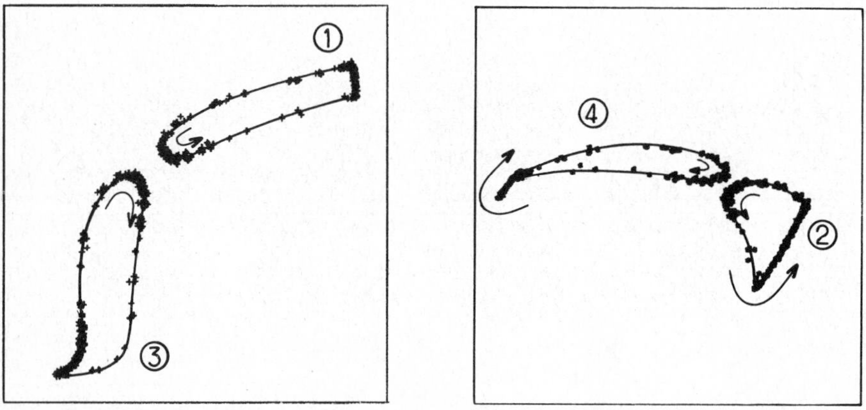

Figure 8: Poincare maps corresponding to the two planes shown fig 7a. Fig 8b is reduced by 5/6 compared to fig 8a

nearly tangent to the trajectories in the outer part of the torus. This particular section (figure 9) unfolds the trajectories at this location, and we can clearly distinguish the wrinkles on the torus.They are very similar to those observed by LANGFORD (15) in its non axisymetric case.

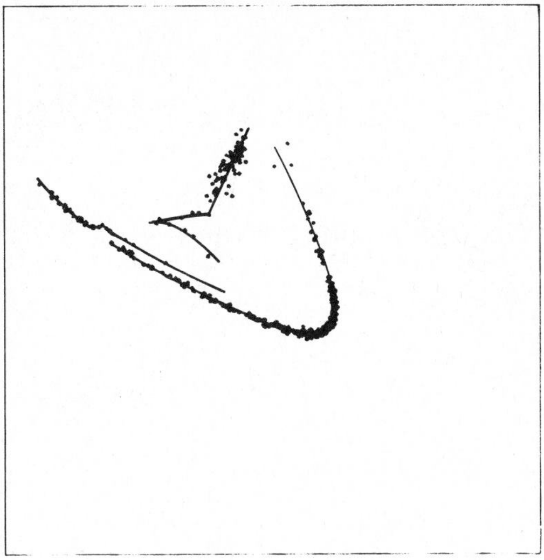

Figure 9: Poincare section between sections 1 and 2 of fig 7.

Conclusions.

On these two examples we have shown how the chemical dynamics observed in the B.Z. reaction illustrate some concepts of the non-linear dynamics. The interest of chemistry in this respect is the facility with which we can vary the constraints on the system and the great variety of the situations obtained. On the other hand the traditional experimental problem of having a precise and strictly constant flow of the chemicals is presently at the point to be solved. Then taking advantage of the very good signal to noise ratio in these experiments we hope to be able soon to produce precise data on the behavior near bifurcations points of codimension two or higher.

REFERENCES

1)-D. RUELLE: Some comments on chemical oscillations. N. Y. Acad. Sci. 35 (1973),66

2)-O. E. ROSSLER Chaos in abstract kinetics, two prototypes. Bull. Math. Biol. 39 (1977) 275

3)-O. E. ROSSLER Chaotic behavior in simple reactions systems. Z. Naturforsch 31a (1976) 259

4)-R. A. SCHMITZ, K. R. GRAZIANI, J. L. HUDSON Experimental evidence of chaotic states in the Belousov - Zhabotinskii reaction. J. Chem. Phys. 67 (1977) 3040

5)-C. VIDAL, J.-.C. ROUX, A. ROSSI, S. BACHELART Etude de la transition vers la turbulence chimique dans la reaction de Belousov - Zhabotinskii. C. R. Acad. Sci Paris 289C (1979) 73

6)-N. H PACKARD, J. P. CRUTCHIELD, J. D. FARMER, R. S. SHAW Geometry from a time series. Phys. Rev. Lett. 45 (1980) 712
7)-F. TAKENS Detecting strange attractor in turbulence. in Lectures Notes in Mathematics 366 (1981), Springer-Verlag.
8)-J.-C. ROUX, A. ROSSI, S. BACHELART, C. VIDAL Representation of a strange attractor from an experimental study of chemical turbulence. Phys. Lett. 77A (1980) 391
9)-J.-C. ROUX, R. H. SIMOYI Observation of a strange attractor. Physica 8D (1983) 257
10)-J.-C. ROUX, J. S. TURNER, W. D. McCORMICK, H. L. SWINNEY Experimental observation of complex dynamics in a chemical reaction. Non Linear Problems Present and Future ed. by A. R BISHOP, D. K. CAMPBELL, B. NICOLAENKO (1982) 409 (North Holland).
11)-J.-C. ROUX, H. L. SWINNEY Topology of chaos in a chemical reaction. Non Linear Phenomena in Chemical Dynamics ed. by C. VIDAL and A. PACAULT (1981) 38 (Springer-Verlag)
12)-J.-C. ROUX, H. L. SWINNEY, R. S. SHAW Strange attractor in a chemical system; a 16mm movie.
13)-J. GUCKENHEIMER On a codimension two bifurcation. Lectures Notes in Mathematics 898 (1980) 99 (Springer-Verlag).
14) W. F. LANGFORD A review of interactions of Hopf and steady state bifurcations. Non Linear Dynamics and Turbulence ed. G. I. BARENBLATT, G. IOOSS, D. D. JOSEPH (1983) 215 (Pitmann)
15)-W. F. LANGFORD This meeting. See also: Numerical studies of torus bifurcations, paper presented at the Conference on numical methods for bifurcation problems. Dortmund August (1983)

* Jean-Claude ROUX
Université de Bordeaux I
Centre de Recherche Paul Pascal
Domaine Universitaire
33405 Talence Cedex
FRANCE

Singularities & Dynamical Systems
S.N. Pnevmatikos (editor)

A SINGULARITY ANALYSIS OF INTEGRABILITY AND CHAOS IN DYNAMICAL SYSTEMS

Tassos C. Bountis

Clarkson College

U. S. A.

The analysis of movable singularities in the complex time plane of n first order, ordinary differential equations has led to a better understanding of the real time behavior of the solutions of dynamical systems. The requirement that these singularities be *poles* with n-1 free constants, i.e. the so-called Painlevé property, has identified many new completely integrable dynamical systems having, of course, no chaotic behavior whatsoever. On the other hand, the violation of the Painlevé property with the introduction of logarithmic terms at *higher orders* in the series expansions has identified many dynamical systems with only "weakly chaotic" behavior. In this paper, I review these results and discuss the methods of singularity analysis with the aid of illustrative examples.

1. INTRODUCTION

It is well-known that a great many problems of applied science often lead to the study of a system of ordinary differential equations of the form

$$\dot{x}_k = f_k(x_j;t) \qquad k,j=1,2,\ldots,n, \tag{1.1}$$

where $(\dot{\ })\equiv d(\ )/dt$. Such systems describe the time evolution of a (finite) set of dynamical variables $x_k(t)$ as they interact among themselves and with their environment according to a well-prescribed force law $\underset{\sim}{f}\equiv(f_1,f_2,\ldots,f_n)$. Eq. (1.1) describes a *Hamiltonian* system if the f_k's are derived from a Hamiltonian function, otherwise it will be referred to as a *non-Hamiltonian* system.

There is a long list of famous physical problems which require the solution of equations of the form (1.1). We shall enumerate here only a few, and refer the interested reader to the literature [1-9] for many more examples: From Hamiltonian Mechanics the motion of stars in the galaxy, the 3-(celestial) body problem, the confinement of plasma in nonuniform magnetic fields and the properties of nonlinear lattices immediately come to mind [1,3-7]. Non-Hamiltonian examples on the other hand include the onset of fluid turbulence in the Navier-Stokes equations, the competition of species in biology, chemical oscillations, relaxation in nonlinear circuits and multiple (electromagnetic) wave interactions in various media [2-4,6-9].

But where is the difficulty in solving eq. (1.1)? Since they describe a *deterministic* system, given an initial state $\{x_k(0); k=1,2,\ldots,n\}$, one should be able, in principle, to determine all later ($t > 0$) or earlier ($t < 0$) states by solving (1.1), if not analytically, at least numerically using some of the advanced present-day computer technology. After all, how unpredictable could the behavior of these solutions or *orbits* turn out to be? Apart from some "pathological" cases, shouldn't most solutions do something reasonable like e.g. oscillate near simple periodic orbits or be attracted to fixed points or limit cycles as $t \to \pm\infty$?

As was recently - and very eloquently! - discussed by J. Ford [10] all of the above simple expectations are actually no more than naive wishful thinking. It takes but the simplest *nonlinear* coupling of the x_k's in (1.1) to give rise to non-zero measure regions of phase space, in which the solutions of (1.1) depend *extremely sensitively* on the choice of initial conditions. These so-called regions of *chaotic* or "stochastic" behavior [1-10] are generic in Hamiltonian systems [11,12] as they are present for "most" nonlinear force laws $\underset{\sim}{f}$ which couple the x_k variables in a non-trivial way. In non-Hamiltonian systems on the other hand, they often form certain highly complicated, so-called *strange attractor* sets on which the motion appears to be exceedingly irregular and unpredictable.

It takes, therefore, only a moment of reflection to realize that in these chaotic (resp. "strangely attracting") regions even the most sophisticated numerical techniques available today cannot accurately follow the solutions of (1.1) for long times. Even the slightest, unavoidable errors of finite-digit arithmetic in regions where nearby solutions separate exponentially in time, would also have to grow exponentially. Thus, in the end, our calculation will tend to "wander" from solution to solution, over the full extent of the chaotic region, rather than follow the precise evolution of one individual trajectory.

This brings us to the main theme of this paper: How can we use mathematical criteria to determine whether a given dynamical system (1.1) has "small" or "large" regions of chaotic behavior within a well-defined part of its available phase space? How does the "size" of chaotic regions depend on the values of certain physical parameters of the problem (masses, coupling constants, Reynolds number) included in the function $f_k(x_j;t)$? And, in which non-trivial examples of (1.1) can one guarantee the complete *absence* of chaos, or strange attractors?

As one might guess, at the present time, we are far from being able to answer completely such ambitious and far reaching questions. Moreover, most mathematical techniques which have met with some success in this direction, have advanced out of step with efforts to provide a rigorous justification for them. Nevertheless, it is the opinion of this author that a search for useful and correct (if non-rigorous) mathematical criteria of chaotic behavior must continue, in view of their important applications to a great many fields of applied science.

In this review, I would like to concentrate on one such mathematical criterion and describe its results to date on a number of illustrative examples. Including myself, there have been several other researchers who have worked on this criterion and I shall try to do justice to their valuable contribution by briefly describing their results here. The interested reader will, of course, be well advised to consult the rather complete set of references listed at the end of this paper.

The mathematical criterion I am referring to can be described as the analysis of *movable singularities* of the solutions of (1.1) in the *complex* time (t-) plane. By "movable" I mean initial condition dependent, in the sense that the actual *location* of a singularity in the (complex) t-plane, $t = t_0$, will always be an arbitrary constant to be specified by the initial conditions $x_k(0)$ of each orbit. The importance of these movable singularities for the cases n = 1,2 in (1.1) has been amply and rigorously demonstrated in the mathematical literature by the classical work of Fuchs, Painlevé, Gambier and others [13,14]. In some sense, the work described here is an attempt to extend these results to higher order (n > 2) systems, as yet, however, without the rigor and completeness of cases n=1 and 2.

Thus, I shall adopt in this paper a physicist's point of view and describe first, in section 2, the success of the (movable) singularity analysis in identifying *completely integrable* systems (1.1), which exhibit no chaotic behavior whatsoever. In fact, it is in this realm of completely integrable systems that the only rigorous justifications of singularity analysis have so far been provided. After defining what I mean by complete integrability for Hamiltonian as well as non-Hamiltonian systems, I will venture to show how it is connected with the *Painlevé property* , i.e. that the solutions of (1.1) have no (movable) singularities other than *poles*, near which they possess Laurent series expansions

$$x_k(t) = (t-t_0)^{p_k} \sum_{r=0}^{\infty} a_r^{(k)} (t-t_0)^r, \qquad (1.2)$$

with (some negative) integers p_k and n-1 free constants $a_r^{(k)}$ to be specified by the initial conditions.

In section 3, I will relax the conditions of the Painlevé property and allow logarithmic terms to enter in the asymptotic expansions of $x_k(t)$ near a singularity at $t = t_0$. We will thus discover in many Hamiltonian systems that if logarithms enter in the *higher* order terms of (1.2) the chaotic regions are much *smaller* than they are in comparable systems, whose asymptotic expansions contain $\log(t-t_0)$ terms already at leading order. A similar situation appears to exist in non-Hamiltonian systems as well, where many *partially integrable* examples have been found by requiring that the (movable) singularities of the solutions be all *pole like* with $\log(t-t_0)$ terms entering only in the higher orders of the asymptotic expansions.

The importance of logarithms arising in higher and higher order terms in (1.2) is examined in more detail in section 4, on the simplest Hamiltonian systems exhibiting chaotic behavior: a one-parameter family of periodically driven, anharmonic oscillators. Finally, in section 5, the results of other researchers will be mentioned and a more comprehensive discussion will be given of the singularity analysis, its achievements to date and its future prospects for becoming a useful method for studying chaotic properties of dynamical systems.

2. COMPLETE INTEGRABILITY AND THE PAINLEVÉ PROPERTY

While it is true that the rigorous and exhaustive work of Fuchs, Painlevé, Gambier et. al [13,14] established the relevance of the Painlevé property for first and second order eq. (1.1), it was Sonya Kowalevskaya who first used it in 1888 [15] to completely integrate a dynamical system of physical significance: the rotating rigid body with moment of intertia ratios 1:1:2.

With a widely growing interest in dynamical systems and nonlinear evolution equations in the 1970's these classical results were revived in a somewhat unexpected way: Ablowitz, Ramani and Segur

[16,17] discovered that several of the 6 Painlevé *irreducible* (i.e. not solvable by the known transcendents) equations arise after similarity transformations, in nonlinear evolution equations which are *exactly solvable* by the Inverse Scattering Transform method [18].

Following this discovery, Segur [19] and Bountis, Segur and Vivaldi [20], inspired by the work of Kowalevskaya, demonstrated the usefulness of the Painlevé property by completely integrating the Lorenz equations [19] and several examples of Hamiltonian systems [20], respectively. As the notion of complete integrability is more firmly established in Hamiltonian systems I shall begin by discussing them first and turn to non-Hamiltonian systems in the second part of this section.

A. Hamiltonian Systems

Taking $n = 2N$, even, in (1.1) and denoting by $p_k \equiv x_{N+k}$ the momentum variable conjugate to x_k, $k=1,2,\ldots,N$, we can write equations (1.1) as

$$\dot{x}_k = \partial H/\partial p_k, \quad \dot{p}_k = -\partial H/\partial x_k, \qquad k=1,2,\ldots,N \tag{2.1}$$

provided, of course, that the f_k's are such that the Hamiltonian function $H = H(x_j, p_j; t)$, $j = 1,2,\ldots,N$, exists. We then say that (2.1) are the equations of motion of an N degree of freedom Hamiltonian system.

In the case of the (N=)1 degree of freedom Hamiltonian

$$H(x,p) = p^2/2 + V(x) \tag{2.2}$$

describing the motion of a unit mass particle in the potential $V(x)$, (2.1) yields the second order ordinary differential equation (o.d.e.) for $x(t)$

$$x = -\partial V/\partial x = -x + x^2 \tag{2.3}$$

with $V(x) = x^2/2 - x^3/3$, giving rise to bounded oscillations for $x(0) \varepsilon [-0.5,1]$ and $p(0) = 0$. The general solution of this problem is well known: Making use of the fact that $H = E = \text{const}$ is an *integral* of the motion, eq. (2.3) can be integrated once to yield the final quadrature

$$t - t_0 = \int dx\,[E-V(x)]^{-1/2}, \tag{2.4}$$

where t_0, E are the two arbitrary constants specified by the initial conditions.

More generally, an N degree of freedom autonomous and bounded Hamiltonian system, possessing N independent, single-valued integrals of the motion

$$F_k(x_j,p_j) = E_k = \text{const}, \quad j,k=1,2,\ldots,N, \qquad (2.5)$$

in *involution*, i.e. with vanishing Poisson brackets $\{F_k,F_j\} = 0$, will be called *completely integrable*, in the sense that it can be integrated by quadratures like (2.4). This is the content of the celebrated Liouville-Arnol'd theorem [11], which also proves that *all* solutions of such a system are quasiperiodic and lie on N-dimensional tori coordinatized by Action-Angle variables (I_k,ϕ_k), $k=1,2,\ldots,N$ [1,3,7,11].

Consider now among all such completely integrable Hamiltonian systems, those which also satisfy the following two requirements: their integrals (2.5) are *rational* functions of the phase space coordinates x_j, p_j and their tori are part of compact, *complex* tori on which the motion is linear. These form the class of the so-called *algebraically* completely integrable Hamiltonian systems, for which Adler and van Moerbeke have proved that all solutions must *necessarily* possess the Painlevé property [21,22]; and although the sufficiency condition is not yet proven, it is certainly expected to be true [23]. Finally, the precise connection between algebraic complete integrability and the Painlevé property has been recently elucidated and explained in detail in the work of Yoshida [24].

From the viewpoint of direct singularity analysis one starts by examining the leading order behavior of the solutions of (2.3) near all possible movable singularities at $t=t_0$

$$x(t) \sim C\tau^p(\log\tau)^q, \quad \tau \to 0 \ (\tau \equiv t-t_0) \qquad (2.6)$$

where t_0 - the location of the singularity in the complex t plane- is the *first free* constant of an expansion starting with (2.6). Inserting (2.6) in (2.3) determines p,q uniquely: $p=-2$, $q=0$; $C=6$. We now look for the second free constant in this expansion by substituting for x in (2.3)

$$x = 6\tau^{-2} + \ldots + a_r\tau^{-2+r}, \quad \tau \to 0, \qquad (2.7)$$

cf. (1.2), and equating terms proportional to a_r. This readily yields

$$(r+1)(r-6)a_r = g_r(a_j), \quad 0 \leq j < r. \qquad (2.8)$$

A free constant is expected to arise at every r value, which makes the coefficient of a_r in (2.8) vanish. This happens at $r=-1$, which corresponds to the free location of the singularity, and $r=6$: whence, the second free constant will be a_6, provided, of course, that the r.h.s. of (2.8) also vanishes at r=6. If $g_6(a_j) \neq 0$ then logarithmic terms would have to be introduced at that order to capture the second free constant [25,26].

In the case of (2.3), of course, $g_6(a_j) = 0$ and (2.7) becomes the Laurent series expansion near the poles of the Jacobi elliptic functions, which are the known, exact solutions of (2.3) [27]. Had we introduced, however, so much as a simple, periodic, driving force F cost, and considered instead of (2.3),

$$\ddot{x} = -x + x^2 + F\cos(\tau + t_0), \quad (\tau \equiv t - t_0), \qquad (2.9)$$

terms of the form $F\cos t_0$ and $F\sin t_0$ would make (as $\tau \to 0$) $g_6(a_j) \neq 0$ in (2.8), giving rise to movable logarithmic singularities in (2.7) at r=6!

There is, of course, a world of difference between eq. (2.3) and (2.9) for $F \neq 0$; we know everything about (2.3) but very little about (2.9). Because of the periodic force term F cos t a second frequency has been introduced in the problem, producing with the x-oscillations an infinity of resonances, which are *nonlinearly* coupled by the x^2 term in (2.9). No global (in t) solutions are known which remain valid over a *range* of initial conditions and new types of trajectories can be observed in phase space with wild and exotic properties: they are located near *unstable* periodic orbits [1-3,7,11,12] and their irregular and *chaotic* behavior is best evidenced by their intersections with the Poincaré map $(x, \dot{x})$ at $t = 2\ell\pi$ (ℓ integer) see fig. 1 below.

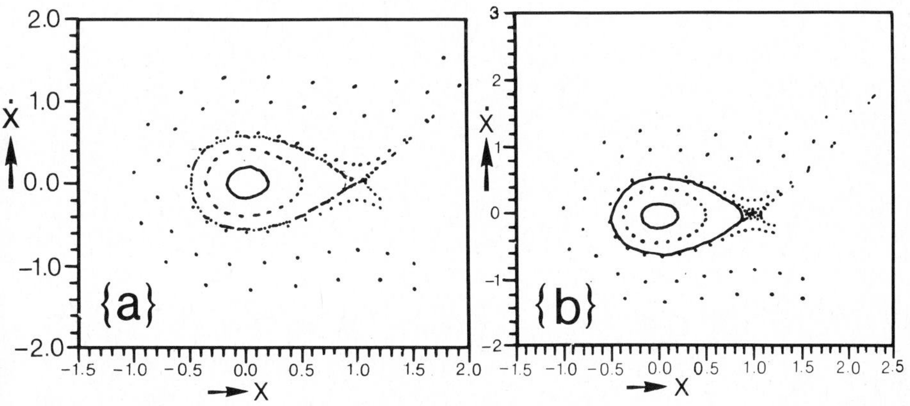

Figure 1. Surface of Section Polots of (2.9) for: (a) F=0, (b) F≠0; Note the chaotic region forming near the unstable fixed point in (b).

Thus, we may call Hamiltonian systems, like (2.9) *non-integrable* (note that its Hamiltonian is *not* a constant of the motion) since they do not satisfy the conditions of the Liouville-Arnol'd theorem for complete integrability. We shall return to their fascinating chaotic properties when we discuss logarithmic singularities in the next section. Let us now resume our discussion of completely integrable Hamiltonian systems:

Assuming the sufficiency of the Painlevé property for complete integrability, many new, completely integrable Hamiltonian systems have been identified to date [20,24] by the following program: Write for the leading order behavior of the solutions of (2.1) near a (movable) singularity at $\tau \equiv t - t_0 = 0$,

$$x_k \sim \tau^{n_k} (\log\tau)^{n_k'}, \quad p_k \sim \tau^{m_k} (\log\tau)^{m_k'}, \quad \tau \to 0, \qquad (2.10)$$

If the nonlinear terms in the equations of motion yield $n_k' = m_k' = 0$ and n_k, m_k integers (some negative) for all $k = 1,2,\ldots,N$, we have a candidate for complete integrability. Expanding further eq. (2.10)

$$x_k = \tau^{n_k} \sum_{r=0}^{\infty} a_r^{(k)} \tau^r, \quad p_k = \tau^{m_k} \sum_{r=0}^{\infty} b_r^{(k)} \tau^r, \qquad (2.11)$$

substitute (2.11) in (2.1) and derive a set of *linear* equations for $a_r^{(k)}$ and $b_r^{(k)}$

$$\underset{\approx}{M}(r) \underset{\sim}{C}_r = \underset{\sim}{g}_r (a_j^{(k)}, b_j^{(k)}) \qquad 0 \le j < r \qquad (2.12)$$

where $C_r^{(k)} \equiv a_r^{(k)}$ and $C_r^{(k+N)} \equiv b_r^{(k)}$ for $k=1,2,\ldots,N$ and $\underset{\approx}{M}(r)$ is a $2N \times 2N$ matrix which also contains the parameters of the problem: masses, spring constants etc.

In analogy with the one degree of freedom example (2.8), in order to have $2N - 1$ arbitrary coefficients in (2.11) we must require that the equation

$$\det \underset{\approx}{M}(r) = 0, \qquad (2.13)$$

possess $2N-1$ positive integer roots r (and $r = -1$). In addition the rhs of (2.12) must satisfy a *compatibility* condition analogous to $g_r(a_j) = 0$ in (2.8): If $\underset{\approx}{M}_j(r)$ is the matrix $\underset{\approx}{M}(r)$ with its jth column replaced by g_r, $\det \underset{\approx}{M}_j(r) = 0$ at the same roots r as (2.13).

All these conditions yield constraints for the parameters of the problem, which have been shown to lead always to algebraically completely integrable systems. Among the most famous Hamiltonians for which completely integrable cases have been identified by the above method are: The Hénon-Heiles system, coupled anharmonic oscillators [20], the Toda lattice [20,22,24] and geodesic motion on SO(4) [21].

Finally, two remarks are in order here: first, even if some of the n_k, m_k powers in (2.10) are *not* integers but rationals, e.g. $n_k = s_k/q_k$, the system could still be algebraically completely integrable (with $X_k \equiv x_k^{q_k}$) provided the roots of (2.13) are all integers. Secondly, we emphasize that for algebraic complete integrability the Painlevé property must be satisfied near every possible singularity (2.10) even though for some (but not all) of them, more than one of the 2N integer roots of (2.13) are negative.

B. Non-Hamiltonian Systems

In dynamical systems whose equations of motion

$$\dot{x}_k = f_k(x_j;t), \qquad k,j=1,2,\ldots,n$$

are not derived from a Hamiltonian, no analogue of the Liouville-Arnol'd theorem exists and thus even the definition of complete integrability is not clear. The methods of singularity analysis, however, are still applicable and, in fact, turn out to suggest such a definition as we shall see below.

Segur [19] was the first to work out the singularity analysis for a very famous non-Hamiltonian system: the Lorenz equations [28]

$$\dot{x}_1 = \sigma(x_2-x_1), \quad \dot{x}_2 = -x_2 + \rho x_1 - x_1x_3, \quad \dot{x}_3 = -bx_3+x_1x_2, \tag{2.14}$$

which are known to exhibit wildly chaotic behavior over large ranges of parameter values [29]. Segur was able to determine that (2.14) possesses the Painlevé property in the following 4 cases:

(i) $\sigma=0$: Equations (2.14) become linear and integration is immediate,

(ii) $\sigma=1/2$, $\rho=0$, $b=1$: Two integrals can easily be found here

$$x_2^2 + x_3^2 = C_1e^{-2t}, \quad x_1^2 - x_3 = C_2e^{-t} \tag{2.15}$$

with the aid of which (2.14) is reduced to a final quadrature completed in terms of elliptic functions,

(iii) $\sigma=1$, $\rho=1/9$, $b=2$: In this case, there is one simple integral $x_1^2 - 2x_3 = C\exp(-2t)$, which reduces (2.14) to the *second* of the 6 irreducible (second order) transcendental equations of Painlevé [13,14], while with

(iv) $\sigma=1/3$, $b=0$ (any ρ), eq. (2.14) lead, also with the aid of one integral, to the *third* Painlevé transcendental equation.

Recently, many other third (and higher) order non-Hamiltonian systems of physical significance like (2.14), have been studied by the methods of singularity analysis [30-32]. These systems include the Lotka-Volterra equations, the Rikitake model and several 3-wave interaction systems. New, completely integrable cases of these systems have been discovered by requiring that they possess the Painlevé property [30-32]. Thus, as a result of our analysis, we have been led to define as completely integrable a non-Hamiltonian system, which fulfills any *one* of the following 3 conditions:

(a) Either it can be reduced with the aid of n-1 integrals to a final quadrature like the Lorenz case (ii) above,

(b) Or, it can be transformed by local changes of variables to a set of *linear* o.d.e.'s (with variable coefficients) as in the case of the Lotka-Volterra equations

$$\dot{x}_1=Cx_1x_2+x_1x_3, \quad \dot{x}_2=Ax_2x_3+x_2x_1, \quad \dot{x}_3=Bx_3x_1+x_3x_2, \tag{2.16}$$

with $B = -1/(A+1)$, $C = -1-1/A$ [30,31],

(c) Or, it can be reduced to one of Painlevé's second (or higher) order transcendental equations, as with cases (iii) and (iv) above of the Lorenz equations.

It must be said, of course, that any definition of complete integrability bears to some extent a reflection of the state of our mathematical ability to integrate a set of equations, like the o.d.e.'s (2.14) of the case at hand. In that sense, the definition offered above should be seen more as an attempt to define our present state of knowledge and hopefully serve to pave the way for future investigations. As for more rigorous results on the connection between integrability and the Painlevé property in non-Hamiltonian systems, this author is only aware of the work of P. Winternitz and co-workers [33,34]: They find systems of o.d.e.'s (1.1) which have nonlinear superposition principles, well-defined Lie Algebraic structure and are linearizable, whence their Painlevé property follows immediately.

3. LOGARITHMIC SINGULARITIES AND "WEAKLY" CHAOTIC BEHAVIOR

In section 2, we saw that there are two ways by which logarithmic terms can enter in the series expansions near a movable singularity: One is already at *leading order*, cf. (2.10), and one is at higher orders through either a violation of the g_r-compatibility conditions [like $g_r(a_j) \neq 0$ in (2.8)] or a multiplicity of the roots of eq. (2.12) [16,17]. In all examples, where these two cases could be compared we have found that the latter one identified dynamical systems with significantly "weaker" chaotic properties.

A. "Weakly" Chaotic Hamiltonian Systems

While analyzing the singularities of the solutions of the Quartic Lattice Hamiltonian

$$H_{QL} = \frac{1}{2}(p_1^2 + p_2^2) + \frac{1}{4}(q_1^4 + q_2^4 + \varepsilon(q_1-q_2)^4) \qquad (3.1)$$

it was noted that only for $\varepsilon = 1/4$ this system possessed pole-like expansions in which the 3 arbitrary coefficients (the 4th is t_0) entered at positive integer order r. These expansions were explicitly carried out [25] and only powers of $\log\tau$ terms were found at higher orders. Studying numerically the intersections of many orbits of (3.1) with surfaces of section it was clearly demonstrated that at that particular value $\varepsilon = 1/4$ the chaotic regions of (3.1) became *minimal*. On the contrary, studying the surfaces of section of the Hénon-Heiles Hamiltonian

$$H_H = \frac{1}{2}(p_1^2 + p_2^2 + Aq_1^2 + Bq_2^2) - q_1^2 q_2 + \varepsilon q_2^3 \qquad (3.2)$$

at the special value $\varepsilon = -2/3$, where again the roots of (2.12) are positive integers, but $\log\tau$ terms appear at *leading order* [with $\log\log\tau$ at higher orders] no appreciable change in the large scale size of the chaotic regions was observed [25] (at 70% of the escape energy).

I then turned to a more detailed study of Hamiltonian systems, whose potential is "perturbed" by an additive linear term, introducing a simple additive constant in the force equations. Starting with a 3-particle Toda lattice with the Painlevé property, the addition of linear terms yielded a (non-integrable) Hamiltonian

$$H_{TL} = \frac{1}{2}(p_1^2 + p_2^2 + p_3^2) + e^{q_1-q_2} + e^{q_2-q_3} - q_1 + q_3 \qquad (3.3)$$

which is much *less* chaotic (at equal total energy H=E) than other perturbations of the Toda lattice like the variation of the masses [35].

Finally, I compared Hamiltonians whose axially symmetric potential is "perturbed" by the same linear term [26]

$$H_1 = \frac{1}{2}(p_1^2 + p_2^2 + q_1^2 + q_2^2) + \frac{1}{4}(q_1^2 + q_2^2)^2 + q_2 = E \qquad (3.4a)$$

$$H_2 = \frac{1}{2}(p_1^2 + p_2^2) + \log(q_1^2 + q_2^2 + 1) + q_2 = E \qquad (3.4b)$$

but whose analytic structure is totally different: H_1 has pole-like singularities and only $\log\tau$ terms at higher orders whereas H_2 has $\log\tau$ at lowest order [with log $\log\tau$ terms at higher orders]. I found at equal energies E that the main chaotic regions of H_1 were considerably *smaller* than those of H_2, see fig. 2 below.

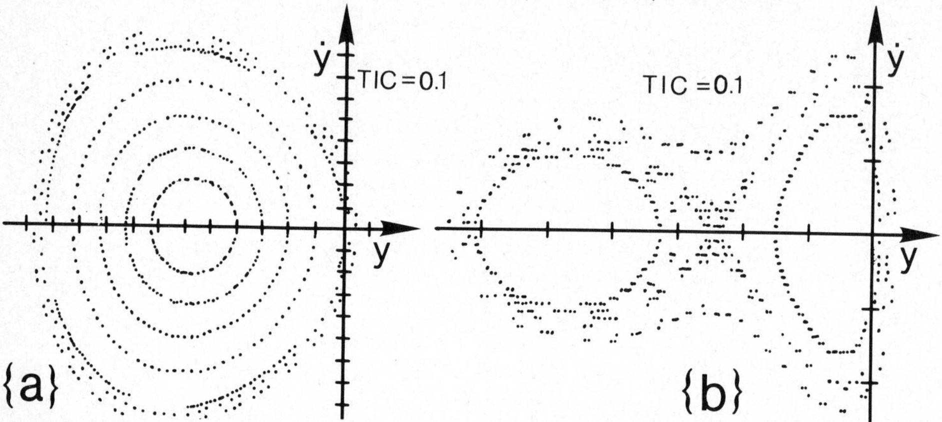

Figure 2. Surfaces of Section of (a) H_1 and (b) H_2 at one energy unit above the minimum. Note the large scale chaotic region in (b).

B. Partially Integrable Non-Hamiltonian Systems

It is interesting to ask whether pole-like singularities (with n-1 free constants) and only $\ell og\tau$ terms at higher orders, can lead to a weaker form of chaotic behavior in *non-Hamiltonian* systems. Our evidence so far suggests that this is indeed the case, at least in a number of simple but non-trivial models. Consider, for example, the Lorenz system (2.14) with:

(a) $b = 2\sigma$: In this case one finds one integral $x_1^2 - 2\sigma x_3 = C\exp(-2\sigma t)$, which as $t\to\infty$ restricts the motion on a two-dimensional paraboloid, on which no chaotic behavior is possible, and

(b) $b=1$, $\rho=0$: Here one also finds one integral $x_2^2 + x_3^2 = C\exp(-2t)$, but no other such simple integral (polynomial in the x_k's and analytic in t) appears to exist.

Such partially integrable cases have been discovered in several non-Hamiltonian systems of physical significance like the Rikitake dynamo [30,32]

$$\dot{x}_1 = -\mu x_1+\beta x_2+x_2x_3, \quad \dot{x}_2 = -\mu x_2-\beta x_1+x_1x_3, \quad \dot{x}_3 = -x_1x_2 + \alpha \qquad (3.5)$$

for (a) $\beta=0$, $\alpha\mu\neq0$ and (b) $\mu=0$ (any α,β) and the 3-Wave Interaction

$$\dot{x}_1 = \gamma x_1+\delta x_2+x_3-2x_2^2, \quad \dot{x}_2 = \gamma x_2-\delta x_1+2x_1x_2, \quad \dot{x}_3 = -2x_3-2x_3x_1, \tag{3.6}$$

for $\gamma=-1$ and any δ. In this latter case we only found the integral $x_1^2+(x_2+\delta/2)^2 + x_3 = C\exp(-2t)$.

From the viewpoint of singularity analysis, what all these partially integrable cases have in common is that their solutions have pole-like singularities with 2 free constants at non-negative integer orders r, cf. (2.12), at which $\log\tau$ terms appear in the expansions. For example, in the $\gamma=-1$ case of (3.6) above these expansions are found to be [30]

$$x = \frac{1}{\tau} - \frac{1}{2} + a_1\tau+\ldots, \quad x_2 = \frac{\delta}{2} -\delta\tau^2\log\tau+b_2\tau^2+\ldots, \quad x_3 = \frac{-1}{\tau^2} + \frac{1}{\tau} + C_0+\ldots$$

where b_2 and a_1 (or C_0) are the 2 free constants.

Finally, we studied an nth order generalization of a system due to Rössler [30]

$$\dot{x}_1 = \sum_{j=1}^{n} x_j, \quad \dot{x}_n = 1+x_1x_n, \quad \dot{x}_k = \sum_{j=1}^{n} a_{kj}x_j, \quad k = 2,3,\ldots,n-1, \tag{3.7}$$

which is known to exhibit wildly chaotic behavior over some range of its parameter values for n=3 [7]. We found that the requirement that (3.7) possess pole-like singularities (with $\log\tau$ terms at higher orders) leads to the condition

$$\underset{\approx}{B} \cdot \underset{\sim}{a} = \underset{\sim}{0}, \qquad \underset{\sim}{a} \equiv (1, a_{2,n},\ldots,a_{n-1,n})^T, \tag{3.8}$$

$\underset{\approx}{B} \equiv (b_{kj})$, which is *identical* with that of the existence of n-1 integrals of (3.7) of the form

$$\sum_{j=1}^{n-1} b_{kj}x_j = C_k\exp(\lambda_k t), \qquad k = 1,2,\ldots,n-1, \tag{3.9}$$

where λ_k's are the eigenvalues and $\underset{\sim}{b}_k \equiv (b_{k,1},\ldots,b_{kn-1})^T$ the eigenvectors of the matrix A with elements $\underset{\approx}{A}_{kl} = 1$ and $A_{kj} = a_{jk}$, $j=2,3,\ldots,n-1$, $k=1,2,\ldots,n-1$.

4. LOGARITHMS IN DRIVEN ANHARMONIC OSCILLATORS

We saw in the previous section that poles at leading order with logarithmic terms at higher orders can distinguish cases with weakly chaotic behavior. The question then naturally arises: Does it matter at which order the logarithms enter in the series expansions? Is it true - as one might expect from the above analysis - that the *higher* the order at which the painleve property is destroyed, the *less* chaotic the overal motion of the system? That this indeed appears to be the case, is the result of the numerical experiments described in this section.

Consider the following one-parameter families of anharmonic potentials

$$V_1(x) = \frac{1}{2}x^2 - A\frac{2k-1}{4k-1}x^{(4k-1)/(2k-1)} \qquad (4.1a)$$

$$k = 1,2,3,\ldots$$

$$V_2(x) = \frac{3}{2}x^2 - \frac{2k-1}{6k-1}x^{(6k-1)/(2k-1)} \qquad (4.1b)$$

where A in (4.1a) is a constant and k a positive integer parameter, see fig. 3. A unit mass particle driven by an oscillatory force

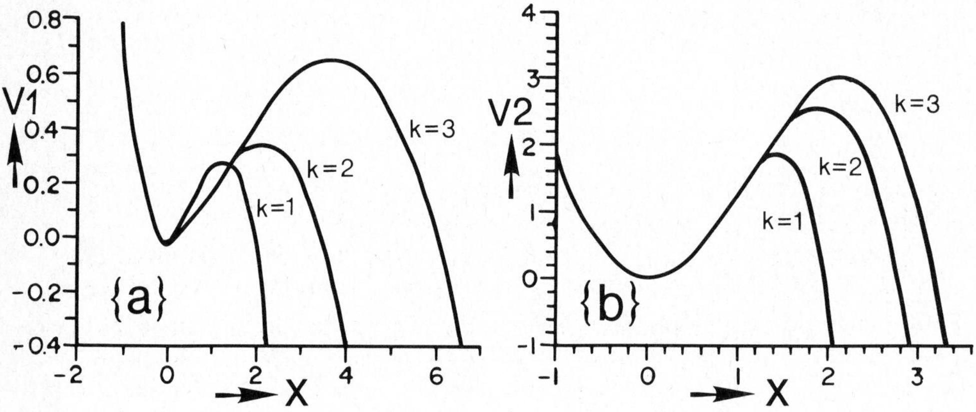

Figure 3. The potentials $V_1(x)$, $V_2(x)$ for k = 1,2,3.

moves in these two potentials, according to the equations

$$(-\partial V_1/\partial x=) \quad \ddot{x} = -x + Ax^{2k/(2k-1)} + F\cos 2t \qquad (4.2a)$$

$$(-\partial V_2/\partial x=) \quad \ddot{x} = -3x + x^{4k/(2k-1)} + F\cos 2t \qquad (4.2b)$$

respectively, where F is also a fixed constant.

Analyzing the singularities of (4.2a,b), as was done in section 2, one finds that the two nonlinear terms determine respectively the leading order behaviors

$$(a) \quad x = c_1\tau^{2-4k}+\ldots, \qquad (b) \quad x = c_2\tau^{2(1-2k)/(2k+1)}+\ldots, \qquad (4.3)$$

as $t-t_0 \equiv \tau \to 0$. Eq. (4.2a) therefore has pole-like singularities while (4.2b) has algebraic branch points, Substituting now $x = C\tau^p + a_r\tau^{p+r}$ in eq. (4.2) to find the order at which the *second* free constant enters we obtain in both cases

$$(r+1)(r+2p-2)a_r = g_r(a_j), \quad 0 \leq j < r \qquad (4.4)$$

cf. (2.8), where p=2-4k for (4.3a) and p=2(1-2k)/(2k+1) for (4.3b).

It is clear from (4.4) that the second free constant enters at r=2-2p. Owing to the presence of the driving term, however, in (4.2), the r.h.s. $g_r(a_j) \neq 0$ at that order (see section 2). Thus logarithmic terms will in general accompany the arbitrary coefficient a_r at r = 2-2p and the series expansions in the two cases will look like:

$$x = C_1\tau^{2-4k} + \ldots + a_r\tau^{8k}\log\tau + \ldots \qquad (4.5a)$$

$$x = C_2\tau^{2(1-2k)/(2k+1)} + \ldots + a_r\tau^{2(4k-1)/(2k+1)}\log\tau + \ldots \qquad (4.5b)$$

where C_1, C_2 are fixed constants, whereas the a_r is free to be specified by the initial conditions in each case.

Note in (4.5a) that as k increases (k=1,2,3,...) the leading order singularity of (4.2a) remains a pole, while the logarithms enter at higher and higher orders. On the other hand, the analytic structure of the solutions of (4.2b), near a novable singularity at $t=t_0$, does not change much with increasing k: As is evident from (4.5b), as $k\to\infty$, x develops a pole $\to \tau^{-2}$ at lowest order with logarithms entering at $r\to 6$, cf. (2.7)-(2.9).

Let us now turn to the *real* time behavior of the solutions of these two models to see whether it might possibly reflect the difference of their singularities in the complex time plane. To do so, we integrate equation (4.2) numerically for several initial conditions and plot the intersections of the orbits with the x, $\dot{x}$ surface of section at $t = \ell\pi$ (ℓ=1,2,3,...) as in section 2.A. The results shown in figure 4 below, clearly, justify our expectations: The chaotic regions of (4.2a) decrease in size as k increases whereas those of (4.2b) remain practically unaffected.

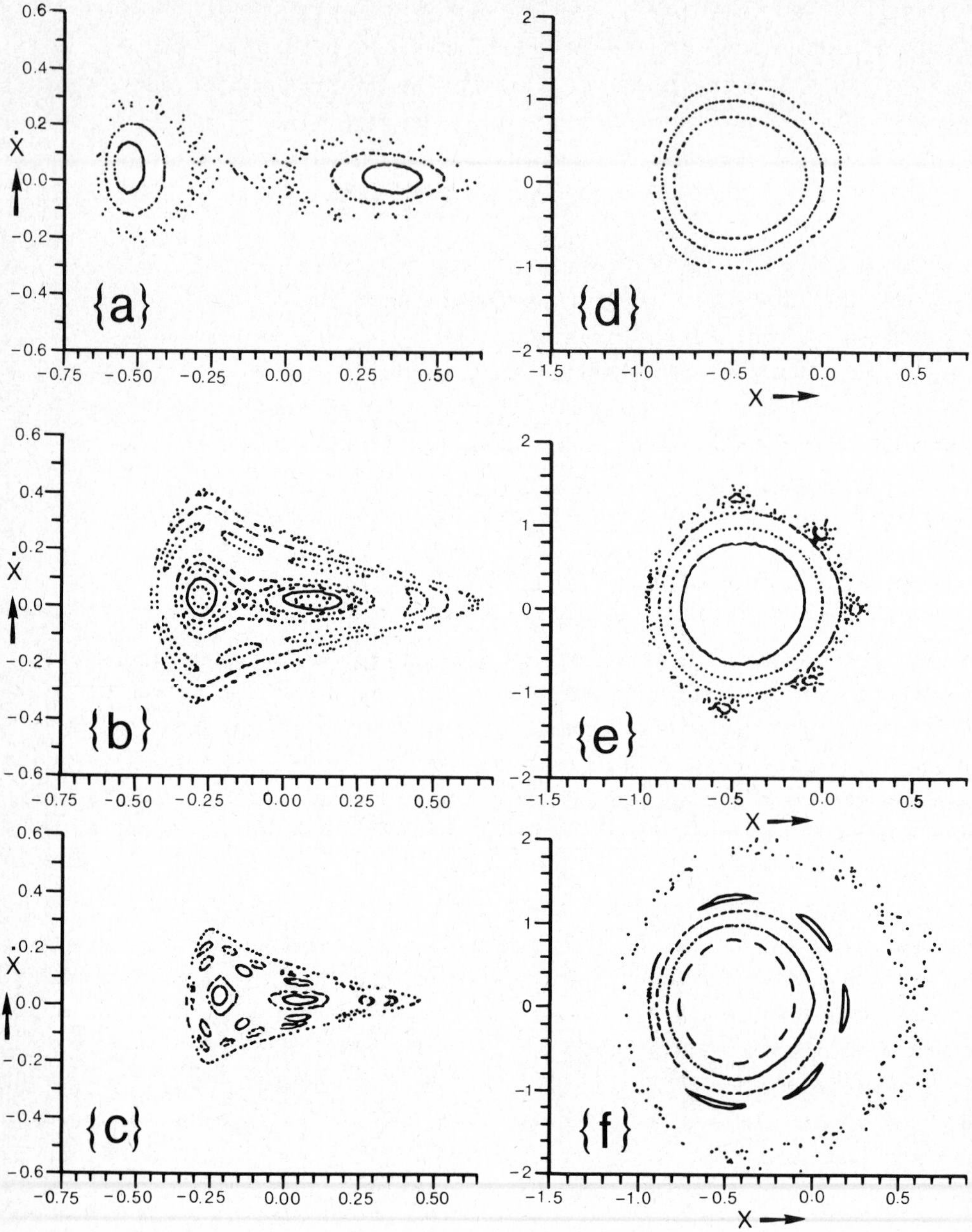

Figure 4. Surfaces of section of: (a)-(c) Equation (4.2a) at A=10/13, F=0.5 and (d)-(f) of Equation (4.2b) at F = 0.5. The main chaotic regions decrease in (a)-(c) as k = 1,2,3,.. whereas they do not significantly change in (d)-(f) as k = 1,2,3,... .

5. DISCUSSION AND RESULTS OF OTHER GROUPS

What I have tried to show in this paper is: (a) that there appears to exist a close connection between the *real* time behavior of systems of o.d.e.'s (1.1) and the singularities of their solutions in the *complex* t-plane, and (b) that this connection is not only limited to the case of completely integrable dynamical systems.

To be sure, the case of complete integrability is the one that is so far best understood - in terms of the Painlevé property - as it also closely parallels the success of the corresponding analysis in exactly solvable (partial differential) nonlinear evolution equations [16-18,36]. What is even more interesting, however, is that singularity analysis can go beyond identifying completely integrable systems: As the results of sections 3 and 4 of this paper indicate, a "controlled" violation of the Painlevé property with the introduction of logarithmic terms at higher orders in the series expansions can identify "weakly chaotic" behavior in Hamiltonian as well as non-Hamiltonian systems.

Still, the identification of the *type* of singularities present in a dynamical system is by itself only the first step in analyzing the analytic structure of its solutions in the complex t-plane. There remains the important task of actually *locating* these singularities, finding how "densely" they are distributed in the complex plane, how close they are to the real axis, etc. In one of the first such investigations to be performed, U. Frisch and R. Morf discovered a connection between the proximity of these singularities to the real t-axis and the turbulent bursts of the solutions of a nonlinear Langevin equation in a regime of intermittency [37]. This connection was further elucidated and studied by M. Tabor and J. Weiss [38] on the Lorenz equations, using a numerical algorithm developed by Y.F. Chang and G. Corliss [39].

Chang, Tabor and Weiss then studied the analytic structure of the Hénon-Heiles Hamiltonian (3.2) [40,41]. They found, over a large range of parameter values, $0 < \varepsilon < 16$, that the solutions of (3.2) possess *natural boundaries* in the complex t plane with a remarkable "self-similar" structure: every singularity near the real t-axis appears to have a double spiral of other singularities around it forming a sequence of ever decreasing isosceles triangles! In fact, these authors observed the existence of such "self-similar" natural boundaries in a *class* of Hamiltonian systems including (3.2), the Toda lattice and the rotating rigid body, when some of the powers

n_k, m_k of the leading order terms, cf. (2.11), are *complex* numbers [41].

Returning to the subject of complete integrability in Hamiltonian systems, I would like to mention here some very interesting results obtained by A. Ramani, B. Grammatikos and B. Dorizzi [42-44]: They extended the Painlevé property as stated in this paper, and introduced the "weak-Painlevé" concept, allowing for *rational* powers $(t-t_0)^{n/r}$ in their series, where r=p (p odd) or r=p/2 (p even) and p is determined by the leading order terms near the singularity. These researchers proved explicitly that all of their weak-Painlevé cases are completely integrable by actually *deriving* the integrals of the motion as second (or higher) degree polynomials in the velocities [44]. Whether - and in what way - these results extend the class of algebraically completely integrable Hamiltonian systems of Adler and van Moerbeke [21,22], remains still an open question.

We have seen, therefore, that there is a compelling amount of evidence, obtained by several research groups in the last few years, which suggests that many properties of the real time motion of dynamical systems can be deduced from their (movable) singularities in the complex t-plane. However, even though these efforts have produced many new and interesting results, it is the opinion of this author that the real power of the singularity analysis has not yet been fully put to use. For example, a study of the singularities of quasiperiodic solutions in "regular" regions, as opposed to those of homoclinic orbits in chaotic regions, may permit a better understanding of the "break-up" of invariant tori as certain parameters (or initial conditions) are varied.

Finally, the applicability of the singularity analysis to an arbitrary - in principle - number of coupled first order o.d.e.'s, makes it an appealing method for studying *multidimensional* problems (e.g. many degree-of-freedom Hamiltonian) systems, for which very few other methods are presently available. These research objectives are currently being pursued, and investigations are in progress [45] whose main aim is to formulate and use specific criteria for studying the chaotic properties of dynamical systems based on the analytic structure of their solutions in the complex time plane.

6. ACKNOWLEDGEMENTS

Most of my recent work reported in this paper, was supported in part by the D.O.E. grant DE-AC02-83ER13052. Useful discussions with M. Adler, E. Trubowitz, M. Kruskal and P. Winternitz on a number of mathematical points are also acknowledged here. Finally, I would like to thank Mr. A. Al-Humadi and T. Thomas for their help with the numerical calculations and the preparation of the figures.

7. REFERENCES

[1] A. Lichtenberg and M. Lieberman, "Regular and Stochastic Motion", Springer (1983).

[2] J. Guckenheimer and P. Holmes, "Nonlinear Oscillations, Bifurcations of Vector Fields and Dynamical Systems", Springer (1983).

[3] S. Jorna, ed., "Topics in Nonlinear Dynamics", A.I.P. Conf. Proc. Vol. 46, A.I.P. (1978).

[4] W. Horton, L. Reichl, V. Szebehely, ed., "Long-Time Prediction in Dynamics", J. Wiley & Sons (1982).

[5] G. Casati and J. Ford, ed., "Stochastic Behavior in Classical and Quantum Hamiltonian Systems", Lect. Notes in Phys. 93, Springer (1979).

[6] L. Garrido, ed., "Dynamical Systems and Chaos", Sitges VII Conf. Proc., Lecture Notes in Phys. 179, Springer (1983).

[7] R.H.G. Helleman, in "Fundamental Problems in Statistical Mechanics", Vol. 5, ed. E.G.D. Cohen, North Holland (1981).

[8] H. Haken, ed., "Chaos and Order in Nature", Springer-Verlag (1981).

[9] A.S. Pikovskii and M.I. Rabinovich, Math. Phys. Rev., Vol. 2, Sov. Sci. Rev. C, Harwood (1981).

[10] J. Ford, "How Random is a Coin Toss?", Physics Today, April 1983.

[11] V.I. Arnol'd, "Mathematical Methods of Classical Mechanics", Springer (1978).

[12] R. Abraham and J. Marsden, "Foundations of Mechanics", Benjamin (1978).

[13] E. Hille, "Ordinary Differential Equations in the Complex Domain", Wiley Interscience (1976).

[14] E.L. Ince, "Ordinary Differential Equations", Dover (1956).

[15] S. Kowalevskaya, Acta Math 12, 177 (1889); also Acta Math. 14, 81 (1889).

[16] M.J. Ablowitz, A. Ramani and H. Segur, Lett. al Nuovo Cimento, 23 (9), 333 (1978).

[17] M.J. Ablowitz, A. Ramani and H. Segur, J. Math. Phys. 21 (4), 715; also J. Math. Phys. 21 (5), 1006 (1980).
[18] M.J. Ablowitz and H. Segur, "Solitons and the Inverse Scattering Transform", SIAM Series Appl. Math. (1982).
[19] H. Segur, Lectures at International School of Physics "Enrico Fermi", Varenna, Italy (1980).
[20] T. Bountis, H. Segur and F. Vivaldi, Phys. Rev. A 25, 1257 (1982).
[21] M. Adler and P. v.Moerbeke, Invent. Math. 67, 297 (1982).
[22] M. Adler and P. v.Moerbeke, Comm. Math. Phys. 83, 85 (1982).
[23] M. Adler, private communication.
[24] H. Yoshida, "Necessary Conditions for the Existence of Algebraic First Integrals" I and II, Cel. Mech., to appear (1984).
[25] T. Bountis and H. Segur, in A.I.P. Conf. Proc. Vol. 88, 279 (1982).
[26] T. Bountis in Am. Inst. Astron. Astrophys. Conf. Proc. Reprint 82-1443, AIAA (1982).
[27] H.T. Davis, "Introduction to Nonlinear Differential and Integral Equations", Dover (1962).
[28] E. Lorenz, J. Atmos. Sci. 20, 130 (1963).
[29] C. Sparrow, "The Lorenz Equations", Springer (1983).
[30] T. Bountis et. al, "On the Complete and Partial Integrability of Non-Hamiltonian Systems", preprint subm. for publ.
[31] T. Bountis, M. Bier and J. Hijmans, Phys. Lett. 97A, 11 (1983).
[32] W.-H. Steeb, A. Kunick and W. Strampp, J. Phys. Soc. of Japan, 52 (8), 2649 (1983).
[33] R. Anderson, J. Harnad and P. Winternitz, Physica 4D, 164, (1982).
[34] S. Snider and P. Winternitz, "Classification of Systems of Nonlinear O.D.E.'s with Superposition Principles", preprint (Univ. de Montreal, 1983).
[35] G. Casati and J. Ford, Phys. Rev. A 12, 4, 1702 (1975).
[36] J. Weiss, M. Tabor and G. Carnevale, J. Math. Phys. 24, 522 (1983).
[37] U. Frisch and R. Morf, Phys. Rev. A 23 (5), 2673 (1981).
[38] M. Tabor and J. Weiss, Phys. Rev. A 24, 2157 (1981).
[39] Y.F. Chang and G. Corliss, J. Inst. Math. Applics. 25, 349 (1980).
[40] Y.F. Chang, M. Tabor and J. Weiss, J. Math. Phys. 23 (4), 531 (1982); see also J. Weiss in same vol. as Ref. 25.

[41] Y.F. Chang et al, Physica 8D, 183 (1983).

[42] A. Ramani, B. Grammatikos and B. Dorizzi, Phys. Rev. Lett. 49 (21), 1539 (1983).

[43] A. Ramani, B. Grammatikos and B. Dorizzi, J. Math. Phys. 24 (9), 2282 and 2289 (1983).

[44] B. Grammatikos, B. Dorizzi and A. Ramani, "Hamiltonians with High-Order Integrals and the Weak-Painleve Concept", preprint.

[45] T. Bountis, M. Bier and T. Wright, "A Singularity Analysis of the Transition to Chaos in Hamiltonian Systems", in preparation.

* Tassos C. BOUNTIS

Clarkson College
Department of Mathematics
& Computer Science
Potsdam, New York 13676
U. S. A.

Singularities & Dynamical Systems
S.N. Pnevmatikos (editor)

BIFURCATIONS IN HAMILTONIAN DYNAMICAL SYSTEMS

George Contopoulos

European Southern Observatory & University of Athens
West Germany Greece

SUMMARY

There are many infinite sequences of period doubling bifurcations of families of periodic orbits in conservative Hamiltonian systems. In systems of two degrees of freedom, the bifurcation ratio has the universal value $\delta=8.72$. In rotating systems these infinite bifurcations are followed by inverse bifurcations, forming an infinity of bubbles. In systems of three degrees of freedom (that cannot be reduced to a 2-D system plus an independent oscillation) it seems that there are no infinite sequences of bifurcations. The bifurcation sequences terminate either by complex instability or by inverse bifurcation.

1. INFINITE BIFURCATIONS IN HAMILTONIAN SYSTEMS

The existence of infinite sequences of period doubling bifurcations of families of periodic orbits in dissipative systems has been well established since the work of Feigenbaum (1978) and Coullet and Tresser (1978). The ratio of successive intervals between bifurcations approaches asymptotically the universal number $\delta=4.67$. The first indication that in conservative systems the bifurcation ratio is different was provided by Benettin et al. (1980). In systems of two degrees of freedom the universal ratio is $\delta=8.72$. This ratio was verified subsequently by several authors, both empirically (Greene et al. 1981, Bountis 1981, Contopoulos 1983a) and theoretically (Widom and Kadanoff 1982).

One of the first examples of successive period doubling pitchfork bifurcations in a Hamiltonian system was provided by Contopoulos (1970). In this paper several bifurcations of reasonant periodic orbits were found in the Hamiltonian

$$H=\frac{1}{2}\ (\dot{x}+\dot{y}+Ax^2+By^2)-\ \varepsilon xy^2=h, \qquad (1)$$

where $A=1.6, B=0.9, h=0.00765$ and the variable parameter was ε . An example of such bifurcations is given in Fig. 1 .

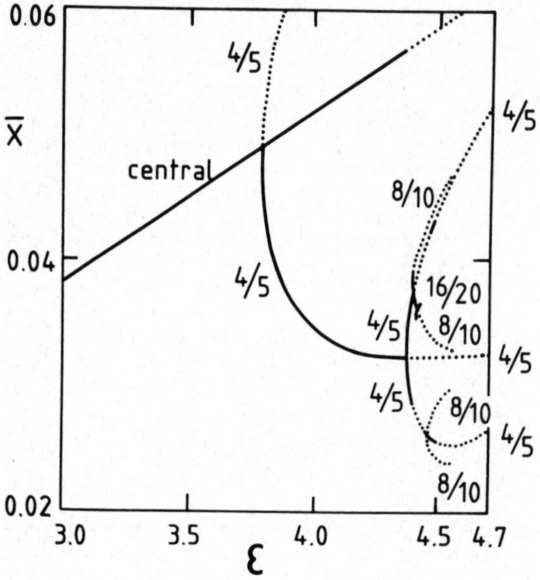

Fig. 1. Characteristics of resonant families of periodic orbits of types 4/5, 8/10,....emanating from the central family. The orbits of all these families intersect perpendicularly the x-axis. The coordinate $\bar{x}=A^{\frac{1}{2}}x$ is given as function of ε; (———) stable (.....) unstable families.

We noticed that the successive bifurcations are closer and closer to each other as ε increases. This led Ford (private communication) to suggest that there would be an infinity of bifurcations with an accumulation point ε_{max} (Contopoulos 1963). However at that time we did not calculate the bifurcation ratio. In retrospect we find that this ratio is close to the universal number $\delta=8.72$.

The first case in which we did calculate the bifurcation ratio was the system (1) with $A=B=\varepsilon=1$ and variable h (Contopoulos and Zikides 1980). These befurcations are not period doubling pitchfork bifurcations, but change the stability character or the <u>same</u> (central) family (Fig. 2) (stable to unstable and vice-versa). In this case the accumulation point is B_∞ and the bifurcation ratio is 9.22 . This ratio was considered by Benettin et al. (1980) to be close to the universal value 8.72, but it was later realized that is it quite different. In fact Heggie (1983) proved that the bifurcation ratio along the <u>same</u> family is not universal but depends on the particular system considered. In our case the ratio 9.22 is equal to $exp(\pi/\sqrt{2})$.

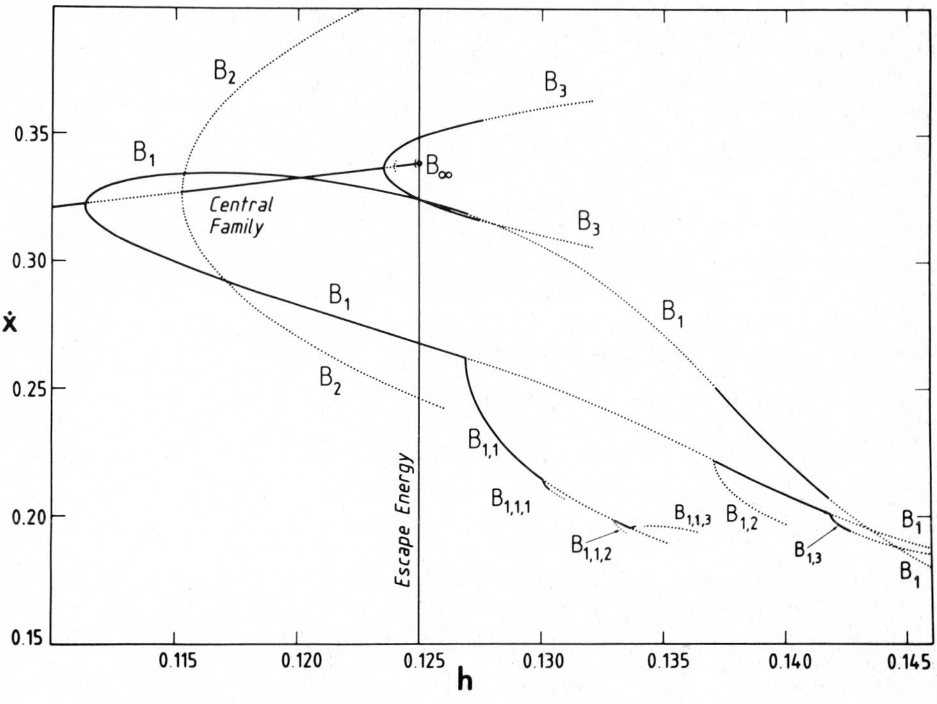

Fig. 2. Characteristics of various families bifurcating from the central family (x=0). The coordinate ẋ is given as function of the energy h; (———) stable (....) unstable families.

In Fig. 2. We see that the stable families bifurcating from the central family, $B_1, B_3, \ldots$ undergo higher order pitchfork bifurcations. Such are the second order families $B_{1,1}, B_{1,2}, \ldots$, the third order families $B_{1,1,1}$, $B_{1,1,2}$,..., etc. The ratios of the intervals between bifurcations of successive orders are consistent with the universal ratio $\delta=8.72$. Therefore both kinds of bifurcations (universal and non-universal) coexist in the same system.

2. THE APPEARANCE AND DISAPPEARANCE OF INFINITE BIFURCATIONS IN ROTATING SYSTEMS

We found that in Hamiltonian systems representing rotating galaxies all bifurcations are followed by inverse bifurcations, so that the characteristics are closed bubbles (Contopoulos 1973a,b). The sequences of infinite bifurcations are also followed by inverse sequences, thus forming infinite bubbles, one inside the other. A theoretical explanation of this phenomenon (Contopoulos 1983d) is based as the fact that far away from the system its potential is approximately that of a point mass. In the case of a point mass, in a rotating frame of reference, the bifurcations from the families of circular periodic orbits do not extend to infinity but form closed bubbles. Thus the same happens to systems approximating the point-mass model at large distances. Examples of sequences of infinite bubbles are given in Figs. 3c,d below. We consider the case of a rotating galaxy composed of an axisymmetric background and a bar perturbation, with Hamiltonian

$$H=\frac{1}{2}\left(\dot{r}^2+\frac{J_o^2}{r^2}\right)-\frac{1}{1+(r^2+1)^{\frac{1}{2}}}+\varepsilon r^{1/2}(16-r)\cos 2\vartheta-\Omega_s J_o=h. \qquad (2)$$

Here (r,ϑ) are polar coordinates, $\dot{r}$ is the radial velocity, J_o is the angular monentum, Ω_s is the angular velocity of the system, h is the value of the energy in the rotating system (Jacobi constant) and ε is a measure of the strength of the bar. When $\varepsilon=0$ we have an axisymmetric galaxy and the only stable families of periodic orbits are the families x_1 and x_4 of circular orbits (direct and retrograde). There are also resonant bifurcating families of various types n/m (making n radial oscillations during m rotations around the center) but all of them are unstable. When ε is small many resonant families (in particular those with $m=1$) are stable but do not undergo higher order bifurcations (Family 3/1, Fig. 3a). As ε increases (Fig. 3b) the family 3/1 generates a bubble 6/2. As ε increases further higher order bubbles are formed until an infinite sequence is generated

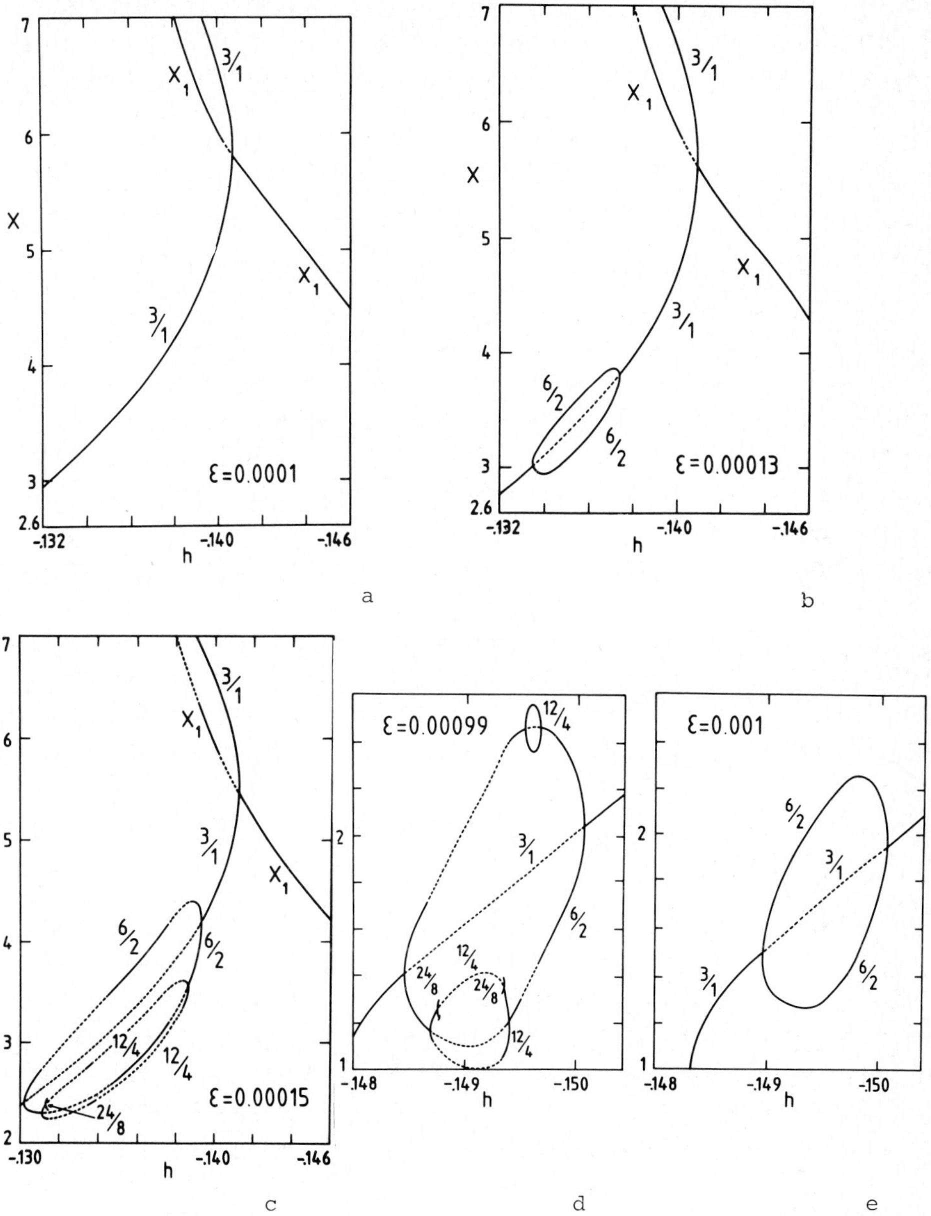

Fig. 3. The generation and disappearance of sequences of bubbles in a rotating barred galaxy (2) as the strength of the bar ε increases. x is the distance of an orbit from the center perpendicularly to the bar as function of the energy h. (———) stable, (....) unstable families (ε=0.0001 corresponds to a bar of density 10% of the background).

(Fig. 3c). In this case the bifurcation ratio is close to $\delta=8.72$, both on the right and left ends of the bubbles.

This infinite sequence of bubbles continues to exist as ε increases until beyond $\varepsilon=0.00099$ (Fig. 3d). However for $\varepsilon=0.001$ (Fig.3e) only one bubble remains and for a little larger ε even this bubble disappears. This example shows how an infinite sequence of bubbles is generated as the perturbation increases. It shows also that an infinite sequence of bifurcations may disappear for still larger perturbations.

In non-rotating systems we may also have bubbles, even an infinity of bubbles one inside the other, but in many cases the various bifurcating families extend to infinity. It seems that such sequences appear whenever we have a transition from ordered to stochastic motion in a dynamical system.

3. SYSTEMS OF THREE DEGREES OF FREEDOM

In dissipative systems the number of degrees of freedom is of minor importance. In the corresponding Poincaré maps the volumes are contracting, but there is one direction of slowest contraction, so that after many iterations the system becomes effectively one-dimensional (Feigenbaum 1980).

However in conservative Hamiltonian systems the volumes are preserved and the dimension of a system may be quite important. For this reason we made recently several investigations of bifurcations in systems of three degrees of freedom (Contopoulos 1984). It seems that in generic 3-D systems the sequences of period doubling bifurcations terminate after a finite number of bifurcations (Contopoulos 1973d). One mechanism that terminates such sequances is complex instability. Complex instability of periodic orbits appears when the eigenvalues of the variational eqations are complex outside the unit circle. A stable orbit has two pairs of complex eigenvalues on the unit circle. A transition to complex instability appears when the two pairs of eigenvalues collide and then leave the unit circle.

This kind of instability does not appear in 2-D systems, or in degenerate 3-D systems composed of a 2-D system plus an independent oscillation. The onset of complex instability is not followed by bifurcation of another family of periodic orbits, as in other types of transition to instability. Therefore when complex instability appears the sequence of bifurcations terminates. This type of instability is very common, both in 3-D Hamiltonian systems and in 4-D conservative mappings. However it does not seem that all sequences of bifurcations terminate in this way.

Another type of termination of a sequence of bifurcations is through an inverse bifurcation. In Fig. 4 we show a direct and an inverse bifurcation. We consider a stable family that becomes unstable as the control parameter η passes through a critical value, producing by bifurcation another family of equal or double period. If the new family exists on the side where the original family is unstable, then it is stable and the bifurcation is direct (Fig. 4a). But if the new family close to the bifurcation point extends bakwards (i.e. in the direction where the original family is stable), then it is unstable and it is inverse. Further away from the

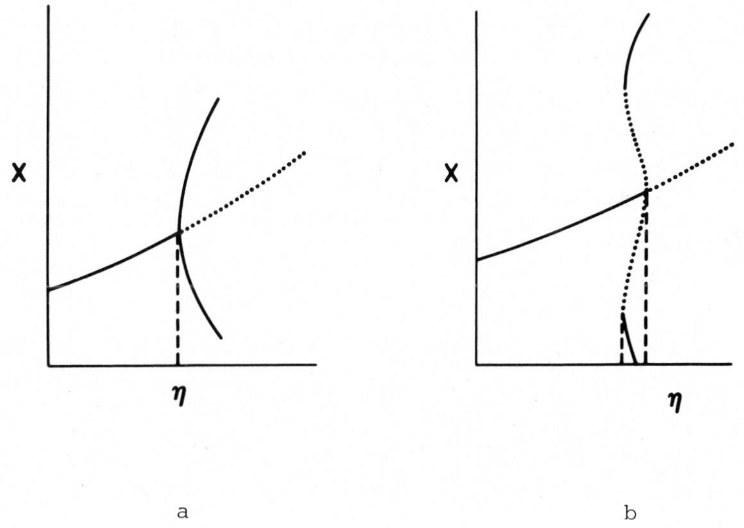

Fig. 4. A direct (a) and an inverse (b) bifurcation; (———) stable (....) unstable families.

original family the new family may become stable (Fig. 4b), but there is no continuous transition from the first stable family to the next. Thus the sequence of (stable) bifurcations terminates at the inverse bifurcation.

It seems that this phenomenon is quite general. A preliminary study of a 4-D conservative mapping by MacKay (1982) and Bountis also failed to provide sequences of infinite bifurcations. On the other hand Janssen and Tjon (1983) described a case of a 4-D mapping where it seems that infinite bifurcations do appear for a certain range of the parameter space. However inverse bifurcations often extend over a small interval of values of the parameter, as in the case of Fig. 4b, and therefore they can be easily missed in numerical investigations, unless one explores very carefully the behaviour of the various families near the bifurcation point.

At any rate it is of great theoretical and practical interest to study several further cases of bifurcations in Hamiltonian systems of three degrees of freedom, and in conservative 4-D mappings. Such investigations are presently in progress.

If the termination of the bifurcation sequences is a general phenomenon then the importance of infinite sequences of bifurcations is rather limited in conservative Hamiltonian systems.

References

Benettin, G., Cercignani, Galgani, L, and Giorgilli, A.: 1980, Lett. Nuovo Cimento 28, 1.

Bountis, T.: 1981, Physica 3D, 577.

Contopoulos, G.: 1970, Astron. J. 75, 96; 108.

Contopoulos, G.: 1973, in Tapley, B.D. and Szebehely, V. (eds) "Recent Advances in Dynamical Astronomy", Reidel, Dordrecht-Holland, p.177.

Contopoulos, G.: 1973a, Lett. Nuovo Cimento 37 , 149.

Contopoulos, G.: 1973b, Physica 8D, 142.

Contopoulos, G.: 1973c, Lett. Nuovo Cimento 38, 257.

Contopoulos, G.: 1973d, Astrophys. J. 275 , p. 511 .

Contopoulos, G.: 1984, preprint.

Contopoulos, G. and Pinotsis, A.: 1984, Astron. Astrophys.(in press).

Contopoulos, G. and Zikides, M.: 1980, Astron. Astrophys. 90, 108.

Coullet, P. and Tresser, J.: 1978, J. Phys. (Paris), 5C, 25.

Feigenbaum, M.J.: 1978, J. Stat. Phys. 19 , 25.

Feigenbaum, M,J.: 1980, Los Alamos Science 1 , 4 .

Greene, J.M., MacKay, R.S., Vivaldi, F. and Feigenbaum, M.J.: 1981, Physica 3D, 468.

Heggie, D.: 1983, Celes. Mech. 29, 207.

Janssen, T. and Tjon, J.A.: 1983, J.Phys. 16A, 567.

MacKay, R.S.: 1982, PhD Thesis, Princeton University.

* George CONTOPOULOS

European Southern Observatory
Garching b. München,W.Germany
&
University of Athens
Athens, Greece

Singularities & Dynamical Systems
S.N. Pnevmatikos (editor)
© Elsevier Science Publishers B.V. (North-Holland), 1985

A GUIDE TO THE HÉNON-HEILES HAMILTONIAN

D.L. Rod* and R.C. Churchill
Univ. of Calgary, Canada
Hunter College, U.S.A.

1. INTRODUCTION

The Hénon-Heiles Hamiltonian was introduced in 1964 as a model problem for investigating conservation laws in axial-symmetric potentials [11]. At low positive energies numerical investigations suggested integrability, but, as the energy increased past 1/12, chaos appeared. In the intervening years considerable effort has been expended in attempts to understand the system, and in 1979 the present authors, together with G. Pecelli, published a survey of those results which could be established with complete mathematical rigor [5]. The present note is both a summary and up-date of that survey, citing references but omitting proofs.

2. BASICS

The Hénon-Heiles Hamiltonian is the real analytic function $H:R^4 \to R$ given by

$$(1)\qquad H(x,y) = \tfrac{1}{2}|y|^2 + V(x),\ x = (x_1,x_2),\ y = (y_1,y_2) \in R^2,$$

where the (Hénon-Heiles) potential $V:R^2 \to R$ is defined by

$$(2)\qquad V(x) = \frac{1}{2}|x|^2 + \frac{1}{3}x_1^3 - x_1x_2^2;$$

the actual system of equations is

$$(3)\qquad \begin{aligned} \dot{x}_1 &= \frac{\partial H}{\partial y_1} = y_1, & \dot{y}_1 &= -\frac{\partial H}{\partial x_1} = -x_1 - x_1^2 + x_2^2 \\ \dot{x}_2 &= \frac{\partial H}{\partial y_2} = y_2, & \dot{y}_2 &= -\frac{\partial H}{\partial x_2} = -x_2 + 2x_1x_2. \end{aligned}$$

The Hamiltonian is invariant under the time-reversing symmetry

(4) $R_1 : (x_1,x_2,y_1,y_2) \to (x_1,x_2,-y_1,-y_2),$

under the reflectional symmetry

(5) $R_2 : (x_1,x_2,y_1,y_2) \to (x_1,-x_2,-y_1,y_2),$

and under the symmetry given by a rotation through the angle $(2\pi/3)$

(6) $$\begin{cases} R_3 : (x_1,x_2,y_1,y_2) \to \\ (-\frac{1}{2}x_1-(\sqrt{3}/2)x_2,(\sqrt{3}/2)x_1-\frac{1}{2}x_2,-\frac{1}{2}y_1-(\sqrt{3}/2)y_2,(\sqrt{3}/2)y_1-\frac{1}{2}y_2). \end{cases}$$

By conservation of energy we have $\frac{d}{dt}H(x(t),y(t)) = 0$ for any solution $(x,y) = (x(t),y(t))$ of (3); the constant $h = H(x(t),y(t))$ is the *energy* of the solution. Obviously such a solution must lie on the *energy surface* $\sum_h = H^{-1}(\{h\}) \subset R^4$. The topological nature of $\sum_h$ can be seen from Figure 1, where the level curves of V are plotted (the points a and b in that figure will be defined in the next paragraph). Indeed, the form of (1) shows that for $(x,y) \in \sum_h$ we have $V(x) \le h$, hence x must

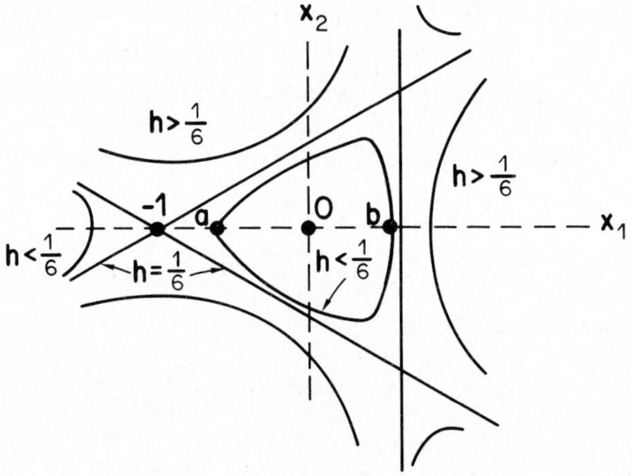

Figure 1

lie within the *Hill's region* (or *potential well*) $\sigma_h = \{x \in R^2 : V(x) \le h\}$ *corresponding to* h. Using (1) it then becomes evident that $\sum_h \simeq \sigma_h \times S^1/\sim$, where the quotient occurs since each circle over a boundary point of σ_h must be collapsed to a point.

For $0 < h < 1/6$ the hyperplane $x_2 = 0$ of R^4 intersects $\sum_h$ in an oval S projecting to the line segment $[a,b]$ of Figure 1. Using this oval as a cross-section of the flow on $\sum_h$, Hénon and Heiles numerically studied the resulting Poincaré mapping $\rho: S \to S$. Using x_1 and y_1 as coordinates on S, the result was Figure 2, roughly for all $0 < h < 1/12$, which certainly suggests integrability. However, chaos ensued as h passed $1/12$, and this has yet to be explained at a rigorous level.

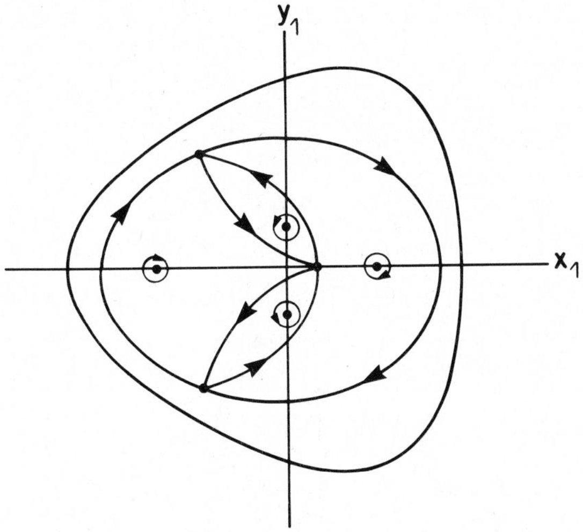

Figure 2

The actual Hamiltonian studied by Hénon and Heiles was

$$H(u,v) = \frac{1}{2}|v|^2 + \frac{1}{2}|u|^2 + u_1^2 u_2 - \frac{1}{3} u_2^3, \tag{7}$$

which is equivalent to (1,2) through the canonical transformation $x_1 = -u_2$, $x_2 = u_1$, $y_1 = -v_2$, $y_2 = v_1$. If we observe that (7) agrees through third order terms with the three-particle Toda lattice Hamiltonian

$$\begin{cases} H_T(u,v) = \tfrac{1}{2}|v|^2 \\ + \frac{1}{24}\{\exp(2(u_2-\sqrt{3}u_1)) + \exp(2(u_2+\sqrt{3}u_1)) \\ + \exp(-4u_2)\} - \frac{1}{8}, \end{cases} \tag{8}$$

and then note that the system associated with (8) admits the function

$$F(u,v) = 8v_1(v_1^2-3v_2^2) + (v_1+\sqrt{3}v_2)\exp(2(u_2-\sqrt{3}u_1)) \tag{9}$$
$$+ (v_1-\sqrt{3}v_2)\exp(2(u_2+\sqrt{3}u_1)) - 2v_1\exp(-4u_2)$$

as an independent integral, the impression of integrability at low energies for (1,2) becomes understandable. Of greater interest is the evolution to chaos as the energy increases.

In the literature the parameter-dependent function

$$H(x,y) = \tfrac{1}{2}|y|^2 + \tfrac{1}{2}|x|^2 + \lambda x_1^3 - x_1x_2^2 \tag{10}$$

is also referred to as the Hénon-Heiles Hamiltonian, since it reduces to (1,2) when $\lambda = 1/3$.

3. CRITICAL POINTS AND ELEMENTARY PERIODIC ORBITS

The vector field (3) vanishes only at points $(a,0) \in R^4$ where $\nabla V(a) = 0$, and $\nabla V(a) = 0$ only for $a_0 = (0,0)$, $a_1 = (-1,0)$, $a_2 = (1/2,-\sqrt{3}/2)$ and $a_3 = (1/2,\sqrt{3}/2)$. The critical points of the Hénon-Heiles vector field are therefore $A_j = (a_j,0)$, $j = 0,1,2,3$. The first coincides with the bounded component of $\sum_0$, and the other three, which can be identified using (6), lie on $\sum_{(1/6)}$. A_0 has repeated eigenvalues $\pm i$, and the remaining three points have eigenvalues ± 1 and $\pm i\sqrt{3}$. A theorem of Moser [15] allows for a complete description of the flow near A_1, A_2 and A_3.

At energies $0 < h < 1/6$, elementary geometrical arguments give the existence of energy h periodic orbits $(\Pi_j,\dot{\Pi}_j)$ of (3) which project to the bounded component of the Hill's region σ_h as shown in Figure 3, $j = 1,\ldots,8$ [5]. As the energy passes $1/6$ three new elementary energy h periodic orbits $(\Pi_j,\dot{\Pi}_j)$ appear, $j = 9,10,11$, projecting as in Figure 4, whereas the first three periodic orbits vanish [5].

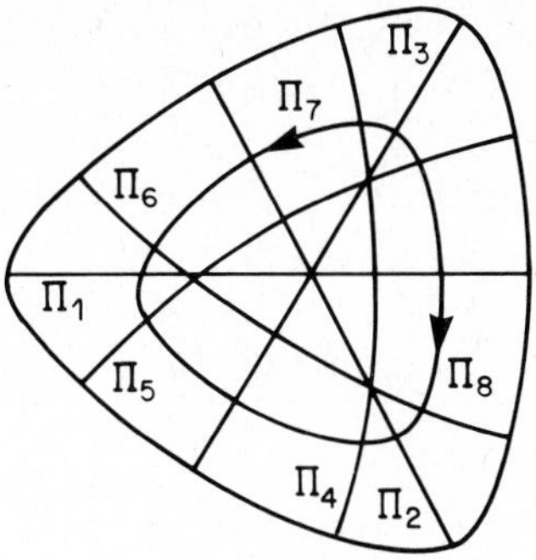

Figure 3

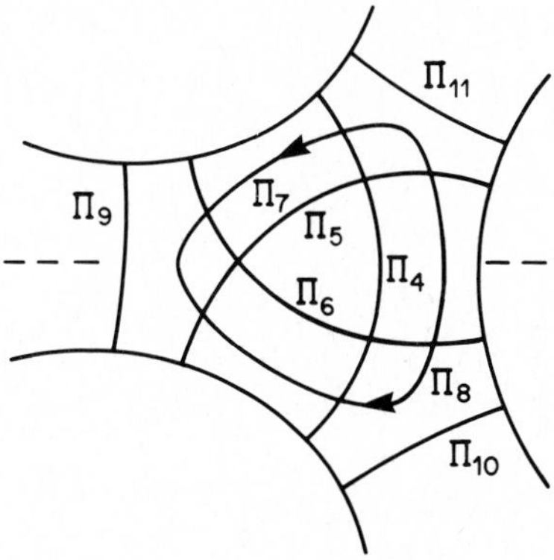

Figure 4

4. THE NORMAL FORM AND REDUCED HAMILTONIANS

Consider an analytic Hamiltonian $H:U \to R$, where $U \subset R^4$ is an open neighborhood of 0. Write

$$H = H_2 + H_3 + H_4 + \dots ,$$

where each H_j is a homogeneous polynomial of degree j in the variables x_1, x_2, y_1, y_2, and assume $k \geq 2$. We say H is in *normal form through terms of order* k if all Poisson brackets $\{H_2, H_j\}$ vanish for $j = 2,\dots,k$ (this is automatic for $k = 2$). For

(11) $$H_2(x,y) = \tfrac{1}{2}|x|^2 + \tfrac{1}{2}|y|^2$$

this is the case if and only if each H_j, for $j = 2,\ldots,k$, can be expressed as a polynomial in the "variables" (see [3])

$$\text{(12)} \qquad \begin{cases} W_1 = 2(x_1x_2 + y_1y_2) \\ W_2 = 2(y_1x_2 - x_1y_2) \\ W_3 = x_1^2 - x_2^2 + y_1^2 - y_2^2 \\ W_4 = x_1^2 + x_2^2 + y_1^2 + y_2^2. \end{cases}$$

The Hénon-Heiles Hamiltonian can be transformed to a Hamiltonian in normal form through order k for any $k \geq 2$. The (non-unique) result, for $k = 6$, is (see [3])

$$\text{(13)} \qquad H = H_2 + H_3 + H_4 + H_5 + H_6 + \ldots ,$$

where

$$\text{(14)} \qquad \begin{cases} H_2 = \tfrac{1}{2}W_4 \\ H_3 = H_5 = 0 \\ H_4 = \frac{1}{48}(7W_2^2 - 5W_4^2) \\ H_6 = \frac{1}{64}(-\frac{67}{54}W_4^3 - \frac{28}{9}W_3^3 + \frac{28}{3}W_1^2W_3 - \frac{7}{18}W_2^2W_4); \end{cases}$$

this form is useful for studies of formal second integrals (this is discussed in detail in [16]) and quantization (e.g. see [13]).

For each $h > 0$ let $S_h \subset R^3$ denote the 2-sphere

$$\text{(15)} \qquad S_h = \{w = (w_1,w_2,w_3) \in R^3 : |w|^2 = 4h^2\}.$$

If for $w \in S_h$ we identify the tangent space $T_w(S_h)$ with $\{u \in R^3 : \langle u,w\rangle = 0\}$, where $\langle\ ,\ \rangle$ is the usual inner product, then the mapping $\omega_R(w):T_w(S_h) \times T_w(S_h) \to R$, given at $u,v \in R^3$ by

$$\text{(16)} \qquad \omega_R(w)(u,v) = -(16h^2)^{-1}\langle w,u \times v\rangle,$$

defines a symplectic structure on S_h. If $K:S_h \to R$ is analytic, i.e. if K is the restriction to S_h of some analytic function $K^e:U \to R$, where $U \subset R^3$ is an open neighborhood of S_h, then this symplectic structure associates with K the "Hamiltonian system"

$$\text{(17)} \qquad \dot{w} = 4w \times \nabla K^e(w)$$

on S_h (see [3]).

For $j = 4,6$ the functions H_j of (14) define analytic functions $K_j^e: R^3 \to R$ as follows: Write H_j as $H_j(W_1,W_2,W_3,W_4)$ and set $K_j^e(w_1,w_2,w_3) = H_j(w_1,w_2,w_3,2h)$. The mappings $K_4 = K_4^e|S_h$ and $K_4 + K_6 = (K_4^e + K_6^e)|S_h$ are called the *first and second reduced Hamiltonians of* (13,14), while (17), first with $K^e = K_4^e$, and then with $K^e = K_4^e + K_6^e$, gives the *first and second reduced systems* of the Hénon-Heiles problem. These reduced entities inherit symmetries corresponding to (4), (5) and (6) (see [3]). The adjective "reduced" is used because (S_h, ω_R) can be identified with the usual reduced space of $(H_2^{-1}(\{h\}), \varphi_t(h))$, where H_2 is as in (11) and $\varphi_t(h)$ is the restriction of the flow of H_2, regarded as defining a free and proper S^1-action on $H_2^{-1}(\{h\})$ [9]. With this identification the first and second reduced Hamiltonians are precisely the Hamiltonians induced by H_4 and $H_4 + H_6$ on this reduced space [3].

The reduced Hamiltonians are useful for studying the flow of (3) in a neighborhood of the origin. In particular, nondegenerate critical points of K_4 and $K_4 + K_6$ correspond to low energy periodic orbits, and (to some extent) the stability status of these points governs that of the orbits [3].

5. RESULTS

At sufficiently low energies $h > 0$ the elementary periodic orbits $(\Pi_j, \dot{\Pi}_j)$, for $j = 2,...,8$, can be identified with the fixed points of the Poincaré mapping $\rho: S \to S$ as in Figure 5 (where $(\Pi_j, \dot{\Pi}_j)$ is abbreviated as Π_j) [3]. Moreover, $(\Pi_1, \dot{\Pi}_1)$ can be identified with the boundary of the oval. As the figure suggests, for such energies $(\Pi_j, \dot{\Pi}_j)$ is elliptic stable for $j = 1,2,3,7,8$ and hyperbolic for $j = 4,5,6$ [3].

As h increases the stability status of $(\Pi_j, \dot{\Pi}_j)$, for $j = 1,2,3$, changes infinitely often. More precisely, there are three increasing positive sequences $\{h_\ell\}$, $\{k_\ell\}$, $\{h'_\ell\}$, converging to $1/6$ and satisfying $h_\ell < k_\ell < h'_\ell < h_{\ell+1}$, such that for $j = 1,2,3$ we have $(\Pi_j, \dot{\Pi}_j)$ elliptic for $h \in (h_\ell, k_\ell) \cup (k_\ell, h'_\ell)$, with elliptic stability holding for almost all such h, and $(\Pi_j, \dot{\Pi}_j)$ hyperbolic for $h \in (h'_\ell, h_{\ell+1})$ [6]. Since elliptic stability precludes ergodicity, a particular consequence is that $\rho = \rho_h: S \to S$ cannot be ergodic for almost all

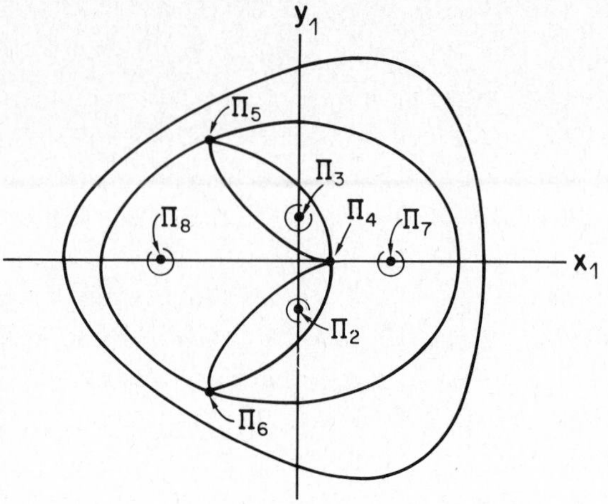

Figure 5

$h \in (h_\ell, k_\ell) \cup (k_\ell, h'_\ell)$, but this does not exclude the possibility of "ergodic seas" (or "zones of instability").

The stability status of $(\Pi_j, \dot{\Pi}_j)$ for $j = 4,5,6,7,8$ remains an open problem for higher energies. Numerical evidence suggests hyperbolicity at all energies for $j = 4,5,6$, and a switch to hyperbolicity at a fairly low energy for $j = 7,8$.

In contrast, for $j = 9,10,11$ the stability status of $(\Pi_j, \dot{\Pi}_j)$ is known: Each is hyperbolic for all $h \in (1/6, \infty)$ [17]. Moreover, for such j and h each $(\Pi_j, \dot{\Pi}_j)$ is "isolated" in $\sum_h$ in the following sense: If $K \subset \sum_h$ is a compact neighborhood of $(\Pi_{11}, \dot{\Pi}_{11})$ projecting into the first quadrant in Figure 1 or 4, then $(\Pi_{11}, \dot{\Pi}_{11})$ is the unique invariant set of the flow of (3) contained within K [4]. Using the symmetry (6), analogous statements obtain for $(\Pi_9, \dot{\Pi}_9)$ and $(\Pi_{10}, \dot{\Pi}_{10})$.

As regards homoclinic orbits, such exist for $(\Pi_j, \dot{\Pi}_j)$ at energies $1/6 < h < \infty$ when $j = 9,10,11$. Moreover, with the possible exception of a discrete set of energies in $(1/6, \infty)$, these homoclinic orbits are nondegenerate, thereby allowing an embedding of the Smale horseshoe into the dynamics, and precluding the existence of a second global analytic integral [7]. More recently it was shown with complex analytic techniques that the flow has no second analytic integral in any neighborhood of the origin [19,20] (also see [10]).

Concerning heteroclinic orbits between the $(\Pi_j, \dot{\Pi}_j)$, there is an $h_0 \geq 1/6$ such that these exist for $j = 9,10,11$ for any $h \in (h_0, \infty)$. A consequence is the following "pathological" behavior of the flow at such energies: There are disjoint neighborhoods N_{j-8} of $(\Pi_j, \dot{\Pi}_j)$ in $\sum_h$, for $j = 9,10,11$, such that for any bi-infinite sequence $\{\varepsilon_k\}_{k=-\infty}^{\infty}$ with $\varepsilon_k = 1,2$ or 3 and $\varepsilon_{k+1} \neq \varepsilon_k$ there is a solution of (3) of energy h which passes from N_k to N_{k+1} in the manner prescribed by the sequence. Moreover, if the sequence is periodic, the solution will be periodic [7].

At low positive energies there are undoubtedly heteroclinic orbits connecting all $(\Pi_j, \dot{\Pi}_j)$, $j = 4,5,6$, as is evident from Figure 5, but this awaits proof. Rigorously establishing the non-degeneracy of such heteroclinic orbits would then explain the chaos arising in $\rho = \rho_h : S \to S$ as h increases past $1/12$.

For λ near $-1/3$ one can establish the existence of non-degenerate homoclinic orbits in the system associated with (10) at all sufficiently low positive energies [12] (also see [2]). A study of the reduced Hamiltonian of that system for all λ is found in [8] (also see [1], [14] and [18]).

REFERENCES

[1] M. Braun: On the applicability of the third integral of motion, J. Differential Equations 13(1973)300-318.

[2] R.C. Churchill: On proving the non-integrability of a hamiltonian system, in The Riemann problem, complete integrability and arithmetic applications, (G. and D. Chudnovsky, Eds.), Springer Lecture Notes in Mathematics 925, Springer-Verlag, New York, 1982, 103-122.

[3] R.C. Churchill, M. Kummer & D.L. Rod; On averaging, symmetry and reduction in hamiltonian systems, J. Differential Equations 49(1983)359-414.

[4] R.C. Churchill, G. Pecelli & D.L. Rod: Isolated unstable periodic orbits, J. Differential Equations 17(1975)329-348.

[5] R.C. Churchill, G. Pecelli & D.L. Rod: A survey of the Hénon-Heiles hamiltonian with applications to related examples, in

Como conference proceedings on stochastic behavior in classical and quantum hamiltonian systems, (G. Casati and J. Ford, Eds.), Springer Lecture Notes in Physics 93, Springer-Verlag, New York, 1979, 76-136.

[6] R.C. Churchill, G. Pecelli & D.L. Rod: Stability transitions for periodic orbits in hamiltonian systems, Archive for Rational Mechanics and Analysis 73(1980)313-347.

[7] R.C. Churchill & D.L. Rod: Pathology in dynamical systems III: analytic hamiltonians, J. Differential Equations 37(1980)23-38.

[8] R. Cushman: Geometry of the bifurcations of the normalized reduced Hénon-Heiles hamiltonian, Proc. R. Soc. Lond. A 382 (1982)361-371.

[9] R. Cushman & D.L. Rod: Reduction of the semisimple 1:1 resonance, Physica D 6(1982)105-112.

[10] J.J. Duistermaat: Non-integrability of the 1:1:2 resonance, preprint.

[11] M. Hénon & C. Heiles: The applicability of the third integral of motion; some numerical experiments, Astronom. J. 69(1964) 73-79.

[12] P. Holmes, Proof of non integrability for the Hénon-Heiles hamiltonian near an exceptional integrable case, Physica D 5 (1982)335-347.

[13] C. Jaffé & W.P. Reinhardt: Uniform semiclassical quantization of regular and chaotic dynamics on the Hénon-Heiles surface, J. Chem. Phys. 77(1982)5191-5203.

[14] M. Kummer: On resonant nonlinearly coupled oscillators with two degrees of freedom, Comm. Math. Phys. 48(1976)53-79.

[15] J. Moser: On the generalization of a theorem of Liapounoff, Comm. Pure Appl. Math. 11(1958)257-271.

[16] J. Moser: Lectures on hamiltonian systems, Memoirs AMS 81, American Mathematical Society, Providence, 1968.

[17] D.L. Rod, G. Pecelli & R.C. Churchill: Hyperbolic periodic orbits, J. Differential Equations 24(1977)329-348 and 28(1978)163-165.

[18] F. Verhulst: Discrete symmetric dynamical systems at the main resonances with applications to axisymmetric galaxies, Phil. Trans. R. Soc. Lond. A 290(1979)435-465.

[19] S.L. Ziglin: Branching of solutions and nonexistence of first integrals in hamiltonian mechanics I, Functional Analysis and

its Applications 16(1982)181-189.

[20] S.L. Ziglin: Branching of solutions and the nonexistence of first integrals in hamiltonian mechanics II, Functional Analysis and its Applications 17(1983)6-17.

David L. Rod
Department of Mathematics
and Statistics
University of Calgary
Calgary, Alberta
Canada T2N 1N4

Richard C. Churchill
Department of Mathematics
Hunter College
695 Park Avenue
New York, New York
U.S.A. 10021

* Author's research supported in part by the Natural Sciences and Engineering Research Council of Canada, Grant A8507.

Singularities & Dynamical Systems
S.N. Pnevmatikos (editor)

SOLITONS IN NONLINEAR ATOMIC CHAINS

Stéphanos N. Pnevmatikos
University of Dijon
France

ABSTRACT

We have studied the dynamical behavior of large amplitude displacements on a discrete monatomic chain whose first and second nearest neighbor particles interact following an anharmonic pair polynomial potential of cubic and/or quartic type. Kink and breather nontopological soliton solutions are found to exist in the continuum limit and envelope solitons or dark solutions are obtained without any continuum restriction for their carrier waves. Numerical simulations are extensively presented showing the solitonlike behavior of these solutions during their propagation and interactions on the discrete chain. New, interesting nonlinear phenomena are observed due to competition between first and second nearest neighbor's coupling (splitting or blow-up of solitons). Discreteness effects are discussed.

I. INTRODUCTION

The wide use of digital computers and the progress in nonlinear mathematics has enabled us to take into account nonlinearity in physical mechanisms. Over the last twenty years, the soliton concept has emerged as widely accepted paradigm [1,2] for exploring and modelling the nonlinear dynamics of the real world. Solitons were found in biological systems [3-9], neurophysical models for nerve axon pulse propagation [10,11], in nonlinear optical fibers [12-15], in theories of commensurate-incommensurate structural phase transitions and central peak phenomena [16-19]. Solitary waves are also important for the nonlinear plasma dynamics [20-24], electronic transmission lines with nonlinear elements [25-28], Josephson junction circuits [29-34], oceanic wave propagation [35,36] and dynamical behavior of anharmonic atomic lattices [37-48], to mention just a few examples. This article covers a very special area of soliton applications, that is the dynamics of one dimensional (1-D) atomic lattices with nonlinear interatomic (pair) potentials.

In order to take into consideration lattice nonlinearity in atomic chain models two different approaches are used : α)In the first one, we consider models where atoms, coupled with harmonic springs, are plunged in minima of a nonlinear periodic substrate potential representing the rest of lattice influence on the atomic chain in question. The classical example of such nonlinear substrate potential is the sinusoidal function. This type of model is described in the continuum limit by a Sine-Gordon (S-G) type equation [49] whose solitary wave solutions are examples of the so-called topological soliton which describe transitions between degenerate equilibrium configurations (figure 1a). The characteristic property of a topological soliton is the possibility of being a stable localized entity independent of time (static atomic excitation) [50-53]. β)In the second one, particles are coupled with nonlinear interatomic strings without any presence of substrate potential (figure 1b). The classical example of such a system is a chain of particles coupled with harmonic plus cubic nonlinear forces [37]. The continuum limit of this model, in the lowest approximation , gives the famous Korteweg de Vries (KdV) equation [42], which admits nontopological soliton solutions, e.g. collective, localized in space, excitations, which, in order to be stable, must have a non-zero velocity.

The purpose of this article is double : α) To give a brief review of some important fundamental results about nontopological solitons in atomic 1D lattice and β) To present new theoretical and numerical results in this area.

Following the early numerical study (1950) of Fermi, Pasta and Ulam [37] and their unexpected results, wave propagation in a monatomic chain of particles, with nonlinearly elastic interaction between nearest neighbors, was extensively studied by Kruskal and Zabusky [38-42,54]. They showed that the nonlinear longitudinal motion of a monatomic chain can be described in terms of a new interesting progressive, localized in space, wave with a characteristic "wavelength" much greater than the interatomic spacing, which is solution of the classical KdV equation [55]. Zabusky and Kruskal observed in numerical simulations that two solitary waves emerge by a mutual collision without deforming or losing their stability or changing their velocities. The only evidence of their collision is that of a possible

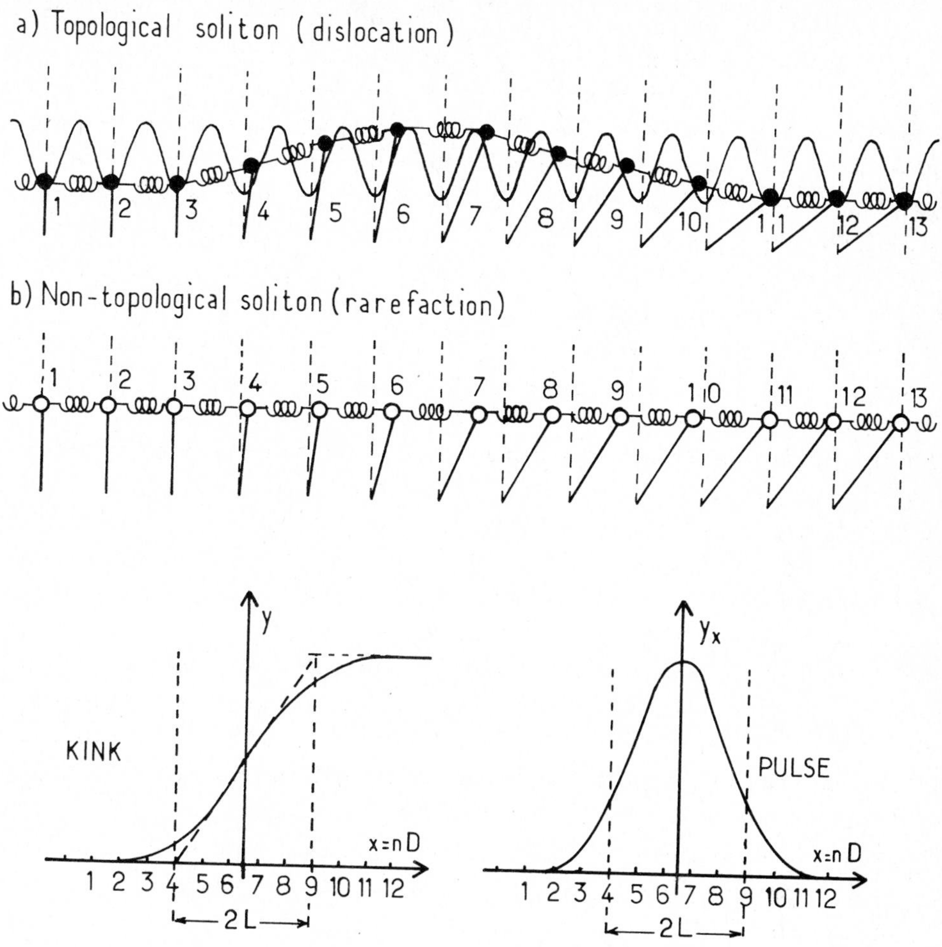

Amplitude of displacement from equilibrium position

Derivative of y(x,t) (or relative displacement $r_n = y_n - y_{n-1}$)

Figure 1 : Topological and nontopological localized excitations.

phase shift. This integrity was reminiscent of elementary particle bahavior and so they named solitary waves, having this remarkable property, "solitons". [56]

In 1967, Gardner, Green, Kruskal and Miura [57] showed how to solve the initial value problem for the KdV equation exactly and to write down the most general solution containing solitons and continuum states, and, Toda introduced a prototype discrete atomic 1D model, with nonlinear exponential interparticle coupling, which is completely integrable [48]. Since then many integrable Hamiltonian systems have been used to describe atomic chains and several results of physical interest were obtained through these mathematical models. However, realistic lattices are non integrable discrete systems with many physical perturbative parameters (impurities, surfaces, damping, discretisation, multiparticle composition per unit cell, long range order interactions, thermal fluctuations, etc ..). For this reason, physicists are soon oriented to this direction. Results on the presence of impurities and interfaces in nonlinear lattices and their influence on the soliton propagation have already been obtained [58-62]. Also, some disordered nonlinear models have been suggested [63,64]. Generalizations on higher dimensionality have been tried [65-70]. In order to study long range order interactions [71,72], a model with first and second nearest neighbors nonlinear coupling has been suggested [73]. The existence of a temperature gradient in non-integrable atomic systems has been seen [46,47,74-78] and experimental results have been explained using soliton theories [3,21, 79-82]

The outline of the paper is as follows. In section II the atomic model and the nonlinear interatomic coupling are described and also general details of the computer simulations are given. In section III kink and breather solitons are obtained in the continuum limit and a comparison between, Bq and KdV type equations are presented. In section IV envelope solitons are studied for which non continuum limit type restrictions are imposed for their carrier wave. In section V a discussion of the second neighbor atom's coupling role is presented both for kink or envelope type solitons. In this section, some numerical observations concerning new nonlinear nonsoliton phenomena are described. In section VI discreteness effects are discussed. In section VII a new nonlinear model is put forward where the extension of the previous results is attempted. In section VIII the main results on a diatomic nonlinear chain are discussed. Finally, in section IX discussions and perspectives are given.

II. THE MODEL

The analytical study of solids, with all the real parameters taken into account, is a very difficult problem. For this reason, we are very often forced to use simplified 1-D models, which allow us to focus our attention on the study of other analytical difficulties, for example that of nonlinearity. The 1-D lattice may be applicable to crystals with a filamentary structure, or to biological and organic macromolecular systems with longitudinal dynamical behavior.

Let us now consider a monatomic chain described by the following general Hamiltonian :

$$H = \sum_n \frac{1}{2} M\dot{y}_n^2 + \sum_{\ell,n} V_\ell (y_{n+\ell} - y_n) \qquad (2.1)$$

where y_n denotes the longitudinal displacement, from its equilibrium position, of the n-th particle with mass M and velocity $\dot{y}_n$. The first term in the Hamiltonian (2.1) represents the kinetic energy of particles and the second one the total potential energy of the chain, where $n, \ell = 1,2,3\ldots, N$ and N is considered large enough $(N\to\infty)$ in order to avoid any boundary effect. This chain is, of course, part of a 3-D lattice, but in a first approximation we ignore the coupling with other chains, and substrate potential or external forces are not considered.

Different types of realistic nonlinear potentials have been used [83-86]. Here, we consider a general potential of polynomial type which can approximate several realistic potentials by the convenient choice of its parameters :

$$V(r_{n,\ell}) = G_i\left(\frac{1}{2} r_{n,\ell}^2 + \frac{A_i}{3} r_{n,\ell}^3 + \frac{B_i}{4} r_{n,\ell}^4\right) \qquad (2.2.)$$

where $r_{n,\ell} = y_n - y_\ell$ is the relative displacement between the n and ℓ particles. G_i, A_i, B_i are the potential parameters whose values depend on the lattice and the index i indicates the vicinity order of the particles n and ℓ. The major advantage of this choice is that, for the potential (2.2) we are able to obtain exact analytical results under some conditions. In this paper, in order to keep calculations simple, we focus our attention on the case where only first and second nearest neighbors are non-zero (see figure 2), i.e. $G_i, A_i, B_i = 0$ if $i > 3$ (with $G_1 > 0$ only, while A_i, B_i can be positive or negative). Nevertheless, the coupling between first and second neighbors seems to be sufficient to describe qualitatively even more general situations [72].

In order to respect the stability of the lattice we choose potential's parameters in such a way that the following stability conditions must be satisfied $(\forall_n)$:

$$\frac{\partial V_{tot}}{\partial y_n} = V'(y_n-y_{n-1})+(V'(y_{n+1}-y_n)+V'(y_n-y_{n-2})+V'(y_{n+2}-y_n)=0 \quad (2.3)$$

$$\frac{\partial^2 V_{tot}}{\partial y_n^2} = V''(y_n-y_{n-1})+V''(y_{n+1}-y_n)+V''(y_n-y_{n-2})+V''(y_{n+2}-y_n)>0 \quad (2.4)$$

where primes denote derivatives with respect to y_n. The first condition defines the equilibrium positions of the system and the second insures the stability of these configurations.

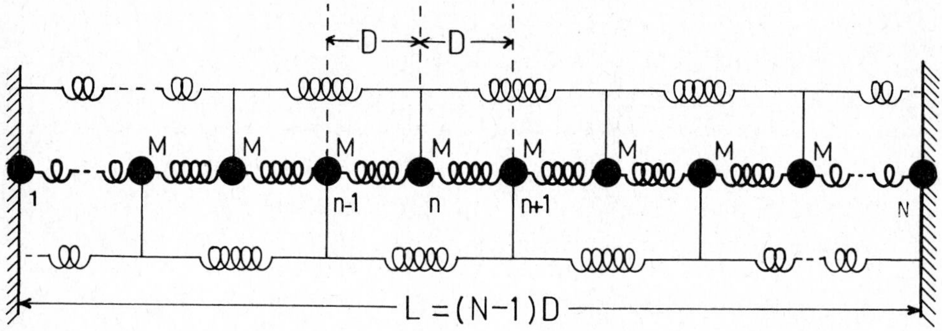

Figure 2: The monatomic chain model. Springs represent first and second nearest neighbors coupling. M is the mass parameter and D is the equilibrium interatomic spacing. Boundaries are fixed.

Following the previous assumptions the lattice's equations of motion are given by :

$$\begin{aligned} M\frac{d^2y_n}{dt^2} &= G_1(y_{n+1}-2y_n+y_{n-1})+G_2(y_{n+2}-2y_n+y_{n-2}) + \\ &+ G_1A_1\left\{(y_{n+1}-y_n)^2-(y_n-y_{n-1})^2\right\} + G_2A_2\left\{(y_{n+2}-y_n)^2-(y_n-y_{n-2})^2\right\}+ \\ &+ G_1B_1\left\{(y_{n+1}-y_n)^3-(y_n-y_{n-1})^3\right\} + G_2B_2\left\{(y_{n+2}-y_n)^3-(y_n-y_{n-2})^3\right\} \end{aligned} \quad (2.5)$$

$$n=1,2,3,...,N$$

The greatest difficulty that prevents us from calculating exact solutions for this system of N differential equations is the nonlinear coupling between them. Thus, it is impossible to obtain analytical results on the discrete model without making some approximations.

Nevertheless, it is possible to study numerically the chain of figure 2 [45,87,88]. All of the numerical results presented here have been taken on chains with 200-1000 particles and fixed boundary conditions (except for the dark solitons) :

$$y_1 = y_2 = y_{N-1} = y_N = 0,\ \dot{y}_1 = \dot{y}_2 = \dot{y}_{N-1} = \dot{y}_N = 0 \qquad (2.6)$$

The numerical method used consists in solving, with a fourth order Runge-Kutta subroutine, the Newtonian equations of motion (2.5) of the discrete chain with a time integration step dt varying between 0.01-0.2. A permanent control of the total energy of the chain enables us not to take into account results where this energy changes over 10^{-2}% of its initial value. For a numerical study we need a set of initial conditions that define displacements $y_n(0)$ and velocities $\dot{y}_n(0)$ for t = 0. Two types of initial conditions are usually considered : α) Arbitrary wave paquets (Gaussian disturbances, step or triangular functions, localized oscillations, etc..) [39,88,89,90]. β) Wave profiles which approximate the exact solutions of the system. Here, in order to study the solitonic motion of the monatomic chain of figure 2, we use solitary wave type initial conditions calculated approximately in the continuum limit as follows.

III. SOLITONS IN THE CONTINUUM LIMIT

3-1. Continuum limit approximation

Limiting our attention to the study of widely spatially extended waves, where every variation in space and time is slow ($\partial/\partial x \sim \varepsilon$, $\partial/\partial t \sim \varepsilon$), functions $y_{n\pm1}(t)$ and $y_{n\pm2}(t)$ can be expanded in Taylor series around y(x,t), with x=nD (D is the equilibrium atomic spacing) :

$$y_{n\pm1} = y \pm D y_x + \frac{D^2}{2} y_{xx} \pm \frac{D^3}{6} y_{xxx} + \frac{D^4}{24} y_{xxxx} + O(\varepsilon^5) \qquad (3.1)$$

$$y_{n\pm2} = y \pm 2D y_x + 2D^2 y_{xx} \pm \frac{4}{3} D^3 y_{xxx} + \frac{2}{3} D^4 y_{xxxx} + O(\varepsilon^5) \qquad (3.2)$$

where $y_x = \partial y/\partial x$, $\partial/\partial x \sim O(\varepsilon)$ and $O(\varepsilon^5)$ terms are neglected. Using relations (3.1) and (3.2) we transform the discrete equation (2.5) into the following continuum form [73] :

$$y_{tt} = c_0^2\, y_{xx} + p_0\, y_x y_{xx} + q_0\, y_x^2 y_{xx} + h_0\, y_{xxxx} \qquad (3.3)$$

with $c_0^2 = \frac{D^2}{M}(G_1+4G_2)$, $p_0 = \frac{2D^3}{M}(A_1G_1+8A_2G_2)$ (3.4)

$$h_0 = \frac{D^4}{12M}(G_1+16G_2) \quad , \quad q_0 = \frac{3D^4}{M}(B_1G_1+16B_2G_2)$$

c_0 is the velocity of sound (velocity of plane waves which are solutions of the linearised version of the equation 3.3), y_{xxxx} is the main term representing the dispersion mechanism due to the discretization of the lattice [54]. The presence of this term is fundamental in order to balance nonlinearity and consequently to allow steady soliton propagation.

3-2. Solitons of Bq type

The equation (3.3) is directly equivalent to the following Generalised Boussinesq (G-Bq) equation :

$$u_{tt} - c_0^2 u_{xx} - p(u^2)_{xx} - q(u^3)_{xx} - h_0 u_{xxxx} = 0 \tag{3.5}$$

$$\text{with} \quad u=y_x \quad , \quad p=p_0/2 \quad \text{and} \quad q=q_0/3 \tag{3.6}$$

For $B_1=B_2=q=0$ and for $A_1=A_2=p=0$ we obtain respectively the classical Boussinesq (Bq) equation [91] and its modified version (M-Bq) [45]. Using a simple method [45] we calculate one-soliton solutions of equation (3.5) and its particular cases. Finally, going back to the initial function $y(x,t)$ we obtain the following kink type solitary wave solutions :

(i) G-Bq , $(p \neq 0, \ q \neq 0)$:

$$y(x,t) = \pm 2\sqrt{\frac{6h_0}{q_0}} \arctan\left\{P_1 \tanh\left[\frac{x-vt}{L_1} + x_1\right]\right\} \tag{3.7}$$

$$\text{with} \quad P = \left[\frac{\sqrt{p_0^2+6(v^2-c_0^2)q_0} \mp p_0}{\sqrt{p_0^2+6(v^2-c_0^2)q_0} \pm p_0}\right]^{1/2} \tag{3.8}$$

(ii) Bq , $(p \neq 0 \ , \ q = 0)$:

$$y(x,t) = s\,\frac{6\sqrt{h_0(v^2-c_0^2)}}{p_0} \tanh\left[\frac{x-vt}{L_1} + x_1\right] \tag{3.9}$$

(iii) M-Bq , $(p = 0 \ , \ q \neq 0)$:

$$y(x,t) = \pm 2\sqrt{\frac{6h_0}{q_0}} \arctan\left\{\exp\left[\frac{2}{L_1}(x-vt) + x_1\right]\right\} \tag{3.10}$$

In the previous expressions the parameter :

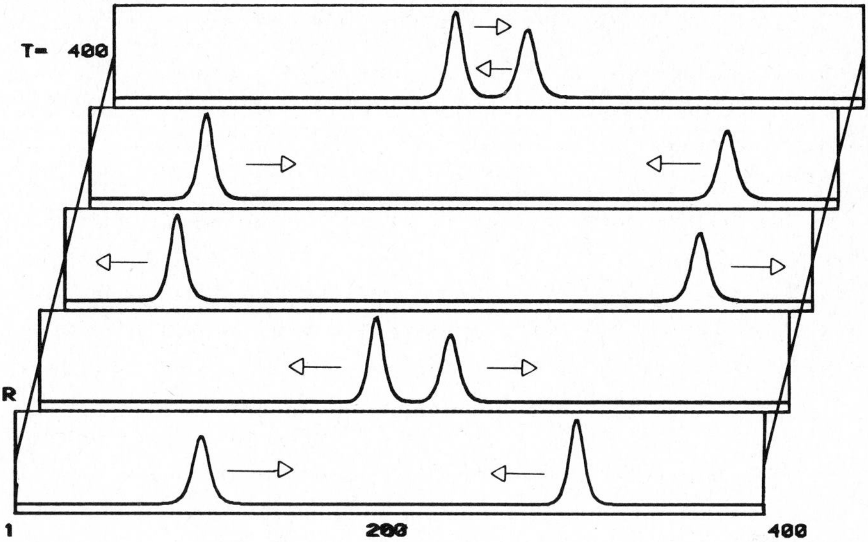

Figure 3 : Pulse-Pulse soliton type head-on collision on a discrete monatomic chain ($M=D=1$, $G_1=1$, $A_1=0.25$, $B_1=0.5$, $G_2=0.1$, $A_2=B_2=0$). The initial conditions ($T=0$) have exact forms of the G-Bq type solutions calculated in the continuum limit, where $v_1=1.193$, $L_1=6.106$, $x_1=100$, (R_{max} = $0.117D$) and $v_2=-1.196$, $L_2=5.338$, $x_2=300$, (R_{max}= $0.147D$).

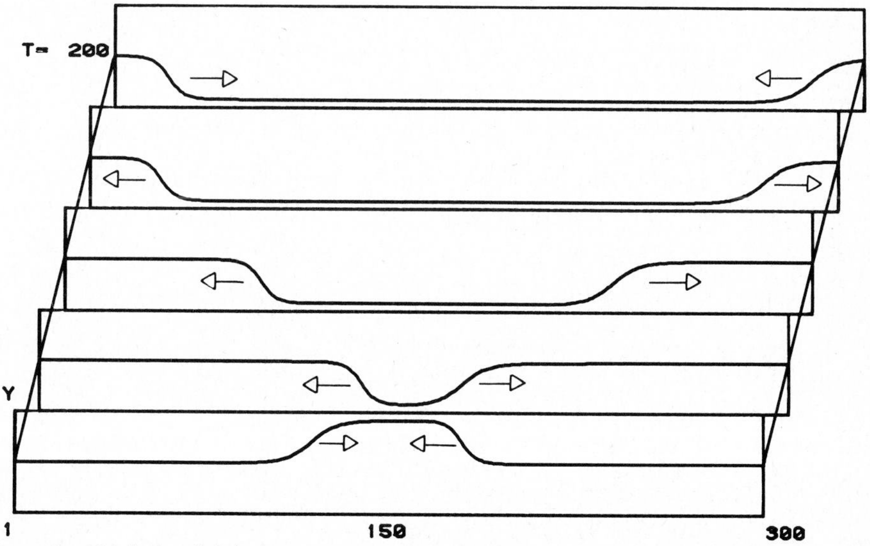

Figure 4 : Kink-Antikink soliton type head-on collision on a discrete monatomic chain ($M=D=1$, $G_1=B_1=1$, $A_1=G_2=A_2=B_2=0$). The initial conditions ($T=0$) have exact formsof the M-Bq type solutions calculated in the continuum limit, where $v_1=1.001$, $L_1=6.453$, $x_1=120$, ($R_{max}=0.063D$) and $v_2=-1.003$, $L_2=3.724$, $x_2=180$, ($R_{max}=-0.108D$)

$$L_1 = 2\sqrt{h_0/(v^2-c_0^2)} \tag{3.11}$$

measures the spatial extension of the wave on the lattice (half-width in half-height), x_1 gives the initial position of the wave center and v is the wave velocity. The last two parameters are arbitrary. Finally, $s=1$ if $h_0 > 0$ and $s=-1$ if $h_0 < 0$ (see table I).

In figures 3 and 4 we are describing the propagation and the elastic collision between two solitons of G-Bq and M-Bq type respectively. For figure 3, we use a relative displacement representation ($r_n = y_n - y_{n-1} \approx Dy_x$), while, in figure 4, we present the absolute displacement $y_n(t)$ for each particle n.

3-3. Solitons of KdV type

The relation (3.3) can be transformed into a very well known family of equations whose the most known is the Korteweg de Vries (KdV) equation [55]. But, in order to obtain KdV equations, we must make a supplementary approximation concerning variations in time. Thus, doing the Galilean transformation :

$$\xi = \alpha(x-c_0t) \quad , \quad \tau = \beta t \quad , \quad u = \gamma y_\xi \tag{3.12}$$

with

$$\alpha = p_0/(6h_0q_0)^{1/2} \; , \; \beta = p_0^3/[2c_0(6q_0)^{3/2}\, h_0^{1/2}], \gamma = (q_0/6h_0)^{1/2} \tag{3.13}$$

and $\partial/\partial\xi \sim 0(\varepsilon)$, $\partial/\partial\tau \sim 0(\varepsilon^{5/2})$, we obtain the following Generalized -Korteweg de Vries (G-KdV) equation [92] :

$$u_\tau + 6uu_\xi + 6u^2u_\xi + u_{\xi\xi\xi} = 0 \tag{3.14}$$

as for the Bq type equations, we easily obtain one-soliton solutions for the equation (3.14) and its particular cases. Finally, taking into account relations (3.12) and (3.13) we arrive at the following kink soliton solutions :

(i) G-KdV , ($p_0 \neq 0$, $q_0 \neq 0$) :

$$y(x,t) = \pm 2\sqrt{\frac{6h_0}{q_0}}\, \arctan\left\{ P_2 \tanh\left[\frac{x-vt}{L_2} + x_2\right]\right\} \tag{3.15}$$

$$\text{with} \quad P_2 = \left[\frac{\sqrt{p_0^2+12q_0c_0(v-c_0)}\mp p_0}{\sqrt{p_0^2+12q_0c_0(v-c_0)}\pm p_0}\right]^{1/2} \tag{3.16}$$

(ii) KdV , ($p_0 \neq 0$, $q_0 = 0$) :

$$y(x,t) = s\,\frac{6\sqrt{2h_0c_0(v-c_0)}}{p_0}\,\tanh\left[\frac{x-vt}{L_2}+x_2\right] \tag{3.17}$$

(iii) M-KdV , $(p_0 = 0\ ,\ q_0 \neq 0)$:

$$y(x,t) = \pm 2\sqrt{\frac{6h_0}{q_0}}\,\arctan\left\{\exp\left[\frac{2}{L_2}(x-vt)+x_2\right]\right\} \tag{3.18}$$

Here too, as for the Bq-type solitons, the parameters :

$$L_2 = 2\sqrt{h_0/2c_0(v-c_0)} \tag{3.19}$$

measures the wave's spatial extension on the lattice and x_2, v are two arbitrary parameters indicating the wave position (at the center) and the soliton velocity. Finally, $s = \pm 1$ as for the Bq case.

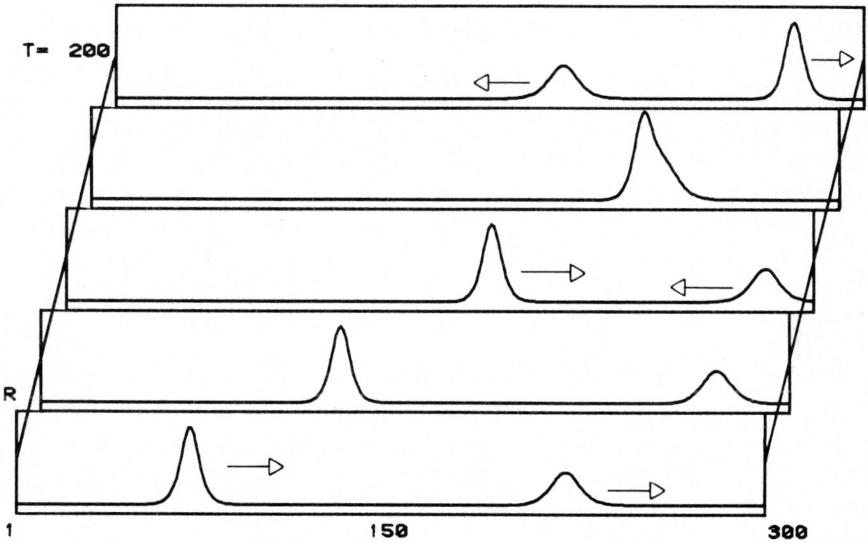

Figure 5 : Pulse-Pulse soliton type collision on a discrete monatomic chain $(M=D=1,\ G_1=1,\ A_1 = 0.5,\ B_1 = G_2=A_2=B_2=0)$. *Initial conditions* $(T=0)$ *are exact solutions of the corresponding KdV equation, obtained in the continuum limit, where* $v_1=1.007$, $L_1=4.879$, $x_1=70$, $(R_{max}=0.041D)$ *and* $v_2=1.003$, $L_2=7.454$, $x_2=220$, $(R_{max}=0.018D)$.

In figure 5 we have presented the propagation and the elastic collision between two KdV solitons (relative displacements). The numerical error on the total energy is 10^{-5}% of the initial conditions's value.

3-4. Comments about kink solitons solutions.

There is a characteristic difference between the equations (3.5) and (3.14). The second one is not invariant to the transformation $t \to -t$. The destruction of this time symmetry was realised for the KdV type equations during the Galilean transformation (3.12) and brings as result that KdV solitons cannot propagate simultaneously in both directions. On the contrary, this restriction is not present for Bq type solitons. Generally, equation (3.5) is a better model than (3.14) to describe nonlinear dynamical evolution in lattices. However, KdV equations are used in a wide range by physicists and this is due to the fact that these are completely integrable systems for which we know an infinite number of conservation laws [92] and we are able to calculate all the periodic or localized solutions using very efficient methods such as Inverse Scattering Transform [57,93], the method of Hirota [94], Bäclund Transformations [95], etc. On the other hand, KdV solitons are going to be identical to Bq solitons when $v \to c_0$. Thus, for the width parameter we have (see fig. 6) :

$$L_1 = 2\sqrt{h_0/(v^2-c_0^2)} = 2\sqrt{h_0/(v+c_0)(v-c_0)} \xrightarrow[v\to c_0]{v\neq c_0} L_2 = 2\sqrt{h_0/2c_0(v-c_0)}$$

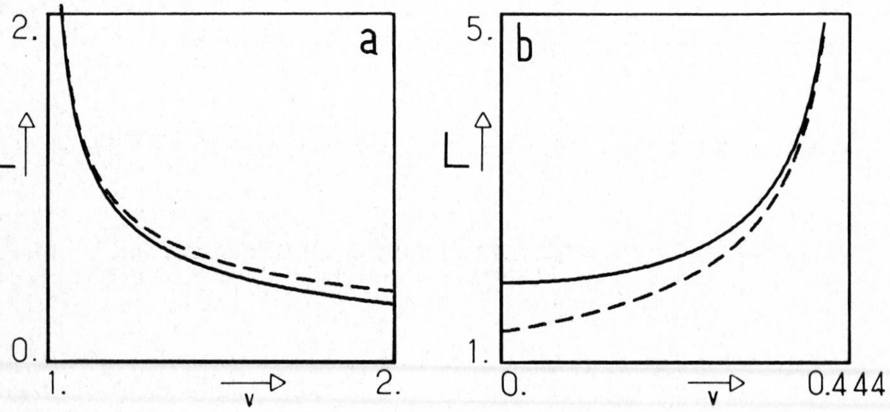

Figure 6 : Spatial extension's parameter (L_1 and L_2) versus the kink's velocity (v) for Bq (full line) and KdV (dashed line) solitons. In figure 6a, we have : $G_1=1$, $G_2=0$, $M=D=1$, from where $c_0 = 1$ and $h_0 > 0$. In figure 6b, we have $G_1=1$, $G_2=-0.2$, $M=D=1$ from where $c_0 = 0.444$ and $h_0 < 0$.

Simultaneously, if $v \to c_0 \Rightarrow P_1 \to P_2$ and the amplitudes of the solutions (3.9) and (3.17) are going to be equal. On the contrary, if the velocity v is very different to the speed of sound c_0 the equations of KdV and Bq type have very different behavior (see also section V).

The sign (+) before these solutions means that soliton produces rarefaction of the lattice and the sign (-) respectiveley means compression. On a cubically anharmonic lattice solitary waves can not be simultaneously compressional and rarefactive (see table I). This particularity is also observed in water wave experiments [96] and in

TABLE I

POTENTIAL'S PARAMETERS			SOLITON	Excitation produced on the lattice	VELOCITY
$h_0 > 0$	$p = 0$	$q = 0$	NO		SUPERSONIC $V > C_0$
		$q > 0$	YES	rarefaction or compression	
		$q < 0$	NO		
	$p > 0$	$q = 0$	YES	rarefaction	
		$q > 0$	YES	rarefaction or compression	
		$q < 0$	NO		
	$p < 0$	$q = 0$	YES	compression	
		$q > 0$	YES	rarefaction or compression	
		$q < 0$	NO		
$h_0 < 0$	$p = 0$	$q = 0$	NO		SUBSONIC $V < C_0$
		$q > 0$	NO		
		$q < 0$	YES	rarefaction or compression	
	$p > 0$	$q = 0$	YES	compression	
		$q > 0$	NO		
		$q < 0$	NO		
	$p < 0$	$q = 0$	YES	rarefaction	
		$q > 0$	NO		
		$q < 0$	NO		

electrical L-C line simulations [28] which are systems also described by the same type of equations. Water or electrical (without second neighbors coupling) solitons are always "rarefactive". Any "compression" (negative pulse) in these systems is immediately decomposed in ripples. On the contrary, lattices with Toda, Morse or Lennard-Jones potentials only allow compressional solitary waves [84,85,97]. For the interatomic potential considered here, the particular conditions allowing stable soliton solutions are indicated in table I.

We have extensively studied through numerical simulations the discrete chain which is initially excited by several solitary wave (kink) initial conditions and we have obtained the following conclusions : "Long wavelength" kink solitary wave solutions obtained in the continuum limit propagate stably on the discrete chain conserving :

1) Their initial shape
2) Their velocity v ("group velocity")
3) Their total energy : $E_{sol} = E_{pot} + E_{kin} = \text{const.}$
4) Their "mass" : $M_{sol} = M\sum_n \frac{y_n - y_{n-1}}{D} = \frac{M}{D}\sum_n r_n$ (✲)

and entered in mutual collision reappeared invariants (see figures 3,4,5) except a possible "phase shift". Obviously, these properties give to solitons particle like character, that has begun to be justified theoretically recently [98].

The remarkable stability of lattice solitons comes from the perfect balance between two competitive mechanisms of the chain : the dispersion and the nonlinearity. This physical reality is very well transported to the equation (3.3) obtained in the continuum limit. If we omit nonlinear terms ($A_1=A_2=B_1=B_2=0$) we obtain the harmonic version of the equation. Here, only dispersive mechanisms are activated and each vibrational lattice motion is extended and dispersed. In equation (3.3) dispersion is represented by the fourth derivative term $h_0 y_{xxxx}$, which, however, remains small, because spatial derivatives are smooth in the continuum limit. Annulating dispersion ($h_0=0$) the imposed mechanism is the nonlinearity that is going to make the wave fall abruptly and dissolve it.

Considering small but finite amplitude relative displacements $r_n = y_n - y_{n-1}$, we are limiting to weakly nonlinear motion. In this case, weak nonlinearity can be balanced by small dispersion and steady solitonlike motion results.

✲ *and eventually other quantities.*

3-5. Breather solution

For the quartic potential case ($G_1, B_1 > 0$ and $A_1 = A_2 = 0$) the G-KdV equation (3.14) obtains the following modified form :

$$u_\tau + 6u^2 u_\xi + u_{\xi\xi\xi} = 0 \tag{3.20}$$

with

$$\xi = x - c_0 t \qquad h_2 = \frac{h_0}{2c_0^2}$$

$$\tau = h_2 c_0 t \tag{3.21}$$

$$u = y_\xi / \eta \qquad \eta = \sqrt{\frac{6h_0}{q_0}}$$

the equation (3.20) admits a supplementary, new, solitary wave solution [99], the exact form of which is given by the Inverse Scattering Transform as follows :

$$u(\xi,\tau) = -4\beta \,\mathrm{sech}\,\Psi \left[\frac{\cos\Phi - \frac{\beta}{\alpha}\sin\Phi \tanh\Psi}{1 + (\frac{\beta}{\alpha})^2 \sin^2\Phi \,\mathrm{sech}^2\,\Psi} \right] \tag{3.22}$$

where $\Phi = 2\alpha\xi + \delta T - \phi_0$ and $\gamma = 8\beta\,(3\alpha^2 - \beta^2)$

$$\Psi = 2\beta\xi + \gamma T + \Psi_0 \qquad \delta = 8\alpha(\alpha^2 - 3\beta^2) \tag{3.23}$$

Integrating (3.22) with respect to ξ and considering the initial variables we obtain :

$$y(x,t) = \pm A_m \arctan\left\{ \frac{\beta}{\alpha}\,\mathrm{sech}\left[\frac{x - v_e t}{L_e} + x_e\right] \sin\left[\frac{x - v_0 t}{L_0} + x_0\right] \right\}$$

with $A_m = -2\eta = -2\sqrt{\dfrac{6h_0}{q_0}}$ (3.24)

$$v_e = c_0 - 2h_0(3\alpha^2 - \beta^2)/c_0 \;, \qquad L_e = 1/2\beta$$

$$v_0 = c_0 - 2h_0(\alpha^2 - 3\beta^2)/c_0 \;, \qquad L_0 = 1/2\alpha \tag{3.25}$$

α , β are two arbitrary parameters that are restricted to remain small in order to respect the continuum limit conditions. The solution (3.24) generated on the atomic chain represents a solitary wave entity propagating in one direction with its amplitude breathing. In figure 7 we present the propagation and the soliton-type collision between a breather and a pulse soliton initial condition of M-KdV type studied numerically on the considered discrete monatomic chain.

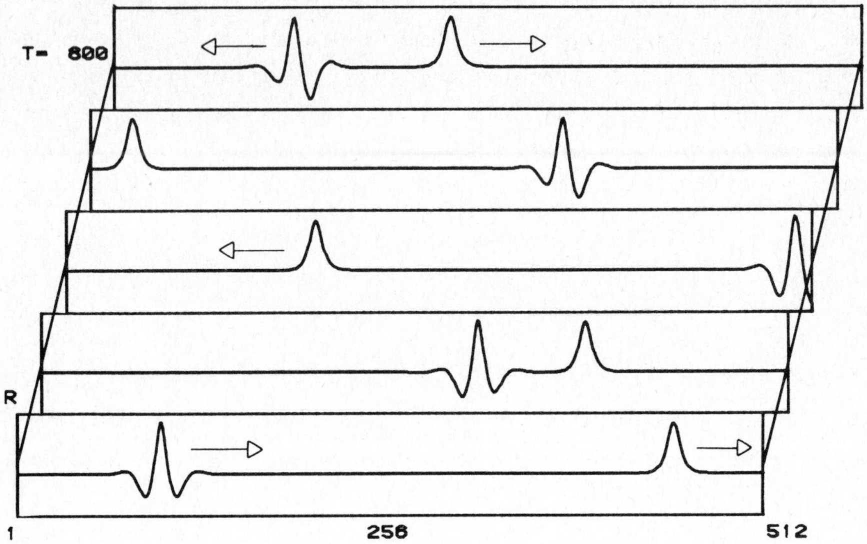

Figure 7 : Breather-Pulse soliton-type collision on a discrete monatomic chain ($M=D=1$, $G_1=B_1=1$, $A_1=G_2=A_2=B_2=0$). *Initial conditions* ($T=0$) *are exact solutions of the corresponding M-KdV equation obtained in the continuum limit, where* $\alpha=0.1$, $\beta=0.07$, $v_2=0.996$, $v_0=1.001$, $x_e=x_0=100$, ($R_{max}=0.112D$) *and* $v_2=1.003$, $L_2=3.727$, $x_2=450$, ($R_{max}=0.108D$).

For the small amplitude limit ($\beta<<1$), and supposing $\beta/\alpha<<1$, the breather expression (3.22) is transformed into an envelope soliton excitation which can be an exact solution of a cubic Nonlinear Schrödinger equation (NLS) describing also approximately the initial model [99]. NLS equations can be obtained also for a KdV (or Bq) and a G-KdV (or G-Bq) equations by looking for solutions of type :

$$u = F\, e^{j\phi} \qquad (3.26)$$

and by using the perturbative technique of the derivative expansion method [100]. But, the envelope soliton solutions obtained will also always be restricted by the continuum limit, which means that envelopes and oscillations (carrier wave) must be indispensably wide. However, it is possible to avoid restrictions concerning the carrier wave by a very interesting method developed in the next section.

IV. DISCRETE SOLITONS

In this section we are studying soliton solution with small and widely extended envelope, modulating oscillations of arbitrary wave vector (discrete carrier wave). These solitons are obtained using a method employed first by Tsurui [101] and modified and extended by Flytzanis et al. [102]. Here, we examaine only the quartic potential case ($A_1=A_2=0$) where no derivative expansion techniques are needed.

4-1. Discrete regime

For the quartic potential case the equation of motion are written as follows :

$$M\frac{d^2y_n}{dt^2}=G_1(y_{n+1}-2y_n+y_{n-1})+G_2(y_{n+2}-2y_n+y_{n-2})+$$
$$+G_1B_1\left\{(y_{n+1}-y_n)^3-(y_n-y_{n-1})^3\right\}+G_2B_2\left\{(y_{n+2}-y_n)^3-(y_n-y_{n-2})^3\right\} \tag{4.1}$$

$$n=1,2,3,\ \ldots,N$$

We look for solutions of the type :

$$y_n(t)=F_n(t)\exp[\,j(\gamma nD-\delta t)\,]+\text{c.c.} \tag{4.2}$$

where the exponential term represents the arbitrary wavevector oscillatory part of the solution modulated by the function $F_n(t)$ which also has a slowly varying phase. The abbreviation c.c. represents the complex conjugate part of the solution.

Replacing the function (4.2.) in the equation (4.1) and neglecting fast oscillation terms (☼) of type $\exp[\pm 3j(\gamma nD-\delta t)]$ we obtain the following equation for $F_n(t)$:

$$M(\ddot{F}_n-2j\delta\dot{F}_n-\delta^2F)=$$

$$=G_1(F_{n+1}e^{j\gamma D}-2F_n+F_{n-1}e^{-j\gamma D})+G_2(F_{n+2}e^{j2\gamma D}-2F_n+F_{n-2}e^{-j2\gamma D})+$$
$$+3G_1B_1\left\{(F_{n+1}e^{j\gamma D}-F_n)^2(F^*_{n+1}e^{-j\gamma D}-F^*_n)-(F_n-F_{n-1}e^{-j\gamma D})^2(F^*_n-F^*_{n-1}e^{j\gamma D})\right\}+ \tag{4.3}$$
$$+3G_2B_2\left\{(F_{n+2}e^{j2\gamma D}-F_n)^2(F^*_{n+2}e^{-j2\gamma D}-F^*_n)-(F_n-F_{n-2}e^{-j2\gamma D})^2(F^*_n-F^*_{n-2}e^{j2\gamma D})\right\}$$

asterisk denotes complex conjugate. In order to calculate the function $F_n(t)$ we pass now to the continuum limit.

4-2. Continuum regime

Limiting our attention to the small amplitude and wide extension

---

☼*This approximation concernes only the function* $F_n(t)$

envelopes :

$$F \sim \varepsilon \quad , \quad F_x \sim \varepsilon^2 \quad , \quad F_{xx} \sim \varepsilon^3 \quad , \quad F_t \sim \varepsilon^2 \tag{4.4}$$

and using the Taylor expansions :

$$F_{n\pm 1} = F \pm DF_x + \frac{D^2}{2} F_{xx} + 0(\varepsilon^4) \tag{4.5}$$

$$F_{n\pm 2} = F \pm 2DF_x + 2D^2 F_{xx} + 0(\varepsilon^4) \tag{4.6}$$

the equation (4.3) becomes :

$$j(F_t + S_1 F_x) + S_2 F + \frac{1}{2}(S_3 F_{tt} + S_4 F_{xx}) + S_5 |F|^2 F = 0 \tag{4.7}$$

where $S_1 = \frac{D}{\delta} \sin(\gamma D) \left[\frac{G_1}{M} + 4 \frac{G_2}{M} \cos(\gamma D) \right]$

$$S_2 = \frac{1}{2\delta} \left\{ \delta^2 - 4\sin^2\left(\frac{\gamma D}{2}\right) \left[\frac{G_1}{M} + 4 \frac{G_2}{M} \cos^2\left(\frac{\gamma D}{2}\right) \right] \right\} , \quad S_3 = -\frac{1}{\delta} \tag{4.8}$$

$$S_4 = \frac{D^2}{\delta} \left[\frac{G_1}{M} \cos(\gamma D) + 4 \frac{G_2}{M} \cos(2\gamma D) \right]$$

$$S_5 = -\frac{24}{\delta} \sin^4\left(\frac{\gamma D}{2}\right) \left[\frac{G_1 B_1}{M} + 16 \frac{G_2 B_2}{M} \cos^4\left(\frac{\gamma D}{2}\right) \right]$$

containing terms of order less than order $O(\varepsilon^4)$. Several methods are available [102] for solving equation (4.7). Here, we use the more simple ones.

4-3. Cubic Nonlinear Schrödinger (NLS) equation.

Considering the following dispersion relation :

$$\delta^2 = 4 \sin^2\left(\frac{\gamma D}{2}\right) \left[\frac{G_1}{M} + 4 \frac{G_2}{M} \cos^2\left(\frac{\gamma D}{2}\right) \right] \tag{4.9}$$

concerning the frequency δ and the wavevector γ of the exponential part of function (4.2.), ((4.9)is nothing other than the dispersion relation of the harmonic monatomic chain [45]) ,we eliminate the term $S_2 F$ in the equation (4.7). Then, the following Galilean transformation :

$$\xi = x - v_g t \qquad (\partial/\partial\xi \sim \varepsilon \ , \ \partial/\partial\tau \sim \varepsilon^2) \tag{4.10}$$

$$\tau = \mu t$$

with

$$v_g = S_1 = \frac{\partial \delta}{\partial \gamma} = \frac{D}{\delta} \sin(\gamma D) \left[\frac{G_1}{M} + 4 \frac{G_2}{M} \cos(\gamma D) \right] \tag{4.11}$$

$$\mu = \frac{\partial^2 \delta}{\partial \gamma^2} = \frac{D^2}{\delta} \left[\frac{G_1}{M} \cos(\gamma D) + 4 \frac{G_2}{M} \cos(2\gamma D) \right] - \frac{v_g^2}{\delta} \tag{4.12}$$

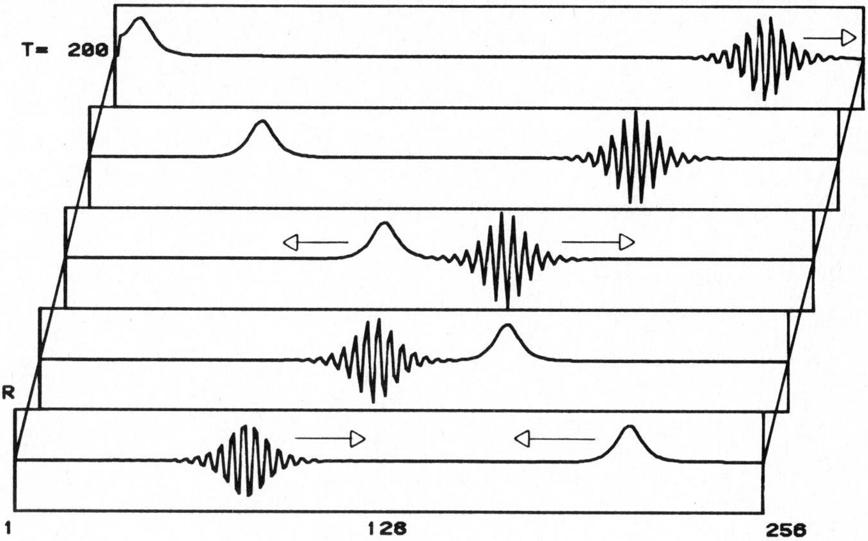

Figure 8 : Pulse-Envelope soliton collision (for $r_n=y_n-y_{n-1}$) on a discrete monatomic chain ($M=D=1, G_1=B_1=1, A=G_2=A_2=B_2=0$) and $\eta_1=0.09$, $\alpha_1=0.01$, $k_1=1.57$, $v_e=0.707$, $L_e= 5.555$, $v_0=0.904$, $L_0=0.637$, ($R_{max}=0.103D$) and $v_2=-1.003$, $L_2=3.724$, ($R_{max}=0.108D$).

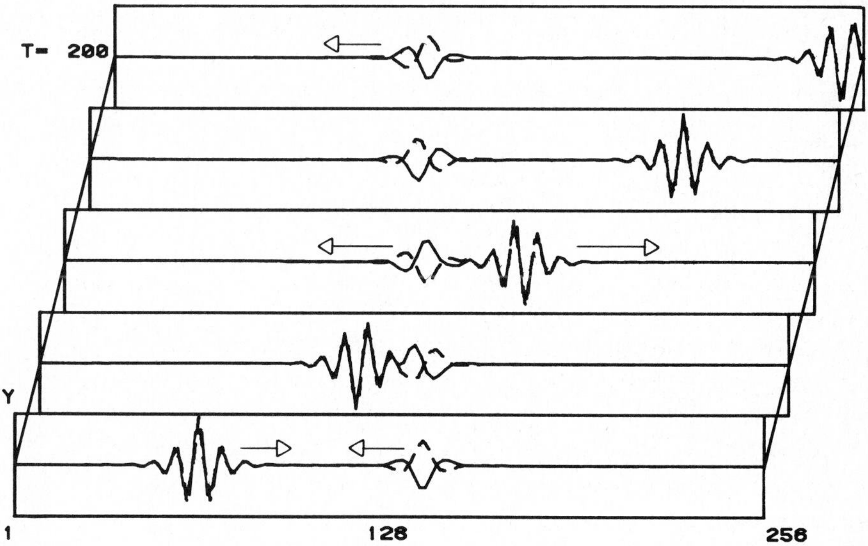

Figure 9 : Envelope-Envelope collision (absolute displacement) on a discrete monatomic chain ($M=D=1, G_1=B_1=1, A_1=G_2=A_2=B_2=0$). The initial conditions ($T=0$) are exact solutions of the corresponding NLS equation , where $\eta_1=0.1$, $\alpha_1=0.01$, $k_1=0.8$, $\omega_1= 0.783$, $v_e=0.921$, $L_e=5$, $v_0=0.979$, $L_0=1.25$, and $\eta_2=0.13$, $\alpha_2=0.01$, $k_2=3.48$, $\omega_2=1.988$, $v_e= -0.168$, $v_0=0.571$, $L_0=0.287$.

leads to the following NLS equation :

$$jF_{\tau} + \frac{1}{2} F_{\xi\xi} + K_0 |F|^2 F = 0 \tag{4.13}$$

with

$$K_0 = S_5/\mu \tag{4.14}$$

and $F_{\tau\tau}$, $F_{\xi\tau}$ neglected. The equation (4.13) is a completely integrable system admitting two different types of solutions [103].

4-4. Envelope Solitons

If $K_0 > 0$, equation (4.13) always allows "bell" envelope solutions which are entities localized in space (see reference 104). Reporting this solution here and putting it into the relation (4.2) we obtain the following envelope soliton :

$$y(x,t) = A_m \operatorname{sech}\left[\frac{x-v_e t}{L_e} + x_e\right] \cos\left[\frac{x-v_0 t}{L_0} + x_0\right] \tag{4.15}$$

with

$$A_m = 4\eta/\sqrt{K_0}, \quad L_e = 1/2\eta, \quad L_0 = 1/(\gamma-2\alpha)$$
$$v_e = v_g - 2\alpha\mu, \quad v_0 = [\delta - 2\alpha v_g + 2\mu(\alpha^2-\eta^2)]/(\gamma-2\alpha) \tag{4.16}$$

where η and α are two small arbitrary parameters. δ, v_g and K_0 are given by the relations (4.9), (4.11) and (4.14). L_e and L_0 measure the width of the envelope and the carrier wave respectively and $k=1/L_0=$ $=\gamma-2\alpha$ is the wavevector of the oscillations of carrier wave ($k\in[-\pi,\pi]$ with γ arbitrary). The frequency ω of these oscillations is given by the formula :

$$\omega = v_0/L_0 = \delta - 2\alpha v_g + 2\mu(\alpha^2 - \eta^2) \tag{4.17}$$

which is to order ε equal to the harmonic frequency δ. v_e and v_0 are respectively the envelope velocity ("group velocity") and the velocity of carrier wave ("phase velocity"). Finally x_e and x_0 are two arbitrary parameters defining the initial positions of the envelope and the carrier wave.

In figure 8, we present a pulse-envelope head-on elastic collision on a discrete quartically anharmonic monatomic chain. While, in figure 9, we present a soliton-type head-on collision between two envelopes with slow (acoustic) and fast (optical) carrier waves. Displacements of odd (solid lines) and even (dashed lines) particles are plotted separately.

4-5. Dark solitons

If $K_0 < 0$, equation (4.13) always allows "dark" soliton (envelope hole) solutions given by the reference 22. These types of solutions are presented in detail elsewhere [88,102]. Here, we present their simplified version, which has the form :

$$y(x,t) = 4\eta \tanh\left[\frac{x-v_e t}{L_e} + x_e\right] \cos\left[\frac{x-v_0 t}{L_0} + x_0\right] \quad (4.18)$$

with

$$L_e = 1/(2\eta\sqrt{|K_0|}) \quad (4.19)$$

and the parameters v_e, v_0, L_0, k and ω are identical to the parameters given for the envelope soliton.

In figure 10, we present the propagation of a dark soliton on a discrete quartically anharmonic monatomic chain with periodic boundary conditions.

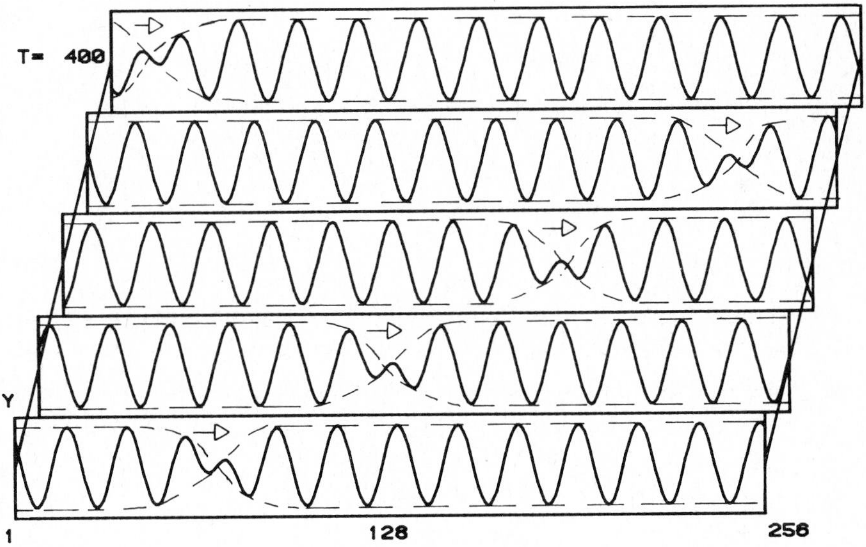

Figure 10 : Dark soliton propagation on a discrete monatomic chain ($M=D=1, G_1=B_1=1, G_2=-0.2, A_1=A_2=B_2=0$). The initial condition ($T=0$) is given by the corresponding NLS equation, where $\eta=0.08$, $\alpha=0.01$, $k=$ $=0.306...$ ($\lambda=20.48$), $\omega=0.147$, $v_e=0.515$, $L_e=11.145$, $v_0=0.479$, $L_0=3.259$, ($R_{max}=0.1D$)

V. THE PHYSICAL MEANING OF SECOND NEIGHBOR COUPLING

Obviously, in real crystal, particles are coupled not only with their first neighbors, but with all the other particles of the lattice. However, the long range order interactions are usually smaller than the short range order coupling. For this reason, physicists have limited their attention to systems where particles are coupled only with their first nearest neighbors. Nevertheless, there are real systems where, at least, a supplementary coupling between next nearest neighbors may be significant (metals), especially when it is competitive with the forces between first neighbors. Moreover, taking into account the supplementary coupling between second nearest neighbors is the first step towards studying long range order interactions [71,72]. On the other hand, this 1-D model may lead to the model of two parallel coupled chains with first nearest neighbor interactions (see section VII and ref. 105).

Looking at expressions (3.4), (4.12) and (4.14) we conclude that, parameters c_0, h_0, p_0, q_0 for kink and breather solitons and parameters μ and S_5 for envelope and dark solitons are quantities depending on lattice constants and essentially on the potential coefficients G_i, A_i, B_i (i=1,2). In order to have soliton solutions we must make a convenient choice of these coefficients. Let us suppose that :

$$D = M = G_1 = 1 \tag{5.1}$$

so the governing parameter will be the coefficinet G_2.

5-1. Kink solitons

For kink soliton solutions of Bq or KdV type we can distinguish three domains of the parameter's G_2 values [73]

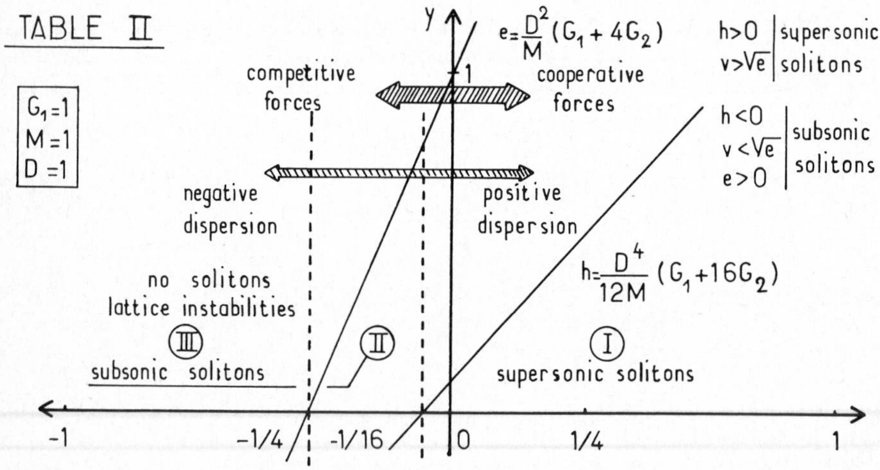

i) DOMAIN I : Here, c_0 is always positive (plane wave solutions of the linearized equation are well defined). Moreover, h_0 (dispersion) is also positive everywhere, so that the soliton velocity v must be greater than the speed of sound c_0, which means that kinks are always supersonic ($v > c_0$). On the other hand, v is also bounded from above, because if $v^2-c_0^2$ is not small enough, then L_1 and L_2 are not sufficiently big and therefore solitons are narrow (discreteness effects). Thus, the faster the soliton is the narrower and the higher it is. The harmonic parts of forces between first and second nearest neighbors are almost everywhere cooperative and where they are competitive [for $G_2 \in (-1/16,0)$] the second neighbor coupling is very small.

In conclusion, there are no qualitative changes on the soliton dynamics, in this domain, through the presence of next nearest neighbor coupling.

ii) DOMAIN II : Here, we have, completely new results due to the competition between first and second neighbor's forces. The speed of sound c_0 is still positive but h_0 (dispersion) is always negative ; thus the velocity v must be less than the speed of sound c_0 and solitons are always subsonic ($v < c_0$). In order to avoid discreteness effects we must consider velocities v very close to the value of c_0 ($c_0^2 - v^2 < \varepsilon$). However, if $c_0 \to 0$, soliton velocity v may be two or four times smaller than c_0 without any dangerous narrowing of the soliton's width. (see § 5-3). In this domain slower solitons are narrower and higher (the opposite of domain I).

iii) DOMAIN III : This is the case where $c_0 < 0$ and $h_0 < 0$ e.g. no plane waves exist in this chain and lattice stability is fragile. Maybe a third nearest neighbor coupling is able to make lattices more stable in this domain [106].

Generally, there are two critical points, for $G_2=-1/16$ where $h_0 = 0$ and for $G_2=-1/4$ where $c_0=0$. However, in the case where $h_0 \to \pm 0$, dispersion still exists in the lattice which comes from higher derivative terms (neglected in the continuum limit).

Chains with competitive coupling between first and second neighbor atoms have a very sensitive lattice stability to the large amplitude solitary excitations, on the contrary, if these interactions are cooperative, lattice stability is stronger.

Finally, where breather solitons are concerned, we can also distinguish three different domains, but breathers can be supersonic or subsonic independent of the sign of h_0 ($v_e, v_0 \simeq c_0 \pm \varepsilon$)

5-2. Envelope solitons

The existence of "bell" envelope solitons or of "dark" solutions depends on the sign of the parameter K_0 (see relation 4.14). In the absence of second nearest neighbor's coupling, if $B_1 > 0$ we have $K_0 > 0$

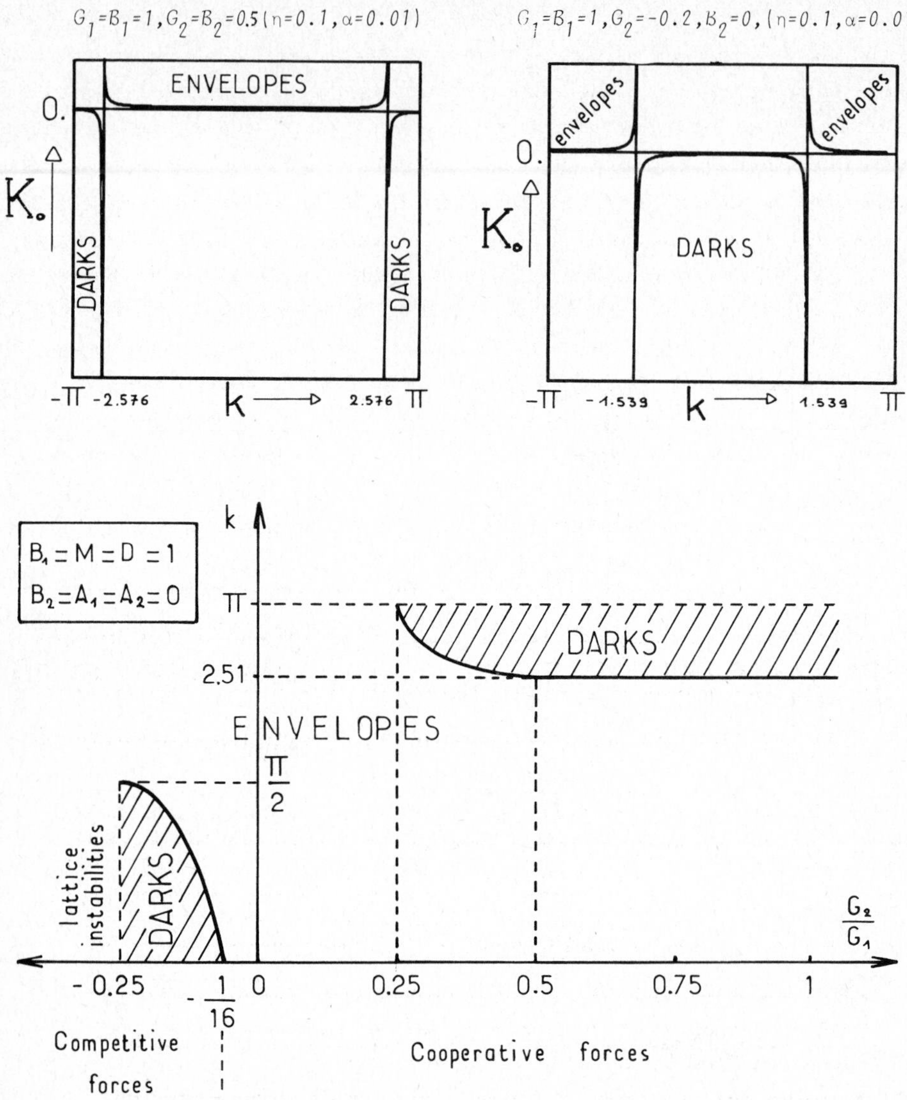

Figure 11 : a,b) Nonlinear coefficient's values, of the NLS equation, versus k. If $K_0 > 0$: envelope ; if $K_0 < 0$: dark.

c) Wave vector-values of switching between dark and envelope solitons versus G_2/G_1 (for the positive values of the first Brillouin Zone).

for any value of the wave vector k, thus only bell-type envelope solitons can be propagated in this chain. If $B_1 < 0$ we always have $K_C < 0$ and only dark solutions can exist. Considering sufficiently strong second nearest neighbor's coupling ($|G_2/G_1| > 0.25$) this situation can be changed. Thus, for example, if $G_1 = B_1 = 1$ and $G_2 = 0.5$, $B_2 = 0$ we can obtain simultaneously envelope and dark solutions for different values of the carrier wave's vector k (figure 11a). A similar situation can be obtained for negative dispersion cases (figure 11b).

It is interesting to study the modulational instability of plane waves [36,107] during this alternation of envelope and dark solution (versus k) resulting from the presence of second neighbor atoms coupling. It is well-known that considering only first neighbors coupling, if envelope are possible, plane wave instability is observed and if dark are possible, plane waves are stable [101]. The most interesting question is : What is the relaxation state in systems where instability is expected ? [108]

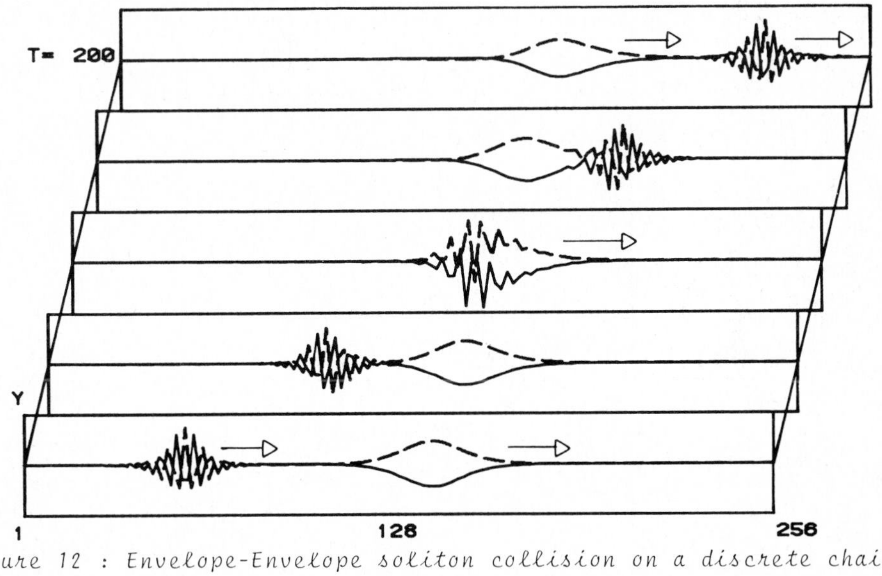

Figure 12 : Envelope-Envelope soliton collision on a discrete chain where kink soliton cannot propagate. (see tableI, $M=D=1, G_1=B_1=1$, $G_2=-0.02, A_1=A_2=B_2=0$). Initial conditions are exact solutions of the equivalent NLS equation. Where $\eta_1=0.1$, $\alpha_1=0.01, k_1=2$, $v_e=0.823$, $L_e=5$, $v_0=0.741$, $L_0=0.5$, and $\eta_2=0.05, \alpha_2=0.01$, $k_2=3.1, v_e=0.037, L_e=10, v_0=$ $=0.646, L_0=0.323$.

Another observation that we must keep in mind is that we can have envelope soliton solutions for a lattice where kinks cannot propagate (see figure 11b). Such a case is presented in figure 12, where, two envelope solitons collide on a discrete quartically anharmonic lattice with first and second nearest neighbor competitive interactions. Absolute displacements of odd (full lines) and even (dashed lines) particles are plotted separately.

5-3. Splitting and blow-up phenomena

In the negative dispersion region (domain II) kink soliton solutions of Bq or KdV type always travel with velocities less than the speed of sound c_0 ($v < c_0$). Decreasing G_2 we obtain lattices where $c_0 \to 0$. Here, it is possible to consider soliton velocities two or four times smaller than the speed of sound without discretization annoyance (solitons are always wide here). In this paragraph, we are studying only cubically nonlinear lattices, with $G_1=1$, $A_1<0$, $G_2<0$ ($G_2 \to -1/4$). But similar interesting phenomena, like those which will be described later, are also possible for other nonlinear potential [109] .

Let us consider the following lattice system :

$$\begin{array}{ll} M=D=1 & c_0 \simeq 0.179 \\ G_1=1, G_2=-0.242 & h_0 \simeq -0.239 \\ A_1=-0.5,\ A_2=0 & p_0 = -1. \end{array} \qquad (5.2)$$

and let us perform numerical experiments with a Bq soliton initial condition varying only its velocity v (see figure 13). We can distinguish three regions of velocity 's values with different soliton or no soliton behavior (table II).

i) REGION A : $v \in [0, c_0/2]$.

Here, the Bq solitary wave initial condition very quickly splits into exactly two soliton propagated in opposite directions (see figures 13b and 13c). The sum of the masses and energies of split solitons are exactly equal to the mass and energy respectively of the initial condition. The more the initial condition's velocity tends to zero, the more the two split solitons are going to be identical (figure 13b). No oscillations are present in the chain after the splitting.

ii) REGION B : $v \in (c_0/2, c_0)$

The Bq soliton initial condition propagate stably in the chain without any loss of energy (figure 13a).

iii) REGION C : $v > c_0$

Kink solitons cannot exist in this chain.

For the moment, we have no satisfactory explanation for this phenomena but we know the following supplementary numerical informations:

TABLE III

For Bq solitons only with $h_0 < 0$ (and $c_0 \ll 1$)

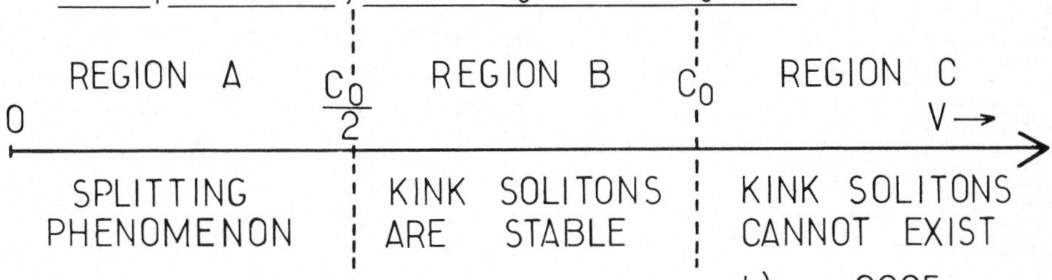

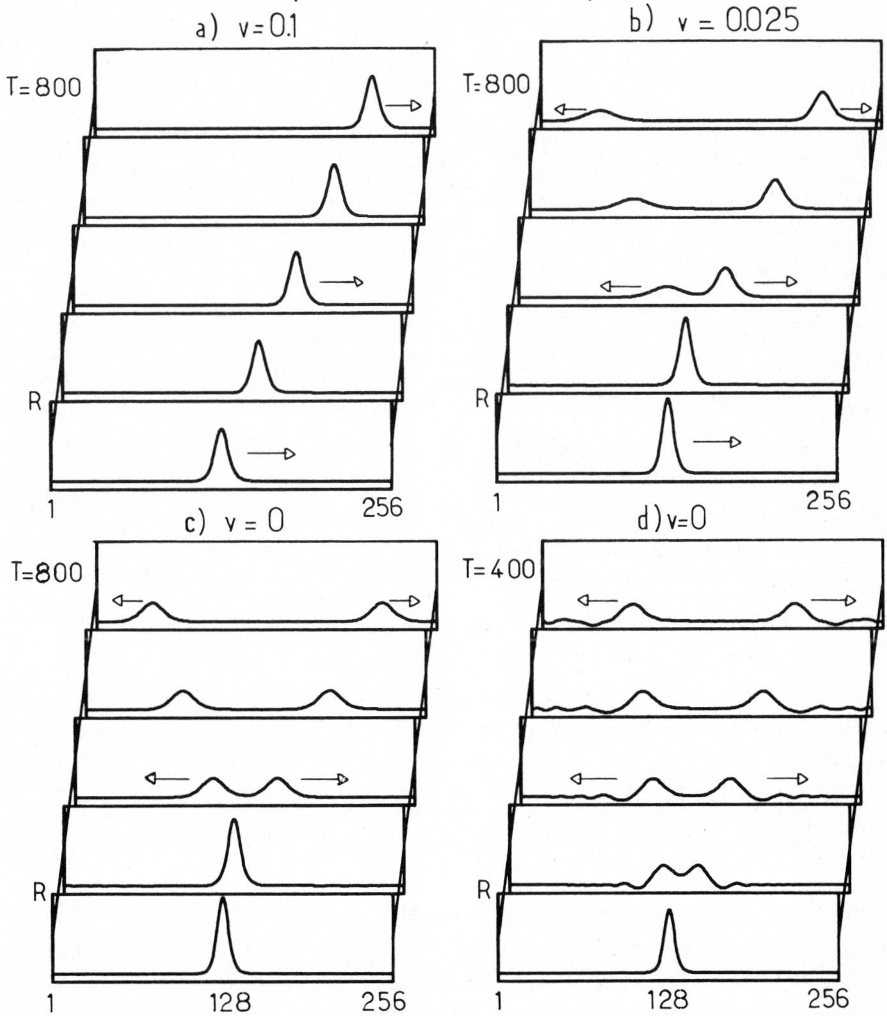

Figure 13: Splitting phenomenon for a Bq initial condition generated on a lattice with $G_1=1, G_2=-0.242, A_1=-0.5, A_2=0$: a) Stable propagation b , c) Splitting, d) Arbitrary (with a pulse form) initial condition. We plot the relative displacements.

1) Exciting the same lattice with an arbitrary initial condition of "tanhϕ"type with zero initial particle velocities, (and also v=0), we obtained results like those we were expecting and they were different from the results of region A (see figure 13d).

2) Exciting the same lattice with a KdV solitary wave initial condition the behavior is quite different. The soliton is blown-up instead of being split into two solitons !

3) The velocities of the two split solitons of figures 13c and 13b seem to satisfy the resonance conditions proposed by Tajiri [110] , valuable for an opposite procedure experiment (three wave interaction). However, contrary to our expectations, when we reverse the time ($t \to -t$), in the numerical experiments 13b and 13c, solitons pass one to the other without reorganising the initial form. Recent theoretical and numerical results will be presented elsewhere [109].

VI. DISCRETENESS EFFECTS

In order to respect the continuum limit approximation's restrictions, the kink soliton solutions of section III must be sufficiently wide. In other words parameters L_1 and L_2 given by the relations (3.11) and (3.19), that measure the spatial extension of Bq and KdV type solitons, must be greater than a critical value ($\sim$3D). Decreasing parameters L_1 and L_2, kinks become narrower and higher (relative displacement amplitude) and their propagation is accompanied by a small amplitude radiation. This radiation is emitted in two phases: Firstly, in a short time, with the beginning of the initial condition's propagation on the lattice, the "continuum" soliton is adapted to the discrete system emitting the excess energy. After that permanent radiation is present behind (for supersonic kinks) or in front (for subsonic kinks) of the main wave. This radiative tail is due to the abrupt restoration of particles to their equilibrium positions (discreteness effects). If the soliton excitation is of large amplitude (large spring elongation or compression), for inertia reasons, particles continue to oscillate even after the passing of the solitary wave. Obviously, this means a continuous rate of energy loss from the soliton which is distributed to the small amplitude oscillations. In figure 14 we present the propagation of a narrow pulse soliton on a discrete monatomic chain.

For envelope soliton solutions, that are calculated using the

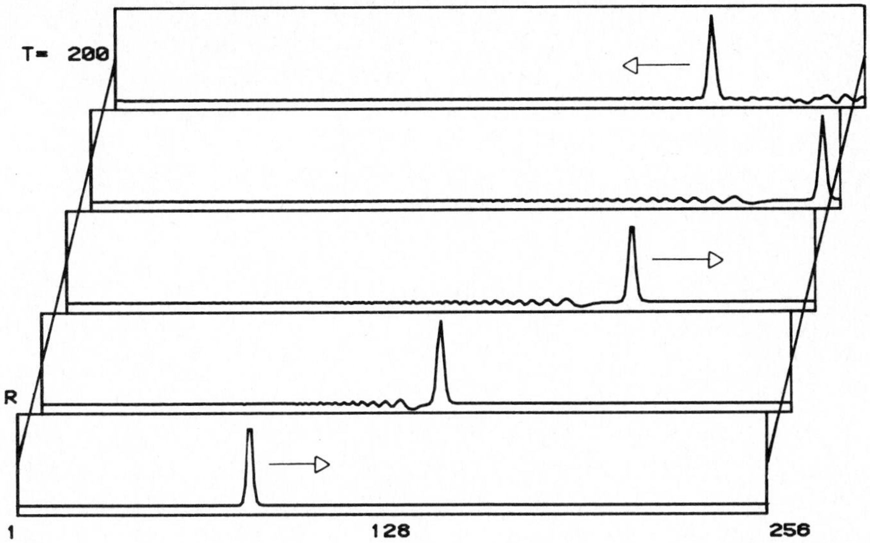

Figure 14 : Evidence of discreteness effects by a narrow Bq soliton pulse (we are plotting the relative displacement) on a discrete monatomic chain ($G_1=A_1=1$, $G_2=0.25$, $B_1=A_2=B_2=0$, $M=D=1$) and $v=1.15$, $L=1.016$, $x=80$, ($R_{max}=0.371D$).

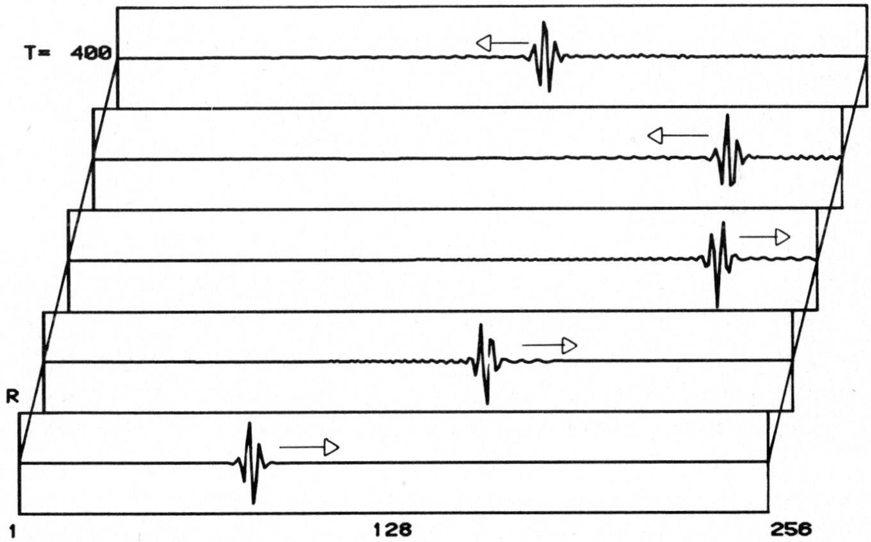

Figure 15: Discreteness effects due to the envelope narrowing. We are plotting the relative displacement ($r_n=y_n-y_{n-1}$), thus, only large amplitude radiation is evident: $G_1=B_1=1$, $A_1=G_2=A_2=B_2=0$, $M=D=1$ and $\eta=0.35$, $\alpha=0.01$, $K=1.57$, $L_e=1.429$, $v_e=0.707$, ($R_{max}=0.4D$).

method of section IV, discreteness effects are only due to the envelope's narrowing. Thus, decreasing the parameter L_e (see relation 4.16) we also have the creation of a radiative tail. However, now this tail is composed of several small amplitude waves with different frequencies which are related to the carrier wave frequency ! In figure 15, we present the propagation of a narrow envelope soliton on a discrete monatomic chain. However, plotting the relative displacement we hide the evidence of small amplitude sound waves [102]. Further work is being undertaken in this direction in connection with recent results obtained for Sine-Gordon discrete soliton [111].

VII. THE MODEL OF PARALLEL CHAINS

The consideration of the supplementary coupling between second nearest neighbor atoms gives us a new possibility to modify the model in figure 2 in a very interesting way. Supposing that all even particles form a chain which is parallel to the initial chain containing now only the odd particles, we obtain a system with two parallel coupled chains (figure 16). The coupling between first neighbors in each chain (previously second neighbor's coupling) and the coupling between respective atoms of the two parallel chains (previously first neighbor's coupling) may be harmonic or anharmonic.

Supposing that transverse motion is negligible compared to the longitudinal motion along each chain we obtain a model which is analytically identical to that if figure 2 (the same equations of motion). Thus, all the results obtained so far are also valid for this new point of view. Moreover, this model allows us to suppose a strong interaction between particles on each chain (previously second neighbors), stronger than coupling between atoms of different chains (previously first neighbors) and to obtain, for some cases, a different dynamical behavior. For example, with this model, we are now able to study stable soliton propagation on a system of parallel chains with a "ϕ^4" nonlinear interatomic potential [88].

The excitation of a solitary wave on only one chain very quickly induces collective motion on the other chain forming a soliton of the coupled system (for appropriate coupling) [88] (see figure 17).

VIII - DIATOMIC CHAIN

In order to simplify our initial atomic model we have limited our

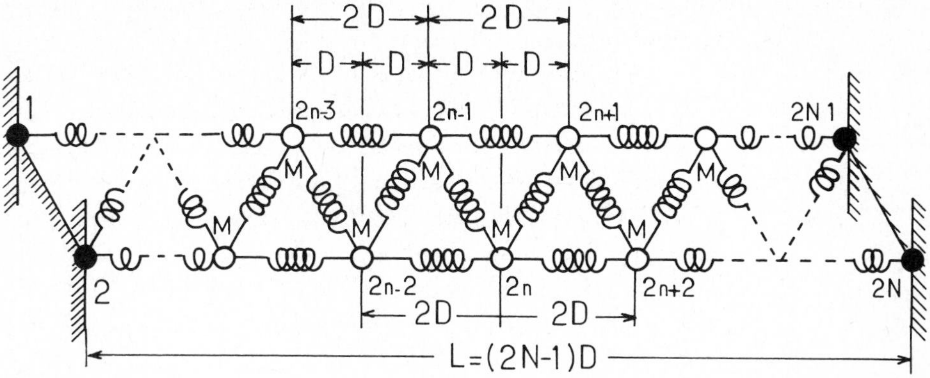

Figure 16 : The two parallel chain model with first nearest neighbor nonlinear coupling (hexagonal lattice). Ends are considered fixed.

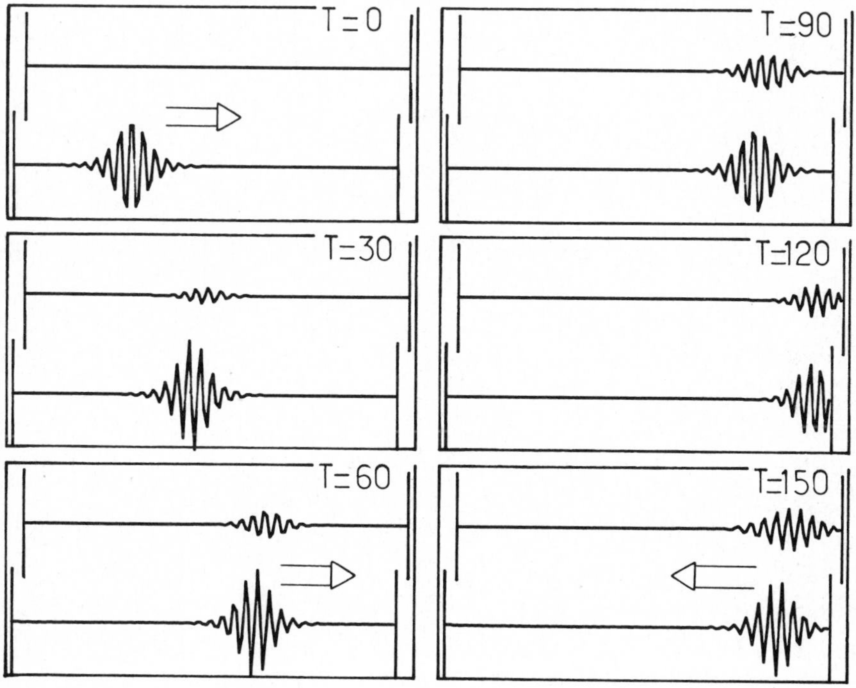

Figure 17 : Chain 1 (lower part of each figure) is excited by an envelope soliton, which is an exact solution of the uncoupled system. The coupling between the two chains (G=0.01, A=B=0) very quickly induces soliton motion on the other chain (upper part of each figure): $G_1=G_2=B_1=B_2=1$, $A_1=A_2=0$, $D=M=1$, $N=256$ and $\eta=0.06$, $k_1=0.8$, $\alpha_1=0.01$.

attention to the monatomic case. However, the majority of solid state systems contain more than one particle per unit cell. On the other hand, physical systems having 1-D dynamical behavior are organic or biological macromolecules. The first natural step to understand the multiparticle chain's anharmonic dynamical bahavior is to study the nonlinear diatomic chain. For this reason, systematic research was undertaken, until 1981, by the author and collaborators [45,88,112, 113], which, accompanied with some recent results obtained in this domain by other poeple[114-119], has today lead to a better understanding of this area. Because, a detailed global description of our results in this domain will be given elsewhere [105], here we give only the main conclusions :

Thus, considering the model in figure 2 with the same type of interaction potential and supposing that all odd particles have mass M_1 and all even particles have mass M_2, we conclude that :

1) Acoustic and optical motion can be studied separately (decoupled) in the continuum limit.

2) For the acoustic mode we have kink solitons where odd and even particles move in phase.

3) For the optical mode and for the quartic interaction potential we have pulse solitons where odd and even particles move out of phase (envelopes with $k \simeq 0$). Optical kinks cannot exist for this potential.

4) Envelope solitons defined for arbitrary k in the first Brillouin zone (as for section IV) and for both modes, can also be propagated on a diatomic chain with quartic type nonlinear interactions between first and second nearest neighbor atoms.

5) Optical kink soliton solutions can be obtained for a ϕ^4 type interatomic potential which may be stable with the assistance of special second neighbor's forces [88]. Further work is actually in progress.

IX. CONCLUSIONS

Nontopological solitons of kink, breather, envelope or dark type can be propagated and elastically interacted on discrete atomic nonlinear and dispersive lattices with a similar anharmonic dynamical behavior as that of the 1-D monatomic chain, presented in section II, whose atoms interact following a polynominal nonlinear interatomic potential of cubic or/and quartic type. This conclusion also seems to be valid when second order cooperative or competitive interactions are added or two different atoms, per unit cell, lattice composition is considered.

Following the procedure of section III and IV it is obvious that long "wavelength" soliton behavior of discrete nonlinear, non-integrable, dispersive atomic chains can be satisfactorily approximated by integrable continuous nonlinear differential equations for which a rich mathematical arsenal is available. Thus kink soliton solutions may be calculated exactly from Bq or KdV type equations. However, I believe Bq equations are better than kdV equations for describing the dynamical behavior of the chain considered here. Envelope and dark exact soliton solutions can be calculated from NLS equations, whose carrier waves may have arbitrary frequencies without any discreteness annoyance. The analytical details are described here for the quartically anharmonic lattice case only. But, similar results are also obtained for cubic or cubic-quartic potentials [102].

A main point in this paper is that a long range order coupling (here second neighbor interactions) may confirm results obtained in systems where only first neighbor coupling is taken into account (supersonic kink solitons,envelope solitons for a quatically anharmonic lattice, etc) or create new situations (subsonic kink solitons, alternation of envelope and dark solitons versus wave vector for the same lattice, etc) or generate new nonsoliton phenomena (resonances, splitting, blow-up, etc).

Solitons can also be stable in models where two parallel nonlinear chains interact harmonically or anharmonically. The excitations of a solitary wave on only one chain very quickly induces motion on the other chain forming a soliton interface if the interchain coupling is sufficient. This means that, in order to describe nonlinear unidirectional dynamical bahavior in 2-D or 3-D lattice systems, we must significantly modify the 1-D soliton theories. In this paper we obtain some analytical results which may be also valid for a pseydo-bidimensional hexagonal lattice (two parallel coupled chains) if transverse motion can be neglected.

Finally, in order to more closely approach the real systems and give a clear answer to the question, if solitons obtained here survive through any physical perturbation, work is in progress concerning the propagation of solitons in disordered lattices and the influence of light dissipation mechanisms.

A considerable effort has been made in this paper to calculate exact soliton solutions. These results may be useful to describe nonlinear dynamics in solids where solitons are already created. However, except perhaps in nonlinear transmission lines, it is very difficult to generate in real systems initial conditions which are exact soliton solutions of these systems. Thus a very interesting question is : what are the qualitative details of an initial wave-paquet which leads to the creation of a kink or breather or envelope or dark soliton of acoustic or optical type ? [39,45,46,78,84, 88]. In order to give a global answer to this physical question studies are in progress [88].

The soliton motion in lattice is related to the determinstic behavior of the nonlinear, non-integrable, conservative systems. However, lattices may simultaneously have stochastic or chaotic behavior. One of the interesting open questions is to identify the physical conditions that determine whether or not a given nonlinear atomic chain, excited by small amplitude waves of periodic or localized type, may exhibit chaotic behavior. This question can be studied in connection with the modulational instability in lattices where envelope or dark solitons are possible.

ACKNOWLEDGMENTS

Many of the original results of this article were obtained in collaboration with Professors M. Remoissenet (O.R.C. lab.-University of Dijon) and N. Flytzanis (Dept of Physics-University of Crete) supported by the official cooperation between the two Universities. The author wishes to thank Dr M. Peyrard for his invaluable discussions during the preparation of this article and also Mr. B. Michaux and Mr. E. Gavalas for technical assistance.

REFERENCES

[1] A. C. Scott, F.Y.F. Chu & D.W. McLaughlin :"The Soliton : A new Concept in Applied Science", Proc. IEEE, 61,(1973),1443-1483.

[2] A.R. Bishop, J.A. Krumhansl & S.E. Trullinger :"Solitons in Condensed Matter: A Paradigm", Physica 1D, (1980), 1-44.

[3] R.H. Enns, B.L. Jones, R.M. Miura & S.S. Rangnekar (eds) : "Nonlinear Phenomena in Physics and Biology", NATO Adv. Study Inst. Series, Vol. 75, (Plenum Press, New York 1981).

[4] A.S. Davydov :"Energy Transfer Along Alpha-Helical Proteins" in : E.Clementi and R.H. Sarma (eds), Structure and Dynamics : Nucleic Acids and Proteins,

(Adenine Press, New-York 1983), 377-387 and references listed therein.

[5] A.S. Davydov :"Theory of Molecular Excitons", (Plenum Press, New-York 1971) and "Bilogy and Quantum Mechanics", (Pergamon, New-York 1982).

[6] J.M. Hyman, D.W.McLaughlin & A.C. Scott :"On Davydov's Alpha-Helix Solitons", Physica 3D (1-2),(1981), 23-44.

[7] A.C. Scott :"Dynamics of Davydov Solitons", Phys. Rev. A. 26(1), (1982),578-595 and references listed therein.

[8] G. Careri, U. Buontempo,F.Carta,E.Gratton & A.C. Scott : "Infrared Absorption in Acetanilide by Solitons", Phys. Rev. Let. 51(4),(1983),304-307.

[9] S. Yomosa :"Soliton Excitations in DNA Double Helices", Phys. Rev. A, 27(4), (1983),2120-2125.

[10] A.L. Hodgkin & A.F. Huxley:"A Quantitative description of Membrane Current and its Applications to Conduction and Excitation in Nerve", J. Physiol,117, (1952),500-544.

[11] A.C. Scott :"Neurophysics",(Wiley-Interscience,New-York(1977) and"The Electrophysics of a Nerve Fibre", Rev.Mod.Phys.47,(1975),487-533 and A.C. Scott & S.D. Lužader :"Coupled Solitary Waves in Neurophysics",Phys. Scripta 20, (1979),395-401.

[12] A. Hasegawa & F.Tappert : "Transmission of Stationary Nonlinear Optical Pulses in Dispersive Dielectric Fibers",Appl. Phys. Let.,"I.Anomalous Dispersion", 23(3),(1973),142-144, and "II.Normal Dispersion", 23(4),(1973),171-172.

[13] L.F. Mollenauer,R.H.Stolen & J.P.Gordon:"Experimental Observation of Picosec-cond Pulse Narrowing and Solitons in Optical Fibers", Phys. Rev. Let. 45(13), (1980), 1095-1098.

[14] A. Hasegawa & Y. Kodama :"Signal Transmissions by Optical Solitons in Monomode Fiber",Proc. IEEE, 69(9),(1981),1145-1150.

[15] D. Anderson :"Variational Approach to Nonlinear Pulse Propagation in Optical Fibers", Phys. Rev. A, 27(6),(1983), 3135-3145.

[16] S. Aubry : "Etude Théorique et Numérique d'un modèle Unidimensionnel pour une Interprétation Unifiée des Transitions Structurales", Thèse d'Etat, Université de Paris VI, 1975.

[17] B.I. Halperin & C.M. Varma :" Defects and the Central Peak Near Structural Phase Transitions", Phys. Rev. B, 14(9), (1976), 4030-4044.

[18] P. Bak :"Commensurate phases, Incommensurate phases and the Devil's Staircase", Progr. Theor. Phys. 45,(1982), 587-629.

[19] T. Riste (ed) : "Ordering in Strongly Fluctuating Condensed Matter Systems", NATO Adv. Stud. Inst. Series, Vol. 50, (Plenum Press, New-York 1980).

[20] V.I. Karpman : "Nonlinear Waves in Dispersive Media", (Pergamon Press, New-York , 1975).

[21] H. Wilhelmsson (ed) : "Solitons in Physics", Phys. Scripta 20(3-4), (1979).

[22] A. Hasegawa :"Plasma Instabilities and Nonlinear Effects",(Springer-Verlag, Berlin 1975).

[23] V.G. Makhankov :"Dynamics of Classical Solitons (in Non-Integrable Systems)", Phys. Rep. 35(1),(1978), 1-128.

[24] B. Kodomtsev :"Phénomènes Collectives dans les Plasmas", (Editions MIR, Moscou 1979 (1976)).

[25] R. Hirota & K. Suzuki : " Theoretical and Experimental Studies of Lattice Solitons in Nonlinear Lumped Networks", Proc. IEEE 61,(1973),1483-1491.

[26] A. Nogushi : "Solitons in a Nonlinear Transmission Line". Elec. and Commun. in Jpn. 57-A, (1974), 9-13.

[27] K. Longgreen & A.C. Scott (eds) : "Solitons in Action", (Academic Press, New-York 1978).

[28] J. Sefrioui : "Dynamique d'impulsions solitons sur des chaînes électriques", Thèse de 3ème Cycle, Université de Dijon, 1984 and references listed therein.

[29] A.C. Scott : "Active and Nonlinear Wave Propagation in Electronics", (Wiley Interscience, 1970).

[30] W.J.Johnson : "Nonlinear Wave Propagation on Superconducting Tunneling Junctions", PhD Thesis, University of Wisconsin,1968.

[31] T. Fulton, R.C. Dynes & P.W. Anderson :" The Flux Shuttle-α Josephson Shift Register employing Single Flux Quanta", Proc. IEEE 61, (1973), 28-35.

[32] S. Reible :"Pulse Propagation on Superconductive Tunnel Transmission Lines", PhD Thesis, University of Wisconsin, 1975.

[33] R.D. Parmentier:"Fluxons in Long Josephson Junctions", in Ref. 27.

[34] P.S. Lomdahl, O.H. Soerensen & P.L. Christiansen : "Soliton Excitations in Josephson Tunnel Junctions", Phys. Rev. B, 25(9), (1982), 5737-5748.

[35] A.R. Osborne & P. Malanotte Rizzoli (eds) : " Topics in Ocean Physics",(North Holland, Amsterdam 1982).

[36] J.P. Boyd : A series of papers in J. Phys. Ocean. 10(1), (1980), 1-11 ; 10(11),(1980),1699-1717 ; 13(3),(1983), 428-466.

[37] E. Fermi, J. Pasta & S. Ulam : "Studies of Nonlinear Problems", Los Alamos Scient. Labor. Report N° LA - 1940, (1955) and in : Collected Works of E. Fermi, Vol. 2, (Univ. of Chicago Press, 1965), 978-988.

[38] N.J. Zabusky :" Synergetic Approach to Problem of Nonlinear Dispersive Wave Propagation and Interaction" in Proc. Supm. Nonlinear P.D.E. (Academic Press 1967), 223-258.

[39] N.J. Zabusky & G.S. Deem : "Dynamics of Nonlinear Lattices. Localized Optical Excitations, Acoustic Radiations and Strong Nonlinear Behavior", J.Comput. Phys. 2, (1967), 126-153.

[40] N.J. Zabusky : "Solitons and Bound States of the Time-Independent Schrödinger Equation", Phys. Rev. 168 (1), (1968), 124-128.

[41] N.J. Zabusky : "Nonlinear Lattice Dynamics and Energy Sharing", J. Phys. Soc. Jpn. (Suppl.) 26, (1969), 196-202.

[42] N.J. Zabusky : "Solitons and Energy Transport in Nonlinear Lattices", Comp. Phys. Commun. 5, (1973), 1-10.

[43] M.A. Collins : "Solitons in Chemical Physics", Adv. Chem. Phys. 53, (1983), 225-340 and references listed therein.

[44] T.Ö. Ogurtani : "Solitons in Solids", Ann. Rev. Mater. Sci. 13, (1983), 67-89 and references listed Therein.

[45] St. Pnevmatikos : " Excitations du type Soliton dans des chaînes atomiques non linéaires", Thèse de 3ème Cycle, Université de Dijon, 1982.

[46] K. Miura : " The Energy Transport Properties of One Dimensional Anharmonic Lattices", PhD Thesis, University of Illinois, 1973.

[47] E.A. Jackson : "Nonlinearity and Irreversibility in Lattice Dynamics", R.M. J. Math. 8(1-2), (1978), 127-196.

[48] M. Toda : " Theory of Nonlinear Lattices", (Springer-Verlag, Berlin 1981 (1978)) and references listed therein.

[49] A. Barone, F. Esposito, C.J. Magee & A.C.Scott :"Theory and Applications of the Sine-Gordon Equation" Riv. Nuov. Cim. 1(2), (1971), 227-267.

[50] A. Seeger : "Solitons in Crystals", in Continuum Models of Discrete Systems, (Univ. of Waterloo Press, Freudenstadt 1979), 253-327.

[51] J.F. Currie : "Classical and Statistical Mechanics of Nonlinear Fields with Applications in Condensed Matter Physics", PhD Thesis, Cornell University 1977.

[52] M. Peyrard & M. Remoissenet :"Solitonlike Excitations in a One-dimensional Atomic Chain with a Nonlinear Deformable Substrate Potential", Phys. Rev. B 26(6), (1982),2886-2899.

[53] M. Remoissenet & M. Peyrard : "Soliton Dynamics in New Models with Parametrized Periodic Double Well and Asymmetric Substrate Potential", Phys. Rev. B, 29(2), (1984).

[54] M.D. Kruskal : "Asymptology in Numerical Computation : Progress and Plans on the Fermi - Pasta - Ulam Problem " in Proc. of the IBM : Scient. Comp. Symp. on Large Scale Problems in Physics, (IBM, New York 1965), 43-62.

[55] D.J. Korteweg & G. de Vries : "On the Change of Form of Long Waves Advancing in a Rectangular Channel and on a New Type of Long Stationary Wave", Philos. Mag. 39, (1895), 422-443.

[56] N.J. Zabusky & M.D. Kruskal : "Interaction of Solitons in A Collisionless Plasma and the Recurrence of the Initial State", Phys. Rev. Let. 15(6),(1965).

240-243.

[57] C.S. Gardner, J.M. Greene, M.D. Kruskal & R.M. Miura : " Method for Solving the KdV Equation", Phys. Rev. Let. 19(19), (1967), 1095-1097.

[58] A. Nakamura : "Interaction of Toda Lattice Soliton with an Impurity Atom", Prog. Theor. Phys. 59(5), (1978),1447-1460.

[59] S. Watanabe & M. Toda : "Interaction of Soliton with an Impurity in Nonlinear Lattice", J. Phys. Soc. Jpn. 50(10), (1981), 3436-3442.

[60] F. Yoshida & T. Sakuma : "Scattering of Lattice Solitons and the Excitations of Impurity Modes", Progr. Theor. Phys. 60(2), (1978), 338-352.

[61] F. Yoshida & T. Sakuma : "Computer-Simulated Scattering of Lattice Solitons from a Mass Interface in a One-Dimensional Nonlinear Lattice", J. Phys. Soc. Jpn. 42(4), (1977), 1412-1417.

[62] N. Yajima : "Reflection and Transmission of Lattice Solitons", Progr. Theor. Phys. 58(4), (1977),1114-1126.

[63] H. Ono : "Wave Propagation in an Inhomogeneous Anharmonic Lattice", J. Phys. Soc. Jpn, 32(2), (1972), 332-336.

[64] F. Kh. Abdullaev & A.A. Abdumalikov : " Dynamics of Solitons in the Disordered Anharmonic Chain", Phys. Stat. Sol. (b), 113, (1982), 685-689.

[65] A.R. Bishop & J.A. Krumhansl : "Mean Field and Exact Results for Structural Phase Transitions in One-Dimensional and Very Anisotropic Two-Dimensional and Three - Dimensional Systems", Phys. Rev. B, 12(7), (1975), 2824-2831.

[66] M.J. Wardrop & D. ter Haar : "The Stability of Three Dimensional Planar Langmuir Solitons", in Ref. 21, 493- 501.

[67] K. Cahill & R.G. Newton : " Soliton-Generating Differential Equations in 3+1 Dimensions", in Ref. 21, 502-504.

[68] P.L. Christiansen & O.H. Olsen : "Ring-Shaped Quasi-Soliton Solutions to the Two and Three Dimensional Sine-Gordon Equation", in Ref. 21, 531-538

[69] V.G. Mankankov : "Computer and Solitons", in Ref 21, 558-562.

[70] D.C. Mattis : "Nonlinear Lattice Dynamics in Two Dimensions", Phys. Rev. B, 27(8),(1983), 5158-5161.

[71] S.K. Sarker & J.A. Krumhansl :"Effect of Solitons on the Thermodynamic Properties of a System with Long-range Interactions", Phys. Rev. B, 23(5),(1981) 2374-2387.

[72] M. Remoissenet & N. Flytzanis : " Solitons in anharmonic chains with long range interactions". in preparation, (1984).

[73] St. N. Pnevmatikos : "Solitons et Couplage entre Seconds Voisins dans des Réseaux Anharmoniques à Une dimension", C.R. Acad. Sc. Paris, t. 296, (1983), 1031-1034.

[74] M. Rich, W.M. Visscher & D.N. Payton III : "Thermal Conductivity of a Two-

Dimensional Two-Branch Lattice", Phys. Rev. A, 4(4), (1971), 1682-1683.

[75] M. Toda : " Solitons and Heat Conduction", in Ref. 21, 424-430.

[76] F. Mokross & H. Büttner :"Thermal Conductivity in the Diatomic Toda Lattice", J. Phys. C, 16, (1983), 4539-4546.

[77] E.A. Jackson, J.R. Pasta & J.F. Waters : " Thermal Conductivity of One Dimensional Lattices", J. Comp.Phys.2, (1968), 207-227.

[78] D.H. Tsai & R.A. MacDonald : "Molecular Dynnamical Study of Second Sound in a Solid Excited by a Stron Heat Pulse", Phys. Rev. B, 14(10), (1976), 4714-4723.

[79] V. Narayanamurti & C.M. Varma :"Nonlinear Propagation of Heat Pulses in Solids", Phys. Rev. Let. 25(16), (1970), 1105-1108.

[80] F.D. Tappert & C.M. Varma : " Asymptotic Theory of self Trapping of Heat Pulses in Solids", 25(16), (1970), 1108-1111.

[81] Kh. G. Boydanova, R.A. Bagautdinov, V.A. Golenishchev - Kutuzov & V.P. Lukomskii : "Observation of an Anomalous Magnetoacoustic Soliton in $KMnF_3$ Crystals", JETP Let. 37(10), (1983), 574-577.

[82] A.R. Bishop & T. Schneider (eds) : " Solitons and Condensed Matter Physics", (Springer Verlag, Berlin 1978).

[83] P. Bocchieri, A. Scotti, B. Bearzi & A. Loinger : " Anharmonic Chain with Lennard-Jones Interaction", Phys. Rev. A, 2(5), (1970), 2013-2019.

[84] T.J. Rolfe, S.A. Rice & J. Dancz : "A Numerical Study of Large Amplitude Motion on a Chain of Coupled Nonlinear Oscillators", J. Chem. Phys. 70(1), (1979), 26-33.

[85] M.K. Ali & R.L. Somorjai : "Quasisoliton Solutions in One-Dimensional Anharmonic Lattices : I. Influence of the Shape of the Pair Potential", J. Phys. A, 12(12), (1979), 2291-2303.

[86] M.A. Collins : "Some Properties of Large Amplitude Motion in an Anharmonic Chain with Nearest Neighbor Interactions", J. Chem. Phys. 77(5), (1982), 2607-2622.

[87] N.J. Zabusky : "Computation : its Role in Mathematical Physics Innovation", J. Comput. Phys. 43(2), (1981), 195-249.

[88] St. Pnevmatikos :"Etude des Solitons dans les Réseaux Diatomiques Nonlinéaires", Thèse d'Etat (in preparation), Université de Dijon, 1984.

[89] Yu.A. Berezin & V.I. Karpman : "Nonlinear Evolution of Disturbances in Plasmas and other Dispersive Media", Sov. Phys. JETP, 24(5),(1967), 1049-1056.

[90] D.F. Strenzwilk :"Shock Profiles Caused by Different end Conditions in One Dimensional Quiescent Lattices", J. Appl. Phys. 50(11),(1979), 6767-6772.

[91] J. Boussinesq : "Théorie des ondes et des remous qui se propagent le long d'un canal rectangulaire horizontal, en communiquant au liquide contenu

dans ce canal des vitesses sensiblement pareilles de la surface au fond" J. Math. Pures Appl. ser.2, Vol. 17, (1872), 55-108.

[92] M. Wadati : " Wave Propagation in Nonlinear Lattice I and II", J. Phys. Soc. Jpn, 38(3), (1975), 673-686.

[93] R.K. Boullough & P.J. Caudrey (eds) : "Solitons", (Springer-Verlag, Berlin 1980).

[94] R. Hirota : " Direct Method in Soliton Theory", in Ref. 93, 157-176.

[95] R.M. Miura (ed) : " Bäcklund transformations", (Springer-Verlag, Berlin 1976).

[96] J.L. Hammack :"Small-Scale Ocean Waves", in ref 35, 278-311.

[97] B.L. Holian & G.K. Straub : "Molecular Dynamics of Shock Waves in One Dimensional Chains", Phys. Rev., 18(4),(1978), 1593-1608.

[98] G. Bowtell & A.E.G. Stuart : "A Particle Representation for KdV solitons" J. Math. Phys. 24(4), (1983), 969-981.

[99] G.L. Lamb Jr : "Elements of Soliton Theory", (Wiley Interscience, New-York 1980).

[100] T. Taniuti & N. Yajima : "Perturbation Method for a Nonlinear Wave Modulation", J. Math. Phys., 10(8), (1969), 1369-1372 and T. Kawahara :" The Derivative Expansion Method and Nonlinear Dispersive Waves", J. Phys. Soc. Jpn, 35(5), (1973), 1537-1544.

[101] A. Tsurui : " Wave Modulations in Anharmonic Lattices", Progr. Theor. Phys. 48(4), (1972), 1196-1203.

[102] N. Flytzanis, St. Pnevmatikos & M. Remoissenet : "Discrete Envelope Solitons in Atomic Systems", in preparation, (1984)

[103] V.E. Zakharov & A.B. Shabat : " Exact Theory of Two-Dimensional Self-Focusing and One-Dimensional Self-Modulation of Waves in Nonlinear Media", Sov. Phys. JETP, 34(1), (1972), 62-69, and "Interaction between Solitons in Stable Medium", Sov. Phys. JETP, 37(5), (1973), 823-828.

[104] H. Segur & M.J. Ablowitz : " Asymptotic Solutions and Conservation Laws for the Nonlinear Schrödinger equation I and II", J. Math. Phys. 17(5), (1976), 710-716.

[105] St. Pnevmatikos, M. Remoissenet & N. Flytzanis : "Soliton Dynamics of Nonlinear Diatomic Lattices", in preparation, (1984).

[106] T. Janssen & J.A. Tjon : "Microscopic Model for Incommensurate Crystal Phases", Phys. Rev. B, 25(6), (1982), 3767-3785.

[107] H.C. Yuen : "Nonlinear Phenomena of Waves on Deep Water", in Ref. 35, 205-234.

[108] D.R. Andersen, S. Datta & R.L. Gunshor : " A Coupled Mode Approach to Modulation Instability and Envelope Solitons", J. Appl. Phys. 54(10), (1983),

5608-5612.

[109] St. N. Pnevmatikos, N. Flytzanis & M. Remoissenet :"Soliton Resonances in Atomic Nonlinear Systems", preprint 1984.

[110] M. Tajiri & T. Nishitani : "Two Soliton Resonant Interactions in One Spatial Dimension : Solutions of Boussinesq Type Equation", J. Phys. Soc. Jpn. 51(11), (1982), 3720-3723.

[111] M. Peyrard & M.D. Kruskal : "Kinks Dynamics in the Highly Discret Sine-Gordon System", to be published, 1984.

[112] St. Pnevmatikos, M. Remoissenet & N. Flytzanis : "Propagation of Acoustic and Optical Solitons in Nonlinear Diatomic Chains", J. Phys. C, 16, (1983), L305 -L310.

[113] St. Pnevmatikos, M. Remoissenet & N. Flytzanis :"Stability of Acoustic and Optical Solitons in a Diatomic Chain", Helv. Phys. Acta, 56, (1983),569 - 574.

[114] J. Tasi : "Initial Value Problems for Nonlinear Diatomic Chains", Phys. Rev. B, 14(6), (1976),2358-2370.

[115] H. Büttner & H. Bilz : "Solitary Wave Solutions in a Diatomic Lattice ", in Ref. 82, 162-165.

[116] P.C. Dash & K. Patnaik : " Solitons in Nonlinear Diatomic Lattices", Progr. Theor. Phys. 65(5).,(1981), 1526-1541.

[117] N. Yajima & J. Satsuma : "Soliton Solutions in a Diatomic Lattice Systems", Progr. Theor. Phys. 62(2), (1979), 370-378.

[118] I. Henry & J. Oitmaa : "Dynamics of a Nonlinear Diatomic Chain", Austr. J. Phys., (1983).

[119] N. Flytzanis : "The Dynamics of a Diatomic Chain on a Parabolic Periodic Substrate", Phys. Let. 85A, (1981), 353-355.

Stéphanos N. Pnevmatikos
Labo. d'Optique du Réseau Cristallin
Faculté des Sciences, Université de Dijon
6, Bd Gabriel, 21100 Dijon, France.

Singularities & Dynamical Systems
S.N. Pnevmatikos (editor)

DECAYING STATES IN QUANTUM SYSTEMS

A. P. Grecos
Université Libre de Bruxelles
Belgium

The problem of exponential decay of quantum unstable systems and its connection with irreversible master equations is discussed.

1. INTRODUCTION.

Various microscopic systems may be characterized as unstable. Either their states decay irreversibly to the fundamental state, the state of lowest energy, or they desintegrate to a number of fragments.Typical examples of such systems are excited atoms decaying by spontaneous emission of photons and radiaoactive nuclei desintegrating by emission of alpha particles. As the energies involved in these processes are low, relativistic effects may be neglected to a first approximation. There are many other examples of unstable systems from different domains of physics. For instance, the overwhelming majority of "elementary" particles in high energy physics have a finite lifetime. However, a satisfactory approach to unstable particles should take into account relativistic invariance and this question is outside the scope of this paper.

Experimentally it is observed that states of unstable systems decay exponentially in time. Concepts such as the lifetime or the width of spectral lines are well-defined and accurately determined by measurements. It is generally admitted that quantum mechanics provides an adequate description of microscopic phenomena and consequently the time evolution of unstable states should follow from this theory.As it is discussed in the next section, the Hilbert space formulation of quantum mechanics due to von Neumann [14] , predicts non-exponential decay. Thus, the concept of an unstable state, for instance an excited atomic state, appears to be ambiguous and, at best, a "useful" approximation.

An unstable system may be regarded as a sybsystem interacting with a field, the combined system being considered as closed. The dynamics is determined by a unitary group, the generator of which is the Hamiltonian H, a self-adjoint operator in a Hilbert space $\mathcal{H}$. A subspace $\mathcal{H}_0 \subset \mathcal{H}$, and the corresponding projection $P : \mathcal{H} \to \mathcal{H}_0$, is taken to characterize the unstable (sub)system. The parameter of the decay are then associated to poles, called resonances of the analytic continuation of the partial resolvent $P(H-z)^{-1}P$ (if such continuation exists). This model is suitable when the unstable system may be thought as becoming a stable one in the limit of a vanishing interaction. In other cases, it might be more convenient to consider unstable states as intermediate states of a scattering system and relate their parameters to the singularities of the analytic S-matrix (see e.g. ref. [2] . A real resonance is just an eigenvalue of H and then a resonance state may be defined as the corresponding eigenvector (in $\mathcal{H}$). It is conceptually interesting to associate resonance states to complex poles in an analogous manner. This cannot be done in the framework of a Hilbert space and generalized vectors need to be introduced as indicated in the third section of the paper.

Decaying systems are similar to the so-called open systems, subsystems interacting with thermal reservoirs, studied in non-equilibrium statistical mechanics [6] . The main difference is that the reservoir is taken to be in a (canonical) equilibrium state while the field is usually in a state with finite energy. Phenomenologically, the irreversible processes taking place when an open system approaches equilibrium are described by some contractive semigroup. The theory of the so-called master equations attempts to derive such semigroups from first principles. Several problems encountered in the derivation and the interpretation of markovian master equations are closely related to those of exponentially decaying states (cf. the review articles [16] and [17]) as it is indicated in the last section.

2. TIME EVOLUTION.

In quantum mechanics, states are represented by vectors in a Hilbert space $\mathcal{H}$, more precisely by unit rays. Observable quantities are represented by self-adjoint operators in $\mathcal{H}$. The expectation value of an observable A for a system in the state f is given by $\langle f,Af\rangle / \langle f,f\rangle$; it is a real quantity determined, in principle, by measurements. Here, $\langle f,g\rangle$ denotes the scalar product in H, antilinear in the first term and linear in the second one. States defined by vectors in H are called pure or vector states, in contrast to the more general class of mixed or statistical states defined by density operators. The latter are positive, trace-class operators on H, normalized (trace norm) to unity $[\rho \geqslant 0,\ \rho \in B_1(H), \mathrm{tr}\rho = 1]$. Then the expectation value $E(A,\rho)$ of an observable A is given by the trace : $E(A,\rho) = \mathrm{tr}A\rho$. Clearly, vector states are the extremal points of the convex set of statistical states ; they are defined by density operators that are one-dimensional projections.

The time evolution of quantum systems is determined by a one-parameter ($t \in \mathbb{R}$) unitary group $U(t) = \exp(-iHt)$ of bounded linear operators on H . For vector states the time-dependence is given by the Schrödinger equation

$$f(t) = U(t)f \quad \Rightarrow \quad i\partial_t f(t) = Hf(t), \quad f \in \mathrm{Dom}(H) \tag{2.1}$$

and for statistical states by the von Neumann equation

$$\rho(t) = U(t)\rho U(-t) \quad \Rightarrow \quad i\partial_t \rho(t) = \mathbb{L}\rho(t) = [H, \rho(t)], \ \rho \in \mathrm{Dom}(\mathbb{L}) \tag{2.2}$$

In most cases, the conservation laws lead to the Hamiltonian, given by a sum of unbounded terms, and its self-adjointness as well as the integration of the equations of motion are generally non-trivial mathematical problems. Using the resolvents of H and $\mathbb{L}$, the solutions of the initial value problem for eqs. (2.1) and (2.2) are

$$f(t) = (2\pi i)^{-1} \int_{\vec{C}} dz \exp(-izt) [H-z]^{-1} f, \quad t \geqslant 0 \tag{2.3}$$

$$\rho(t) = (2\pi i)^{-1} \int_{\vec{C}} dz \exp(-izt) [\mathbb{L}-z]^{-1} \rho, \quad t \geqslant 0 \tag{2.4}$$

The countour $\vec{C}$ is parallel and above the real axis. For $t \leqslant 0$ it suffices to replace $\vec{C}$ by a contour $\overleftarrow{C}$ below the axis. From the definition of the von Neumann operator $\mathbb{L}$ as a commutator, cf. eq. (2.2), it follows that

$$[\mathbb{L}-z]^{-1}\rho = (2\pi i)^{-1} \int_{\vec{\Gamma}} dw \, [H-w]^{-1} \rho \, [H-w+z]^{-1} \tag{2.5}$$

where the contour Γ encloses counterclockwise the spectrum of H (but not the point z). In the remaining of this section and in the next one, vector states are considered. Statistical states that will be discussed in the last section.

Suppose now that the system is initially in the state f ($\|f\|=1$). Then, according to the basic postulates of quantum mechanics, $p(t)=|\langle f,f(t)\rangle|^2$ is the probability to find the system at time t in the same state. Thus, if it is assumed that f(=f(O)) represents the state of an unstable system, p(t) is the probability that at time t this state has not decayed [7] . The question arises whether the quantity

$$a(f;t) = \langle f, \exp(-iHt)f\rangle = (2\pi i)^{-1}\int_{\vec{C}} dz \exp(-izt) \langle f, [H-z]^{-1}f\rangle \quad (2.6)$$

decays exponentially in time, that is, $|a(f;t)|=\exp(-\gamma|t|)$. Clearly, in order to have decay at all, the spectrum of the Hamiltonian must be (at least partly) continuous. However, as the Hamiltonian represents the energy of the system, in most cases, it is an operator bounded from below. It is a consequence of the theory of Sz.-Nagy and Foias on the dilation of contracting semigroups [21] that the last condition precludes exponential decay. Indeed, it may be shown [22] that if $p(t) \leqslant b \exp(-c|t|)$, for all $t\in R$, the spectrum of the Hamiltonian must be the entire real axis. This result implies that there is no state for which p(t) decays exponentially, even for long times. There is an extensive literature as to whether or not it is important for "real" physical systems (see e.g. [7] for various references).

It is often assumed that $H=H_O+\lambda V$ and that the state f is an eigenvector of an unperturbed Hamiltonian H_O corresponding to an eigenvalue embedded in the continuous spectrum. Such an eigenvalue is, in general, unstable and when the interaction λV is extremely weak, by introducing a "rescaled" time $\tau=\lambda^2 t$ the probability p(t) behaves exponentially in the weak coupling limit

$$\lim_{\lambda\to 0} p(\lambda,t) = \exp(-2\gamma|\tau|); \quad \tau = \lambda^2 t \quad (2.7)$$

This result may be proved rigorously in several case and may be extended to more general situations [6, 19] . It gives a precise mathematical meaning to second order perturbative calculations but its physical significance is not always clear. In fact, neither the (real) time interval for which eq. (2.7) is a valid approximation to the exact expression for p(t) can be estimated in a reliable manner, nor its extension to finite, albeit weak, coupling is possible.

The same conclusions apply if instead of a single vector, a subspace is considered. If P is the projection onto the subspace (where dim $P\mathcal{H}$ may be greater than one), the analog of a(f;t), cf. eq. (2.6), is the family of operators

$$V(t) = P \exp(-iHt)\, P = (2\pi i)^{-1} \int_{\vec{C}} dz \exp(-izt)\, P\, [H-z]^{-1} P, \quad t \geqslant 0 \quad (2.8)$$

Again, a necessary condition for V(t) to define a contractive semigroup is that the spectrum of H is the entire real axis [unless a weak coupling limit is taken]. It should be noted that when $\sigma(H) = \mathbb{R}$ and a(f;t) decays exponentially, or V(t) is a semigroup, the state f, or the subspace $P\mathcal{H}$, cannot belong to the domain of H. Furthermore, the system admits then arbitrary lefetimes and this remark indicates that exponential decay does not necessarily leads to the definition of a state of an unstable system because the experimentally observed lifetimes form a discrete rather than a continuous set.

Although the restriction of the time evolution to a subspace does not lead to exponential decay, contributions to eq. (2.8) that depend exponentially in time may exist. These contributions arise from resonances, that is, poles of the analytic continuation of the partial resolvent $P(H-z)^{-1}P$ in the lower half-plane. Indeed, by assumption, H has a continuous spectrum which for $P(H-z)^{-1}P$, as an operator valued function of z, is a cut. Under certain conditions, this function may be continued through the cut in some region of the lower half-plane and be meromorphic there. The real and the imaginary part of a resonance are interpreted, respectively, as the energy and the (inverse) lifetime of an unstable state. Various methods have been develped to prove the existence of resonances and to devise effective ways of computation (see e.g. [18]).

Consider now a set $\{z_r\}$ of resonances. Then, there is a contribution $P\hat{f}(t)$ to $Pf(t)$ with exponential time dependence

$$P\hat{f}(t) = \sum \exp(-iz_r t)\, K_r f; \quad \mathrm{Im}\, z_r \leqslant 0, \quad t \geqslant 0 \tag{2.9}$$

where K_r are operators determined by taking residues in eq. (2.3). It should be remarked that the contribution $P\hat{f}(t)$ need not be a "good" approximation to $Pf(t)$. It may even not permit a probabilistic interpretation because sometimes $\|P\hat{f}(0)\| > \|Pf(0)\|$. By approximating to second order in the coupling parameter λ the various terms in eq. (2.9), a semigroup is obtained, identical to the one derived by taking the weak coupling limit of $V(t)$. For $t \leqslant 0$, an expression analogous to eq. (2.9) may be deduced where the various quantities are defined essentially by replacing the resonances by their complex conjugate. This is a consequence of the symmetry of the Schödinger equation with respect to time inversion. Although $P\hat{f}(t)$ describes to some extent the evolution of the unstable subsystem, the difficulty with a correct interpretation stems from the fact that the question as to what states should be attributed to resonances remains open.

3. RESONANCE STATES.

Formally a resonance state may be constructed, using the pair of projections P and Q, by analogy to an eingenvalue problem. In fact if z_0 is an eigenvalue of an operator H, not necessarily self-adjoint, and if ξ and ξ' are the corresponding right and left eigenvectors, it is easily seen that

$$H\xi = z_0\xi \quad \Rightarrow \quad M(z_0)u = z_0 u, \quad u = P\xi \;;\; Q\xi = -(QHQ-z_0)^{-1}QHPu \tag{3.1}$$

and ($f \to f^*$ denotes the involution in $\mathcal{H}$)

$$H = z \quad = v\, M(z\ ) = z\ v\ , \quad v^* = P\xi'^*; \; Q\xi'^* = -v^*PHQ(QHQ-z_0)^{-1} \tag{3.2}$$

provided that z_0 is not an eigenvalue of QHQ. The operator $M(z)$ is defined [13] by the partial resolvent

$$P(H-z)^{-1}P = [M(z)-z]^{-1}P \quad \Rightarrow \quad M(z) = -PHQ(QHQ-z)^{-1}QHP + PHP \tag{3.3}$$

Eigenvectors corresponding to different eigenvalues are orthogonal,

$<\xi'_1, \xi_2> = 0$ for $z_1 \neq z_2$. It is reasonable to normalize the components u and v to unity, $<u,v> = 1$, but then for the eigenvectors a normalization constant must be introduced ($P\xi = c\,u$, $P\xi' = c'v$)

$$<\xi', \xi> = 1 \quad \Rightarrow \quad \bar{c}'c< v \, , [\, P - M'(z_0)] u> \; = 1 \tag{3.4}$$

where $M'(z_0)$ is the derivative of $M(z)$ at $z = z_0$.

For a self-adjoint operator H, the preceeding relations hold for real eigenvalues and then left and right eigenvectors coincide. In the case of a non-real resonance ($Imz_0 \leqslant 0$), $M(z)$ must be replaced by its analytic continuation $M_+(z)$ through the cut in (the second Riemann sheet of) the lower half-plane. The left and right "eigenvectors" do not coincide anymore and certainly they do not represent vectors in the Hilbert space. In fact, in defining the Q-components, it has been assumed implicitly that the resolvent $(QHQ-z)^{-1}$ has been continued analytically through the cut. The point here is that the "scalar"product $<f,Q\xi>$ must be evaluated by constructing $F(z) = \; <f,(QHQ-z)^{-1}QHP\,\xi>$ and taking its analytic continuation (through the cut) at $z = z_0$ ($Imz_0 \leqslant 0$), that is, $<f,Q\xi>$ $= -F_+(z_0)$ and not just $F(z_0)$. Thus, in the normalization condition, eq. (3.4), the "scalar" product leads to the operator $-PHQ(QHQ-z_0)^{-2}QHP$, which must be interpreted as $M'_+(z_0)$ [and not $M'(z_0)$]. Once this operation is admitted, the formal analogy with an eigenvalue problem is practically complete. For instance, (generalized) eigenvectors corresponding to distinct resonances are "orthogonal" : $<\xi_r, \xi_s> = \delta_{rs}$

The necessity of considering contributions in the subspace $Q\mathcal{H}$ may be seen already in the case where all resonances are real. Then, the subsystem is stable and it is characterized by the subspace spanned by the eigenvectors $\{\xi_r\}$. In terms of the corresponding eigenprojections $\{\Pi_r\}$ of H the evolution of the subsystem is given by the group ($Imz_r = 0$)

$$S(t) = \sum \exp(-iz_r t)\,\Pi_r \qquad ; \qquad H\Pi_r = \Pi_r H = z_r \Pi_r, \tag{3.5}$$

while eq. (2.9) defines only the component PS(t)P that, in general, satisfies no group property. In a certain sense, the P-projection serves to define a "reference space" whereas the projection $\Pi = \Sigma \Pi_r$ defines the state space of the subsystem. To extend this point of view to resonance states, it is necessary to formulate the eigenvalue problem in a larger space.

Generalized vectors are introduced for Hamiltonians (and other operators) having continuous spectrum (cf. ref. [10]). A mathematically rigorous meaning to the eigenvalue problem

$$Hf_\nu = \nu f_\nu \ , \qquad \nu \in \sigma_c(H) \tag{3.6}$$

may be given by constructing a rigged Hilbert space : $\Phi \subset H \subset \Phi^\times$. Here, Φ is a countably normed, nuclear, topological space that is (isomorphic to) a dense subspace of H , and $\Phi^\times$ is its topological dual, the space of continuous (anti)linear functionals on Φ [the notation $<f,\phi>$; $\phi \in \Phi$, $f \in \Phi^\times$ will be used] . Suppose now that is stable under H ; then this operator can be extended to the whole dual space Φ by

$$H\Phi \subset \Phi \quad ; \quad <Hf, \phi> = <f, H\phi>, \qquad \forall \phi \in \Phi \ , \quad \forall f \in \Phi^\times \tag{3.7}$$

The (nuclear) spectral theorem asserts that the set of generalized eigenvectors of the (self-adjoint) operator H, corresponding to real generalized eigenvalues is complete

$$<Hf_\nu, \phi> = \nu <f_\nu, H\phi> \quad ; \quad <f_\nu, \phi> = 0, \ \forall \nu \in \mathbb{R} \ \Rightarrow \ \phi = 0 \tag{3.8}$$

Moreover, if Φ is stable under the time evolution, $\exp(-iHt)\Phi \subset \Phi$, then the generalized eigenvectors can be used to solve the initial value problem of the Schrödinger equation.

Rigged Hilbert spaces have been used in quantum theory in order to justify Diracs formalism (cf. [3,4] and references cited therein). Because, in general, their Hilbert norm is not finite, vectors in $\Phi^\times$ do not lead directly to quantum states with a probabilistic interpretation. Nevertheless, for observables that are represented by operators mapping $\Phi^\times$ into Φ relative probabilities, that is, ratios $<f,Af>/<f,Bf>$ $(f \in \Phi^\times)$, are well defined. Usually, one demands that the rigging is "tight" so that the only solutions of eq. (3.7) are those corresponding to the spectrum of H as an operator in H .

In the case of resonances the idea is to choose the space Φ sufficiently "small", and consequently to have a dual space $\Phi^{\times}$ sufficiently "large", so that eq. (3.7) admits complex eigenvalues. The construction is relatively simple [1] when H is represented by $\mathbb{C}^n \oplus L_2(I, d\omega)$, where I is some interval of the real axis, and QHQ is represented by a multiplication in $L_2(I, d\omega)$: QHQ $\to \omega f(\omega)$. Then QHP is given by a set of functions $v_r(\omega) \in L_2$ $(r=1,\dots,n)$ and PHQ by their complex conjugate, that are assumed to be analytic. It follows that M(z) , an operator on $\mathbb{C}^n$, is a matrix-valued function, analytic in the complex plaine, except for a cut on I . Various exapmles of such Hamiltonians may be presented [12]. Assuming that the interactions $v_r(\omega)$ are holomorphic in some domain G of the complex plane, the matrix M(z) can be continued through the cut in this domain. Let Φ_c be the space of functions holomorphic in G , with the topology defined by a sequence of semi-norms sup $|f(z)|$,$z\in K$, where K is a family of compact domains exhausting G. Then, Φ_c and $\Phi = \mathbb{C}^n \oplus \Phi_c$ are nuclear spaces, the latter being dense in H . Assuming that the interval I is bounded, it may be shown that $H^m \Phi \subset \Phi$ $(\forall m)$; if I is not bounded, Φ_c must be restricted to functions rapidly vanishing at infinity. The dual space $\Phi^{\times}$ is the space of analytic functionals on Φ. Then if eqs. (3.1) and (3.2) hold with M(z) replaced by its analytic continuation $M_+(z)$ through the cut, every solution of

$$M_+(z_r)u_r = z_r u_r \;, \qquad v_r^* M_+(z\;) = z_r v_r^* \;; \qquad \operatorname{Im} z \leqslant 0 \tag{3.9}$$

defines a pair, ξ_r and ξ_r $(\in \Phi^{\times})$, of generalized eigenvectors of H, corresponding to the generalized eigenvalue z . Moreover, generalized eigenprojections $\{\Pi_r\}$ and an associated semigroup S(t) can be defined as in eq. (3.5), but now these operators map Φ into $\Phi^{\times}$.

The semigroup S(t) may be considered as describing the evolution of the unstable states of the subsystem. However, at this point it is necessary to make two remarks concerning this interpretation. Firstly, the space Φ is not stable under exp(-iHt). To define the time evolution it is required to decompose this space in two subspaces $\Phi_{\pm}$ that are stable for t 0. They are constructed by considering the intersection $\Phi_c \cap H_{\pm}$, where H_+ (H_-) are Hardy spaces in the upper (lower) half-plane [8]. Secondly, to treat the eigenprojections as mutually orthogonal, $\Pi_r \Pi_s = \delta_{rs}\Pi_s$, one needs to define the product $<\xi'_r, \xi_s>$ in $\Phi^\times$, and this may be done consistently by introducing the analytically continued operator $M_+(z)$ or its derivative and evaluating them at the resonances [11,20]. The question of associating generalized states to resonances has been discussed in several recent publications [1,5,8,9,15], for Friedrichs-type models and scattering systems.

4. MASTER EQUATIONS

In the context of non-equilibrium statistical mechanics, a master equation denotes an equation describing the evolution of a component of the statistical state of a dynamical system. This component is defined by a projection $\mathbb{P}$ acting on the space of density operators. Thus, if the initial state $\rho(0)$ is the $\mathbb{P}$-subspace, it follows from eq. (2.4) that $(t \geqslant 0)$

$$\mathbb{P}\rho(t) = (2\pi i)^{-1} \int_{C^{\to}} dz \, \exp(-izt) \, [\mathbb{PLP} + \psi(z) - z]^{-1} \, \mathbb{P}\rho(0) \tag{4.1}$$

with the collision operator $\psi(z)$ defined by $(\mathbb{Q}=\mathbb{I}-\mathbb{P})$

$$\mathbb{P}(\mathbb{L}-z)^{-1}\mathbb{P} = [\mathbb{PLP} + \psi(z) - z]^{-1}\mathbb{P}; \qquad \psi(z) = -\mathbb{PLQ}(\mathbb{QLQ}-z)^{-1}\mathbb{QLP} \tag{4.2}$$

By taking the time derivative of eq. (4.1), a non-markovian equation for $\mathbb{P}\rho(t)$, the so-called generalized master equation, is easily deduced [16,17]. It should be noted that the formalism may be extended to apply to more general initial conditions with non-vanishing $\mathbb{Q}$-component.

To describe dissipative processes in dynamical systems one needs to derive an irreversible "markovian" equation, defining a contractive semigroup. Assuming a meromorphic continuation of the partial resolvent $\mathbb{P}(\mathbb{L}-z)^{-1}\mathbb{P}$ in the lower half-plane, through the cut due to the continuous spectrum of $\mathbb{L}$, a theory may be formulated analogous to that of resonances (for details and references see [11,16]. If $\{z_k\}$ is a set of poles of the partial resolvent, and a set $\{\Pi_k\}$ of generalized eigenprojections of $\mathbb{L}$ is defined, a contribution $\hat{\rho}(t)$ to the state is obtained that reads

$$\hat{\rho}(t) = \sum \exp(-iz_k t) \, \Pi_k \, \rho(0), \qquad t \geqslant 0 \tag{4.3}$$

Of course, the possibility of an analytic continuation and the position of the poles do not depend only on $\mathbb{L}$ but also on the projection $\mathbb{P}$. It is generally admitted that the choice of the latter is dictated by physical considerations, although such a justification is not entirely satisfactory as it is based on practical rather than theoretical arguments. As for pure states, $\hat{\rho}(t)$ must be interpreted as a generalized (statistical) state in manner similar to that for resonance states (but a rigorous construction is still lacking).

For decaying systems, as explained in the first section, the Hilbert space is decomposed in a direct sum, $H = H_o + H_2$, and a projection $P : H \quad H_o$ is introduced. The corresponding projection on the space of density operators is

$$\mathbb{P}\rho = P\rho P ; \qquad \rho \in \mathcal{B}_1(H) \tag{4.4}$$

Because of eq. (2.5) and the structure of $\mathbb{P}$, the poles of $\mathbb{P}(\mathbb{L}-z)^{-1}\mathbb{P}$ are simply : $z_k = w_r - w_s$, where $\{w_r\}$ are the resonances of the Hamiltonian. Moreover, the generalized eigenprojections of $\mathbb{L}$ are given in terms of those of H by : $\mathbb{\Pi}_k\rho = \Pi_r \rho \Pi_s$.

In the remaining part of this section some points are briefly discussed concerning the derivation of markovian master equations for weakly coupled systems. The Hamiltonian is written as $H = H_o +$ V, and correspondingly the von Neumann opeator $\mathbb{L} = \mathbb{L}_o + \lambda \mathbb{L}_1$. The projection $\mathbb{P}$ is taken to commute with $\mathbb{L}_o$, as it is the case for the projection defined by eq. (4.4). It is assumed that the various operators depend continuously on the coupling parameter λ and that they may be expanded at least up to second order. These assumptions imply, in particular, that the generalized eigenprojections $\mathbb{\Pi}_k$ of L reduce for $\lambda = 0$ to eigenprojections $\mathbb{P}_k$ of $\mathbb{L}_o$. Between these two sets of projections a one-to-one correspondence can be introduced by an invertible (generalized) transformation $\mathbb{\Lambda}(\lambda)$

$$\mathbb{\Pi}_k(\lambda)\, \mathbb{\Lambda}(\lambda) = \mathbb{\Lambda}(\lambda)\, \mathbb{P}_k ; \qquad \mathbb{P}_k = \mathbb{\Pi}_k(\lambda=0) \tag{4.5}$$

This transformation need not be unique, but the quantity $\mathbb{\Lambda}^{-1}\hat{\rho}(t)$ is always in the $\mathbb{P}$-subspace and, from eqs. (4.3) and (4.5), it satisfies

$$\Lambda^{-1}\rho(t) = \exp(-i\Phi t)\Lambda^{-1}\rho(0) = \sum \exp(-iz_k t)\mathbb{P}_k\Lambda^{-1}\rho(0) \quad (4.6)$$

To second order in λ, the operator Φ can be expressed in terms of the collision operator

$$\Phi = \mathbb{PLP} + \lambda^2 \sum \mathbb{P}_k \Psi_2(\nu_k + i0)\mathbb{P}_k + O(\lambda^3) \quad (4.7)$$

Here, ν_k and $\mathbb{P}_k$ are the eigenvalues and the eigenprojections of $\mathbb{PL}_0\mathbb{P}$ while $\lambda^2\Psi_2(z)$ is the second order approximation of $\Psi(z)$, namely $\Psi_2(z) = -\mathbb{PL}_1\mathbb{Q}(\mathbb{L}_0 - z)^{-1}\mathbb{QL}_1\mathbb{P}$. It defines the generator of a semigroup preserving the positivity of density operators and leaving their traces less than or equal to one.

It is interesting to compare this semigroup to the one obtained by considering a weak coupling approximation for $\mathbb{P}\hat{\rho}(t)$. The expression for $\mathbb{P}\hat{\rho}(t)$ may be written in the form

$$\mathbb{P}\hat{\rho}(t) = \exp(-i\Theta t)\mathbb{AP}\rho(0) \quad (4.8)$$

where the eigenvalues of Θ are identical to those of Φ, that is the set $\{z_k\}$. To second order in λ, Θ is given by

$$\Theta = \mathbb{PLP} + \lambda^2 \sum \Psi_2(\nu_k + i0)\mathbb{P}_k + O(\lambda^3) \quad (4.9)$$

while $\mathbb{A}$ reduces to $\mathbb{P}$ for $\lambda = 0$. Often, it is stated that eq. (4.8) should be a good approximation to $\mathbb{P}\rho(t)$ for long times $(t \to \infty)$. This assertion is incorrect and, as a matter of fact, the only consistent interpretation is to consider $\mathbb{P}\hat{\rho}(t)$ as a contribution to $\mathbb{P}\rho$ with a specific (exponential) time dependence. One problem with the semigroup $\exp(-i\Theta t)$ is that it does not preserve density operators and this difficulty subsists even for weakly coupled systems where Θ is approximated as in eq. (4.9). Taking into account eq. (4.5), it may be shown that Φ and Θ are related by

$$\Phi = \mathbb{X}^{-1}\Theta\mathbb{X}; \qquad \mathbb{X} = \mathbb{P}\mathbb{A}\mathbb{P} \quad (4.10)$$

assuming that the inverse of $\mathbb{X}$ (in the $\mathbb{P}$-subspace) exists. Then, by a perturbation calculation eq. (4.7) is recovered.

The weak coupling limit ($\lambda \to 0$, $t \to \infty$, $\lambda^2 t = \tau$: finite) may be used to define a semigroup $\mathbb{T}(t)$ in the P-sbuspace and to prove that $\lim_{\lambda \to 0} \|\mathbb{P}\exp(-i\mathbb{L}t)\mathbb{P}\rho - \mathbb{T}(t)\rho\|_1 = 0$, where $\|\cdot\|_1$ is the trace norm (cf. [6, 19]). However, this procedure does not lead to a unique $\mathbb{T}(t)$ and supplementary conditions must be imposed to determine the proper generator. For instance, both $\exp(-i\,\Theta_2 t)$ and $\exp(-i\Phi_2 t)$ satisfy the weak coupling limit. By relating the semigroup in the $\mathbb{P}$-subspace to $\hat{\rho}(t)$ and the generalized eigenvalues and eigenprojections of $\mathbb{L}$, a natural construction is proposed that need not be limited to weakly coupled systems only. On the other hand, the possibility of an analytic continuation and the existence of poles are rather strong assumptions that can be verified on relatively simple models. This method may be used in the theory of open systems, but there the poles of the partial resolvent of the von Neumann operator are not directly related to resonances because of the different projection used and the finite temperature of the reservoir.

REFERENCES

[1] H. Baumgärtel : Resonances of perturbed seladjoint operators and their eigenfunctionals. Math. Nachr. 75 (1976) 133-151.

[2] A.I. Baz', Ya.B. Zel'dovich and A.M. Perelomov : Scattering, reactions and decay in nonrelativistic quantum mechanics. (Israel Program for Scientific Translations, Jerusalem, 1979).

[3] N.N. Bogolubov, A.A. Logunov and I.T. Todorov : Introduction to axiomatic quantum field theory (Benjamin, Reading, Mass., 1975).

[4] A. Bohm : The rigged Hilbert space and quantum mechanics. Lecture Notes in Physics 78 (1978), Springer-Verlag.

[5] A. Bohm : Resonance poles and Gamow vectors in the rigged Hilbert space formulation of quantum mechanics. J. Math. Phys. 22 (1981) 2813-2823.

[6] E.B. Davies : Quantum theory of open systems (Academic Press, New York, 1976).

[7] L. Fonda, G.C. Ghirardi and A. Rimini : Decay theory of unstable quantum systems. Rep. Prog. Phys. 41 (1978) 587-631.

[8] M. Gadella : A rigged Hilbert space of Hardy-class functions, applications to resonances. J. Math. Phys. 24 (1983) 1462-1469.

[9] M. Gadella : A description of virtual scattering states in the rigged Hilbert space formulation of quantum mechanics. J. Math. Phys. 24 (1983) 2142-2145.

[10] I.M. Gel'fand and N.Y. Vilenkin : Generalized functions : applications of harmonic analysis (Academic Press, New York, 1964).

[11] A.P. Grecos : Lectures on dissipative processes in dynamical systems, in F.C. Auluck, L.S. Kothari and V.S. Nanda (eds.) : Frontiers of Theoretical Physics (Macmillan, New Delhi, 1978).

[12] A.P. Grecos : Solvable models for unstable states in quantum physics. Adv. Chem. Phys. 38 (1978) 143-171.

[13] J. Howland : "The Livsic matrix in perturbation theory. J. Math.Anal. Appl. 50 (1975) 415-437.

[14] J. von Neumann : Mathematical foundations of quantum mechanics (Princeton Univ. Press, Princeton, 1955).

[15] G. Parravicini, V. Gorini and E.C.G. Sudarshan : Resonances, scattering theory, and rigged Hilbert spaces. J. Math. Phys. 21 (1980) 2208-2226.

[16] O. Penrose : Foundations of statistical mechanics. Rep. Prog.Phys. 42 (1979) 1937-2006.

[17] I. Prigogine and A.P. Grecos : Topics in non-equilibrium statistical mechanics, in G. Toraldo di Francia (ed.) : Problems in the foundations of physics (North-Holland, Amsterdam, 1979).

[18] Sanibel Workshop 1978 : Complex scaling in the spectral theory of the Hamiltonian. Intern. J. Quantum Chem. 14 (1978) n°4.

[19] H. Spohn : Kinetic equations from Hamiltonian dynamics, markovian limits. Rev. Mod. Phys. 53 (1980) 569-615.

[20] E.C.G. Sudarshan, C.B. Chiu and V. Gorini : Decaying states as complex energy eigenvectors in generalized quantum mechanics. Phys. Rev. D18 (1978) 2914-2929.

[21] B. Sz.-Nagy and C. Foias : Harmonic analysis of operators on Hilbert space (North-Holland, Amsterdam, 1970).

[22] D.N. Williams : Difficulty with a kinematic concept of unstable particles, the Sz.-Nagy extension and the Matthews-Salam-Zwanziger representation. Commun. Math. Phys. 21 (1971) 314-333.

A.P. Grecos
Faculté des Sciences, C.P. 231
Université Libre de Bruxelles
B-1050 Bruxelles, Belgique.

List of participants in the International Conference

SINGULARITIES & DYNAMICAL SYSTEMS

Crete, Greece: 30 August - 6 September 83

*

Ralph ABRAHAM .... University of California, Mathematics Board, Santa Cruz, California 95064, U.S.A.

Paulo ALMEIDA .... Universidade de Lisboã, Departamento de Matematica Instituto Tecnico Superior, Lisboa, PORTUGAL.

Rodrigo BAMON .... Universidad do Chile, Departamento de Matematicas Las Palmeros 3425, Santiago, CHILE.

Rachid BEBOUCHI .... Université d'Oran, Département de Mathématiques, B.P. 1524, Es Senia, Oran, ALGERIE.

Imme Van de BERG ...... Université de Strasbourg, Institut de Mathématiques, 7 rue R.Descartes, 67084 Strasbourg, FRANCE.

Tassos BOUNTIS .... Clarkson College, Department of Mathematics, Porsdam, New-York 13676, U.S.A.

Colin BOYD .... University of Warwick, Institute of Mathematics, Coventry CV4 7 AL, Warwick, ENGLAND.

Jonathan BRITT .... Southampton College of Higher Education, East Park Terrace, Southampton SO9 4 WW, ENGLAND.

César CAMACHO .... Instituto de Matemática Pura e Applicada, Estrada Dona Castorina 110, Rio de Janeiro, BRASIL.

Maria-Izabel CAMACHO ... Universidade Federal do Rio de Janeiro, Matemática, CP 68530, Rio de Janeiro, BRASIL.

Ana CASCON .... Université de Dijon, Département de Mathématiques, 214 rue Mirande, 21100 Dijon, FRANCE.

Dominique CERVEAU ..... Université de Dijon, Département de Mathématiques, 214 rue Mirande, 21100 Dijon, FRANCE.

Marc CHAPERON .... Ecole Polytechnique de Paris, Centre de Mathématiques, 91128 Palaiseau, FRANCE.

Alain CHENCINER .... Université Paris VII, Département de Mathématiques, 2 place Jussieu, 75251 Paris, FRANCE.

David CHILLINGWORTH .... Southampton University, Department of Mathematics, Southampton SO9 5 NH, ENGLAND.

George CONTOPOULOS ..... European Southern Observatory, 8046 Garching b. München, West GERMANY.

Costas DAFERMOS .... Brown University, Department of Mathematics, Providence, Rhode Island 02912, U.S.A.

Francine DIENER .... Université d'Oran, Département de Mathématiques, B.P. 1524, Es Senia, Oran, ALGERIE.

Marc DIENER .... Université d'Oran, Département de Mathématiques, B.P. 1524, Es Senia, Oran, ALGERIE.

Paul DOUSSON .... Université de Saint-Etienne, Départ. de Mathématiques, 23 rue Dr P.Michelon, 42123 Saint-Etienne, FRANCE.

Jean-Paul DUFOUR ...... Université de Montpellier, Institut de Mathématiques, Place E.Bataillon, 34000 Montpellier, FRANCE.

Freddy DUMORTIER ...... Limburgs Universitair Centrum, Dept of Mathematics, Universitaire Campus, 3610 Diepenbeek, BELGIUM.

Gerassimos EFTHIMIATOS.. N.R.C.Democritus, Division of Theoretical Physics, Agia Paraskevi, Athens, GREECE.

Ludwig FADDEEV .... Steklov Mathematics Institute, Leningrad Branch, Fontanka 27, 191011 Leningrad, U.S.S.R.

Dietrich FLOCKERZI ..... Universität Wurzburg, Mathematische Institute, AM Hubland, 8700 Wurzburg, West GERMANY.

Nikos FLYTZANIS .... University of Crete, Department of Physics, Heraklion, Crete, GREECE.

Stavros FOURFOULAKIS ... University of Crete, Department of Mathematics, Heraklion, Crete, GREECE.

Jean-Pierre FRANÇOISE .. Université Paris Sud, Département de Mathématiques, 91040 Orsay, FRANCE.

Xavier GOMEZ MONT ..... Universidad National Autonoma de Mexico, Matematicas, Mexico 22 D.F., MEXICO.

Daniel GOROFF .... Institut des Hautes Etudes Scientifiques, 35 Route de Chartres, 91440 Bures-sur-Yvette, FRANCE.

Alkis GRECOS .... Université Libre de Bruxelles, Faculté des Sciences, Physique-Chimie, C.P.231, 1050 Bruxelles, BELGIUM.

Carlos GUTIEREZ .... Instituto de Matemática Pura e Applicada, Estrada Dona Castorina 110, Rio de Janeiro, BRASIL.

Mike IRWIN .... University of Liverpool, Department of Mathematics, P.O.B. 147, Liverpool L 693 BX, ENGLAND.

Nikos KADIANAKIS ...... National Technical University of Athens, Panepistimioupolis Zographou, 15771 Athens, GREECE.

Loucas KANAKIS .... Technical University of Thessaloniki, Panepistimioupolis Thessaloniki, GREECE.

George KORDOULIS ...... Riga Fereou 4, Kessariani, Athens, GREECE.

Christos KOUROUNIOTIS .. King's College London, Department of Mathematics, The Strand, London, ENGLAND.

Martin KRUSKAL .... University of Princeton, Department of Physics, Princeton, NJ 08544, U.S.A.

Nicolaus KUIPER .... Institut des Hautes Etudes Scientifiques, 35 Route de Chartres, 91440 Bures-sur-Yvette, FRANCE.

Rémi LANGEVIN .... Université de Dijon, Département de Mathématiques, 214 rue Mirande, 21100 Dijon, FRANCE.

William LANGFORD ...... University of Guelph, Department of Mathematics, Guelph, Ontario N1G 2W1, CANADA.

Anthony MANNING .... University of Warwick, Institute of Mathematics, Coventry CV4 7 AL, Warwick, ENGLAND.

Jean MARTINET .... Université de Strasbourg, Institut de Mathématiques, 7 rue R.Descartes, 67084 Strasbourg, FRANCE.

Jean-François MATTEI ... Université de Toulouse I, Mathématiques,
Place Anatole France, 31042 Toulouse, FRANCE.

Wellington de MELO ..... Instituto de Matemática Pura e Applicada,
Estrada Dona Castorina 110, Rio de Janeiro, BRASIL.

Robert MOUSSU .... Université de Dijon, Département de Mathématiques,
214 rue Mirande, 21100 Dijon, FRANCE.

Mario de OLIVEIRA ...... Universidade Federal do Rio de Janeiro, Matemática,
CP 68530, Rio de Janeiro, BRASIL.

Bernard d'ORGEVAL ...... Université de Dijon, Département de Mathématiques,
214 rue Mirande, 21100 Dijon, FRANCE.

Maria-Josè PACIFICO .... Universidade Federal do Rio de Janeiro, Matemática,
CP 68530, Rio de Janeiro, BRASIL.

Jacob PALIS .... Instituto de Matemática Pura e Applicada,
Estrada Dona Castorina 110, Rio de Janeiro, BRASIL.

Paris PAMFILOS .... University of Crete, Department of Mathematics,
Heraklion, Crete, GREECE.

Fernand PELLETIER ...... Université de Dijon, Département de Mathématiques,
214 rue Mirande, 21100 Dijon, FRANCE.

Bernard PERRON .... Université de Dijon, Département de Mathématiques,
214 rue Mirande, 21100 Dijon, FRANCE.

Andrew du PLESSIS ...... Aarhus Universitet, Matematisk Institut,
8000 Aarhus C., DENMARK.

Spyros PNEVMATIKOS ..... 45-49 rue Elie Zervou,
11144 Athènes, GRECE.

Stéphanos PNEVMATIKOS .. Université de Dijon, Département de Physique,
6 Boulevard Gabriel, 21100 Dijon, FRANCE.

Jean-Pierre RAMIS ...... Université de Strasbourg, Institut de Mathématiques,
7 rue R.Descartes, 67084 Strasbourg, FRANCE.

Mark ROBERTS .... Southampton University, Department of Mathematics,
Southampton SO9 5 NH, ENGLAND.

Claude-André ROCHE ..... Université de Dijon, Département de Mathématiques,
214 rue Mirande, 21100 Dijon, FRANCE.

Davis ROD .... University of Calgary, Department of Mathematics,
Calgary T2N 1N4, CANADA.

Robert ROUSSARIE .... Université de Dijon, Département de Mathématiques,
214 rue Mirande, 21100 Dijon, FRANCE.

Jean-Claude ROUX .... Université de Bordeaux I, Centre de Recherche P.Pascal,
Domaine Universitaire, 33405 Talence, FRANCE.

John SCHINAS .... Democritus University of Thrace, School of Engineering,
Xanthi, GREECE.

Herbert SPOHN .... Universität München, Theoretische Physik,
Theresien Str. 37, 8 München 2, West GERMANY.

Floris TAKENS .... Universiteit Groningen, Department of Mathematics,
Postbus 800, 900 PH Groningen, THE NETHERLANDS.

René THOM .... Institut des Hautes Etudes Scientifiques,
35 Route de Chartres, 91440 Bures-sur-Yvette, FRANCE.

David TROTMANT .... Université d'Angers, Faculté des Sciences,
2 Boulevard Lavoisier, 49045 Angers, FRANCE.

Costas TZANAKIS .... Université Libre de Bruxelles, Faculté des Sciences, Physique-Chimie, CP 231, 1050 Bruxelles, BELGIUM.

Antonis VASSILIOU ...... University of Crete, Department of Physics, Heraklion, Crete, GREECE.

C.T.C. WALL .... University of Liverpool, Department of Mathematics, P.O.B. 147, Liverpool L 693 BX, ENGLAND.

Bassilis XANTHOPOULOS .. University of Crete, Department of Physics, Heraklion, Crete, GREECE.

Johannes Van ZEIJTS .... Twente University of Technology, Dept of Physics, P.O. 217, 7500 AE, Enschede, THE NETHERLANDS.

SECRETARIAT

University of Crete

Christine ALEVIZOU
Mary ANDRIANAKI
Marine HOURDAKI
Petros LAREDJAKIS

Université de Dijon

Annick CERVEAU
Françoise MOUSSU
Maria PNEVMATIKOU

* UNIVERSITY OF CRETE
Department of Mathematics
Department of Physics
Heraklion, Crete
GREECE

* UNIVERSITE DE DIJON
Département de Mathématiques
Faculté des Sciences
214 rue Mirande, 21100 Dijon
FRANCE